Handbook of Research on E-Planning:
ICTs for Urban Development and Monitoring

Carlos Nunes Silva
Institute of Geography and Spatial Planning
University of Lisbon, Portugal

INFORMATION SCIENCE REFERENCE

Hershey • New York

Director of Editorial Content: Kristin Klinger
Director of Book Publications: Julia Mosemann
Acquisitions Editor: Lindsay Johnston
Development Editor: Elizabeth Ardner
Publishing Assistant: Sean Woznicki
Typesetter: Deanna Zombro
Production Editor: Jamie Snavely
Cover Design: Lisa Tosheff
Printed at: Yurchak Printing Inc.

Published in the United States of America by
Information Science Reference (an imprint of IGI Global)
701 E. Chocolate Avenue
Hershey PA 17033
Tel: 717-533-8845
Fax: 717-533-8661
E-mail: cust@igi-global.com
Web site: http://www.igi-global.com/reference

Copyright © 2010 by IGI Global. All rights reserved. No part of this publication may be reproduced, stored or distributed in any form or by any means, electronic or mechanical, including photocopying, without written permission from the publisher.
    Product or company names used in this set are for identification purposes only. Inclusion of the names of the products or companies does not indicate a claim of ownership by IGI Global of the trademark or registered trademark.

Library of Congress Cataloging-in-Publication Data

Handbook of research on e-planning : ICTs for urban development and monitoring / Carlos Nunes Silva, editor.
   p. cm.
 Includes bibliographical references and index.
 Summary: "This book provides relevant theoretical perspectives on the use of ICT in Urban Planning as well as an updated account of the most recent developments in the practice of e-planning in different regions of the world"- -Provided by publisher.
  ISBN 978-1-61520-929-3 (hbk.) -- ISBN 978-1-61520-930-9 (ebook) 1. City planning--Information technology. 2. City planning--Geographic information systems. 3. City planning--Citizen participation. I. Silva, Carlos Nunes.
  HT166.H362 2010
  307.1'216--dc22
                               2009040554

British Cataloguing in Publication Data
A Cataloguing in Publication record for this book is available from the British Library.

All work contributed to this book is new, previously-unpublished material. The views expressed in this book are those of the authors, but not necessarily of the publisher.

Editorial Advisory Board

Ari-Veikko Anttiroiko, *University of Tampere, Finland*
Roger Caves, *San Diego State University, USA*
Soon Ae Chun, *City University of New York, USA*
Morten Falch, *University of Aalborg, Denmark*
Lech J. Janczewski, *The University of Auckland, New Zealand*

List of Reviewers

Ian Bishop, *The University of Melbourne, Australia*
Angus Whyte, *University of Edinburgh, UK*
Richard E. Klosterman, *University of Akron, USA*
Tomas Ohlin, *Linkoping University, Sweden*
Richard Heeks, *University of Manchester, UK*
Shirin Madon, *London School of Economics and Political Science, UK*
Zorica Nedovic-Budic, *University of Illinois at Urbana-Champaign, USA*
Rina Ghose, *University of Wisconsin-Milwaukee, USA*
Edward J. Malecki, *The Ohio State University, USA*
Hendrick Wagenaar, *Leiden Universiteit, Netherlands*
Anni Dugdale, *University of Canberra, Australia*
Clay Wescott, *Asia Pacific Governance Institute, USA*
Ramona McNeal, *University of Northern Iowa, USA*
James Stewart, *University of Edinburgh, UK*
Gabriel Dupuy, *Université Paris 1 Panthéon Sorbonne, Paris*
Gregory G. Curtin, *University of Southern California, USA*
Richard M. Levy, *University of Calgary, Canada*
Michael McCall, *Universidad Nacional Autonoma de Mexico, Mexico*
Scott McQuire, *University of Melbourne, Australia*
Teal Triggs, *University of the Arts London, UK*
Rana Tassabehji, *University of Bradford, UK*

Ralf Klischewski, *German University in Cairo - GUC, Egypt*
Michael Flamm, *Ecole Polytechnique Fédérale de Lausanne, Switzerland*
Bauke de Vries, *Eindhoven University of Technology, The Netherlands*
Sarah Elwood, *University of Washington, USA*
Georg Aichholzer, *Austrian Academy of Sciences, Austria*

List of Contributors

Abdullah, Alias / *International Islamic University Malaysia, Malaysia*..........435
Abdullah, Muhammad Faris / *International Islamic University Malaysia, Malaysia*..........435
Aikins, Stephen Kwamena / *University of South Florida, USA*..........404
Al-Kodmany, Kheir / *University of Illinois at Chicago, USA*..........143
Arnberger, Arne / *University of Natural Resources and Applied Life Sciences, Austria*..........103
Åström, Joachim / *Örebro University, Sweden*..........237
Baum, Scott / *Griffith University, Australia*..........324
Bolay, Jean-Claude / *Ecole polytechnique fédérale de Lausanne EPFL, Switzerland*..........306
Borning, Alan / *University of Washington, USA*..........36
Bourdakis, Vassilis / *University of Thessaly, Greece*..........268
Camarda, Domenico / *Technical University of Bari, Italy*..........195
Caperna, Antonio / *Università degli Studi Roma Tre, Italy*..........340
Chen, Yun / *University of Salford, UK*..........36
Conroy, Maria Manta / *The Ohio State University, USA*..........218
Deffner, Alex / *University of Thessaly, Greece*..........268
Döllner, Jürgen / *Hasso-Plattner-Institute at University of Potsdam, Germany*..........120
Evans-Cowley, Jennifer / *The Ohio State University, USA*..........218
Granberg, Mikael / *Örebro University, Sweden*..........237
Hamilton, Andy / *University of Salford, UK*..........36
Horelli, Liisa / *Helsinki University of Technology, Finland*..........58
Howell, Angela / *Department of Defense, USA*..........388
Jobst, Markus / *Hasso-Plattner-Institute at University of Potsdam, Germany*..........120
Klessmann, Jens / *Fraunhofer Institute for Open Communication Systems (FOKUS), Germany*...252
Kubicek, Herbert / *University of Bremen, Germany*..........168
Lubanski, Olaf / *Jobstmedia Präsentation Verlag, Austria*..........120
Mahizhnan, Arun / *National University of Singapore, Singapore*..........324
Morgan-Morris, Vanessa / *Constellation Energy and Community College of Baltimore County, USA*..........388
Nummi, Pilvi / *Helsinki University of Technology, Finland*..........80
Pang, Les / *University of Maryland University College, USA*..........388
Rantanen, Heli / *Helsinki University of Technology, Finland*..........80
Reed, Darren J. / *University of York, UK*..........365
Reichhart, Thomas / *University of Natural Resources and Applied Life Sciences, Austria*..........103

Repetti, Alexandre / *Ecole polytechnique fédérale de Lausanne EPFL, Switzerland* 306
Silva, Carlos Nunes / *Institute of Geography and Spatial Planning, University of Lisbon, Portugal* 1
Staffans, Aija / *Helsinki University of Technology, Finland* 80
Velibeyoglu, Koray / *Izmir Institute of Technology, Turkey* 420
Wallin, Sirkku / *Helsinki University of Technology, Finland* 58
Webster, Andrew / *University of York, UK* 365
Wessels, Bridgette / *University of Sheffield, UK* 286
Yigitcanlar, Tan / *Queensland University of Technology, Australia* 15
Zahari, Rustam Khairi / *International Islamic University Malaysia, Malaysia* 435

Table of Contents

Preface xx

Acknowledgment xxv

Section 1
Theory and Methods in E-Planning

Chapter 1
The E-Planning Paradigm-Theory, Methods and Tools: An Overview 1
    Carlos Nunes Silva, Institute of Geography and Spatial Planning, University of Lisbon, Portugal

Chapter 2
Planning Online: A Community-Based Interactive Decision-Making Model 15
    Tan Yigitcanlar, Queensland University of Technology, Australia

Chapter 3
Modelling & Matching and Value Sensitive Design: Two Methodologies for E-Planning
Systems Development 36
    Yun Chen, University of Salford, UK
    Andy Hamilton, University of Salford, UK
    Alan Borning, University of Washington, USA

Chapter 4
The Future-Making Assessment Approach as a Tool for E-Planning and Community
Development: The Case of Ubiquitous Helsinki 58
    Liisa Horelli, Helsinki University of Technology, Finland
    Sirkku Wallin, Helsinki University of Technology, Finland

Chapter 5
Local Internet Forums: Interactive Land Use Planning and Urban Development
in Neighbourhoods 80
    Aija Staffans, Helsinki University of Technology, Finland
    Heli Rantanen, Helsinki University of Technology, Finland
    Pilvi Nummi, Helsinki University of Technology, Finland

Chapter 6
Does Computer Game Experience Influence Visual Scenario Assessment of Urban
Recreational Paths? A Case Study Using 3-D Computer Animation 103
    Arne Arnberger, University of Natural Resources and Applied Life Sciences, Austria
    Thomas Reichhart, University of Natural Resources and Applied Life Sciences, Austria

Chapter 7
Communicating Geoinformation Effectively with Virtual 3D City Models 120
    Markus Jobst, Hasso-Plattner-Institute at University of Potsdam, Germany
    Jürgen Döllner, Hasso-Plattner-Institute at University of Potsdam, Germany
    Olaf Lubanski, Jobstmedia Präsentation Verlag, Austria

Chapter 8
Political Power, Governance, and E-Planning 143
    Kheir Al-Kodmany, University of Illinois at Chicago, USA

Section 2
Citizen Participation in E-Planning

Chapter 9
The Potential of E-Participation in Urban Planning: A European Perspective 168
    Herbert Kubicek, University of Bremen, Germany

Chapter 10
Beyond Citizen Participation in Planning: Multi-Agent Systems for Complex
Decision-Making 195
    Domenico Camarda, Technical University of Bari, Italy

Chapter 11
The E-Citizen in Planning: U.S. Municipalities' Views of Who Participates Online 218
    Maria Manta Conroy, The Ohio State University, USA
    Jennifer Evans-Cowley, The Ohio State University, USA

Chapter 12
Planners Support of E-Participation in the Field of Urban Planning 237
    Mikael Granberg, Örebro University, Sweden
    Joachim Åström, Örebro University, Sweden

Chapter 13
Portals as a Tool for Public Participation in Urban Planning 252
    Jens Klessmann, Fraunhofer Institute for Open Communication Systems (FOKUS), Germany

Chapter 14
Can Urban Planning, Participation and ICT Co-Exist? Developing a Curriculum and an
Interactive Virtual Reality Tool for Agia Varvara, Athens, Greece 268
Vassilis Bourdakis, University of Thessaly, Greece
Alex Deffner, University of Thessaly, Greece

Chapter 15
The Role of Local Agencies in Developing Community Participation in E-Government
and E-Public Services 286
Bridgette Wessels, University of Sheffield, UK

Chapter 16
ICTs and Participation in Developing Cities 306
Alexandre Repetti, Ecole polytechnique fédérale de Lausanne EPFL, Switzerland
Jean-Claude Bolay, Ecole polytechnique fédérale de Lausanne EPFL, Switzerland

Chapter 17
Public Participation in E-Government: Some Questions about Social Inclusion
in the Singapore Model 324
Scott Baum, Griffith University, Australia
Arun Mahizhnan, National University of Singapore, Singapore

Section 3
Innovations and Challenges in Urban Management

Chapter 18
Integrating ICT into Sustainable Local Policies 340
Antonio Caperna, Università degli Studi Roma Tre, Italy

Chapter 19
Architectures of Motility: ICT Systems, Transport and Planning for Complex Urban Spaces 365
Darren J. Reed, University of York, UK
Andrew Webster, University of York, UK

Chapter 20
RFID in Urban Planning 388
Les Pang, University of Maryland University College, USA
Vanessa Morgan-Morris, Constellation Energy and Community College of Baltimore County, USA
Angela Howell, Department of Defense, USA

Chapter 21
E-Planning: Information Security Risks and Management Implications 404
Stephen Kwamena Aikins, University of South Florida, USA

Chapter 22
E-Planning Applications in Turkish Local Governments 420
    Koray Velibeyoglu, Izmir Institute of Technology, Turkey

Chapter 23
GIS Implementation in Malaysian Statutory Development Plan System 435
    Muhammad Faris Abdullah, International Islamic University Malaysia, Malaysia
    Alias Abdullah, International Islamic University Malaysia, Malaysia
    Rustam Khairi Zahari, International Islamic University Malaysia, Malaysia

Compilation of References 455

About the Contributors 503

Index 512

Detailed Table of Contents

Preface xx

Acknowledgment xxv

Section 1
Theory and Methods in E-Planning

Chapter 1
The E-Planning Paradigm-Theory, Methods and Tools: An Overview 1
Carlos Nunes Silva, Institute of Geography and Spatial Planning, University of Lisbon, Portugal

The chapter discusses the relationships between planning theory and the use of information and communication technologies in urban planning. It explores how recent organizational transformations in urban planning, associated with the widespread use of information and communication technologies, are incorporated by different planning theories. It is argued that the way information and communication technologies tools are considered or included by the different planning perspectives is in part responsible for the various forms of e-planning.

Chapter 2
Planning Online: A Community-Based Interactive Decision-Making Model 15
Tan Yigitcanlar, Queensland University of Technology, Australia

The first use of computing technologies and the development of land use models in order to support decision-making processes in urban planning date back to as early as mid 20th century. The main thrust of computing applications in urban planning is their contribution to sound decision-making and planning practices. During the last couple of decades many new computing tools and technologies, including geospatial technologies, are designed to enhance planners' capability in dealing with complex urban environments and planning for prosperous and healthy communities. This chapter, therefore, examines the role of information technologies, particularly internet-based geographic information systems, as decision support systems to aid public participatory planning. The chapter discusses challenges and opportunities for the use of internet-based mapping application and tools in collaborative decision-making, and introduces a prototype internet-based geographic information system that is developed to integrate public-oriented interactive decision mechanisms into urban planning practice. This system, referred as the 'Community-based Internet GIS' model, incorporates advanced information technologies, distance learning, sustainable urban development principles and community involvement techniques in decision-making processes, and piloted in Shibuya, Tokyo, Japan.

Chapter 3
Modelling & Matching and Value Sensitive Design: Two Methodologies for E-Planning
Systems Development 36
    Yun Chen, University of Salford, UK
    Andy Hamilton, University of Salford, UK
    Alan Borning, University of Washington, USA

In this chapter the authors present two methodologies: Modelling & Matching methodology (M&M) and Value Sensitive Design (VSD), which can help address the knowledge gap in the methodologies for designing e-Planning systems. Designed to address the requirements of diverse user groups and multi-disciplinary cooperation for systems development, these two methodologies offer operational guidance to e-Planning systems developers. After the background introduction on e-Planning systems, these two methodologies are described, along with their application in two projects, namely VEPs and UrbanSim. This is followed by suggestions for the further work and conclusions.

Chapter 4
The Future-Making Assessment Approach as a Tool for E-Planning and Community
Development: The Case of Ubiquitous Helsinki 58
    Liisa Horelli, Helsinki University of Technology, Finland
    Sirkku Wallin, Helsinki University of Technology, Finland

As e-planning takes place in a complex and dynamic context, consisting of many stakeholders with a diversity of interests, it benefits from an evaluation approach that assists in the monitoring, supporting and provision of feedback. For this purpose, the authors have created a new approach to e-planning, called the Future-making assessment. It comprises a framework and a set of tools for the contextual analysis, mobilisation and nurturing of partnerships for collective action, in addition to an on-going monitoring and evaluation system. The aim of this chapter is to present and discuss the methodology of the Future-making assessment-approach (FMA) and its application in a case study on e-planning of services in the context of community development, in a Helsinki neighbourhood.

Chapter 5
Local Internet Forums: Interactive Land Use Planning and Urban Development
in Neighbourhoods 80
    Aija Staffans, Helsinki University of Technology, Finland
    Heli Rantanen, Helsinki University of Technology, Finland
    Pilvi Nummi, Helsinki University of Technology, Finland

The Internet is shaking up the expertise and production of knowledge in the planning institution. Digital citizens are searching for information from different places, combining formal and informal sources without apology, and are debating and speaking out on matters. Public planning organisations will be fully stretched to adapt their practices and services to meet these demands. This chapter will present the research results of a project that embarked on gathering and combining local information and knowledge on urban planning on Internet forums. Interactive applications were also developed for these forums to support public participation in ongoing land use planning and development projects in the City of Espoo, Finland. The research results demonstrate how fragmented local, place-based knowledge is, how difficult it is to combine informal and formal information in urban planning, and how inaccessible public data systems still are.

Chapter 6
Does Computer Game Experience Influence Visual Scenario Assessment of Urban
Recreational Paths? A Case Study Using 3-D Computer Animation.. 103
    Arne Arnberger, University of Natural Resources and Applied Life Sciences, Austria
    Thomas Reichhart, University of Natural Resources and Applied Life Sciences, Austria

During the past decades, computer visualizations have been frequently used in urban e-Planning and research. The question arises of whether the degree of experience with the computer during leisure time can have an influence on the assessment of computer-visualized outdoor environment scenarios using visualizations comparable to computer games. The authors used a computer-animated choice model to investigate the influence of computer game experience on respondents' preferences for an urban recreational trail. Static and animated representations of use scenarios were produced with 3-D computer animation techniques. Three social factors were investigated: number of trail users, user composition, and direction of movement: The scenarios were presented to respondents (N = 149), segmented into groups with different computer game experience. The results indicate that the individual experience with computer gaming and the presentation mode influences the evaluation of trail scenarios. Animated trail scenarios seem to be more useful than static ones.

Chapter 7
Communicating Geoinformation Effectively with Virtual 3D City Models.. 120
    Markus Jobst, Hasso-Plattner-Institute at University of Potsdam, Germany
    Jürgen Döllner, Hasso-Plattner-Institute at University of Potsdam, Germany
    Olaf Lubanski, Jobstmedia Präsentation Verlag, Austria

Planning situations are commonly managed by intensive discussions between all stakeholders. Virtual 3D city models enhance these communication procedures with additional visualization possibilities (in opposite to physical models), which support spatial knowledge structuring and human learning mechanisms. This chapter discusses key aspects of virtual 3D city creation, main components of virtual environments and the framework for an efficient communication. It also explores future research for the creation of virtual 3D environments.

Chapter 8
Political Power, Governance, and E-Planning 143
    Kheir Al-Kodmany, University of Illinois at Chicago, USA

Spatial information is a crucial cornerstone of e-planning. This chapter explains the process of constructing a mega geospatial database for the Hajj, the annual Muslim Pilgrimage to Makkah, Saudi Arabia. It discusses the complex process and influence of top-down political power on the comprehensive planning process for the Hajj. It specifically examines how the process led by a politically powerful agency impacted the creation and adoption of a mega geospatial database. The chapter provides transferable and useful lessons on GIS misconceptions and solutions, as well as insight on how and when political power may help in advancing the planning process.

Section 2
Citizen Participation in E-Planning

Chapter 9
The Potential of E-Participation in Urban Planning: A European Perspective................................... 168
Herbert Kubicek, University of Bremen, Germany

Because urban planning affects the living conditions of its inhabitants, most countries, at least western democracies, require some kind of citizen participation by law. The rise of the World Wide Web has led to recommendations to offer participation via the Internet (eParticipation) in various forms. However, many eParticipation applications are not well accepted and fall short of the expectations associated with them. This chapter argues that the electronic mode of participation per se does not change much. Rather, electronic forms of participation have to be embedded in the context of the respective planning processes and participation procedures. If citizens are not interested in participating in an urban planning process, they will not do so just because they could do it via the Internet. Therefore, an analysis of the barriers and deficits of eParticipation has to start with a critical review of traditional offers of participation. Against this background, the forms and methods of electronic participation are described and assessed in regard to expectations and barriers associated with them. It becomes apparent that eParticipation research has still not provided solid knowledge about the reasons for low acceptance of eParticipation tools. This research is largely based on case studies dealing with quite different subject areas. There is also high agreement that electronic tools will not substitute traditional devices for a long time. Instead, they will only complement them. Therefore, online and traditional forms of participation have to be designed as a multi-channel communication system and need to be analyzed against each context together. Accordingly, this chapter starts with summarizing both the institutional context of urban planning and traditional modes of citizen participation and the development and use of technical tools as two backgrounds. Recognizing a certain degree of disappointment with the low use of eParticipation, future eParticipation research should focus on fitting electronic tools better into their context and apply more comprehensive and rigorous evaluation.

Chapter 10
Beyond Citizen Participation in Planning: Multi-Agent Systems for Complex
Decision-Making... 195
Domenico Camarda, Technical University of Bari, Italy

The new complexity of planning knowledge implies innovation of planning methods, in both substance and procedure. The development of multi-agent cognitive processes, particularly when the agents are diverse and dynamically associated to their interaction arenas, may have manifold implications. In particular, interesting aspects are scale problems of distributed interaction, continuous feedback on problem setting, language and representation (formal, informal, hybrid, etc.) differences among agents (Bousquet, Le Page, 2004). In this respect, an increasing number of experiences on multi-agent interactions are today located within the processes of spatial and environmental planning. Yet, the upcoming presence of different human agents often acting au paire with artificial agents in a social physical environment (see, e.g., with sensors or data-mining routines) often suggests the use of hybrid MAS-based approaches (Al-Kodmany, 2002; Ron, 2005). In this framework, the chapter will scan experiences on the setting up of cooperative multi-agent systems, in order to investigate the potentials of that approach on the interaction of agents in planning processes, beyond participatory planning as such. This investigation will reflect on agent roles, behaviours, actions in planning processes themselves. Also, an attempt will be carried out to put down formal representation of supporting architectures for interaction and decision making.

Chapter 11
The E-Citizen in Planning: U.S. Municipalities' Views of Who Participates Online 218
    Maria Manta Conroy, The Ohio State University, USA
    Jennifer Evans-Cowley, The Ohio State University, USA

Municipalities that plan have both a legal obligation and a professional directive to incorporate citizens into the planning process, but garnering sufficient and diverse citizen participation is often a struggle. Online participation tools as a component of e-government provide a potential venue for enhancing the participation process. However, e-government participation raises challenges pertaining to trust, exclusion, and responsiveness. This chapter contributes to our understanding of these issues by analyzing how municipalities in the U.S. view the e-participant. The analysis is based on an ongoing longitudinal study that examines planning department web sites for U.S. cities with 2000 census populations of 50,000 or more. The authors' findings highlight respondents' views of online tools as a means to further efficiency and citizen satisfaction, rather than as a means by which to potentially enhance discussion of community issues.

Chapter 12
Planners Support of E-Participation in the Field of Urban Planning 237
    Mikael Granberg, Örebro University, Sweden
    Joachim Åström, Örebro University, Sweden

The chapter questions what planners really mean when they display positive attitudes toward increased citizen participation via ICTs? Are they aiming for change or the reinforcement of existing values and practices? What are the assumptions that underlie and condition the explicit support for e-participation? In addressing these questions, this chapter draws upon a survey mapping the support for e-participation in the field of urban planning, targeting the heads of the planning departments in all Swedish local governments in 2006. The results show confusing or conflicting attitudes among planners towards participation, supporting as well as challenging the classic normative theories of participatory democracy and communicative planning.

Chapter 13
Portals as a Tool for Public Participation in Urban Planning 252
  Jens Klessmann, Fraunhofer Institute for Open Communication Systems (FOKUS), Germany

In this chapter it will be shown how different general types of portals can be utilized to foster public participation processes in urban and regional planning. First portals and the objectives of their use in the public sector are explained. This happens before the background of different concepts of administrative reform and a transition of government to an electronic manner. Then public participation will be described and different categories thereof are presented. This part forms the basis for the delineation of electronic participation in urban planning. Finally the already introduced general portal types will be applied to distinguish several kinds of participation portals.

Chapter 14
Can Urban Planning, Participation and ICT Co-Exist? Developing a Curriculum and an
Interactive Virtual Reality Tool for Agia Varvara, Athens, Greece.. 268
  Vassilis Bourdakis, University of Thessaly, Greece
  Alex Deffner, University of Thessaly, Greece

One of the recent main problems in urban planning is to find ways in order to employ practical, very broad and commonly used theoretical principles such as participation. An additional issue is the exploitation of the possibilities of new technologies. The process of developing a flexible three-part (common core, public and planners) curriculum in the case of Agia Varvara (Athens, Greece) in the framework of the Leonardo project PICT (2002-2005) showed that ICT (Information Communication Technologies) can help in participation, mainly because it constitutes a relatively simple method of recording the views of both the public and the planners in a variety of subjects (both 'open' and 'closed').

Chapter 15
The Role of Local Agencies in Developing Community Participation in E-Government
and E-Public Services 286
  Bridgette Wessels, University of Sheffield, UK

This chapter discusses the way in which understanding of participation in e-services has evolved through a social learning process within planning and implementation processes. The chapter traces the development of methodologies, partnerships and design constituencies in pilot projects that inform the development of inclusive e-services. It draws on case studies of e-services between 1995 and 2009 to show how planning processes become embedded in cycles of learning and development. E-services involve change in services as well socio-technological change and relate to change in forms of participation. This has led to the development of partnerships to plan and implement e-services and to the development of research and design methodologies that foster participation in the design and use of e-services.

Chapter 16
ICTs and Participation in Developing Cities 306
  Alexandre Repetti, Ecole polytechnique fédérale de Lausanne EPFL, Switzerland
  Jean-Claude Bolay, Ecole polytechnique fédérale de Lausanne EPFL, Switzerland

Developing cities are experiencing substantial gaps in urban planning. They are due to approaches and instruments that do not correspond to the realities of the developing city including the prevalence of informal sector and slums, urban governance problem, and few resources. Information and communication technologies (ICTs) now offer enormous possibilities to use information flows, communication, and land-use models better. ICTs offer solutions that take greater account of informal activities, enable discussions with civil society and Internet forums to take place, etc. ICTs can enhance the planning of developing cities, if conditions are right. The chapter provides a review of the situation in developing cities. It analyses the challenges and potential of using ICTs to improve urban planning. Lastly, it puts forward key conditions for the successful and relevant implementation of ICTs in order to create the best conditions for real technological added value.

Chapter 17
Public Participation in E-Government: Some Questions about Social Inclusion
in the Singapore Model 324
  Scott Baum, Griffith University, Australia
  Arun Mahizhnan, National University of Singapore, Singapore

Singapore's E-government model is considered to be among the best in the world. Over the past decade the Singapore government has constantly developed and re-developed its on-line presence. International comparisons have consistently rated Singapore as one of the most advanced E-government nations. However, despite significant progress towards full E-government maturity, some issues of full public participation remain. It is these issues which this chapter discusses. In particular, it will consider the ways in which a digital divide within the Singapore model has emerged, despite specific policies to address such a problem.

Section 3
Innovations and Challenges in Urban Management

Chapter 18
Integrating ICT into Sustainable Local Policies 340
  Antonio Caperna, Università degli Studi Roma Tre, Italy

This work analyses the Information and Communication Technologies' (hereafter referred to as ICT) phenomenon, the opportunities it offers, the potential problems, and the relationship with local policies. It moves on the actions needed to develop, within the Agenda 21 process, a framework able to define some fundamental features for a new spatial theory in the information age, which will eventually consider Information and Communication Technology not just a simple tool, but a crucial aspect of a sustainable policy, capable, if well addressed, to mitigate various current or emerging territorial challenges such as literacy and education, public participation in the planning process, social and geographical divide, institutional transparency, etc.. This chapter will illustrate a framework able to assist politicians and planners in planning a sustainable development through ICT.

Chapter 19
Architectures of Motility: ICT Systems, Transport and Planning for Complex Urban Spaces 365
    Darren J. Reed, University of York, UK
    Andrew Webster, University of York, UK

This chapter engages with contemporary approaches to urban planning by introducing an analytic strategy rooted in the sociological approach of Science and Technology Studies. By demarcating a 'social frame' and comparing this to the established 'engineering frame' through different 'architectures', the chapter reveals hitherto unrecognised features of the implementation of an intelligent transportation system called BLISS (the Bus Location and Information SubSystem). Through the 'mobilities' conceptual approach, the relationships between various aspects, including the urban space, the experience of passengers, drivers and managers, and component technologies, are revealed as forming an 'assemblage' of conflicting features, that at the same time move toward a form of 'stabilization'. The underlying point, is that we need to engage not only with the technical difficulties of technology implementation in the city, but also with the contingent and experiential processes of those who use, and are affected by such implementations.

Chapter 20
RFID in Urban Planning 388
    Les Pang, University of Maryland University College, USA
    Vanessa Morgan-Morris, Constellation Energy and Community College of Baltimore County, USA
    Angela Howell, Department of Defense, USA

Radio Frequency Identification (RFID) is a significant emerging technology that enables the automation of numerous applications globally. Professions, businesses and industries have integrated this technology into their procedures and it has resulted in great advances in the accuracy of data, operational efficiencies, logistical enhancements and other process improvements. This chapter discusses the application of RFID technology to support the needs and requirements within the realm of urban planning. First, the historic and technical background behind RFID is reviewed. Illustrative examples of its use are presented. Next, the technology's potential is explored in terms of a practical tool for urban planners. Consequently, issues and challenges associated with RFID are identified and considerations to be made when applying the technology are offered. Finally, the outlook for RFID technology is examined as an instrument in urban development and the expected exponential growth of the technology is discussed.

Chapter 21
E-Planning: Information Security Risks and Management Implications 404
    Stephen Kwamena Aikins, University of South Florida, USA

This chapter discusses the security risks and management implications for the use of information technology to manage urban and regional planning and development processes. The advancement in GIS technology and planning support systems has provided the opportunity for planning agencies to adopt innovative processes to aid and improve decision-making. Although studies show that a number of impediments to the widespread adoption these technologies exist, emerging trends point to opportunities for the integration of planning supporting systems with various models to help estimate urban growth, environmental, economic and social impact, as well as to facilitate participatory planning. At the same time, information security infrastructure of most public agencies is reactive-based approach, and security preparedness lags behind vulnerabilities. Drawing on the literature on planning, e-planning and information security, the author argues that the emergence of e-planning as an efficient approach to urban planning and development also poses enormous security challenges that need to be managed to ensure integrity, confidentiality and availability of critical planning information for decision-making.

Chapter 22
E-Planning Applications in Turkish Local Governments 420
Koray Velibeyoglu, Izmir Institute of Technology, Turkey

This chapter examines the pivotal relationship between e-planning applications and their organizational context. It employs various evaluation frameworks by searching explicit and implicit structures behind the implementation process. The study is largely based on the statement that 'the organizational and user dimension of implementation factors more than technical ones, constitute the main obstacles to the improvement of e-planning tools in urban planning agencies'. The empirical part of the study scrutinizes the personal and situational factors of users in the process of implementation, benefits and constraints of an e-planning implementation and planning practitioners' perception of new technologies on urban planning practice and debate. Using a case study research in Turkish local governments, the findings of this study reveal that the organizational and human aspects of high order information systems are still the biggest obstacle in the implementation process.

Chapter 23
GIS Implementation in Malaysian Statutory Development Plan System 435
Muhammad Faris Abdullah, International Islamic University Malaysia, Malaysia
Alias Abdullah, International Islamic University Malaysia, Malaysia
Rustam Khairi Zahari, International Islamic University Malaysia, Malaysia

The chapter presents the current state of GIS implementation in Malaysian development plan system. It offers an overview of GIS implementation worldwide, touching briefly on the history of GIS, planners' early acceptance of the system, factors that promote GIS implementation, level of usage among planners, and factors that impede successful GIS implementation. At the end, the chapter provides a comparison between the state of GIS implementation in Malaysian statutory development plan system with the state of GIS implementation worldwide. The evidence was derived from three main sources: literature, empirical observation of GIS implementation in Malaysia, and a survey conducted in 2008.

Compilation of References 455

About the Contributors 503

Index 512

Preface

Urban planning has experienced numerous changes in its long history but none seems so challenging, for planners and other urban stakeholders, as the methodological revolution associated with the use of information and communication technologies (ICT) in all stages of the planning process. E-Planning, the name given to the new planning paradigm that is emerging in association with the extensive use of information and communication technologies, especially the Internet, geographic information systems and virtual reality technologies, entails a move from a paper based urban planning system, described in this handbook as conventional urban planning, to one based primarily on the integration of various new information and communication technologies and on the interaction of multiple urban stakeholders, referred here as e-planning. Although this is perhaps the most visible difference between conventional urban planning and e-planning, the new urban planning paradigm, this move away from a paper based practice must be seen as more than a simple transfer to a computer system of the traditional paper based routines, requiring also the re-engineering of procedures, the development of a full ICT integrated back office and, most important of all, changes in the nature and purpose of urban planning.

The term e-planning is used here as synonymous of e-government or digital government applied to the field of urban and regional planning. Like these others terms, e-planning is also employed in the literature as a fairly broad multi-dimensional concept (digital terminology is still far from consensus and other definitions of e-planning can be found in the literature). It refers to urban planning, either as part of a hierarchical oriented form of urban government or as an activity co-initiated and co-coordinated by citizens and other private and public stakeholders within the overall urban governance network. In the literature the term e-planning is often employed to refer other more specific subcategories, such as e-urban management, e-urban services, etc., or categories specified according to the dominant technology used, as those associated with the concepts of ubiquitous government and mobile government. Like conventional urban planning, e-planning is also regarded as an interdisciplinary research field.

Based on this broad and multidimensional concept of e-Planning, the purpose of this handbook is to explore the nature and to examine the impacts of the transformations in the urban planning field that result from the use of information and communication technologies in all phases of the urban planning process and to raise new questions for further research. However, it is not intended to be an exhaustive coverage of themes that make up the field of e-planning, since numerous other critical issues were not included. For that reason readers will certainly find at the end that there is much ground yet to be explored and researched on the theory, ethics, methodology and practice of e-planning.

Students, scholars, researchers and practitioners interested to become familiar with new concepts, methods and technologies applied in e-planning, with innovative approaches to improve citizen participation through the Internet, as well as with ground-breaking planning e-tools, will find here an accessible, updated, and research focused reference. Readers will find in these empirical studies practical guidance on how to do cutting edge research on e-planning and useful ideas for the design of new methods of citizen e-participation in urban planning as well.

The handbook is divided into three interrelated sections. The first section deals with theories and methods in e-Planning. The second is devoted to citizen participation in e-Planning. The last section provides an overview of innovations in specific sectors within the urban planning field. It goes without saying that some of the themes of these essays are interrelated and for that reason they could fall into more than one of these three sections. The 23 chapters of this handbook cover a wide range of issues on the theory, methods and tools of e-planning, which make it a useful source for different types of readers. It brings together a collection of multidisciplinary studies, on the many faces of e-Planning, written by 41 distinguished scholars and researchers from leading universities, research institutions, or specialized institutions, from 14 countries, with different perspectives about what e-planning is, representing to some extent the diversity of perspectives and methodologies that can be found in the e-planning field around the world.

Each chapter is divided into six sections: (1) an introduction that provides a general perspective of the chapter and of its main objectives; (2) a background providing broad definitions and discussions of the topic, based on a literature review of the issues discussed, as well as the author's perspective about these issues, controversies, and problems. When appropriate, this section also includes a discussion of solutions and recommendations of the problems presented by the author; (3) a section on future research directions where future and emerging trends are discussed and when appropriate also suggestions for future research within the topic discussed in the chapter; (4) a conclusion with a discussion of the overall coverage of the chapter and concluding remarks; (5) a reference and further readings section, and, finally, (6) a list of terms and definitions applied in the chapter.

Section 1, titled "Theory and Methods in E-Planning," begins with an overview of e-planning followed by seven chapters. These chapters describe and discuss different planning methodologies, based on the use of information and communication technologies in different planning contexts, exploring key facets of the move towards a new paradigm of urban planning.

Chapter 1, "The E-Planning Paradigm-Theory, Methods and Tools: An Overview," serves as the introduction to the book and discusses the relationships between planning theories and the use of information and communication technologies in urban planning. The way information and communication technologies tools are incorporated by the different planning perspectives is considered to be in part responsible for the different forms of contemporary urban planning.

The role of information technologies, particularly internet based geographic information systems, as decision support systems to aid public participatory planning, is examined in the Chapter 2 "Planning Online: A Community-Based Interactive Decision-Making Model". Tan Yigitcanlar also discusses the challenges and opportunities for the use of internet based mapping application and tools in collaborative decision-making, introducing a prototype internet based geographic information system that was developed to integrate public oriented interactive decision mechanisms into urban planning practice.

In "Modelling & Matching and Value Sensitive Design: Two Methodologies for E-Planning Systems Development," Yun Chen, Andy Hamilton, and Alan Borning explore two methodologies which can help address the knowledge gap in the methodologies for designing e-Planning systems. Planned to address the needs of diverse user groups and multi-disciplinary cooperation for systems development, these two methodologies offer operational guidance to e-Planning systems developers.

In Chapter 4, "The Future-Making Assessment Approach as a Tool for E-Planning and Community Development: The Case of Ubiquitous Helsinki," Liisa Horelli and Sirkku Wallin, offer readers an in-dept look at an evaluation approach to be used in e-Planning, called the Future-making assessment (FMA), to assist in the monitoring and provision of feedback in the implementation of e-Planning.

The question of the role of the Internet in the production of planning knowledge is addressed by Aija Staffans, Heli Rantanen, and Pilvi Nummi in the Chapter 5, "Local Internet Forums. Interactive Land Use Planning and Urban Development in Neighbourhoods". The authors describe the results of a research project that tried to gather and to combine local information and knowledge on urban planning through Internet forums. They show that local, place-based knowledge is highly fragmented and that it is difficult to combine informal and formal information and knowledge in urban planning.

The next two chapters look at the use of 3-D images as a method of communicating information in urban planning. In Chapter 6, "Does Computer Game Experience Influence Visual Scenario Assessment of Urban Recreational Paths? A Case Study Using 3-D Computer Animation," Arne Arnberger and Thomas Reichhart explain the results of a study in which they used a computer-animated choice model to investigate the influence of computer game experience on respondents' preferences for an urban recreational trail, concluding that the individual experience with computer gaming and the presentation mode influence the evaluation of trail scenarios. Markus Jobst, Jürgen Döllner and Olaf Lubanski, in "Communicating Geoinformation Effectively with Virtual 3D City Models," focus on Virtual 3-D city models and how they can enhance the communication between different urban stakeholders. The authors discuss key aspects of virtual 3-D city creation, the main components of virtual environments, the framework for an efficient communication, and explore future research for the creation of virtual 3-D environments.

The last chapter in Section 1, "Political Power, Governance, and E-Planning," analyse the construction of a mega geospatial database for the Hajj, the annual Muslim Pilgrimage to Makkah, Saudi Arabia. Kheir Al-Kodmany discusses this complex process, including in his analysis the influence of top-down political power on the planning process for the Hajj. The chapter provides transferable and useful lessons on GIS application in spatial urban planning, as well as insights on how and when political power may help in advancing the planning process.

Section 2, titled "Citizen Participation in E-Planning," with nine chapters, introduces readers to a range of experiences and practices of e-participation, in different countries and contexts, which provides a good illustration of how citizen participation in the urban planning decision-making process is changing, and the type of challenges faced by planners and planning departments.

Herbert Kubicek, through a comprehensive review of the literature, examines in Chapter 9, "The Potential of E-Participation in Urban Planning: A European Perspective," different cases of public participation and argues that the electronic mode of participation by itself will not change the low levels of public participation in urban planning, suggesting that it will be necessary to include these new electronic forms of participation within the formal planning processes and in the respective participation procedures, arguing that if citizens are not interested to participate in the urban planning process, they will not take part only because they could do it via the Internet. The author reveals that ICT tools for citizen participation in urban planning will not substitute the traditional forms of public participation in the near future, arguing that it is necessary to combine both, offering specific recommendations for that.

Next, Domenico Camarda in his chapter "Beyond Citizen Participation in Planning: Multi-Agent Systems for Complex Decision-Making" examines how to set up cooperative multi-agent systems, and discusses the potentials of multi-agent system for complex decision-making in public participation processes in urban planning.

In "The E-Citizen in Planning: U.S. Municipalities' Views of Who Participates Online," Maria Manta Conroy and Jennifer Evans-Cowley examine how online participation tools, regarded as a component of e-government, provide a potential venue for enhancing citizen participation in the urban planning process. However, as e-government participation raises challenges pertaining to trust, exclusion, and responsiveness, the chapter examines how municipalities in the U.S. view the e-participant, concluding that municipal officials view these online tools as a means to advance efficiency and citizen satisfaction, rather than as a means by which to potentially enhance discussion of community issues.

In Chapter 12, "Planners Support of E-Participation in the Field of Urban Planning," Mikael Granberg and Joachim Åström discuss what planners really mean when they display positive attitudes toward increased citizen participation via the use of information and communication technologies, based on a survey about the support for e-participation in the field of urban planning by the heads of planning departments in Sweden, concluding for the existence of confusing or conflicting attitudes among planners towards participation.

Jens Klessmann, in "Portals as a Tool for Public Participation in Urban Planning," looks critically how different types of portals and different kinds of participation portals can be used to encourage public participation processes in urban planning.

After that, in Chapter 14, "Can Urban Planning, Participation and ICT Co-Exist? Developing a Curriculum and an Interactive Virtual Reality Tool for Agia Varvara, Athens, Greece," Alex Deffner and Vassilis Bourdakis examine how information and communication technologies can help in urban participation processes, mainly because it constitutes a relatively simple method of recording the views of both the public and the planners in a variety of subjects.

Bridgette Wessels, in "The Role of Local Agencies in Developing Community Participation in E-Government and E-Public Services," discusses the way in which understanding of participation in e-services has evolved through a social learning process within planning and implementation processes.

In "ICTs and Participation in Developing Cities," Alexandre Repetti and Jean-Claude Bolay provide a review of the use of information and communication technologies for public participation in urban planning, in cities located in developing countries. The authors analyse the challenges and potential of ICT to improve urban planning and public participation, and put forward a number of recommendations for the successful and relevant implementation of ICT in this kind of cities.

In the final chapter of Section 2, "Public Participation in E-Government: Some Questions about Social Inclusion in the Singapore Model," Scott Baum and Arun Mahizhnan examine the case of Singapore's E-government model, which, despite being considered to be among the best in the world, has still important weaknesses in what respects public participation.

Section 3, titled "Innovations and Challenges in Urban Management," with six chapters, explores a number of experiences and innovative practices of urban e-management, in different countries and contexts, as an illustration of the type and extent of the changes going on within the urban planning system.

Antonio Caperna in Chapter 18, "Integrating ICT into Sustainable Local Policies," analyses the role of information and communication technologies in the promotion of sustainable local policies, the opportunities it offers, potential problems, and the relationship with other local policies.

Chapter 19, "Architectures of Motility: ICT Systems, Transport and Planning for Complex Urban Spaces," by Darren J. Reed and Andrew Webster, examines the implementation of an intelligent transportation system called BLISS (the Bus Location and Information Sub-System), and shows that urban planners need to engage not only with the technical difficulties of technology implementation in the city, but also with the contingent and experiential processes of those who use it, and are affected by such implementations.

In "RFID in Urban Planning," Leslie Pang, Vanessa Morgan-Morris, and Angela Howell discuss the application of Radio Frequency Identification (RFID) technology to support the needs and requirements within the realm of urban planning. The authors provide an account of the historic and technical background behind RFID, explore this technology's potential as a practical tool for urban planners and discuss the issues and challenges associated with RFID.

Stephen Aikins, in "E-Planning: Information Security Risks and Management Implications," engages with the security risks and management implications associated with the use of information technology to manage urban and regional planning and development processes, and argues that the emergence of e-planning poses enormous security challenges that need to be managed to ensure integrity, confidentiality and availability of critical planning information for decision-making.

The last two chapters in section three present and discuss the situation of e-Planning in two countries, Turkey and Malaysia. In the penultimate chapter, "E-Planning Applications in Turkish Local Governments," Koray Velibeyoglu explores the critical relationship between e-planning applications and their organizational context. The author shows, based on a case study of Turkish municipalities, that the organizational and human aspects of information systems are still the main obstacle in the implementation of information and communication technologies in urban planning.

In the final chapter, "GIS Implementation in Malaysian Statutory Development Plan System," Muhammad Faris Abdullah, Alias Abdullah and Rustam Khairi Zahari examine the current state of GIS implementation in the Malaysian development plan system comparing it with the state of GIS implementation worldwide.

In sum, the readings in this Handbook of Research on e-Planning provide a well grounded and research focused overview of the emerging e-planning paradigm and will hopefully point readers to future research directions. Ultimately, this collection of essays will ask each reader to reflect on the planning theories that frame the urban planning practice, which methods to use in the preparation and implementation of urban e-plans, how to organise citizen e-participation in urban planning processes, or how to use new information and communication technologies to collect and manage data in different areas of the urban planning process.

For that reason, I hope that this Handbook of Research on e-Planning will assist students, scholars, researchers and practitioners in the field of urban planning and e-government in general, and other public and private stakeholders in the urban arena as well, to advance their understating of e-planning, to identify its main weaknesses and strengths, threats and opportunities, in order to make cities and surrounding areas a better place to live, in a world where more than half of the population now live in urban areas.

Carlos Nunes Silva
Institute of Geography and Spatial Planning
University of Lisbon, Portugal

Acknowledgment

I would like to begin by thanking each of the contributors to this Handbook for the effort devoted to this project and for the excellent work and commitment in the preparation of each chapter. It has been a truly enriching professional experience to work with this group of distinguished colleagues from different disciplines and academic backgrounds.

I also am indebted to all colleagues that accepted to assist me in the review process of several manuscripts, within their field of expertise, and whose names are listed at the front of the book as members of the editorial advisory board and in the list of reviewers, as well as some contributing authors who also served as reviewers. The constructive and helpful comments they offered and their attention to detail helped to improve the quality of each individual manuscript and as a result the overall quality of the Handbook

For the support in the preparation of this publication, I also thank the publishing team at IGI Global, in particular Elizabeth Ardner, with whom I worked most closely throughout the development of the Handbook, for her professionalism and commitment, and Tyler Heath for a similar participation in the initial stage of this project.

Any factual errors or lapses that survived the editorial process are, needless to say, my responsibility.

Carlos Nunes Silva
Editor

Section 1
Theory and Methods in E-Planning

Chapter 1
The E-Planning Paradigm – Theory, Methods and Tools:
An Overview

Carlos Nunes Silva
Institute of Geography and Spatial Planning, University of Lisbon, Portugal

ABSTRACT

The chapter discusses the relationships between planning theory and the use of information and communication technologies in urban planning. It explores how recent organizational transformations in urban planning, associated with the widespread use of information and communication technologies, are incorporated by different planning theories. It is argued that the way information and communication technologies tools are considered or included by the different planning perspectives is in part responsible for the various forms of e-planning.

INTRODUCTION

Contemporary urban planning practice is embedded in a complex and diverse social, political and economic urban world. The implementation of e-Planning, the new urban planning paradigm, requires new concepts, methods, and tools, as happened in the past when other technologies were introduced in this professional field. The history of urban planning, since the end of the nineteenth century, reveals a process of continuous change in the prevailing theories and methodologies, which led to an increasingly complex professional practice (Friedman, 1996; Hall, 2002; Peterson, 2003; Silva, 1994; Talen, 2005; Ward, 2004). The Garden City model, in the formula proposed by Ebenezer Howard (Hall and Ward, 1998, Howard, 1902/2001), or the CIAM discourse on urbanism, steered among others by Le Corbusier (Le Corbusier, 1971; Mumford, 2000), had a vision of planning rather different from the rational planning paradigm that followed it as the main planning paradigm and which framed most of the twentieth century urban planning practice. Rational scientific planning, system theory and the following paradigms, namely the political economy perspective of planning, collaborative or communicative planning, and the various streams of postmodern planning put forward different visions of

DOI: 10.4018/978-1-61520-929-3.ch001

what urban planning is, who benefits from it, and how it should be practiced (Allmendinger, 2002; Faludi, 1973; 1973a; Hillier and Healey, 2008).

Even though information and communication technologies may be seen as neutral technologies, they can certainly be applied to serve different political and social purposes, or to respond to different principles and values (Anttiroiko and Malkia, 2007; Budthimedhee et al., 2002). It is for this reason that the use of information technologies within the rational planning approach has different objectives compared to what happens in collaborative or communicative planning. In the first case, the introduction of information and communication technologies allows planners and planning departments to carry out new actions or to implement conventional practices through new tools, such as geographic information systems, virtual reality technologies, e-participation devices, including public participation GIS applications, among other tools, with the aim of improving conventional decision-making processes. In the second case, the use of similar information and communication technologies tend to be associated with an epistemological turn and in the limit with a change of planning paradigm that goes beyond the basic objective of improving established planning routines.

The provision of better planning and urban management services, more efficient, with lower costs and, at the same time, a more collaborative and participative, transparent and accountable planning decision-making process are some of the basic objectives usually associated with the move from conventional urban planning to e-planning, as the empirical evidence collected in this handbook illustrates. However, this move is not always followed by a paradigm shift from the point of view of planning theory. Neo-positivist as well as post-positivist epistemologies and the corresponding urban planning paradigms do incorporate and use information and communication technologies with different purposes and under different rationales. If e-planning is seen simply as the extensive use of information and communication technologies in urban planning, there will be as much types of e-planning as planning theories that frame them. An e-planning system organized according to the principles of rational planning or system theory will be different from another framed by the principles and by the rationale of collaborative planning or from a third one that combines elements from both in a hybrid and contradictory form of urban planning. For that reason, the term e-planning is frequently used with slightly different meanings, and this handbook is no exception in that respect. In this chapter the term e-planning stands for the new urban planning paradigm, characterized by the extensive use of information and communication technologies in all phases of the urban planning process, within the framework of a post-positivist planning theory.

BACKGROUND

Planning Theory

The questions "who gains from urban planning?" and "what role citizens can play in the planning process?" and, more specifically, "how to handle communication and collaboration between different urban stakeholders?", a discussion renewed due to the new possibilities offered by information and communication technologies, are key points in the planning theory debate. To some extent, the ways these issues have been addressed are responsible for the differences between the various modern and post-modern planning paradigms, which more than well delimited theories must be seen as clusters of perspectives co-existing in time and with strong commonalties.

In the Garden City movement, City Beautiful movement or in the CIAM discourse on Urbanism, for example, urban planning was regarded as urban design (Silva, 2003; 2005; 2005a). This planning as design paradigm was largely dominant until the 1960s, when the rational theory of planning and systems theory became the core paradigm in the field of urban planning. Rational theory and system theory perceive planning as an instrument to administer social processes, contrasting with the previous paradigm (Allmendinger, 2002; Friedman, 1996; Hillier and Healey, 2008). This shift represented also a change on how the nature and purpose of urban planning was considered by professionals in the field, as well as in the methodology and planning tools that should be used. Later, in reaction to rational scientific planning, new approaches emerged, as the political economy perspective, and as a result a new vision of planning in society and new concepts, methodologies and tools were developed, especially in the 1970s and in the 1980s.

In the political economy perspective, issues of power and class struggle are key references on how planners perceive the city and how they think it should be planned. It challenges the individualistic point of view and the importance given to individual choice, contrary to what happens in the rational planning paradigm. In other words, urban land use patterns and the distribution of urban functions within the city are determined not by individual preference but by the capitalist economic system (for a comprehensive revision of this approach, see, among others: Allmendinger, 2002; Friedman, 1996; Hillier and Healey, 2008a). Urban planning is regarded as an instrument, controlled by the dominant social class, for the creation of conditions for capital accumulation and social reproduction at the local level (Fainstein and Fainstein, 1979; Harvey, 1985; Scott and Roweis, 1977). For that reason, e-planning, as the conventional paper based urban planning before, is regarded as an instrument for capital accumulation and for the reproduction and legitimization of capitalist social relations, controlled by the dominant social class.

The theory of communicative or collaborative planning is, perhaps, the most influent planning theory among those that emerged as alternatives to the rational planning paradigm. Like other post-positivist planning paradigms, this perspective regards urban planning as an activity that deals with socially fragmented communities (Allmendinger, 2002; Forester, 1989; Hillier and Healey, 2008b). Compared to previous perspectives of planning (e.g., planning as design, rational planning, system theory or political economy), communicative or collaborative planning recognizes and gives more emphasis to the diversity of values, meanings, and interests within each local community (Healey, 1997). Parallel to this theoretical shift there was also a change in the prevailing discourse on urbanism, a move from the CIAM and its Charter of Athens to the discourse(s) on Urbanism synthesized in the Charter of New Urbanism (CNU, 1998) and in the New Charter of Athens (ECTP, 2003; Silva, 2005b).

Planning is considered to be a communicative or interactive activity among public and private stakeholders within public decision-making processes (Innes, 1995). As a result, what is important in urban planning, from this point of view, is to consider the needs of the various urban stakeholders and to organize communication between them (Innes and Booher, 1999). Planners need to be aware of the values and interests of these different groups of urban stakeholders, and should then identify common interests and visions through the use of appropriate participatory practices within the urban planning decision-making process (Healey, 2003). In order to address the different needs and demands of multiple urban stakeholders, planning practice should give emphasis to communication and collaboration between urban stakeholders more than to the formal procedures, contrary to what was common in the rational planning paradigm. In other words, e-planning, informed by the theory of collaborative or communicative planning will tend to adopt methods and tools that seek to provide a better understanding of how citizens perceive their own condition and will try to articulate these contradictory perspectives.

For that reason, e-Planning should not be regarded as an instrument of urban design control but as an instrument for collective action in the urban arena. From this perspective, the importance of e-planning and its social relevance don't result exclusively from the enforcement of administrative and technical planning procedures, as was the case in planning as design paradigm or in the rational planning paradigm, but also from the fact that it steers communication and collaboration among urban stakeholders in order to articulate the different interests and perspectives. Therefore, for the communicative or collaborative planning perspective, e-Planning is a highly political professional practice and not a value-neutral professional activity as is admitted by neo-positivist perspectives.

However, a number of problems have been pointed out (for example, see Allmendinger, 2002; Fischler, 2000; Miraftab, 2003; Margerum, 2002) questioning the emphasis put on planning procedures and on participatory methods, and the exclusion of critical factors that are responsible for the uneven access to the urban planning decision-making process, such as the unequal distribution of power among the various urban stakeholders, or the argument of Talen and Ellis (2002) in favour of a search for good city form.

There are other post-structuralist and post-modern planning theories, which have in common, among other characteristics, the fact that they are based on a non universal vision of the urban world (for a comprehensive account see, among others: Allmendinger, 2002; Hillier and Healey, 2008b; and Peet, 1998 for a review of these social theories). These and other emerging post-structuralist and post-modern planning theories, which tend to see planning as a complex practice and social values as no longer permanent but in continuous change, will certainly shape how the new e-Planning paradigm will be practiced in the future.

However, despite the growing influence of these new paradigms, rational theory of planning and systems theory continue to influence urban planning. The various types of impact analysis applied in urban planning (e.g., on environmental issues, on urban transport systems, etc.), usually a compulsory component in most urban planning systems, are based on assumptions and principles of system theory and rational theory of planning. The same can be said about similar analysis within e-planning, in which more sophisticated models based on geographic information systems and on other information technologies are usually employed for the description, interpretation, and evaluation of urban structures and processes, as well as urban scenarios, suggesting a situation in which different urban planning paradigms, their assumptions, values, and methodological tools are interrelated in apparently contradictory theoretical perspectives.

E-Planning Tools

In its more basic definition, e-planning is the result of the shift from a paper based urban planning to a system supported primarily by information and communication technologies, as mentioned before (Silva, 2007). It involves also changes in the nature of the planning process, and should not be seen as a simple transfer to a computer system of long-established paper based urban planning routines. This move requires a change in the methods of data collection, storage and analysis, a revision of public participation practices, new mechanisms for the control of planning scenarios, and for monitoring and evaluation of urban development processes, as well as new ethical considerations (Buchanan, 2004; Silva, 2008). It also requires an adequate geospatial database and an integrated back office for online services (Kubicek et al., 2007).

The e-tools applied in e-Planning include, among others, geographic information systems, internet interactive mapping or Internet GIS, virtual reality technologies, data modelling, computer aided design, various types of computer supported collaborative working environments, interactive social media tools (web 2.0) applied to participatory planning as well as to the production of knowledge relevant for planning. Other tools are, for example, public participation geographic information systems (for an overview see, among others: Elwood, 2006; Harrison and Haklay, 2002; Jankowski, 2009; Nedovic-Budic, 2000), the combination of Internet GIS with virtual reality, offering citizens, and other urban stakeholders, better visualizations of the urban plan, and the use of location technologies in the planning process. Web 2.0 tools, such as planning blogs, wikis, chat rooms, discussion forums, virtual communities, mailing lists (see Noveck, 2009, on wiki government; Wyld, 2008, on blogs), are examples, among others, of tools that can be used to promote citizen participation in planning decision-making and more active contributions to the planning process, as well as citizen-to-citizen exchanges about the urban plan. Mobile phones, one of the most important applications of ICT in developing countries, can fulfil an important role in the implementation of citizen e-participation and in other communication components of e-planning (see, among others, chapters 2 and 3 and Section III for an overview of new technologies applied in urban planning; and chapter 9 for e-participation tools).

In sum, a fully developed and accessible e-Planning system provides on-line general information about the planning system, planning laws and regulations, and planning procedures, and contains basic and specialized information on all aspects of the planning system, in its several scales – local, regional and national –, increasing the level of automation in the planning process and improving the conditions for citizen participation. It comprises also the publication of local plans, including 3D-visualization of built and natural environments, technical reports, public participation documents, monitoring and evaluation reports, and urban marketing tools. Urban plans are available in 2-D or 3-D electronic format, allowing citizens and other urban stakeholders to visualize land use proposals through a public participation geographical information system (for an overview of some of these issues, see, among others: Carver et al., 2001; Jankowski, 2009; Sieber, 2006). Detailed information on each parcel of land or building is also available online, with restricted access conditions to sensitive personal information. The system provides ubiquitous access to on-line planning services: pre-application advice, submission of applications, consultation, e-petitions, commentaries, complaints and planning decisions, among others, including the possibility to pay online. It ensures appropriate external and internal security controls, reducing or avoiding data theft (see chapter 21 for security issues and Arnesen and Danielsson, 2007; Janczewski and Portougal, 2007).

The main planning portal, in a typical e-planning system, should include an agenda of events related to urban planning (e.g., public inquires or municipal political boards meetings dealing with planning issues), and surveys to measure citizen satisfaction with e-Planning services, among other functions and online services. If not integrated or linked to a wider municipal web site, this planning portal should also contain general information about the municipality and the respective municipal organization. A section for children and youths, with information for different ages, structured around planning issues, should be incorporated as a stimulus for the participation of young people in urban life (CoE, 2003; UN, 1989). In an increasingly multicultural society, an e-planning system should also have versions in the main non-native languages present in the local community.

However, it is not enough to make information available and to stimulate participation, as the research results of Carver et al. (2001), Madon et al. (2009) and Warschauer (2003), among others, seem to suggest. It is necessary to give information that can be understood by all urban stakeholders, to provide technology that the common citizen can use, to apply information and communication technology for community empowerment (Ghose, 2001), and to apply the principles and norms of the Aarhus Convention on access to information and public participation in urban decision-making processes (UN, 1998), in order to foster e-inclusion (see chapter 17).

E-Planning in Practice

In practice, however, e-Planning systems do not function in the same way in all places, and, as a result of that, major differences exist between developed and developing countries (see chapter 16 in this handbook), or within the same country, where each local e-planning system is unique for the reason that every municipality's economic, social, political, and cultural characteristics are different. For example, in developing countries, chaotic urbanization processes and the failure of conventional paper based urban planning offer complex challenges for e-planning, different from those faced by planning organizations in more developed countries (Bishop et al., 2002). In addition, the way urban planning is organized locally, within the limits allowed by central state, is also responsible for some of the differences encountered.

There are also major distinctions between planning authorities within the same country, associated with the values and aims assigned to the urban planning system and not with the level of ICT infra-structures or with those social, economic, political and cultural characteristics. Geographic information systems and virtual reality technologies (see chapters 6, 7 and 8) are applied to model and simulate the outcomes of an urban plan, replacing paper based forms of representation, under different planning paradigms. Some chapters in this handbook illustrate well this diversity of perspectives within e-Planning. Chapter 2 discusses the use of internet-based interactive mapping and virtual reality technologies in collaborative decision-making. Although the author approaches public participation from a collaborative planning perspective, the role assigned to geographic information systems and other technologies in urban planning modelling may well follow to some extent the rational theory of planning. In chapter 3, the authors propose two methodologies for the design of collaborative e-planning systems, which however can also be used within a neo-positivist or rational scientific perspective of planning, and chapter 4 refers a multi-ontology in future-studies and the different methods that may be used, and presents a participatory planning scheme based on communicative and post-structural planning theories (see, for example, chapter 5).

The evidence available also suggests that the use of new technologies to enhance public participation in urban planning per se does not change much, since it is dependent on the specific context in which the communication between the planning organization and the citizens takes place (see, for example, chapters 9 to 15 in this handbook). In other words, new e-participation tools can be used with a different purpose and rationale, within the rational planning paradigm or system theory or in the framework of the communicative or collaborative perspective of urban planning. For instance, Internet Forums and Planning Portals and other electronic devices applied in urban planning, can potentially be used by different planning approaches (see chapters 5 and 13; and Campagna and Deplano, 2004). However, without the commitment to empower citizens and to share power, by those that hold the political authority to decide on planning matters, the impact of these e-participation tools in the overall urban planning decision-making process will be limited.

In synthesis, the evidence available suggests that the planning theory and the policy that guide the use of the technology is far more important than the type of tool, electronic or conventional, employed in the planning process.

FUTURE RESEARCH DIRECTIONS

Future developments in the e-Planning systems are dependent of external and internal conditions (Anttiroiko and Malkia, 2007; Falch, 2006; Silva, 2007). The overall growth of the ICT sector and the expansion of the information society (e.g., interoperability of ICT, market penetration of specific ICT, broadband universal access, digital literacy, digital divide, etc.) are examples of these external factors. In the second case, the prevailing planning theory, confidence in the system, material and financial conditions experienced by the planning department are some of the factors that will certainly have an effect on the expansion of e-planning systems.

From the point of view of the external conditions, the growth of ICT use among urban stakeholders, including the use of online urban planning services, in spite of its social and geographical unevenness (Graham, 2002; Grimes, 2003; Winden et al., 2004), will increase the number of participants in e-Planning systems and, due to that, will create new demands for planning services and information, a trend that will be further stimulated by broadband connectivity, which will allow the use of more sophisticated online planning services. In addition, the present resistance in the use of information and communication technologies due to weak digital literacy will diminish or disappear as analogical forms of communications are gradually replaced by digital technologies. Parallel to these technological trends that stimulate greater use of online information and online planning services, there is at the political level, in national as well as in local governments all over the world (see examples in Anttiroiko and Malkia, 2007; Garson and Khosrow-Pour, 2008), a trend towards the implementation of digital government projects that will certainly reinforce the ongoing expansion of the information society.

Internal factors are also critical for the future expansion of e-Planning systems. Data integrity protection, the risks associated with the loss of privacy and confidentiality in the transactions between citizens and the e-Planning system and a myriad of other ethical issues are critical factors that urban planning departments have to consider carefully (Aikins, 2008; Dodig-Crnkovic and Horniak, 2007; Melville, 2007; Rowe, 2007), as well as planning professional organizations (Danielson, 2007; Schultz, 2006; Silva, 2007a; 2008). The characteristics of digital archives for future use are also factors that may affect confidence in the system. Security and confidence in the e-planning system are for those reasons two critical factors for its success (see also chapter 21 in this handbook for an analysis of some of these challenges).

Future research on e-planning must examine the way both internal and external conditions affect the development of urban e-planning systems in different geographical contexts. Complexity, rapid technological innovations, security challenges and ethical issues suggest that there is still much to be researched on e-planning, as several chapters in this handbook point out. For example, how successful is e-planning in each of its components? How important are organizational factors for the success of e-planning systems? How should it adapt to particular contexts, in developed as well as in developing countries? How to tackle the digital divide in its various dimensions (economic divide, computer literacy divide, language divide, disability divide, and geographical divide) and the effects it has on e-planning systems? How to deal with security vulnerabilities of urban planning data and to ensure data accuracy, integrity or protection against unauthorized change of data, and confidentiality of critical and sensitive information within the planning system? How does it impact on common citizens? What are the main obstacles in the implementation of an e-planning system? What are the tangible and intangible benefits and costs of e-planning? How should it be organized and implemented according to the different post-positivist planning theories?

There is also scope for further research on each

specific e-tool applied in e-planning, especially how to use them in specific (multi-)cultural environments, as well as ethical dilemmas associated with each of them (Buchanan, 2004; Silva, 2008). The impact of interactive social media tools (web 2.0) on knowledge production, in public participation and in other components of the planning system; the impact of other electronic tools in public participation processes; or the validity of virtual representations as substitutes of the real world in public participation processes are some examples of the numerous directions that research on e-planning must follow.

CONCLUSION

In sum, a fully developed urban e-planning system requires an extensive use of information and communication technologies, in all phases of the planning process, within the framework of a post-positivist planning theory. This extensive use of information and communication technologies has the potential to improve the efficiency, effectiveness and the social, economic and environmental impacts of current urban planning systems. With the new technologies, the urban planning system can help promote citizen empowerment, improve social cohesion, economic competitiveness and environmental sustainability, although it can also create new social exclusions, in some cases coincident with old social, cultural, economic and geographical divides, as did other technologies in the past. However, it will only mean a true shift of planning paradigm if associated with a change of principles and values. On balance, no information and communication technology is as important and determinant for the urban planning system as the planning theory and the policy that guide the use of the technology.

Among the key challenges faced by a standard e-Planning system, it is important to highlight the uneven access to planning information and planning services by the most disadvantaged social groups, the elderly and persons with a disability, as well as the problems associated with broadband access in the most remote rural areas and in the most socially deprived neighbourhoods in the suburbs and in the inner cities. This digital divide in the access to online planning information and planning services limit the effectiveness and sustainability of e-planning systems. If large segments of the population can't access on-line planning services and information or if they are not able to comprehend the information, e-Planning will certainly contribute to increase social exclusion and geographical inequalities. To prevent that to happen, it is necessary to ensure digital literacy and broadband accessibility for all citizens independently of her/his ability, age, gender, ethnicity, or economic capacity.

The growing access and use of information and communication technologies, especially the Internet, although experienced unevenly by the different urban stakeholders, implied a democratization of knowledge on urban issues, which affect the dominant position of planners within the urban planning process, in all phases of the process. This shift in the way knowledge on urban issues is produced, both formal and informal knowledge, on how the urban planning process is organized and on how urban plans are implemented, monitored and evaluated, creates new challenges for urban planners, for citizens in general, and for other urban stakeholders as well, challenges that need to be addressed by planning theory and planning ethics as well as through new urban planning methods and e-tools.

REFERENCES

Aikins, S. K. (2008). Practical measures for securing government networks. In Garson, G. D., & Khosrow-Pour, M. (Eds.), *Handbook of Research on Public Information Technology* (pp. 386–394). New York: Information Science Reference.

Almendinger, P. (2002). *Planning theory*. Basingstoke, UK: Palgrave Macmillan.

Anttiroiko, A.-V., & Malkia, M. (Eds.). (2007). *Encyclopedia of Digital Government*. Hershey, PA: Idea Group Publishing.

Arnesen, R. R., & Danielsson, J. (2007). Protecting citizen privacy in digital government. In Anttiroiko, A.-V., & Malkia, M. (Eds.), *Encyclopedia of Digital Government* (pp. 1358–1363). Hershey, PA: Idea Group Publishing.

Bishop, I. D., Barry, M., McPherson, E., Nascarella, J., Urquhart, K., & Escobar, F. (2002). Meeting the Need for GIS Skills in Developing Countries: The Case of Informal Settlements. *Transactions in GIS, 6*(3), 311-326.

Buchanan, E. A. (2004). *Readings in virtual research ethics. Issues and controversies*. Hershey, PA: Information Science Publishing.

Budthimedhee, K., Li, J., & George, R. V. (2002). e-Planning: a snapshot of the literature on using the World Wide Web in urban planning. *Journal of Planning Literature, 17*(2), 227–246. doi:10.1177/088541202762475964

Campagna, M., & Deplano, G. (2004). Evaluating geographic information provision within public administration websites. *Environment and Planning. B, Planning & Design, 31*, 21–37. doi:10.1068/b12966

Carver, S., Evans, A., Kingston, R., & Turton, I. (2001). Public participation, GIS, and cyber democracy: evaluating on-line spatial decision support systems. *Environment and Planning. B, Planning & Design, 28*, 907–921. doi:10.1068/b2751t

CNU. (1998). *Charter of the New Urbanism*. San Francisco: Congress for the New Urbanism.

CoE. (2003). *European Charter on the Participation of Young People in Local and Regional Life*. Strasbourg, France: Council of Europe.

Danielson, P. (2007). Digital morality and Ethics. In Anttiroiko, A.-V., & Malkia, M. (Eds.), *Encyclopedia of Digital Government* (pp. 377–381). Hershey, PA: Idea Group Publishing.

Dodig-Crnkovic, G., & Horniak, V. (2007). Ethics and privacy of communications in the e-Polis. In Anttiroiko, A.-V., & Malkia, M. (Eds.), *Encyclopedia of Digital Government* (pp. 740–744). Hershey, PA: Idea Group Publishing.

ECTP. (2003). *The New Charter of Athens*. Bruxells, Belgium: European Council of Town Planners.

Elwood, S. (2006). Participatory GIS and community planning: restructuring technologies, social processes and future research in PPGIS. In S. Balram & Suzana Dragićević (Eds.), Collaborative Geographic Information Systems (pp. 66-84). Hershey, PA: Idea Group Publishing.

Fainstein, N. I., & Fainstein, S. S. (1979). New debates in urban planning: the impact of Marxist theory within the United States. *International Journal of Urban and Regional Research, 3*, 381–403. doi:10.1111/j.1468-2427.1979.tb00796.x

Falch, M. (2006). ICT and the future conditions for democratic governance. *Telematics and Informatics, 23*, 134–156. doi:10.1016/j.tele.2005.06.001

Faludi, A. (1973). The rationale of planning theory. In Faludi, A. (Ed.), *A Reader in Planning Theory* (pp. 35–53). Oxford, UK: Pergamon Press.

Faludi, A. (1973a). What is planning theory? In Faludi, A. (Ed.), *A Reader in Planning Theory* (pp. 1–10). Oxford, UK: Pergamon Press.

Fischler, R. (2000). Communicative Planning Theory. A Foucaldian Assessment. *Journal of Planning Education and Research, 19*(4), 358–368. doi:10.1177/0739456X0001900405

Forrester, J. (1989). *Planning in the face of power*. Berkeley, CA: University of California Press.

Friedmann, J. (1996). Two centuries of planning theory: An Overview. In Mandelbaum, S. J., Mazza, L., & Burchell, R. W. (Eds.), *Explorations in planning theory* (pp. 10–29). New Brunswick, CT: Centre for Urban Policy Research.

Garson, G. D., & Khosrow-Pour, M. (Eds.). (2008). Handbook of Research on Public Information Technology (Vol. 1 & 2). New York: Information Science Reference.

Ghose, R. (2001). Use of Information Technology for Community Empowerment: Transforming Geographic Information Systems into Community Information Systems. *Transactions in GIS, 5*, 141–163. doi:10.1111/1467-9671.00073

Graham, S. (2002). Bridging urban digital divides? Urban polarization and information and communications technologies. *Urban Studies (Edinburgh, Scotland), 39*(1), 33–56. doi:10.1080/00420980220099050

Grimes, S. (2003). The digital economy challenge facing peripheral rural areas. *Progress in Human Geography, 27*(2), 174–193. doi:10.1191/0309132503ph421oa

Hall, P. (2002). *Cities of tomorrow: An intellectual history of urban planning and design in the twentieth century*. Oxford, UK: Blackwell.

Hall, P., & Ward, C. (1998). *Sociable Cities: The Legacy of Ebenezer Howard*. Chichester, UK: John Wiley & Sons.

Harrison, C., & Haklay, M. (2002). The potential of public participation geographic information systems in UK environmental planning: appraisals by active publics. *Journal of Environmental Planning and Management, 45*(6), 841–863. doi:10.1080/0964056022000024370

Harvey, D. (1985). On planning the ideology of planning. In Harvey, D. (Ed.), *The Urbanization of capital* (pp. 165–184). Oxford, UK: Blackwell.

Healey, P. (1997). *Collaborative Planning: shaping places in fragmented societies*. Basingstoke, UK: Palgrave.

Healey, P. (2003). Collaborative planning in perspective. *Planning theory, 2*(2), 101-123.

Hillier, J., & Healey, P. (Eds.). (2008). *Critical essays in planning theory* (Vol. 1). Aldershot, UK: Ashgate.

Hillier, J., & Healey, P. (Eds.). (2008). *Critical essays in planning theory* (Vol. 2). Aldershot, UK: Ashgate.

Hillier, J., & Healey, P. (Eds.). (2008). *Critical essays in planning theory* (Vol. 3). Aldershot, UK: Ashgate.

Howard, E. (2001). *Garden Cities of To-morrow*. New York: Books for Business.

Innes, J. E. (1995). Planning theory's emerging paradigm: communicative action and interactive practice. *Journal of Planning Education and Research, 14*, 183–189. doi:10.1177/0739456X9501400307

Innes, J. E., & Booher, D. E. (1999). Consensus building as role playing and bricolage: toward a theory of collaborative planning. *Journal of the American Planning Association. American Planning Association, 65*, 9–26. doi:10.1080/01944369908976031

Janczewski, L., & Portugal, V. (2007). Managing security clearances within government institutions. In Anttiroiko, A.-V., & Malkia, M. (Eds.), *Encyclopedia of Digital Government* (pp. 1196–1202). Hershey, PA: Idea Group Publishing.

Jankowski, P. (2009). Towards participatory geographic information systems for community-based environmental decision making. *Journal of Environmental Management, 90*(6), 1966–1971. doi:10.1016/j.jenvman.2007.08.028

Kubicek, H., Millard, J., & Westholm, H. (2007). Back-Office integration for online services between organizations. In Anttiroiko, A.-V., & Malkia, M. (Eds.), *Encyclopedia of Digital Government* (pp. 123–130). Hershey, PA: Idea Group Publishing.

Le Corbusier (1971). *La Charte d'Athènes*. Paris: Seuil.

Madon, S., Reinhard, N., Roode, D., & Walsham, G. (2009). Digital Inclusion Projects in Developing Countries: Processes of Institutionalization. *Information Technology for Development, 15*(2), 95–107. doi:10.1002/itdj.20108

Margerum, R. D. (2002). Collaborative Planning Building Consensus and Building a Distinct Model for Practice. *Journal of Planning Education and Research, 21*, 237–253. doi:10.1177/0739456X0202100302

Melville, R. (2007). Ethical dilemmas in online research. In Anttiroiko, A.-V., & Malkia, M. (Eds.), *Encyclopedia of Digital Government* (pp. 734–739). Hershey, PA: Idea Group Publishing.

Miraftab, F. (2003). The Perils of Participatory Discourse: Housing Policy in Post-apartheid South Africa. *Journal of Planning Education and Research, 22*, 226–239. doi:10.1177/0739456X02250305

Mumford, E. (2000). *The CIAM discourse on Urbanism, 1928-1960*. Cambridge, MA: The MIT Press.

Nedovic-Budic, Z. (2000). Geographic Information Science Implications for Urban and Regional Planning. *URISA Journal, 12*(2), 81–93.

Noveck, B. S. (2009). *Wiki Government. How technology can make government better, democracy stronger, and citizen more powerful*. Washington, DC: Brookings Institution Press.

Peet, R. (1998). *Modern geographical thought*. Oxford, UK: Blackwell.

Peterson, J. A. (2003). *The birth of city planning in the United States, 1840-1917*. London: The John Hopkins University Press.

Rowe, N. C. (2007). Cyber Attacks. In Anttiroiko, A.-V., & Malkia, M. (Eds.), *Encyclopedia of Digital Government* (pp. 271–276). Hershey, PA: Idea Group Publishing.

Schultz, R. A. (2006). *Contemporary Issues in Ethics and Information Technology*. Hershey, PA: Information Science Publishing.

Scott, A. J., & Roweis, S. T. (1977). Urban planning in theory and practice: a reappraisal. *Environment & Planning A, 9*(10), 1097–1119. doi:10.1068/a091097

Sieber, R. (2006). Public Participation Geographic Information Systems: A Literature Review and Framework. *Annals of the American Association of Geographers, 96*(3), 491–507. doi:10.1111/j.1467-8306.2006.00702.x

Silva, C. N. (1994). *Política Urbana em Lisboa, 1926-1974*. Lisbon: Livros Horizonte.

Silva, C. N. (2003). Urban utopias in the twentieth century. *Journal of Urban History, 29*(3), 327–332. doi:10.1177/0096144203029003011

Silva, C. N. (2005). Charter of Athens. In Caves, R. W. (Ed.), *Encyclopedia of the City* (pp. 52–53). London: Routledge.

Silva, C. N. (2005a). City Beautiful. In Caves, R. W. (Ed.), *Encyclopedia of the City* (pp. 69–70). London: Routledge.

Silva, C. N. (2005b). New Charter of Athens. In Caves, R. W. (Ed.), *Encyclopedia of the City* (pp. 328–329). London: Routledge.

Silva, C. N. (2007). e-Planning. In A.-V. Anttiroiko & M. Malkia (Eds.), Encyclopedia of Digital Government (pp. 703-707). Hershey, PA: Idea Group Publishing.

Silva, C. N. (2007a). Urban Planning and Ethics. In Rabin, J., & Berman, E. M. (Eds.), *Encyclopedia of Public Administration and Public Policy* (2nd ed.). New York: CRC Press / Taylor & Francis Group. doi:10.1201/NOE1420052756.ch410

Silva, C. N. (2008). Research Ethics in e-Public Administration. In Garson, G. D., & Khosrow-Pour, M. (Eds.), *Handbook of Research on Public Information Technology* (Vol. 1, pp. 314–322). New York: Information Science Reference.

Talen, E. (2005). *New Urbanism and American Planning: the conflict of cultures*. London: Routledge.

Talen, E., & Ellis, C. (2002). Beyond Relativism. Reclaiming the search for good city form. *Journal of Planning Education and Research, 22*(1), 36–49. doi:10.1177/0739456X0202200104

UN. (1989). *Convention on the Rights of the Child*. New York: United Nations.

UN. (1998). *Convention on access to information, public participation in decision-making and access to justice in environmental matters (Aarhus Convention)*. New York: United Nations.

Ward, S. V. (2004). *Planning and Urban Change*. London: Sage Publications.

Warschauer, M. (2003). Social capital and access. *Universal Access in the Information Society, 2*, 315–330. doi:10.1007/s10209-002-0040-8

Winden, W. V., & Woets, P. (2004). Urban broadband Internet policies in Europe: a critical review. *Urban Studies (Edinburgh, Scotland), 41*(10), 2043–2059. doi:10.1080/0042098042000256378

Wyld, D. C. (2008). Blogging. In Garson, G. D., & Khosrow-Pour, M. (Eds.), *Handbook of Research on Public Information Technology* (pp. 81–93). New York: Information Science Reference.

ADITIONAL READING

Arnstein, S. R. (1969). A Ladder of Citizen Participation. *Journal of the American Institute of Planners, 35*, 216–224.

Beatley, T. (1994). *Ethical Land Use. Principles of Policy and Planning*. Baltimore: The John Hopkins University Press.

Beauregard, R. A. (1989). Between modernity and postmodernity: the ambiguous position of US Planning. *Environment and Planning. D, Society & Space, 7*, 381–395. doi:10.1068/d070381

Booher, D. E., & Innes, J. E. (2002). Network power in collaborative planning. *Journal of Planning Education and Research, 21*, 221–236. doi:10.1177/0739456X0202100301

Borja, J., & Castells, M. (1997). *Local and global: management of cities in the information age*. London: Earthscan.

Brail, R. K., & Klosterman, R. E. (Eds.). (2001). *Planning Support Systems: Integrating Geographic Information Systems, Models, and Visualization Tools*. Redlands, CA: ESRI, Inc.

Brennan, L. L., & Johnson, V. E. (2004). *Social, Ethical and Policy Implications of Information Technology*. Hershey, PA: Information Science Publishing.

Campbell, H., & Marshall, R. (1999). Ethical frameworks and planning theory. *International Journal of Urban and Regional Research, 23*, 464–478. doi:10.1111/1468-2427.00208

Campbell, S., & Fainstein, S. S. (2003). *Readings in planning theory*. Oxford, UK: Blackwell.

Castells, M. (1977). *The urban question: a Marxist approach*. London: Edward Arnold.

Castells, M. (2000). The Information Age: Economy [nd Ed.). Oxford, UK: Blackwell.]. *Society and Culture, 1*, 2.

Caves, R. W. (Ed.). (2005). *Encyclopedia of the City*. London: Routledge.

Craig, W. J., Harris, T. M., & Weiner, D. (Eds.). (2002). *Community Participation and Geographic Information Systems*. London: Taylor and Francis.

Davidoff, P. (1965). Advocacy and Pluralism in Planning. *Journal of the American Institute of Planners, 31*, 331–338.

Davidoff, P., & Reiner, T. A. (1973). A choice theory of planning. In Faludi, A. (Ed.), *A Reader in Planning Theory* (pp. 11–39). Oxford, UK: Pergamon Press.

Dear, M. J. (1986). Postmodernism and planning. *Environment and Planning. D, Society & Space, 4*, 367–384. doi:10.1068/d040367

Fainstein, N. I., & Fainstein, S. S. (1979). New debates in urban planning: the impact of Marxist theory within the United States. *International Journal of Urban and Regional Research, 3*, 381–403. doi:10.1111/j.1468-2427.1979.tb00796.x

Forester, J. (1993). Understanding planning practice: an empirical, practical and normative account. In Forester, J. (Ed.), *Critical theory, public policy, and planning practice* (pp. 15–35). Albany, NY: State University of New York Press.

Friedmann, J. (1973). The transactive style of planning. In Friedmann, J. (Ed.), *Retracking America: A theory of transactive planning* (pp. 171–193). New York: Doubleday.

Graham, S., & Marvin, S. (2001). *Splintering Urbanism: Networked Infrastructures, Technological Mobilites and the Urban Condition*. London: Routledge. doi:10.4324/9780203452202

Grant, J. (2006). *Planning the Good Community. New Urbanism in Theory and Practice*. London: Routledge.

Harper, T. L., & Stein, S. M. (1995). A classical liberal (libertarian) approach to planning theory. In Hendler, S. (Ed.), *Planning Ethics: A Reader in Planning Theory, Practice and Education* (pp. 11–29). New Brunswick, CT: Center for Urban Policy Research.

Harvey, D. (1973). *Social justice and the city*. London: Edward Arnold.

Healey, P. (1992). 'A planner's day: knowledge and action in communicative practice. *Journal of the American Planning Association. American Planning Association, 58*, 9–20. doi:10.1080/01944369208975531

Hendler, S. (Ed.). (1995). *Planning Ethics: A Reader in Planning Theory, Practice and Education*. New Brunswick, CT: Center for Urban Policy Research.

Imrie, R. (1996). *Disability and the city*. London: Paul Chapman.

King, A. D. (2004). *Spaces of Global Cultures. Architecture, Urbanism, Identity*. London: Routledge.

Lindblom, C. E. (1973). The Science of Muddling Through. In Faludi, A. (Ed.), *A Reader in Planning Theory* (pp. 151–169). Oxford, UK: Pergamon Press.

Mandelbaum, S. J., Mazza, L., & Burchell, R. W. (Eds.). (1996). *Explorations in planning theory*. New Brunswick, CT: Centre for Urban Policy Research.

McLoughlin, J. B. (1969). The guidance and control of change: physical planning as the control of complex systems. In McLoughlin, J. B. (Ed.), *Urban and Regional Planning: A system approach*. London: Faber and Faber.

Monclús, J., & Guàrdia, M. (Eds.). (2006). *Culture, Urbanism and Planning*. Aldershot, UK: Ashgate.

Mossberger, K., Tolbert, C. J., & McNeal, R. S. (2007). *Digital citizenship: the Internet, Society, and Participation*. Cambridge, MA: MIT Press.

Mumford, L. (1961). *The City in History: Its origins, its transformations, and its prospects*. London: Secker and Warburg.

Newman, P., & Thornley, A. (2005). *Planning World Cities. Globalization and Urban Politics*. Basingstoke, UK: Palgrave-Macmillan.

Nyerges, T. L., & Jankowski, P. (2007). Participatory Geographic Information Science. In Anttiroiko, A.-V., & Malkia, M. (Eds.), *Encyclopedia of Digital Government* (pp. 1314–1318). Hershey, PA: Idea Group Publishing.

Rowe, N. C. (2007). Trust in digital government. In Anttiroiko, A.-V., & Malkia, M. (Eds.), *Encyclopedia of Digital Government* (pp. 1572–1576). Hershey, PA: Idea Group Publishing.

Sandercock, L. (1998). *Towards cosmopolis*. Chichester, UK: John Wiley.

Sandercock, L., & Forsyth, A. (1992). Feminist theory and planning theory: the epistemological linkages. *Planning Theory*, 7-8, 45–49.

Taylor, N. (1998). *Urban planning theory since 1945*. London: Sage.

Thornley, A., & Rydin, Y. (Eds.). (2003). *Planning in a Global Era*. Aldershot, UK: Ashgate.

van den Berg, L., van der Meer, A., van Winden, W., & Woets, P. (2006). *E-Governance in European and South African Cities. The Cases of Barcelona, Cape Town, Eindhoven, Johannesburg, Manchester, Tampere, The Hague and Venice*. Aldershot, UK: Ashgate.

Verma, N. (1996). Pragmatic Rationality and Planning Theory. *Journal of Planning Education and Research*, 16, 5–14. doi:10.1177/0739456X9601600102

von Lucke, J. (2007). Portals for the public sector. In Anttiroiko, A.-V., & Malkia, M. (Eds.), *Encyclopedia of Digital Government* (pp. 1328–1333). Hershey, PA: Idea Group Publishing.

Walters, G. (2002). *Human Rights in an Information Age. A philosophical analysis*. Toronto, Canada: University of Toronto Press.

KEY TERMS AND DEFINITIONS

Digital Literacy: The capacity to make effective use of information and communication technologies.

E-Inclusion: Means equal access, for all citizens, to planning information and planning services through information and communication technologies, independently of her/his ability, age, gender, ethnicity, or economic capacity.

E-Planning: The application of e-government principles to urban planning or, in other words, the extensive use of information and communication technologies in all phases of the urban planning process, within the framework of a post-positivist planning theory.

Information and Communication Technologies (ICT): Include local computer networks, the Internet, electronic mail, digital television, mobile communications, etc.

Planning Portal (or Planning Gateway): The access point in the Internet, where information on urban planning issues is available and where online planning services are provided.

Public Participation Geographical Information System (PPGIS): A Web GIS facility that allows the viewing of plan proposals. It enables citizens and other urban stakeholders to participate actively in the planning process.

Chapter 2
Planning Online:
A Community-Based Interactive Decision-Making Model

Tan Yigitcanlar
Queensland University of Technology, Australia

ABSTRACT

The first use of computing technologies and the development of land use models in order to support decision-making processes in urban planning date back to as early as mid 20th century. The main thrust of computing applications in urban planning is their contribution to sound decision-making and planning practices. During the last couple of decades many new computing tools and technologies, including geospatial technologies, are designed to enhance planners' capability in dealing with complex urban environments and planning for prosperous and healthy communities. This chapter, therefore, examines the role of information technologies, particularly internet-based geographic information systems, as decision support systems to aid public participatory planning. The chapter discusses challenges and opportunities for the use of internet-based mapping application and tools in collaborative decision-making, and introduces a prototype internet-based geographic information system that is developed to integrate public-oriented interactive decision mechanisms into urban planning practice. This system, referred as the 'Community-based Internet GIS' model, incorporates advanced information technologies, distance learning, sustainable urban development principles and community involvement techniques in decision-making processes, and piloted in Shibuya, Tokyo, Japan.

INTRODUCTION

Urban planners and designers have employed wide range of techniques in order to plan, model and simulate the outcome of urban planning and development processes. In the field of urban planning, physical models of large scale plans have been created with three aims in mind: to aid decision-making processes; to democratise planning processes, and; to support dissemination of planning ideas among a wider community. In the past, planners have employed physical models developed as paper and cardboard

DOI: 10.4018/978-1-61520-929-3.ch002

plans and supported with perspective drawings and photomontages in order to promote a design solution or a proposed development plan (Doyle et al., 1998:138). Whilst still employed, such methods are started to be used less extensively due to the emergence of increasingly more powerful desktop computing with constantly improving graphic capabilities, available at modest financial cost to individuals and planning agencies. The increase in computing power and information technology is also facilitating the advancement of geographic information systems (GIS), multimedia and virtual reality software that are being used in many disciplines which are concerned with modelling or analysis aspects of urban environments (Doyle et al., 1998:138). At present, particularly internet and GIS are becoming the most popular means for involving community in the planning process, at least in democratic places that encourage public participation. During the last decades, a range of innovative technologies are developed that offer different ways of modelling and representing built form and associated urban information with real-time interaction over the world-wide-web (in short web). This chapter, therefore, investigates the potentials of interactive mapping, virtual reality technologies and, especially, Internet GIS for the visualisation, modelling and analysis of urban environments and phenomena. This chapter presents a review of the capabilities of the technologies and how they are applied in planning and designing liveable urban environments. The chapter advocates the idea that new information technology based on the web offers excellent means for involving the public in the planning process. The main opportunity here is the use of Internet GIS as a communication and interaction vehicle between different interest groups such as planners, decision-makers and most importantly the general public.

The following background section of this chapter discusses the potential of GIS in urban planning and its importance in supporting public participation. The section reviews computer supported collaborative work systems for group decision-making in urban planning, underlines the significance of virtual reality and 3D GIS visualisation, and highlights the necessity of utilising Internet GIS for public participatory planning. The main thrust of this chapter introduces a new online planning support system, referred as a 'Community-based Internet GIS' model that is piloted in Shibuya, Tokyo, Japan. Following to underlining future trends in online planning, the chapter concludes by advocating and determining the potential contributions of the web-based participatory planning support systems as e-planning mechanisms of the urban planning discipline.

BACKGROUND

Planning is a future oriented activity, strongly conditioned by the past and present. As Friedmann (1987:38) highlights, it links "scientific and technical knowledge to actions in the public domain". Ideally, it proceeds via public discourse between all groups and individuals interested in and affected by urban development and management activities pursued by the public and/or private sectors. In practice, such "comprehensive sharing of information and decision-making is rarely found" (Nedovic-Budic, 2000:81). Planners have always sought tools to establish the missing piece of planning (public participation) in order to enhance the analytical, problem solving, and decision-making capabilities of the planning mechanism. And now new technologies are promising opportunities in improving the efficiency of planning and two way communication between planners and the public (Al-Kodmany, 2007). Throughout the last century, urban and regional planning has been subject to a rapid technological change. Advancement of information technology especially that of the internet has urged the development of network-based support systems (wired or wireless) such as GIS, facility management and automated mapping (AM/FM), Internet GIS, virtual reality (VR) and various types of groupware. As Batty (1996) predicted these tools are now contributing to the automation of the planning process, reduction of the planning time, and increase in the opportunity for public participation in urban planning.

Beginning in the late 1950s, planners started to develop and use computerised models, planning information systems, and decision support systems to improve performance and precision of urban areas (Klosterman, 2008). GIS had been introduced to planning for the first time in 1966 by the Harvard Laboratory for Computer Graphics and Spatial Analysis, Harvard University (GIS Development, 2002). However, effective and efficient use of geographic information technologies dates back to 1980s. Since then computer graphics, including digital maps, satellite images and aerial photos are used significantly in the planning practice. GIS are an appropriate technology for dealing with urban problems and they have revolutionised the way that spatial data is generated stored, analysed and disseminated. The output of a GIS system, which is information in either/both in visual and/or attribute form, helps people to manage what they know, by making it easy to organise, manipulate and apply to planning problems (Longley et al., 2001). GIS technology integrates common database operations such as query and statistical analysis with the unique visualisation and geographic analysis benefits offered by intelligent maps. With this technology not only existing situations and impacts can be displayed on cartographic representations of areas, but also projected patterns and usages of environments can be modelled, simulated, their outcomes envisaged, and local expertise are transferred.

Over the past two decades significant attention has been given to the adoption of GIS and land information systems to promote local authorities' planning services and information delivery to the public (Nedovic-Budic, 2008). Planning departments have been on the forefront of GIS use among local government agencies to incorporate new tools and technologies for sound and sustainable urban development (French and Wiggins, 1990; Juhl, 1993; French and Skiles, 1996; DeMers, 2005). Presently, planners all around the world are applying geographic information technologies in all aspects of the planning process, including data collection and storage, data analysis and presentation, planning and policy-making, communication with the public and decision-makers, and policy implementation and administration (Cho, 2005). GIS technology is most commonly used for comprehensive planning, zoning, land use inventories, site suitability assessments, and socio-demographic analysis, and is generally used for mapping purposes (Budic, 1993; 1994; Harris and Elmes, 1993; Nedovic-Budic, 2000; Pamuk, 2006). Presently, GIS dominate the use of computers in spatial analysis, thus shifting the attention from the previous non-routine or strategic models to the much more routine type operations such as spatial database management and map overlays. Figure 1 shows how GIS and related modelling technologies fit within the urban planning process. Indeed, this is the kind of structure that Harris (1989, 1991) refers to as a decision support system (DSS) or more accurately a planning support system (PSS), which links a variety of computer supported decisions at different stages of the planning process (Batty, 1995).

During the last years planners started to make much more use of the Internet as a medium of interaction and developed web-based platforms for online planning. Online planning – sometimes referred to as internet-assisted urban planning or e-planning – is a new frontier for the planning discipline. It creates a new platform for planning operations and processes, and increases the opportunity for public participation. Online planning offers access opportunity to a seamless record of the progress and approval of planning proposals and policies (Shiode, 2000; McGinn, 2001; Peng and Tsou, 2003). Online information technologies are essential tools for urban planning as they support information sharing and participatory decision-making in the planning

Figure 1. GIS and urban planning process (Batty, 1995:9)

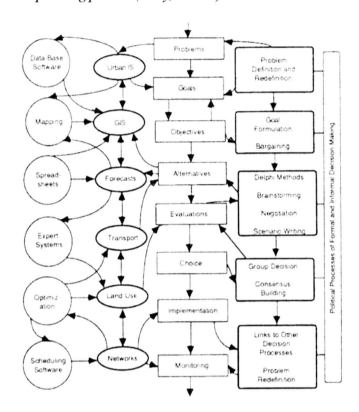

process. Popular online planning technologies include: public participatory GIS, computer supported collaborative work environment, 3D GIS, virtual reality, and internet mapping. Implications of these technologies on the planning practice are discussed below.

Public Participatory GIS

In the information era, an urban plan is seen as a process as much as a design outcome, if not more. The reason is clear: too many plans had accomplished nothing other than occupying space on a shelf. If more people were involved in developing a plan, it is more likely that this plan will appropriately address issues that are important to the community (Sieber, 2007). Plans that have engaged many people have the support that is needed to bring the plans to fruition (Craig, 1998:394). Since the 1960s, citizens have increasingly shared their views on the planning issues and they have become an important part of the urban planning process. Community focused or public participatory planning is introduced to planning after the rise of this attention. Public participatory planning briefly is a local planning process that provides governmental assistance to communities as they address opportunities and challenges. It aims to develop a community-based process with broad citizen participation in order to build local capacity to plan for sustainable urban development and to benefit from the insights, knowledge, and support of local residents (Kelly, 2004).

Public participation is a vital constituent of the planning process. Numerous urban researchers have investigated computer technology and social issues and questioned whether technology aids public participation and democracy or diminishes it. As reported by Craig (1998) such technology provides individuals and smaller organisations with relevant information, thereby reducing monolithic decision-making and redirecting necessary resources to them. Various IT tools, including GIS, support public participation and advocate communities' requests (Sieber, 2006). However, these tools cannot directly change the decision-making processes from top-down to bottom-up, but they raise awareness and build a capacity for the public to reorient this structure. For that purpose the concept of Public Participatory GIS (PPGIS) is introduced by the National Centre for Geographic Information and Analysis (NCGIA) as the Research Initiative 19 entitled 'the Social Implications of How People, Space and Environment are Represented in GIS' in 1993. PPGIS deals with community's demand on the spatial and non-spatial issues, addresses a specific scale, involves public participation in the second cybernetic sense, is pluralistic, inclusive and non-discriminatory, includes qualitative and quantitative data, and it is for and by the community (Yigitcanlar, 2002).

Following to NCGIA's initiative for almost a decade PPGIS has been a hot issue for scholars and practitioners from both urban planning and geographic information science disciplines (Ghose, 2007). According to Kingston et al. (2000) a succeeding PPGIS should concern following issues: PPGIS should provide equal access to data and information for all sectors of the community; it should have the capability to empower the community by providing the necessary data and information which matches the needs of the community who are, or potentially, participating; and a high degree of trust and transparency needs to be established and maintained within the public realm to give the process legitimacy and accountability. However, another as important issue is the need for an efficient collaborative work environment for PPGIS systems to become fully operational.

Computer Supported Collaborative Work Environment

In the broad sense, a computer supported collaborative work environment (CSCW) is a computer software, hardware and network environment that functions to provide a means for human collaboration. More specifically, CSCW is considered as a software system that supports a group of decision-makers engaged in a common decision task by providing access to the same shared environment and information. It is often considered synonymous with 'groupware'. During the last few decades, computers have mainly been used to design urban plans and store information for urban management. However, recent developments in computer, multimedia and networking technologies now provided the opportunity for technicians, policy-makers and the public to be able to participate in group decision-making sessions by using groupware and computers on wide-area networks including Internet. Actual groupware functionality is typically described as in terms of the time and place of interaction (Figure 2). The dimensions of time and place and ability of groupware to equate them is a function of technology itself (Lococo and Yen, 1998). The ability to collaborate productively in real-time with other associates, regardless of location, is

Figure 2. Groupware functionality (Khoshaflan and Buckiewicz, 1995:83)

	Same Place	Different Places
Same Time *Synchronous*	Electronic Meetings Team Rooms Group Decision Support Systems Electronic Whiteboards	Video Conferencing Teleconferencing Screen Saving Document Sharing Electronic Whiteboards
Different Times *Asynchronous*	Shared Containers Mailboxes Electronic Bulletin Boards Virtual Rooms, Kiosks Document Management Systems	Electronic Mail Workflow Form Flow Messaging Routing & Notification

invigorating and beneficial for management effectiveness (Figure 3).

Research in spatial decision (or planning) support systems has a rather long tradition (Densham, 1991; Goodchild and Densham, 1995; Yigitcanlar and Gudes, 2008). Projects related to collaborative spatial decision support systems (CSDSS) focused on overcoming the limitations of single user GIS (Armstrong, 1994; Densham et al., 1995). CSDSS benefit from both GIS and CSCW systems for supporting public participation in planning and assisting communities in local decision-making (e.g. environmental decisions). As pointed out by Couclelis and Monmonier (1995:92), CSDSS have their roots in operations research and the approach is fully applicable to the planning discipline. For instance, planning initiatives by Faber (1995, 1997) offer a good example of an approach in which the architecture of a CSDSS is based on the integration of groupwork tools and GIS tools for sustainable urban development (Sarjakoski, 1998; Kain and Soderberg, 2008). More recent examples of CSDSS include INDEX (www.crit.com) and CommunityViz (www.communityviz.org) that are advanced yet easy-to-use GIS software designed to help people visualise, analyse, and communicate about important community planning decisions. These new GIS customisation examples are particularly useful when public is involved in decision-making as these applications include invaluable visual analysis modules such as 3D and virtual reality.

3D GIS and Virtual Reality

There has been a considerable effort to push the capabilities of urban GIS from simple two dimensional (2D) mapping towards three dimensional (3D) visualisation of the built environment (Faust, 1995). This has often been achieved through the linkage of computer aided design (CAD) technologies to GIS databases (Liggett et al., 1995). Beyond this, there has been a demand in using virtual reality techniques to produce 3D solid geometry models for interactive exploration and interrogation. The practical implementations of these techniques are achieved by using virtual reality modelling language (VRML) with 3D models (Doyle et al., 1998:143). The interest in 3D GIS has been substantial from both application and theoretical perspectives. Traditionally, virtual reality systems did not have much analytical functions that are applicable to urban environments. At present, to overcome this disadvantage

Figure 3. Groupware applications (Lloyd, 1994:39)

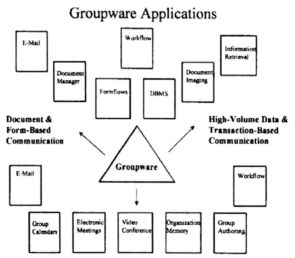

of VRML, GIS software has been linked with contemporary VRML technologies. VRML has also for the capacity in enabling the use of 3D models in web browsers (Sarjakoski, 1998:386). By benefiting from this technology, it is possible to share various instances of 'virtual cities' or digital representations of real-world urban locations on the web. There are websites that contain basic graphic interfaces for geographic data, but there are only few that integrate virtual reality technologies with the spatial databases held within GIS so as to create a true virtual city – an effective simulation environment for planners.

Many of the commercial GIS vendors realised the importance and opportunities of 3D GIS based on virtual reality technologies. For example, Environmental Systems Research Institute (ESRI) launched a powerful 3D module (3D Analyst) to be used with ArcGIS software (ESRI, 2003; 2004). This product provides 3D visualisation capabilities within the GIS itself allowing a degree of modelling and analysis in 3D and the ability to exploit a VRML on the desktop. Other good examples of more recent tools are the CommunityViz's SiteBuilder 3D module (www.communityviz.org) and INDEX 3D module (www.crit.com). These good examples reveal that virtual reality and 3D GIS visualisation are on the verge of changing the practice of urban planning and design. Instead of presenting citizens with abstract maps and descriptive text to explain, analyse and debate design ideas and urban processes, planners are now able to present explicit photo textured information of what their city will look like after a proposed change. Although there are promising stand alone 3D GIS applications, it is crucial that these applications to be placed in the internet environment to be accessible by all interest groups.

Internet Mapping

Urban planners and designers recognise that local knowledge is a key element in developing appropriate and effective solutions to community design and planning problems. In an environment in which computer and internet technologies are becoming common, it is important that planners develop ways to harness these technologies to work effectively with the public (Batty, 1998). Computer visualisation and web-based systems offer a new opportunity to support and facilitate public participation and democratic decision- and plan-making, which leads to the establishment of an e-planning mechanism. As city administrations and planning departments increasingly understand their roles as public oriented service centres, they have become interested in providing mapping services to the public through internet. This includes "providing wide and easy to use access to spatially referenced data" – e.g. demographic, environmental, economic data (Andrienkoa et al., 1999:425-426). Accessing the web and other information technologies empowers neighbourhood organisations by providing them with more information to use in their own deliberations. Whether these deliberations are internal or external, more and reliable information makes them powerful in the decision-making process. Internal deliberations help the community set its priorities, and "[i]nformation used in external deliberations helps the organisation better argue its case" (Craig, 1998:403). Shared information is valuable to citizens and local interest groups (e.g. environmental organisations, local companies) to get involved in the development and decision processes.

Since the first webpage with an interactive map was put up by Xerox Palo Alto Research Centre (PARC) as an experiment in interactive information retrieval in 1993 (Harder, 1998), many sites serving maps appeared on the web. Plewe (1997) discusses the benefits of 'GIS Online' and refers to many good practices that distribute geographic information. Among them, some sites provide publicly accessible geographic data and geo-spatially referenced information, where some have only restricted access for project stakeholders. Many other data portals provide commercial or free-of-charge geo-data, such as Microsoft's

TerraServer (http://terraserver.homeadvisor.msn.com/default.asp), Teale Data Centre (www.gislab.teale.ca.gov), Map Quest (www.mapquest.com), and Google Maps (http://maps.google.com.au). Planners and designers, who traditionally use maps and related information, have found internet and intranet technologies useful mainly because they reduce the cost of data management and information distribution (Tripathy, 2002). With the advent of GIS and internet technologies in urban planning, the conventional intricacies to get solutions in time and position have been improved.

The recent years have witnessed an explosive development of internet, which has now become an important means for acquiring and disseminating information. Most GIS vendors and some commercial spatial data providers have realised that the web is the next generation GIS platform, providing a powerful medium for geographic information distribution, as well as a particularly lucrative new market to exploit (Doyle et al., 1998). Correspondingly, a number of Internet GIS solutions for deploying maps have been developed. Plewe (1997) provides a review of the integration of Internet mapping and GIS, and Huang and Lin (1999) discuss the possible services that such integration could offer. Internet has also recently been seized upon as a new powerful vehicle with which to engage the public. Visionaries look forward to future in which internet-based systems will be used to inform citizens of developments in their environment and allow citizens to interact with professionals and policy-makers in decision-making processes. Because many of the issues that affect local government are land or property based, Internet GIS play a major role within the "movement towards internet participation" (Reeve et al., 2002:49).

There is a vast literature on successful experiments in Web-based PPGIS (see Reeve et al., 2002). For example, Jankowski and Stasik (1997) introduce an Internet GIS that makes collaborative spatial decision-making via public participation possible. Keisler and Sundell (1997) present an integrated geographic multi-attribute utility system and an extensive survey of applications and research issues for geographic information technology applications in urban planning. In recent years, GIS have begun to appear on the web ranging from simple demonstrations and references to GIS use, to more complex online GIS and spatial decision support systems. The level of sophistication in online GIS systems in providing access to a variety of GIS functionality and data is improving constantly (Doyle et al., 1998). Thus, "the previous criticisms of GIS being an elitist technology (Pickles, 1995) [is] no longer valid" in the same context (Kingston et al., 2000: 109-110).

GIS and internet are ever evolving technologies and hold great potential for public use, allowing wider involvement in environmental and urban decision-making (Innes and Simpson, 1993). As Kingston et al. (2000:109-110) state "the rise of the Internet and the web over the past decade has created many opportunities for its use in local, regional and national democratic processes". One area where the web can be used to great advantage is for the enhancement of participatory democracy in local environmental decision-making. The web generates a new public sphere supporting interaction, debate, new forms of democracy and cyber cultures which feedbacks to support a renaissance in the social and cultural life of cities, and as a result, the web helps communities in shaping their future (Yigitcanlar and Gudes, 2008). In the developed world, almost every city now has a presence on the web which offers the potential to deliver public goods and services through this relatively new media (see Reeve et al., 2002; Sacramento City, 2002; Southern Tier, 2002; Yamato City, 2002; Caloundra City, 2009). The following section introduces one of the good practices, 'Shibuya Community-based Internet GIS Project', of online planning that empowers decision-making process by using innovative techniques and tools in urban and environmental planning.

SHIBUYA COMMUNITY-BASED INTERACTIVE DECISION-MAKING PROJECT

While local governments, urban planners and community agencies frequently use the web to offer information to community members, it has been less common for these entities to use internet as a medium for two way communication. Until recently, the use of spatial data with web technology has largely been limited to the provision of information to the public through 'image-maps' rather than any actual public participation. It is crucial to create systems and tools that would allow people to become both receivers and providers of information. To meet this purpose a new Internet GIS model referred as a 'Community-based Internet GIS Approach (CIGA)' has been developed. It is one of only a few projects that involve the public returning spatial data as well as consuming it. The model advances the prospect of allowing citizens to comment, using web-based maps, upon government actions and development proposals.

CIGA is an online collaborative decision-making system which enables various users, such as the public, technicians and decision-makers to obtain and share information interactively on their environs at different levels. CIGA has been developed to support the urban planning process with community needs in mind (Figure 4). Since most urban planning problems have spatial and multi-dimensional characteristics and tend to be increasingly complex, the inclusion of community-oriented decisions through the provision of web-based innovative applications will be greatly beneficial (Yigitcanlar, 2008). CIGA offers just this alternative perspective and encourages the facilitation of users to participate in the problem solving and decision-making stages of the participatory planning process via online interactive sources. It is also meant to be a mechanism to undertake studies on sustainable urban development, to identify planning and community goals, to draw up planning guidelines and criteria, and to collect and store data in the GIS environment. Moreover, further steps of urban planning, collaboration, negotiation and consensus building are integrated in the system. Thus, a powerful 'systems' methodology is used in this model to serve local needs, preserve urban ecosystems and

Figure 4. Community-based Internet GIS approach (Yigitcanlar, 2008:350)

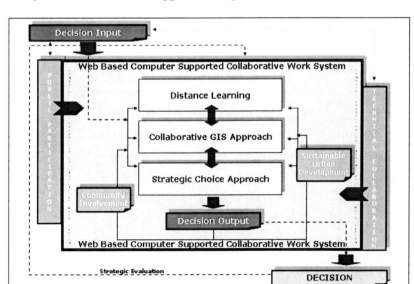

enhance sustainable urban development (Yigitcanlar, 2008).

The system architecture includes a CSCW and specific applications to support collaboration. The system constructs a trio of communication, collaboration and coordination among all of the participants and contains forms and common gateway interfaces that the user can interact with to provide information and feedback. As Doyle et al. (1998) and Brown (1999) point out, integrating multimedia forms with Internet GIS has numerous advantages and potential for the creation of successful learning environments (Yigitcanlar, 2008). Video and audio transmissions are, therefore, supported and intelligent maps are supplied for the users by internet map servers (Figure 5).

The multitude of factors that must be taken into account by decision-makers turns urban planning into a daunting system. CIGA aims to raise the awareness on these factors among the public, to give them essential knowledge and empower them to be able to deal with problems in their environment. CIGA contains distance learning and problem solving tools to give collaborators a basic background on the relevant issues to their decisions (Yigitcanlar, 2008). CIGA is constructed on the basics of distance learning and problem solving logics by utilising CSCW with advanced information technologies (Figure 6).

Social interaction is an important theme of this model. Online decision support systems can not be fully successful without an interaction between the collaborators. CIGA is, therefore, designed to use traditional digital community building tools to encourage social interaction on the Web. These traditional tools will include newsgroups, email lists, and bulletin boards. CIGA is comprised of 'problem' and 'solution' stages. Both of the stages are connected through a series of five sections and fourteen steps (Figure 7). The problem stage contains various e-learning tools to raise the awareness of the public on system and planning issues. Moreover, at this stage, the system endeavours to provide basic education about environmental systems, planning and decision-making in a collaborative manner. This stage also assists the public to better understand, forecast and visualise the consequences of alternative scenarios (Yigitcanlar, 2008). The solution stage accommodates mechanisms to develop the alternative scenarios, evaluate and make strategic decisions. Spatial decision support systems are often designed

Figure 5. System architecture of CIGA (Yigitcanlar, 2008:353)

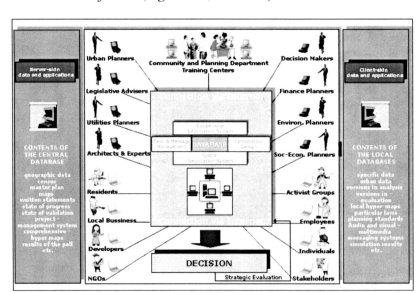

Figure 6. System overview of CIGA (Yigitcanlar, 2008:355)

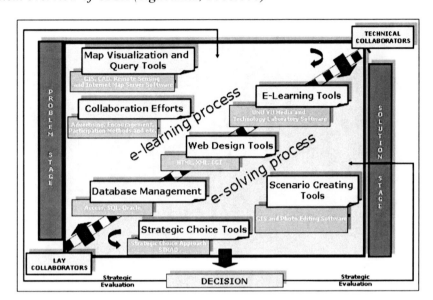

specifically to address semi-structured problems, however, they often lack support for collaborative decision-making, and for this reason, CIGA specifically accommodates 'Collaborative GIS' (Faber, 1997; Sapient, 2000) and 'Strategic Choice Approach' (Friend and Hickling, 1997; Friend, 2002) as decision support tools (Yigitcanlar, 2008).

As a prototype Internet GIS model, CIGA aims to integrate a public-oriented interactive decision support system for the urban planning process. The system components of CIGA have recently been developed and this Internet GIS model is piloted in Shibuya, Tokyo. By placing CIGA 'online' within a website, which incorporates an interactive map as an interface to the information, interested parties can obtain information on existing and previous developments, land-use, ownership, contact details in order to collaborate and voice an opinion and so on (for more information on CIGA, see Yigitcanlar, 2002; Yigitcanlar and Okabe, 2002; Yigitcanlar, 2008).

This project was online for six months between May and November 2002. In this online decision-making exercise participants discussed sustainability issues within, Shibuya which covered five major issues and problems of energy, open spaces, conservation, pollution, urbanisation and development. Only stage one of CIGA was applied in this pilot study as the development of the second stage still continues. Once it is completed it will be tested on several pilot projects (i.e. Brisbane and Gold Coast, Australia). During the May and November 2002 period over 60 people logged on to this website. At the moderated discussion forum 37 people communicated very actively and 24 participated rather passively. The major reason behind this low level of participation was the language barrier as this database and website was developed in English instead of the local language of Japanese (Yigitcanlar, 2008). The URL of the webpage of the database is: http://kanagawa.csis.u-tokyo.ac.jp (Figure 8). When users first enter the site, after an initial welcome window, they are prompted to fill in a profile. This was seen as an essential part of the system design as it could be used to build up a database of users to help validate responses and analyse the type of people who were using the system. The main page contains; map, legend, tool buttons, and query frame.

Figure 7. System flow chart of CIGA (Yigitcanlar, 2008:356)

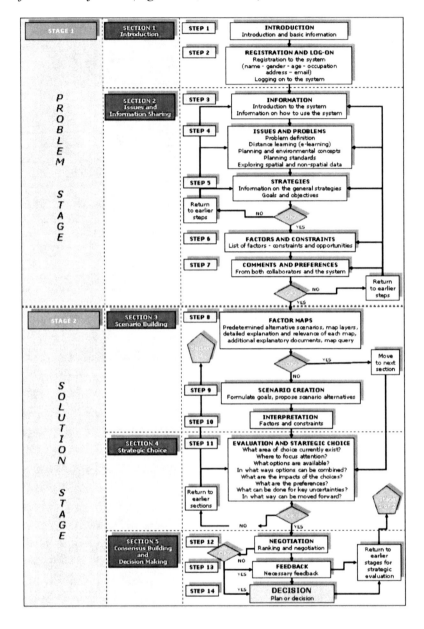

Similar to what Marker et al. (2002) talked about the design and conception of socio-technical systems for online citizen participation, this pilot project has shown that public participation should not be solely technology-driven but also should be oriented towards the basic principles of cooperative planning approaches. This is known as the 'new planning culture', and, among other things, allows: (1) participation at an early stage; (2) assures an equal opportunity to participate; (3) remains open with respect to both process and results; (4) assures communication and dialogue; (5) integrates multiple perspectives; and (6) allows moderation by neutral third parties. On the other hand, the realities of city politics cannot be ignored; that is, participation procedures usu-

Figure 8. Shibuya city internet GIS pilot project website (Yigitcanlar, 2008:360)

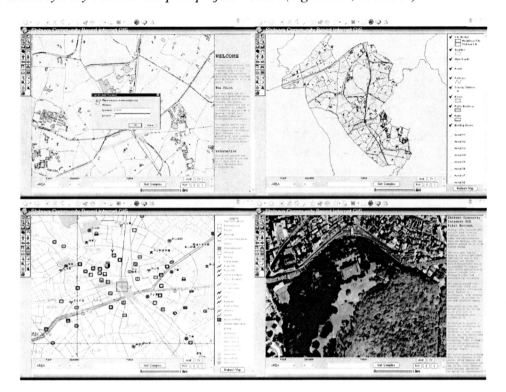

ally cannot achieve more than is allowed by the existing context of communication and power relationships (Yigitcanlar, 2008).

The experience with the pilot study has shown that, as Talen and Shah (2007) argue, online urban information systems and GIS have the potential to catch public interest in environmental planning and also in preparation of urban development plans. The system demonstrates how it is possible to combine information and provide a greater insight into Shibuya's urban and environmental situation. The system also communicates urban and environmental information to a potentially wide ranging audience in an easy to use and intuitive manner through a variety of ways including an online discussion platform. The ability to instantaneously update the database and profile users online was seen as one of the most useful advantages of the system over the traditional techniques. The online system allows people to use the system regardless of time and place. The public do not need to attend a meeting at a particular time or place which is often the single most inhibiting factor in participating via traditional methods. The system allows faster collation of results from log files and the website can be used to disseminate results and feedback (Yigitcanlar, 2008).

Clearly there is a need to improve public access to urban and environmental information, and encourage collaborative decision-making. Online urban and environmental information systems like the Shibuya CIGA project offer a variety of tools and technologies to assist the management and use of urban and environmentally related data and information, and hence can be used as a tool to help to achieve the sustainable urban development goal. If we are to strive to achieve the most efficient form of a sustainable information society, urban and environmental information systems, their tools and results should attract the interest of as wide an audience as possible (Yigitcanlar, 2008). This pilot study has shown us the role of internet, public participation and the development of Internet GIS are important factors in taking this potential further.

FUTURE RESEARCH DIRECTIONS

Decision support systems have evolved rapidly over the last two decades from stand alone or limited networked solutions to online participatory solutions (Hopkins, 2001). One of the major enablers of this change is the fastest growing areas of GIS technology development that relates to the use of internet as a means to access, display and analyse geospatial data remotely. World-wide many federal, state and particularly local governments are designing to facilitate data sharing using interactive internet map servers. This new generation decision or planning support systems, interactive internet map server, is the solution for delivering dynamic maps and GIS data and services via the web, and providing PPGIS opportunities to a wider community (Carver, 2001; Jankowski and Nyerges, 2001). These planning support systems provide a highly scalable framework for GIS web publishing, Web-based PPGIS or Internet GIS, which meets the needs of corporate intranets and demands of worldwide internet access (Craig, 2002). The establishment of Internet GIS provides spatial data access through a support centre or a GIS portal to facilitate efficient access to and sharing of related geospatial data (Yigitcanlar et al., 2003). As more and more public and private entities adopt Internet GIS technology, the importance and complexity of facilitating geospatial data sharing is growing rapidly (Carver, 2003; Yigitcanlar and Gudes, 2008). This chapter focused on the online public participation dimension of the GIS technology to aid local decision-making and planning. The continuum of advance development of planning support technologies along with provision of democratic platforms for the public in decision-making will eventually lead planning to go online and a true e-planning to be realised (Yigitcanlar 2005; 2006).

There are few important developments that are likely to be the hallmark of the online planning support systems within the coming years. One of them is combining virtual reality with Internet GIS, in order to give the participants much clearer visual opportunities in a specific case study. Another one is using Internet GIS in large scale planning programs rather than local ones such us: planning metropolitan or large regional level projects and plans. Furthermore, combining enterprise resource planning systems with GIS seems to be a very useful trend for the success of the e-government and e-planning processes. Additionally increasing use of 'location based services' such as: routing, locations, and tracking services, seems to be new rising trends. Few examples of these trends can be seen in the USA and Australia, where parents use location-based services to track their children's daily movement patterns (Yigitcanlar and Gudes, 2008). Finally, the Google Earth service accelerates the use of GIS among a wider population rapidly, as it makes GIS a common and useful tool for community and individuals.

CONCLUSION

In its short history, Internet GIS has proven to be a powerful tool to develop flexible and versatile functions and to deliver rich information content to the users through the web. As a result, interactive mapping or Internet GIS has developed rapidly over the past few years resulting in the migration of GIS functionality to the web (Plewe, 1997). This migration has taken elements of GIS to mass audiences. Without owning the individual software, the Internet GIS technology has opened new paths for disseminating, sharing, displaying, and processing spatial information on internet. Web-based solutions provide a low cost, efficient way to deliver map products to users. Internet GIS based solutions for urban planning helps the public in a significant way to participate in decision-making processes. Placing urban planning 'online', serving maps on the web for public participatory planning and Internet GIS model (i.e. CIGA) have the potential to provide numerous contributions to the urban planning discipline by supplying tremendous support to the public. To recapitulate, these contributions are:

- Deliberating inclusions of social issues compared to other technology implementations and provision of equal access to data and information for the users over internet.
- Promoting of a healthy balance of urban growth, conservation of natural resources and cultural identity for communities.
- Bridging the gap between decision-makers, technicians and the public by enhancing citizen consciousness and encouraging individuals to share their opinions.
- Accommodating an equitable representation of diverse views of stakeholders and communities.
- Preserving contradictions and endorsing the role of individuals and communities in the creation and evaluation of development plans and strategies by promoting citizen participation in urban planning.
- Offering powerful tools to provide information and support decision-making in planning, sustainable urban development and management.
- Promoting the integrated management of resources based on the sensitivity and needs of local communities and helps to construct a healthier relationship between individuals.
- Reinforcing the appropriate use of analytical tools and datasets in policy-making and planning to predict and examine the possible short and long term outcomes on the environment and to empower the urban planning process overall.

All community groups, non-profit organisations, local authorities and planning agencies should make every effort to adapt recent information technologies to policy-making and planning processes. They need to understand its strengths and limitations so that they can view it from a critical perspective. Placing urban planning 'online' and integrated community-based methodologies to urban planning are new and important approaches. Furthermore, they are effective considerations that show local authorities, planning agencies, communities and the public, a strategic way to further develop local economy, preserve socio-cultural and environmental values, and shape their neighbourhoods.

REFERENCES

Al-Kodmany, K. (2007). Creative approaches for augmenting two-way spatial communication and GIS. *GIS Development*, *3*(12), 1–9.

Andrienkoa, G., Andrienkoa, N., Vossa, H., & Carter, J. (1999). Internet mapping for dissemination of statistical information. *Computers, Environment and Urban Systems*, *23*(1), 425–441. doi:10.1016/S0198-9715(99)00044-7

Armstrong, M. (1994). Requirements for the development of GIS-based group decision-support systems. *Journal of the American Society for Information Science American Society for Information Science*, *45*(9), 669–677. doi:10.1002/(SICI)1097-4571(199410)45:9<669::AID-ASI4>3.0.CO;2-P

Batty, M. (1995). Planning Support Systems and the New Logic of Computation. *Regional Development Dialogue*, *16*(1), 1–17.

Batty, M. (1996). Planning, late-20th-century style. *Environment and Planning. B, Planning & Design*, *23*(1), 1–2.

Batty, M. (1998). Digital planning. In Sikdar, K., & Rao, K. (Eds.), *Computers in urban planning and urban management* (pp. 13–30). New Delhi: Narosa.

Brown, I. (1999). Developing a Virtual Reality User Interface for Geographic Information Retrieval on the Internet. *Transactions in GIS*, *3*(3), 207–220. doi:10.1111/1467-9671.00018

Budic, Z. (1993). GIS Use Among South-eastern Local Governments - 1990/1991 Mail Survey Results. *Journal of Urban and Regional Information Systems Association*, *5*(1), 4–17.

Budic, Z. (1994). Effectiveness of Geographic Information Systems in Local Planning. *Journal of the American Planning Association. American Planning Association*, *60*(2), 244–263. doi:10.1080/01944369408975579

Caloundra City. (2009). *Strategic City Plan of the Caloundra City*. Retrieved on March 13, 2009, from http://maproom.caloundra.qld.gov.au

Carver, S. (2001). Public participation using web-based GIS. *Environment and Planning. B, Planning & Design*, *28*(1), 803–804. doi:10.1068/b2806ed

Carver, S. (2003). The Future of Participatory Approaches Using Geographic Information: Developing a Research Agenda for the 21st Century. *URISA Journal*, *15*(1), 61–71.

Cho, G. (2005). *Mastering geographic information science: technology, applications and management*. New York: John Wiley and Sons.

Couclelis, H., & Monmonier, M. (1995). Using SUSS for resolve NIMBY: How spatial understanding support systems can help with the 'not on my back yard' syndrome. *Geographical Systems*, *2*(1), 83–101.

Craig, W. (1998). The Internet Aids Community Participation in the Planning Process. *Computers, Environment and Urban Systems*, *22*(4), 393–404. doi:10.1016/S0198-9715(98)00033-7

Craig, W. (2002). *Community Participation and Geographic Information Systems*. London: Taylor and Francis.

DeMers, M. (2005). *Fundamentals of geographic information systems*. New York: John Wiley and Sons.

Densham, P. (1991). Spatial decision support system. In Maguire, J., & Rhind, D. (Eds.), *Geographical information systems: Principles and applications* (pp. 403–412). London: Longman.

Densham, P., Armstrong, M., & Kemp, K. (1995). *Collaborative spatial decision making, Scientific Report of the NCGIA Initiative 6 Specialist Meeting, NCGIA Technical Report 90-5*. National Center for Geographic Information and Analysis, UCSB., California.

Development, G. I. S. (2002). *GIS Development History: Milestones of GIS*. Retrieved on May 2, 2002, from http://www.gisdevelopment.net/history/1960-1970.htm

Doyle, S., Dodge, M., & Smith, A. (1998). The Potential of Web Based Mapping and Virtual Reality Technologies for Modelling Urban Environments. *Computers, Environment and Urban Systems*, *22*(2), 137–155. doi:10.1016/S0198-9715(98)00014-3

ESRI. (2002). *Architecture of ArcIMS*. Retrieved on June 29, 2004, from http://www.esri.com/software/arcims/architecture.html

ESRI. (2003). ArcGIS 3D Analyst: Three Dimensional Visualization, Topographic Analysis, and Surface Creation. *ESRI White Paper*, 1-16.

ESRI. (2004). *ArcGIS 3D Analyst*. Retrieved on June 29, 2004, from http://www.esri.com/software/arcgis/arcgisxtensions/3danalyst

Faber, B. (1995). Extending electronic meeting systems for collaborative spatial decision making. In M. Densham & K. Kemp (Eds.), *Collaborative spatial decision-making: Scientific Report for the Initiative 17 Specialist Meeting, NCGIA Technical Report 95-14*. National Center for Geographic Information and Analysis, UCSB, California.

Faber, B. (1997). Active response GIS: An architecture for interactive resource modelling. In *GIS'97 Annual Symposium on Geographic Information Systems*, Vancouver, BC, March.

Faust, N. (1995). The virtual reality of GIS. *Environment and Planning. B, Planning & Design*, *22*(1), 257–268. doi:10.1068/b220257

French, S., & Skiles, A. (1996). Organizational Structures for GIS Implementation. In M. Salling (Ed.), *URISA '96 Conference*, July 27-August 1, 1996, Salt Lake City, Utah (pp. 280-293).

French, S., & Wiggins, L. (1990). California Planning Agency Experiences with Automated Mapping and Geographic Information Systems. *Environment and Planning B*, *17*(4), 441–450. doi:10.1068/b170441

Friedmann, J. (1987). *Planning in the Public Domain: From Knowledge to Action*. Princeton, NJ: Princeton University Press.

Friend, J. (2002). *Stradspan: New Horizons in Strategic Decision Support*. Retrieved on May 2, 2002, from http://www.btinternet.com/~stradspan/program.htm

Friend, J., & Hickling, A. (1997). *Planning Under Pressure, The Strategic Choice Approach*. Oxford, UK: Butterworth Heinemann.

Ghose, R. (2007). Politics of Scale and Networks of Association in PPGIS. *Environment & Planning A*, *39*(1), 1961–1980. doi:10.1068/a38247

Goodchild, M., & Densham, P. (1995). *Spatial decision support systems*. Scientific Report of the NCGIA Initiative 17 Specialist Meeting, NCGIA Technical Report 95-14. National Center for Geographic Information and Analysis, UCSB, California.

Harder, C. (1998). *Serving Maps on the Internet*. Redlands, California: Environmental Systems Research Institute Inc.

Harris, B. (1989). Beyond Geographic Information Systems: Computers and the Planning Professional. *Journal of the American Planning Association. American Planning Association*, *55*(1), 85–90. doi:10.1080/01944368908975408

Harris, B. (1991). Planning Theory and the Design of Planning Support Systems. In *Second International Conference on Computers in Planning and Management*, 6-8 July, 1991, Oxford.

Harris, T., & Elmes, G. (1993). GIS Applications in Urban and Regional Planning: The North American Experience. *Applied Geography (Sevenoaks, England)*, *13*(1), 9–27. doi:10.1016/0143-6228(93)90077-E

Hopkins, L. (2001). *Urban Development: The Logic of Making Plans*. New York: Island Press.

Huang, B., & Lin, H. (1999). GeoVR: a web-based tool for virtual reality presentation from 2D GIS data. *Computers & Geosciences*, *25*(1), 1167–1175. doi:10.1016/S0098-3004(99)00073-4

Innes, J., & Simpson, D. (1993). Implementing GIS for planning. *Journal of the American Planning Association. American Planning Association*, *59*(1), 230–236. doi:10.1080/01944369308975872

Jankowski, P., & Nyerges, T. (2001). *Geographic information systems for group decision-making: Towards a participatory geographic information science*. London: Taylor and Francis.

Jankowski, P., & Stasik, M. (1997). Design considerations for space and time distributed collaborative spatial decision making. *Journal of Geographic Information and Decision Analysis*, *1*(1), 1–12.

Juhl, G. (1993). Government Agencies Let Their Hair Down about GIS. *Geographical Information Systems*, *3*(7), 20–26.

Kain, J., & Soderberg, H. (2008). Management of complex knowledge in planning for sustainable development: the use of multi-criteria decision aids. *Environmental Impact Assessment Review*, *28*, 7–21. doi:10.1016/j.eiar.2007.03.007

Keisler, J., & Sundell, R. (1997). Combining multi-attribute utility and geographic information for boundary decisions: an application to park planning. *Journal of Geographic Information and Decision Analysis*, *1*(2), 101–118.

Kelly, S. (2004). *Community Planning: How to Solve Urban and Environmental Problems*. New York: Rowman and Littlefield Publishers.

Khoshaflan, S., & Buckiewicz, M. (1995). *Introduction to groupware, workflow, and workgroup computing*. London: John Wiley and Sons.

Kingston, R., Carver, S., Evans, A., & Turton, I. (2000). Web-based public participation geographical information systems: an aid to local environmental decision-making. *Computers, Environment and Urban Systems*, *24*(1), 109–125. doi:10.1016/S0198-9715(99)00049-6

Klosterman, R. (2008). Urban future strategies: concepts and tools for a new urban management. In 2008 International Seminar on Future City, October 24, 2008, National Hanbat University, Daejeon, Korea.

Liggett, R., Friedman, S., & Jepson, W. (1995). Interactive design/decision making in a virtual urban world: Visual simulation and GIS. In *1995 ESRI User Conference*, ESRI, California.

Lloyd, P. (1994). *Groupware in the 21st century: Computer supported collaborative working toward the millennium*. New York: Greenwood.

Lococo, A., & Yen, D. (1998). Groupware: Computer Supported Collaboration. *Telematics and Informatics*, *15*(1), 85–101. doi:10.1016/S0736-5853(98)00006-9

Longley, P., Goodchild, M., Maguire, D., & Rhind, D. (2001). *Geographic Information Systems and Science*. New York: Wiley.

Märker, O., Hagedorn, H., & Trénel, M. (2002). *Internet-Based Public Consultation: Relevance – Moderation – Software*. Retrieved February 25, 2009 from http://www.ercim.org/publication/Ercim_News/enw48/maerker.html

McGinn, M. (2001). *Getting Involved in Planning*. Edinburgh, UK: Scottish Executive Development Department.

Nedovic-Budic, Z. (2000). Geographic Information Science Implications for Urban and Regional Planning. *URISA Journal*, *12*(2), 81–93.

Nedovic-Budic, Z. (2008). ICTs to support urban planning strategies. In 2008 International Seminar on Future City, October 24, 2008, National Hanbat University, Daejeon, Korea.

Pamuk, A. (2006). *Mapping global cities: GIS methods in urban analysis*. Redlands, CA: ESRI Press.

Peng, Z., & Tsou, M. (2003). *Internet GIS: distributed GIS services for the internet and wireless networks*. New York: John Wiley and Sons.

Pickles, J. (1995). *Ground truth: the social implications of geographical information systems*. New York: Guilford Press.

Plewe, B. (1997). *GIS On-Line: Information, retrieval, mapping and the Internet*. Santa Fe, NM: On Word Press.

Reeve, D., Thommason, E., Scott, S., & Simpson, L. (2002). Engaging Citizens: The Bradford Community Statistics Project. In S. Wise, Y. Kim, & C. Openshaw (Eds.), *GISRUK: GIS Research UK 10th Annual Conference*, 3rd-5th April 2002, University of Sheffield, Sheffield, UK (pp. 49-51).

Sacramento City. (2002). *Sacsites: Sacramento's Business and Development Resource*. Retrieved on May 2, 2002, from http://www.maps.cityofsacramento.org/website/sacramentoed/ed.htm

Sapient. (2000). *Smart Places: Collaborative GIS Approach*. Retrieved on May 2, 2002, from http://www.saptek.com/smart

Sarjakoski, T. (1998). Networked GIS for Public Participation – Emphasis on Utilizing Image Data. *Computers, Environment and Urban Systems*, *22*(4), 381–392. doi:10.1016/S0198-9715(98)00031-3

Shiode, N. (2000). Urban Planning, Information Technology, and Cyberspace. *Journal of Urban Technology*, *7*(2), 105–126. doi:10.1080/713684111

Sieber, R. (2006). Public Participation Geographic Information Systems: A Literature Review and Framework. *Annals of the American Association of Geographers*, *96*(3), 491–507. doi:10.1111/j.1467-8306.2006.00702.x

Sieber, R. (2007). Spatial Data Access by the Grassroots. *Cartography and Geographic Information Science*, *34*(1), 47–62. doi:10.1559/152304007780279087

Southern Tier. (2002). *Southern Tier Regional Planning and Development Board: Community GIS*. Retrieved on May 2, 2002, from http://www.southerntierwest.org/st/cgis/html/locgovgis1.htm

Talen, E., & Shah, S. (2007). Neighbourhood Evaluation Using GIS: An Exploratory Study. *Environment and Behavior*, *39*(5), 583–615. doi:10.1177/0013916506292332

Tripathy, G. (2002). *Web-GIS Based Urban Planning and Information System for Municipal Corporations – A Distributed and Real-Time System for Public Utility and Town*. Retrieved on May 2, 2002, from http://www.gisdevelopment.net/application/urban/overview/urbano0028pf.htm

Yamato City. (2002). *Online Urban Master Plan of the Yamato City*. Retrieved on May 2, 2002, from http://www.city.yamato.kanagawa.jp/t-soumu/TMP/e/index.html

Yigitcanlar, T. (2002). Community Based Internet GIS: A Public Oriented Interactive Decision Support System. In S. Wise, Y. Kim, & C. Openshaw (Eds.), *GISRUK: GIS Research UK 10th Annual Conference*, 3rd-5th April 2002. University of Sheffield, Sheffield, UK (pp. 63-67).

Yigitcanlar, T. (2005). Is Australia Ready to Move Planning to an Online Mode? *Australian Planner*, *42*(2), 42–51.

Yigitcanlar, T. (2006). Australian local governments' practice and prospects with online planning. *URISA Journal*, *18*(2), 7–17.

Yigitcanlar, T. (2008). A public oriented interactive environmental decision support system. In Wise, S., & Craglia, M. (Eds.), *GIS and Evidence-Based Policy Making* (pp. 347–366). London: Taylor and Francis.

Yigitcanlar, T., Baum, S., & Stimson, R. (2003). *Analyzing the Patterns of ICT Utilization for Online Public Participatory Planning in Queensland* (pp. 5–21). Australia: Assessment Journal.

Yigitcanlar, T., & Gudes, O. (2008). Web-based public participatory GIS. In Adam, F. (Ed.), *Encyclopedia of Decision Making and Decision Support Technologies* (*Vol. 2*, pp. 969–976). London: Information Science Reference.

Yigitcanlar, T., & Okabe, A. (2002). *Building Online Participatory Systems: Towards Community Based Interactive Environmental Decision Support Systems*. Tokyo: United Nations University, Institute of Advanced Studies.

ADDITIONAL READING

Bédard, Y., Rivest, S., & Proulx, M. (2006). (SOLAP), Concepts, architectures, and solutions from a geomatics engineering perspective. In *Data Warehouses and OLAP: Concepts, Architecture, and Solutions*. Spatial On-Line Analytical Processing.

Berndt, D. J., Hevner, A. R., & Studnicki, J. (2003). The catch data warehouse: Support for community health care decision-making. *Decision Support Systems*, *35*(3), 367–384. doi:10.1016/S0167-9236(02)00114-8

Bharati, P., & Chaudhury, A. (2004). An empirical investigation of decision-making satisfaction in web-based decision support systems. *Decision Support Systems*, *37*(2), 187–197.

Birkin, M., Clarke, G., Clarke, M., & Wilson, A. (1996). *Intelligent GIS: Location decisions and strategic planning*. Cambridge, UK: GeoInformation International.

Caldeweyher, D., Zhang, J., & Pham, B. (2006). OpenCIS-Open source GIS-based web community information system. *International Journal of Geographical Information Science*, *20*(8), 885–898. doi:10.1080/13658810600711378

Cromley, E. K., & Mclafferty, S. L. (2003). GIS and public health. *Health & Place*, *9*(1), 279–292.

Croner, C. M. (2003). Public Health, GIS and the Internet. *Public Health*, *24*(1), 57–80.

Healey, P. (2003). Collaborative planning in perspective. *Planning Theory*, *2*(2), 101–112. doi:10.1177/14730952030022002

Keenan, P. B. (2006). Spatial decision support systems: A coming of age. *Control and Cybernetics*, *35*(1), 9–23.

Kelly, N. M., & Tuxen, K. (2003). WebGIS for monitoring sudden oak death in coastal California. *Computers, Environment and Urban Systems*, *27*(1), 527–547. doi:10.1016/S0198-9715(02)00065-0

Peng, Z. R. (2001). Internet GIS for public participation. *Environment and Planning. B, Planning & Design*, *28*(6), 889–905. doi:10.1068/b2750t

Richards, T., Croner, C., Rushton, G., Brown, C., & Fowler, L. (1999). Geographic Information Systems and public health: Mapping the future. *Public Health Reports*, *114*(4), 359–360. doi:10.1093/phr/114.4.359

Sarjakoski, T. (1998). Networked GIS for public participation and GIS on utilizing image data. *Computers, Environment and Urban Systems*, *22*(1), 381–392. doi:10.1016/S0198-9715(98)00031-3

Shim, J. P., Warkentin, M., Courtney, J. F., Power, D. J., Sharda, R., & Carlsson, C. (2002). Past, present, and future of decision support technology. *Decision Support Systems*, *33*(2), 111–126. doi:10.1016/S0167-9236(01)00139-7

Turban, E. (1993). *Decision support and expert systems: Management support systems*. Upper Saddle River, NJ: Prentice Hall.

Velibeyoglu, K., & Yigitcanlar, T. (2008). Information communication and technology for e-regions. In Khosrow-Pour, M. (Ed.), *Encyclopedia of Information Science and Technology* (2nd ed., Vol. 4, pp. 1944–1949). Hershey, PA: Information Science Reference.

Yigitcanlar, T. (2009). Planning for smart urban ecosystems: information technology applications for capacity building in environmental decision making. *Theoretical and Empirical Researches in Urban Management Journal*, *3*(12), 5–21.

Yigitcanlar, T., Han, H., & Lee, S. (2008). Online environmental information systems. In Adam, F. (Ed.), *Encyclopedia of Decision Making and Decision Support Technologies* (Vol. 2, pp. 691–698). Hershey, PA: Information Science Reference.

Yigitcanlar, T., & Saygin, O. (2008). Online urban information systems. In Adam, F. (Ed.), *Encyclopedia of Decision Making and Decision Support Technologies* (Vol. 2, pp. 699–708). Hershey, PA: Information Science Reference.

Yigitcanlar, T., & Velibeyoglu, K. (2008). Strengthening the knowledge-base of cities through ICT strategies. In Cartelli, A., & Palma, M. (Eds.), *Encyclopedia of Information and Communication Technology* (Vol. 2, pp. 728–734). Hershey, PA: Information Science Reference.

KEY TERMS AND DEFINITIONS

Decision Support Systems: A class of computerized information systems that support decision making activities.

Environmental Information System: An extended geographic information system that serves as a tool to capture, save and present spatial, time-related and content-specific data and at the same time to describe the condition of the environment in terms of negative impacts and risks.

Geographical Information System: A system for managing spatial data and associated attributes. It is a computer system capable of integrating, storing, editing, analyzing, and displaying geographically-referenced information.

Interactive Internet Map Server: An online mapping utility which enables users who may not be familiar with GIS to view and interact with online GIS.

Location-Based Service: A service provided to the subscriber based on their current geographic location. This position can be known by user entry or a global positioning system receiver. Most often the term implies the use of a radiolocation function built into the cell network or handset that uses triangulation between the known geographic coordinates of the base stations through which the communication takes place.

Online Environmental Information System: A web-based system for making environmental information transparent and instantaneously accessible to the public.

Planning Support Systems: Interactive computer-based systems designed to help decision-makers process data and models to identify and solve complex problems in large scale urban environment and make decisions.

Public Participatory GIS: A GIS technology that is used by members of the public, both as individuals and grass-root groups for participation in the public processes (i.e. data collection, mapping, analysis and decision-making) affecting their lives.

Web-Based GIS: Also known as 'Internet GIS'; is a new technology that is used to display and analyze spatial data on the internet. It combines the advantages of both internet and GIS. It offers public a new means to access spatial information without owning expensive GIS software.

Web-Based Public Participatory GIS: An online application of GIS that is used for increasing public access to information and active participation in the decision-making process and is an important improvement over existing public and decision-maker power relationships.

Chapter 3
Modelling & Matching and Value Sensitive Design:
Two Methodologies for E-Planning Systems Development

Yun Chen
University of Salford, UK

Andy Hamilton
University of Salford, UK

Alan Borning
University of Washington, USA

ABSTRACT

In this chapter the authors present two methodologies: Modelling & Matching methodology (M&M) and Value Sensitive Design (VSD), which can help address the knowledge gap in the methodologies for designing e-Planning systems. Designed to address the requirements of diverse user groups and multi-disciplinary cooperation for systems development, these two methodologies offer operational guidance to e-Planning systems developers. After the background introduction on e-Planning systems, these two methodologies are described, along with their application in two projects, namely VEPs and UrbanSim. This is followed by suggestions for the further work and conclusions.

INTRODUCTION

Cities are dynamic living organisms that are constantly evolving. Thus city planning has always been difficult. Today our rapidly changing society makes the job of predicting future needs of city dwellers, and those who depend on the services cities provide, even more problematic. Particular problems include: transport, pollution, crime, conservation and economic regeneration. Thus in addressing the complex problems of city planning it is not sufficient just to be concerned with the physical structure of the city; the interplay of intangible economic, social and environmental factors needs to be considered as well.

Those involved in the sophisticated art of city planning use a variety of tools. Many authors argue that Information Communication Technologies

(ICTs) offer the potential to improve the current situation in the consultative urban planning process, in particular the Internet, Geographic Information Systems, and Virtual Reality. The rapid development of these technologies provides new opportunities to improve the planning processes and make better use of resources (Dollner et al., 2006). Recently it has become clear that in order to provide more efficient city management and administration, it is important to design holistic systems that integrate information from city authorities, utility and transport system providers etc (Hamilton et al. 2005; Knapp et al., 2009). E-Planning systems are a new variety of information system for use in urban planning, with the goal of making the urban planning process more effective and efficient with appropriate ICTs. Nowadays, many countries around the world intend to modernise the planning process or have already initiated such movement (ODPM, 2004).

An e-Planning system is a complex information system that involves integration of various new technologies and interaction among multiple user groups. However, the vast literature that covers computer use for e-Planning systems arguably devotes insufficient attention to gathering requirements of different user groups and to the system design process. Effective development methodologies are needed to ensure that new technologies are matched to required functionality, and that successful e-Planning systems are implemented. In addition, the technical problems of creating an integrated ICT infrastructure to support e-Planning systems development are considered not to be the most difficult challenge (Curwell & Hamilton, 2003). Greater concern is expressed over social, human and activity issues, which should be addressed in a system development process.

In her PhD research, Chen has identified significant gaps in current knowledge regarding appropriate development methodologies for complex system development that involves multiple stakeholders, such as e-Planning systems (Chen, 2007). It also shows that the individual Information System Development Methodology (ISDM) does not offer a total solution that address both 'soft' and 'hard' perspectives in the system development process. Hence, this chapter is intended as a contribution toward the development of hybrid methodologies that can provide sufficient support to e-Planning systems designers. This chapter concentrates on how multiple stakeholders can be involved in the design process of e-Planning systems and how system designers can work together to identify goals in terms of cities designing via appropriate system development methodologies; so that the systems produced can enhance the planning process.

The chapter starts with the general investigation of e-Planning systems from its definition, current applications for both professionals and general public, and future trends. These investigations indicate challenges of the development of these systems. The potential solutions to address these challenges are discussed in the second part of this chapter, which presents two approaches that we believe will be useful for e-Planning systems development: Modelling & Matching, and Value Sensitive Design. The intended outcome of the application of both M&M and VSD is improved planning systems, in terms of better alignment to planning needs.

The Modelling & Matching (M&M) methodology (Chen, 2007) aims to address the need to develop systems that satisfy the requirements of stakeholders who display wide variance in terms of domain knowledge and motivations. In addition, it focuses on the appropriate process to facilitate cooperative system development engaging multi-disciplinary professionals, and to integrate technical design into the knowledge management/structuring process. M&M brings together many approaches and bridges the design/implementation gap by providing different design models with immediate feedback from stakeholders that is so lacking at present.

Value Sensitive Design (VSD) is a theoretically grounded approach to the design of technology that attempts to account for human values in a principled and comprehensive manner throughout the design process (Borning et al., 2005). Key features of the methodology are its interactional perspective, tripartite methodology, and emphasis on indirect as well as direct stakeholders.

Why link M&M and VSD? Both development systems address stakeholders concerns but are focused on different aspects. M&M is particularly focused on requirements capture and collaborative work. The purpose behind this is the production of systems that reflect the way stakeholders currently work, and could work more effectively with the support of an IT system to perform particular functions. For example, through involvement with local stakeholders M&M can help to design systems where pedestrians crossing outside schools are considered as a matter of course. However M&M does not specifically help to find answers to questions such as "why are we planning cities for cars when perhaps we should be planning for other types of transport?". Such questions are value laden and very much the focus of VSD. VSD helps to examine values and make choices, whether this is for low impact cars, highly efficient public transport, or any other solution in which values conflict.

However the two systems are complementary. Stakeholder views are closely associated with values. The requirement for pedestrian crossings is value driven in terms of the value we put on life and prevention of loss of life. In this respect we could see the difference between the two systems in terms of accepted values (M&M) and value conflict or changing values (VSD). Some reflections on applications of these methodologies in the real world will illustrate this point.

Finally, the chapter concludes with a discussion of directions for future work, including additional applications and the potential for coordinating the application of these two methodologies, based on reflections from case studies discussed in the chapter.

BACKGROUND TO E-PLANNING SYSTEMS

A city is not only a space, but also involves cultural, social and environmental aspects. As a result, urban planning is a multi-disciplinary process. Due to the complicated and iterative characteristics of the urban planning process, urban plans have to be flexible, allowing the incorporation of influences from all stakeholders of cities. This is one reason that the urban planning process should be transparent and accomplished through a participatory process involving multiple user groups, including government, citizens, urban planners, private sectors (e.g. real estate companies, forest industries, and investors) and affected communities (e.g. environmental non-governmental organizations, and transport and utility service providers). It was agreed in Lisbon by the Council of the European Union that 'information technology can be used to renew urban and regional development'.

In the early 1990's, software products for urban planning were mainly used for mapping, storing information and, in a lesser extent for analysis (Fischer and Nijkamp, 1993). With the evolution of computer techniques, conventional urban planning can be reconceived (Hamilton et al. 1998). In recent years, advances in visualisation techniques, data modelling and the development of new standards have given the potential for the creation of a new generation of e-Planning systems. For example, a 3D city model can be integrated with GIS systems, so that it is not only possible to have a visual effect of what the city is like, but also, to understand the information behind each building, each street, and each parcel of land in the city (Wang & Hamilton, 2009; Knapp et al., 2009). The rapid development of ICT provides new opportunities to improve the planning processes and make better use of resources. However, it is argued that ICTs can only become e-Planning systems for social development if they address the complex challenge of improving the lives of the most needy (Ghose, 2001).

In this section, the background to e-Planning systems will be discussed, in terms of its definition, the state of the art, future trends, and challenges of development.

Definition and Characteristics of E-Planning Systems

E-Planning, as an important part of eGovernment, can enable easy access to high quality information, guidance, and services that support and assist planning applicants, and streamline means of sharing and exchanging information amongst key stakeholders (The Planning Service, 2004). The principal aims of e-Planning are to enable more people to become involved in planning; to increase openness, efficiency and effectiveness; and to arrange the delivery of planning services to meet citizens' needs.

There is no doubt that urban planning is a complicated process, and a comprehensive method of collecting and utilising ideas would help in making better decisions. An e-Planning system is constructed via negotiations between various social groups, which include professionals, decision-makers, developers, special interest groups, citizens, and other stakeholders. E-Planning systems should facilitate the social cohesion and inclusion in the planning and policy making process.

E-Planning systems act as 'information media', which should be 'more concise, more regionally specific and more focused on delivery' (Mugumbu, 2000). This characteristic of e-Planning systems also informs the selection of its component technologies, which can be summarised in three perspectives:

- **Geo-spatial analytic ability:** Since the process of urban planning is a geographic-related process, geo-spatial information-based technologies are a focus of most e-Planning systems. In the drive for information-rich urban environments, it is significant that an estimated 80% of actions taken by municipal authorities are supported by geo-spatial information.
- **Technologies interoperability:** New methods and tools for system dynamics modelling, together with increasing computing power, are opening up new possibility in addressing urban development. Standards such as e-GIF, SDI and SIS for e-Planning systems development can ensure the reliability, credibility, security and interoperability of these new technologies and boost international and national cooperation (Song, et al., 2009).
- **Cost-efficiency and feasibility:** Concerning limited time and money, taking advantage of massive amounts of e-Planning system from various techniques to make timely and critical decisions requires significant computing power at 'reasonable' cost (Rowe et al., 2005). Efficiency is defined as the ability to achieve desirable goals by managing limited resources. It asks developers and designers to consider how to make best use of all the available resources (Ranke et al., 2005).

E-Planning systems offer considerable opportunity for early and rapid change to the future delivery of planning services, with an emphasis on electronic delivery (The Planning Service, 2004). In addition, a successful e-Planning system should have positive effects on social inclusion, allowing people to become more effectively involved in the planning process. In order to achieve these purposes, e-Planning systems need to be accessible, comprehensive and interactive.

Current E-Planning System Applications

Many local governments in Europe have attempted to apply new ICTs for e-Planning activities. For example, Stuttgart City has developed a system called „Nachhaltiges Bauflächenmanagement" (sustainable land use management system), which is intended to support the strengthening of inner-city development based in context of sustainable urban development. Also, novel modelling and representation technologies can contribute to presentation of planning options and communication with stakeholders. It is expected that, especially urban planners and the government can profit from the use of e-Planning systems to facilitate planning work in the following perspectives:

- **Provision of information:** Geo-spatial databases and digital maps are more often used to provide environmental and planning information. The use of GIS eases the analysis of complex planning data and contributes to a better quality of information.
- **Scenarios / simulations / forecast of planning outcomes:** ICTs facilitate the creation of different planning alternatives as well as the forecast of their outcomes ((Song, et al., 2009). Thus planners are able to select options that are forecast to better support sustainability.
- **Visualising:** Visualising planning issues with the help of 3D VR modelling contributes to a better forecast of planning outcomes such as visual impact, noise and wind. Furthermore, 3D modelling can become 4D (time embedded) or include other 'dimensions' that can be managed and visualised in the same consistent model (e.g. costs and environmental impacts) (Aouad, et al., 2006).
- **Disseminating information / planning participation:** ICTs are used for disseminating planning information via the Internet and facilitating the public to comment on local and strategic urban plans, thus promoting transparency of activities and public awareness of planning and sustainability issues. In this way, planning participation can be supported so that a better collaboration between public administration and civil society can be achieved.
- **Networking / contacts:** ICTs are also used for professional communication (i.e. professional contacts and networks) to make it easier and faster.

Compared with e-Planning systems design for professional use, the design of an e-Planning system for the general public presents additional challenges. PPGIS (Public Participation GIS) is one example. PPGIS is a field of research that focuses on the use of GIS by non-experts and occasional users. It is about empowering GIS various user groups (including general public) and enabling them to use the technology purposefully to capture their local knowledge and advance their goals (Sester, et al., 2009). These potential users tend to have a diverse range of computer literacy, cultural backgrounds, experiences and knowledge. These aspects require that PPGIS systems are accessible and easy to use. Although the usability of GIS products has improved immensely, they still require users to have considerable technical skills to operate them. This presents major obstacles to non-expert users in terms of navigating an interface that embeds a language and concepts that support the system's architecture rather than the user's language (Haklay 2002).

To enable the accessibility of advanced technologies to a wider public, current research in e-Planning systems has opened areas of research, such as integrating appropriate visualisation or multimedia to improve the usability of GIS for occasional and non-specialist users, and adopting user-centred design and Human Computer Interaction (HCI) approaches to e-Urban planning projects (Haklay and Tobon, 2003). In practice, however, planning procedures and instruments are taking the potentials of innovative ICTs into account slowly. Community mapping, PPGIS and XPlanGML are examples of emerging approaches that aim to build ICTs into planning procedures in order to improve coordination or facilitate new forms of interaction, but that do not reflect a mainstream yet (Balram and Dragićević 2006; Sester, et al., 2009). Most of them have been developed and tested in the lab, but few of them are widely used in cities. Furthermore, many failures of these systems ultimately occur due to the complexity of e-Planning systems and failures of meeting basic requirements of their stakeholders. This illustrates that we need an approach that can contribute substantially to the way in which we understand how e-Planning systems can operate as a social-technical object. To develop an e-Planning system, it is doubly important to use a transparent design approach to understand and engage with multiple stakeholders so that the system can be designed for them.

The next subsection further describes the challenges in E-Planning system development, to inform the potential solutions in the following sections.

E-Planning Systems: Future Trends and Development Challenges

'eGovernment is not just about enabling services to be made available over the Internet – it is about total revolution as to how services are processed internally and how State Agencies interact with each other to provide citizen and business focused integrated services.' (Fitzpatrick, 2005)

As a result, e-Planning is not just about 'service delivery', but also about 'interaction'. In addition, the trend of e-Planning systems evolution increases its complexity, since it involves the integration of various new technologies and interaction among multiple user groups. There are two primary challenges for its development due to this complexity.

First, this kind of system has a wide range of users.

In February 2008, the Population Division of the United Nations reported that more than half of the world's population would be living in urban areas by the end of the year. To cope with this rapid urbanisation, there are predictions that cities will become more violent and socially exclusive. In 1990s, the 'digital divide' quickly became one of the political and academic 'hot-topics'. Crucially, the groups most likely to be 'digitally excluded' continue to be synonymous with those who can be characterised as being already socially excluded – especially in terms of income and socio-economic status (Landorf, 2009). In order to avoid the 'digital divide' and to encourage the general public to participate into the urban planning process, new kinds of governance and decision-making process involving a large variety of stakeholders are essential. Hence, understanding stakeholders has a significant role to play in the pursuit of effective e-urban planning, and achieving its goals.

In Europe, more than 80% of Europeans live in cities, and urban planning affects every one of its stakeholders, who may have divergent interests. The complexity of the interaction among these stakeholders may result in their relationship in synergy, conflict, or negotiation. In addition, different stakeholders may use e-Planning systems in a large number of application areas. Accommodating such a wide spectrum of needs and effective use of technologies to facilitate interaction among various stakeholders is a challenge, since it is difficult to arrive at a shared view on e-Planning System specification. The literature concerned with computer use for e-Planning systems generally does not fully address the problem of gathering requirements from disparate user groups and integrating these requirements into the system design process.

Second, the complexity of urban processes requires multi-disciplinary cooperation to identify goals in terms of cities designing.

Generally speaking, urban planning involves various factors and is extremely uncertain. As information systems to facilitate the urban planning processes, e-Planning systems face the difficulty of understanding concrete requirements which may result in the gap between initial understanding and final functional specification. To avoid the mismatch between supplied technologies and demand requirements in e-Planning systems development process is a challenge. Experts from different areas (e.g. urban planning, geo-spatial system, and government policy) need a well-formulated collaborative environment and process to work together for the system development. In addition, like most information systems, the development of e-Planning systems is an incremental process. With the continual process of urban planning and developments in the available techniques, requirements for e-Planning systems are increasing as well, such as: more specific tools in terms of functionality to solve particular problems (Vircavs, 2006); more attractive interfaces and visualisation in terms of high efficiency of usage (Dollner et al., 2006), and much better geo-spatial applications in terms of public accessibility (Onsrud and Craglia, 2003), etc. The question of how to select these technologies to match the demand of the uses also needs to be considered carefully. A proper Information System Development Methodology (ISDM) would be highly beneficial, which combines both 'soft' (with a social and human focus) and 'hard' (with a technical development focus) ISDMs into a 'Multi-methodology'.

'Multi-methodology' is not the name of a single methodology, or of a specific way of combining methodologies. Rather, it refers in general to utilizing a plurality of methodologies or techniques, both qualitative and quantitative, within a real-world intervention. The purpose is to generate a richer and more effective way of addressing the problem at hand. We argue that to develop complex systems such as e-Planning systems, a multi-methodology is the most appropriate approach. There are two arguments to support this statement: First, all real-world problems existing in the complex urban planning process have personal, social and technical dimensions. Combining methodologies to deal with all these characteristics are therefore likely to be more effective. Second, a typical IS development passes through several stages, from an initial exploration and appreciation of the situation, through analysis and assessment, to implementation and action. Individual methodologies and techniques have their strengths and weakness with regard to these various stages.

To summarise, a well-understood and feasible development methodology can aid in the design and implementation of successful e-Planning systems, so that they are more likely to meet the different requirements of various user groups. Such methodologies should devote significant attention to issues of human values and the political context in which these systems are deployed, in addition to more traditional engineering concerns. In the rest of the chapter, two methodologies will be introduced and discussed as potential solutions to address challenges discussed above, namely Modelling & Matching (M&M) and Value Sensitive Design (VSD). These two methodologies are intended to address challenges from different perspectives. M&M aims to provide wider access to involved stakeholders in systems development process and to encourage the cooperation among multi-disciplinary system developers. VSD focuses more on value issues to ensure a human-centred design approach.

Methodology (1): Modelling & Matching

In order to address the challenges discussed above, the Modelling & Matching methodology (M&M) was designed to model system functionality in terms of stakeholders' needs, to communicate this information effectively to system developers for a shared view, and to matching the requirements with available technologies in terms of interoperability and feasibility. As suggested in the literature, to make the most effective contribution in dealing with the richness of e-Planning systems development, it is desirable to go beyond using a single methodology to generally combining several methodologies, in whole or in part, and possibly from different paradigms. M&M

combines elements of UML, Soft Systems Methodology, and rapid development methodology, together with User Requirements Engineering, Human Factors Engineering and Usability Evaluation, and embeds them into a five-step process that reflects the human-centred approach. It has been successfully adopted in a multi-partner, geographically distributed system development project (the Rosensteinviertel Project, a 100 million Euro regeneration project in Stuttgart), as part of Virtual Environment Planning Systems project (VEPs) (E109, INTEREG 3b, €4.7 million, 2004 -2008), and enabled partners in different countries in Europe to design and implement a planning system co-operatively (Knapp et al, 2009). M&M is currently used in the Sustainable Regeneration project (SURegen) (UK Government EPSRC funded, £2.5 Million, 2008-2012) to design a regeneration decision-support system, the 'Regeneration Workbench'. In this section, M&M's theoretical framework, practical methodology, and application will be discussed briefly, to illustrate its potential to address the challenges of e-Planning system development discussed above.

M&M: The Theoretical Framework

Based on the particular characteristics of e-Planning systems development, we suggest that the possible approaches to e-Planning systems development can be conducted through four processes: Modelling Process, Matching Process, Iteration process and Evaluation Process. In the theory these processes are developed in terms of specific characteristics of e-Planning systems, and thus have significant implications for the effective e-Planning systems development.

The four-process framework is underpinned by both objectivist ('hard') viewpoints (Hirschheim and Klein, 1989) and interpretivism ('soft') viewpoints (Hirschheim, 1992) to ensure that social context, stakeholders, activities and technologies are all covered during the whole development process. It is proposed based on three conjectures about e-Planning systems that come out of the literature on urban planning and ISDMs. The first conjecture is that understanding the complexity of various stakeholders' interaction and relationships in the wide social context is the first step to develop e-Planning systems. This conjecture implies the importance of modelling the complexity and providing 'soft' conceptual models to manage the complexity. The second conjecture is that technology reuse and selection are important in e-Planning systems development. For designing e-Planning systems, there is a compromise in deciding between available technologies and desired system functionality by stakeholders, remembering the feasibility and interoperability are key points. In addition, 'hard' technical models are essential for technology reuse. The third conjecture is that rapid prototyping is useful both to decompose the incremental development process and to continually involve stakeholders. Users can describe what they currently do, but they find it difficult to visualise how they might use a computer system or how the whole tasks might be re-engineered unless they are provided with the concrete system. Figure 1 below illustrates the general layout of the framework:

Urban planning can be considered as a process involving several stakeholders, and the role of the information system is to solve the urban problems or at least some of them (Laurini, 2001). For constructing an information system, the first step is to model its context and user needs. The key to modelling is to find the right abstractions that explain the desired aspect and the careful weeding out of irrelevant details. For e-Planning systems development, models are used for two purposes. First, different models can be produced to bridge the gap between the initial e-Planning system boundary and the desired system functionality by stakeholders. To bridge the gap the following aspects have to be considered:

- **Focus on Social Context:** the real-life scenarios, the involved stakeholders, and

Figure 1. A four-process framework

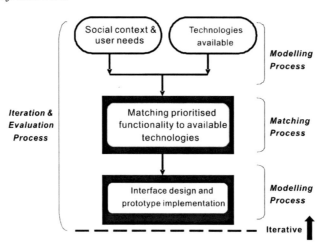

- other social information gathered before beginning the development.
- **Focus on Activities:** a high-level description of the application's fundamental functions from stakeholders' perspective, without reference to technology or how the stakeholders will actually perform them.
- **Focus on Stakeholders and Representational Context:** how stakeholders will see, think about, and interact with the application.

The second way of using models in the framework is to model technologies for reuse, which happens after the matching between available technologies and user models in terms of functionality, feasibility and interoperability. ICT solutions promise to provide an effective and efficient means of streamlining the urban planning process, driving down the costs of delivering services as well as providing a more convenient channel of communication for all involved stakeholders (Laurini, 2001). However, e-Planning solutions, like all large-scale ICT systems and solutions, typically have high implementation costs and inherent production risks, both of which are barriers to effective e-Planning systems development in the face of failing projects and falling budgets (Ranke et al., 2005). In addressing these challenges, the concepts of open standards, open source, solution reuse and collaborative solution development promise to help (Rowe et al., 2005). Choice should lead to the technologies that best suit and balance the needs of involved stakeholders, taking into account price, quality, ease of use, support and protection of legacy investments. Functionality, feasibility and interoperability are three key guidelines to undertake this matching process.

The modelling and matching process takes place in the iterative cycle with evaluation activates. This will split the complicated system functionality into different development stages and provide system users with the opportunity to immediately assess the success or failure of the prototypes (in an prioritised order), in contrast to lengthy design sessions whose result may be validated only late in the development process. An e-Planning system is a complicated system involving multiple stakeholders, whose interests might be different and even conflict. Hence, implementing all user requirements in one development cycle is difficult. The key concept in the iteration process refers to rapid prototyping. It means that the prototype system with high-prioritised functions is to be produced first to provide basic system ideas to its users for early system evaluation.

In summary, the main concept of this framework is the separation of requirement elicitation and system modelling into views reflecting the peculiarities of the e-Planning system. This results in models of sub-problems, which are then matched and combined to an overall model of the system. As a result, modelling and matching processes are two main processes in the proposed framework, complemented by iteration process and evaluation process. Due to the distinctive role of modelling and matching processes in this framework, the methodology proposed based on the framework is called Modelling and Matching (M&M), which will be discussed in the section below.

M&M: A Practical Methodology

There is a considerable gap between the theory and the practice of information system development. Because of this gap, problems could occur which may lead to the failure of the system development (Lindgaard, 1994). To be a practically effective methodology, M&M has distinct procedures for action, and addresses practical issues identified from the real case study (Hamilton et al., 2005), such as 'cooperation and communication among development partners', 'results presentation and sharing', 'combination between user requirements and technical development', 'evolved requirements', and 'system test'. In addition, analysis of the case study implied that, besides stakeholder involvement, closer collaboration is also needed among development teams early in the process and throughout the data collection phase, as well as a greater understanding of each other's tasks, roles and responsibilities. As a result, the basic principle of M&M is the use of different models throughout the development process for knowledge sharing and cooperation among development teams and stakeholders. The intent is to address a problem with most current ISDMs: with few exceptions (e.g. Maguire, 1997; Beyer and Holtzblatt, 1998), these do not address how to present findings from each step of the process in a consistent, structured, and accessible manner, which can then be used internally with development teams and externally with various stakeholders. The methodology is elucidated succinctly below.

There are five stages in the 'M&M' methodology for modelling and matching process, shown in Figure 2.

As shown in Figure 2, the essence of Stage One of the process involves interviews and meetings with various stakeholders to gain an understanding of the problem situation, which is represented by the use of 'rich pictures' and CATWOE techniques in Soft Systems Methodology (SSM) (Checkland, 1981). The three most important components of a

Figure 2. Modelling and matching process of M&M

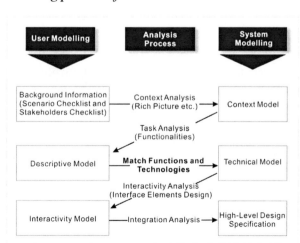

rich picture are structure, process, and concerns, which are used to identify two main aspects of the human activity system—primary tasks and concerns. Key tasks can be identified by the links among actors, i.e., the flow of information or documents. Main concerns are depicted as the think bubbles. It is obvious that the rich picture is intended to be a broad, fine-grained view of the problem situation.

CATWOE represents six elements for system analysis, which are:

1. **Customers:** the victims/beneficiaries of the purposeful activity.
2. **Actors:** those who are involved in the activities.
3. **Transformation process:** the purposeful activity transforming an input into an output.
4. **Weltanschauung:** the view of the world that makes the root definition meaningful in context.
5. **Owners:** those who can stop the activity.
6. Environmental constraints that affect the situation.

In a project with multiple primary and secondary user groups and a complex situation such as e-Planning system development, it is essential to spend time clarifying the 'big picture' to develop a shared understanding of the project with its various stakeholders. Stage One in M&M could help retain the big picture while also yielding easy insight into the details when gathering user requirements as they emerge during the analysis. The big picture is then summarised in the Context Model, which expresses the features of the situation and its stakeholders.

Stage Two encompasses task analysis. Like Stage One, it is concerned with system definition; but unlike that stage, it is driven by the needs of external stakeholders rather than any inherent purpose of the e-Planning system development project. The outcomes from stage two are presented in the Descriptive Model, which tends to focus inwardly on use cases and stakeholders interacting with the system. The combination of Context Model and Descriptive Model can yield a more balanced view of system definition and provide a systematic development path from high-level context analysis down into an object-oriented implementation.

In Stage Three, the matching process occurs, which facilitates the establishment of a feasible and interoperable development solution against the system functionality identified in previous stages. Existing soft methodologies (e.g. SSM) focus almost entirely on the tasks and activities of user requirement gathering with little or no mention of the contributions that technical software developers could bring to the system functionality. Likewise, certain aspects of hard development models, such as object-oriented models, are good at facilitating technical software developers to do system modelling, but it is not obvious what user requirement gathering could benefit from. Matching process in M&M aims to bridge the gap between soft and hard methodologies.

There are three activities in the matching process, namely:

1. Prioritising use cases for rapid prototyping to continually involve stakeholders for quick feedback on user requirements;
2. Balancing functional and non-functional requirements to build quality into the system;
3. Considering available technologies/data/tools and technical standards in terms of system functionality for effective system development.

The technical solution is then defined by the object-oriented language (i.e., Unified Modelling Language) in the Technical Model, which needs to be handed over to technical developers for realisation.

Stage Four concerns the interface design issues based on proposed technologies. The interface elements and their arrangement are presented in the Interactivity Model by storyboards. The final stage of M&M (i.e. Stage Five) is an integration process to construct the high-level design specification based on outcomes produced from above four stages.

The Application of M&M

M&M has been applied in an EU-funded project, called VEPs. VEPs is a collaborative project with eight academic and industry partners located across Europe, alongside associated planning authorities (VEPs, 2006). It aims to improve the knowledge base on the potential of ICTs for territorial development in the North West European (NWE) region, specifically regarding the use of ICT for e-Planning, consultation and communication of citizens' views on planning issues.

In the project, we used M&M to understand user requirements and to improve the design of the system (Chen et al., 2007). The process started with the detailed stakeholders' investigation and a series of workshops and semi-structured interviews with identified stakeholders.

To define use cases for Rosensteinviertel case study, stakeholders' workshops were held with participants from different user groups, including planners, members of the public, members of local authorities, and architects. In these workshops, professionals and non-professionals discussed their requests on web-based participation tool. Findings from these workshops were summarised into the scenario checklist, which is a 'one-page' document used to capture and define various stakeholders involved in the proposed system and to state their motivations and expectations for the proposed system. A series of semi-structured interviews were also conducted with representatives of each stakeholder group. The stakeholders' checklist was used for interviews, which includes five sections, namely Basic Characteristics, Knowledge and Experience, Motivation and Expectation, System Requirements, and Desired Data.

Through these individual interviews and group workshops, stakeholders started to be involved in the design process. The analysis represented an initial background exercise to understand who was involved during the e-Planning system design process. It also outlined the nature of the involvement played by each stakeholder for the e-Planning systems. From the findings, 'dominant stakeholders' and their requirements were identified and prioritised as 'must have' functions for further technical consideration in prototyping process.

Under the guidance of the Matching and Technical Modeling stages of M&M, two prototypes were developed for interactive participation concerning urban regeneration in Rosensteinviertel (Stuttgart, Germany), the location of one of the demonstration projects in VEPs, which address requirements of dominant stakeholders. They were developed rapidly for quick feedback from those stakeholders. Two techniques were adopted to evaluate the methodology after the development of these prototypes, namely questionnaires to end users and interviews with project partners.

In the evaluation using the questionnaire, the test website was distributed to twenty citizens in UK. In addition, ten pairs of questionnaires were brought to attendees in the workshop at Urban Data Management System International Conference 2006 (UDMS2006) held in Aalborg. The results of these questionnaires demonstrate that the first rapid prototype for the Rosensteinviertel demonstration project was effective as an e-Planning system. Simultaneously, the evaluation indicates two weaknesses of the development. The first is the geo-spatial interface design and the second is the long-term effective evaluation regarding the social inclusion and selflessness issues.

In the evaluation using the interview technique, two interviews were conducted with team members in the Rosensteinviertel project. The highlights in using the methodology observed in the interview include:

1. The methodology presents very precise ideas and development plans for project cooperation;
2. The methodology provides profound structure and control for development process;
3. The methodology documents back-up cognitions of stakeholders' profiles, stakeholder requests and technical must-be;
4. The methodology encourages the good communication among partners;
5. The methodology is effective in combining 'hard' issues with 'soft' issues during development process to avoid technique bias;
6. The methodology integrates the evaluation plan into the process to facilitate the rapid prototyping, which does help for iterative development process.

During the application, several factors emerged as important for using M&M to design e-Planning system to be realised in practice in the advocated manner. The factors emerged from the reflection of the case study, with some successfully demonstrated in practice and others requiring further consideration:

Stakeholder investigation and prioritisation. Stakeholders are groups or individuals who have a stake in, or expectation of, a project's performance. In the case of urban planning, these include the public, professionals (e.g. urban planner, architects etc.), local authorities, and managers of different sectors etc. The relationship among stakeholders is complex. Frequent conflicts between stakeholders revolve around long-term versus short-term objectives, cost efficiency versus need, quality versus quantity, and control versus independence. It is important to undertake stakeholder investigation and prioritisation at the early stage of e-Planning system development process. Some techniques are useful to understand the complexity among stakeholders, e.g. scenario checklist, stakeholder checklist, and rich picture. In addition, according to the prioritisation of stakeholders, requirements can be categorised as 'Must Have', 'Should Have', 'Would Have', and 'Might Have' for rapid prototyping. The rapid prototyping process can enhance the involvement of stakeholders in the development process via quick feedback to them.

The contribution of 'Mediator' in the development process. This factor addresses the difficulty of using the methodology in practice. The research demonstrated the value gained throughout the process by the expert knowledge provided by the mediator among project teams. This role is not typical within a project team, but in this case the contribution was clearly beneficial in supporting the level of understanding of members of the team from both 'soft' requirements gathering and 'hard' technical development perspectives. The stakeholders also recognised the potential value of this role from the inception of the project. The guiding role provided by the advisor's experience and general understanding, coupled with their technical expertise, clearly improved the team's ability to engage with various stakeholders and to effectively interact with the design across the project lifecycle.

Consideration of human values in the development process. Although this project represents progress toward delivering effective e-Planning systems, there were a number of aspects that require additional research, for example the difficulty of involving target users in the interface design. This methodology did not demonstrate the concrete realization of how to achieve this aim. In addition, a greater emphasis is required for the wider issues of human values across the design life cycle. M&M by its nature has a limited provision for these issues, and needs to address them via other methodologies. The desired methodologies should have the potential to take values of all stakeholders into account in the design process, for both user interface design and long-term social inclusion in urban planning process. One suitable methodology is VSD, which is described in the next section.

Methodology (2): Value Sensitive Design

As we have discussed, e-Planning systems are complex and involve interaction among numerous stakeholders. Particularly when such systems involve substantial public participation, fundamental value conflicts among the stakeholders may make it hard to make progress. Even with more back-room systems, value hotspots can significantly undermine system acceptance. Setting aside issues of conflict and difficulties of system adoption, e-Planning system design and deployment often raise challenging issues around such basic values as fairness, democratic participation, and transparency.

Value Sensitive Design (VSD) (Friedman et al., 2006a) is an approach to the design of technology that accounts for human values in a principled and comprehensive manner throughout the design process, and is intended to address questions of just these sorts. This section provides an overview of the VSD theory and methodology, followed by a case study of its application and extension in an e-Planning domain, namely urban simulation and public participation. In addition to the urban planning domain, VSD and closely related theory and methods have been used successfully to conceptualise value tensions and corresponding design trade-offs in information systems (Friedman et al., 2006a, Miller et al., 2007), to understand the value-oriented user experience of specific technologies (Friedman et al., 2006b, Kahn et al., 2006), and to design new or redesign existing technologies in response to value analyses and user experience (Flanagan et al., 2005, Kahn et al., 2006, Sengers et al., 2005).

Overview of VSD

Key features of VSD are its interactional perspective, tripartite methodology, and emphasis on direct and indirect stakeholders.

VSD is an interactional theory: values are viewed neither as inherent in a technology, nor as simply transmitted by social forces independent of the characteristics of the technology. Rather, people and social systems affect technological development, and technologies shape (but do not rigidly determine) individual behavior and social systems.

VSD employs a tripartite methodology, consisting of conceptual, empirical, and technical investigations. These investigations are applied iteratively and integratively, with results from new investigations building on and integrating earlier ones. Conceptual investigations comprise philosophically informed analyses of the central constructs and issues under investigation. For example, how does the philosophical literature conceptualize certain values and provide criteria for their assessment and implementation? How should we engage in trade-offs among competing values in the design, implementation, and use of information systems? Empirical investigations focus on the human response to the technical artifact, and on the larger social context in which the technology is situated, using quantitative and qualitative methods from the social sciences. Technical investigations focus on the design and performance of the technology itself. Technical investigations can involve either retrospective analyses of existing technologies or the design of new ones.

A third key aspect of VSD is its focus on both direct and indirect stakeholders. Direct stakeholders are people who interact with the system or its output directly, while indirect stakeholders are people who do not directly touch the system but are nonetheless substantively affected by its use. Traditional user-centred design methods are primarily concerned with the end users of the technology being developed, i.e., the direct stakeholders. User-centred design is certainly a welcome reorientation from technology-focussed development, but has the potential to ignore the values and needs of people who may be strongly affected by the system being designed but are not the immediate users. These considerations are quite relevant to the e-Planning domain, since urban planning decisions usually affect all of the residents in the region, not just the users of the e-Planning system.

UrbanSim Background

UrbanSim (Borning et al., 2008a, Waddell & Borning 2004) is a simulation system for projecting the impacts of alternative policies and transportation infrastructure projects over periods of 20-30 years. In many regions throughout the United States and worldwide, there is great concern about such issues as traffic congestion, resource consumption, lack of sustainability, and sprawl. Elected officials, planners, and citizens grapple with these difficult issues as they develop and evaluate alternatives for major land use and transportation decisions, such as building a new rail line or freeway, establishing an urban growth boundary, or changing incentives or taxes. UrbanSim's purpose is to inform decision-making about such alternatives by projecting their long-term consequences for land use, housing costs, transportation utilization, environmental impacts, and related outcomes.

UrbanSim is implemented as a set of interacting component models that simulate different actors or processes within the urban environment. The system is open source, and freely available for download from the project website at http://www.urbansim.org. As of May 2009, it has been applied operationally in Detroit, Houston, and the Puget Sound region in Washington State (Seattle and surrounding cities). The UrbanSim group has also worked with other agencies in applying UrbanSim in the urban areas around Eugene, Honolulu, Salt Lake City, and San Francisco. There have also been research and pilot applications in such diverse regions as Amsterdam, Burlington, Durham, El Paso, Melbourne, Paris, Phoenix, Tel Aviv, and Zurich. There is an active users group, including significant representation from regional planning agencies.

Results from UrbanSim simulations are presented in the form of indicators. The overall goals of the alternate plans, and which impacts are particularly important to consider, are reflected in the choice of UrbanSim indicators and how they are described and interpreted. As used in the planning literature (Gallopín 1997), an indicator is a variable that conveys information on the condition or trend of an attribute of the system, taking on a specific value at a given time. Examples of indicators in UrbanSim include the population density in different neighbourhoods, the ratio of car trips to bus trips for the region, and the projected cost of land per acre in different parts of the region, each under different possible scenarios and for 30 successive years. Indicator values may then be displayed as tables, graphs, charts, or maps.

Applying VSD to Interaction Design for UrbanSim

Our value sensitive interaction design work to date has concentrated on UrbanSim's indicators, specifically:

1. how to select which indicators to provide;
2. which indicators to use in presenting the results of a particular simulation;
3. how to provide useful and fair technical documentation for the indicators;
4. how to integrate this technical documentation with support for value advocacy;
5. more generally, how to better support public participation in urban planning using results from urban simulations.

As described in the previous section, UrbanSim is beginning to be used operationally to inform important regional planning decisions. However, over a decade of work has preceded this deployment. In recognition of this very long development time, we adopted the strategy of laying the foundations for public participation in advance of wide-scale public deployment, with the goal of having interaction designs ready as the system is put into such use. This strategy is described in more detail in references (Friedman et al., 2008) and (Borning 2009).

Our interaction design involves providing descriptive technical information alongside a range of organizational perspectives. This is implemented as a set of web pages, divided into Technical Documentation and Indicator Perspectives, as well as infrastructure to compute, display, and browse the indicator values. The Technical Documentation provides a description of the available indicators (Borning et al., 2005). It includes a categorized list of all available indicators, with each indicator name linked to a separate page for that indicator. Each of these indicator description pages in turn includes a definition, discussion of how to interpret the results, the code used to compute the indicator value, test cases, and other elements. The Indicator Perspectives Framework, on the other hand, provides a mechanism for different partner organizations to present their own perspectives on major land use and transportation issues, on which indicators are most important, and on how best to evaluate alternative scenarios of land use and transportation. Each perspective includes one or more web pages that present this information, including links back to specific indicators in the Technical Documentation.

Following the VSD methodology, our work included conceptual, technical, and empirical investigations. In our conceptual investigations, we made a sharp distinction between explicitly supported values (i.e., ones that we explicitly want to support in the interaction design) and stakeholder values (i.e., ones that are important to some but not necessarily all of the stakeholders). We identified fairness (and more specifically freedom from bias), accountability, and support for democratic process as explicitly supported values. Stakeholder values cover a wide range, including clean air, economic development, equity, low taxes, walkable neighbourhoods, and many more. In turn, as part of supporting a democratic society, we decided that the system should not a priori favor or rule out any given set of stakeholder values, but instead should allow different stakeholders to articulate the values that are most important to them, and evaluate the alternatives in light of these values. We identified comprehensibility, and subsequently legitimation (Habermas 1979, 1984) and transparency, as key instrumental values as well. Making this distinction between explicitly supported and stakeholder values helped us address in a principled way questions of which values to consider, as well as questions of designer bias (are we privileging our own views and values?).

Our technical investigations included the design of the SQL queries initially used to implement the different indicators, and a testing framework for these queries (including placing the test cases as runnable components of the documentation). Subsequently, for efficiency we replaced the SQL queries with Python code, and finally with expressions written in a domain-specific programming language custom-designed for writing model variables and indicators (Borning et al., 2008b).

Finally, our empirical investigations included semi-structured interviews with urban modellers to help assess the technical documentation (Borning et al., 2005), and semi-structured interviews with 20 Seattle residents to assess the overall design consisting of Technical Documentation alongside the Indicator Perspectives (Friedman et al., 2008). A key finding in this study was that more participants viewed the system as a whole as supporting a legitimate decision-making process than any of the individual components, or even a subset consisting of several components.

FUTURE RESEARCH DIRECTIONS

This section deals with the recommendations for further research in this area. Future work should focus on more extensive application and possible integration of the two methodologies discussed above in the light of evaluation of their applications. Its aim is to improve the validity and generalization of e-Planning system design methodologies in the following ways:

Further evaluation of these methodologies: Extend the richness of the experience of these methodologies through the development of additional e-Planning systems and integrate them into the geospatial infrastructure workflow of the municipality, to evaluate their effectiveness in the long term. Future research should aim at integrating the developed e-Planning systems into a municipality's existing geo-data infrastructure with existing legacy systems, taking into account data security and network security issues, as data will be requested from outside the municipality's network.

Possible integration of these methodologies: Since these two methodologies have compatible goals, they have potential to be integrated for a more effective development process. The theoretical framework and practical methodology of M&M can be evolved by integrating the tripartite investigations in VSD, which could provide concrete guidance for deeper understanding of human values in user requirements elicitation and common application interface development. In addition, the focus of VSD on both direct and indirect stakeholders can help M&M profile users who will be affected by the system to clarify their requirements and address values in a long-term period.

As discussed in the previous section, M&M is currently being applied in a UK Government EPSRC funded project, called SURegen, for designing the Regeneration Workbench. By integrating the elements in VSD into M&M, we intend to use the combined methodology to assist technical developers to communicate with multiple stakeholders to clarify their requirements, and to support cooperative work among multi-disciplinary partners in the system development process for the project.

CONCLUSION

In this chapter, we have illustrated how e-Planning systems present a number of challenges for IS development and have proposed two methodologies (i.e. M&M and VSD) to enable effective e-Planning system development.

The design of the M&M methodology focuses on multiple user groups, the wide variety of user needs and the complexity in urban planning process. Initial evaluation of the methodology following its application in the VEPs indicates that it facilitates the documentation and analysis of user profiles and requests together with technical requirements, and encourages good communication among partners in the real development project. The evaluation also suggests that some areas of work remain, such as the involvement of users in the interface design and the integration with existing prototypes. As discussed in the 'Future Work' section, one aspect of future work will be more extensive application and evaluation of the methodology following adjustments to address the issues raised in the initial evaluation.

VSD is a principled, theoretically-grounded approach to the design of technology that seeks to account for human values throughout the design process. As we have argued in this chapter, the e-Planning domain is a political and value-laden one, making VSD an appropriate method for this domain. We illustrated its application with results from applying it to interaction design for UrbanSim. VSD is intended to be used in concert with other design methods, making a coordinated use of M&M and VSD an attractive prospect. The possible combination of these two methodologies is suggested in the 'Future Work' section.

This chapter describes the M&M and VSD theoretical background and has shown their potential to be effective in a real-world situation. In addition, to address the range of future urban issues, it would be more effective to combine these two methodologies to address both complex stakeholder issues and values. It is expected that the ongoing work in SURegen will provide more reflections on the combination of these two methodologies in the near future.

REFERENCES

Aouad, G., Lee, A., & Wu, S. (2006). *Constructing the Future: nD Modelling*. New York: Taylor & Francis.

Balram, S., & Dragićević, S. (2006). *Collaborative Geographic Information Systems*. Hershey, PA: Idea Group Inc.

Beyer, H., & Holtzblatt, K. (1998). *Contextual Design: Defining Customer-Centered Systems*. New York: Academic Press.

Borning, A., Friedman, B., Davis, J., Gill, B., Kahn, P., Kriplean, T., & Lin, P. (2009). Public participation and value advocacy in information design and sharing: Laying the foundations in advance of wide-scale public deployment. *Information Polity, 14*(1-2), 61–74.

Borning, A., Friedman, B., Davis, J., & Lin, P. (2005). Informing public deliberation: Value sensitive design of indicators for a large-scale urban simulation. In *Proc. ECSCW 2005* (pp. 449-468.)

Borning, A., Ševčíková, H., & Waddell, P. (2008b). A Domain-Specific Language for Urban Simulation Variables. In *Proceedings of the 9th Annual International Conference on Digital Government Research*, Montréal, Canada, May 2008.

Borning, A., Waddell, P., & Förster, R. (2008a). UrbanSim: using simulation to inform public deliberation and decision-making. In Chen, H. (Eds.), *Digital Government: Advanced Research and Case Studies*. Berlin: Springer-Verlag.

Checkland, P. (1981). *Systems Thinking, Systems Practice*. Chichester, UK: Wiley.

Chen, Y. (2007). *Modelling And Matching: A Methodology For Complex ePlanning Systems Development*. PhD thesis, University of Salford, UK.

Chen, Y., Kutar, M., & Hamilton, A. (2007). Modelling and Matching: A Methodology for ePlanning System Development to Address the Requirements of Multiple User Groups. In D. Schuler (Ed.), Online Communities and Social Computing, HCII 2007 (LNCS 4564, pp. 41–49). Berlin: Springer-Verlag.

Curwell, S., & Hamilton, A. (2003). *The roadmap (final deliverable) for the INTELCITY one-year RTD roadmap project (2002-2003)*. Funded by the EU DG Research Information Society Technologies (IST) Programme (IST-2001-37373).

Dollner, J., Baumann, K., & Buchholz, H. (2006). Virtual 3D City Models as Foundation of Complex Urban Information Spaces. In the Proceeding of the 11th International Conference on Urban Planning and Regional Development in the Information Society. 13th – 16th February 2006, Vienna, Austria.

Fischer, M., & Nijkamp, P. (1993). *Geographic information systems, spatial modelling and policy evaluation*. New York: Springer-Verlag.

Flanagan, M., Howe, D., & Nissenbaum, H. (2005). Values at play: Design tradeoffs in socially-oriented game design. In *Proc. CHI 2005* (pp. 751-760). New York: ACM Press.

Friedman, B., Borning, A., Davis, J., Gill, B., Kahn, P., Kriplean, T., & Lin, P. (2008). Laying the Foundations for Public Participation and Value Advocacy: Interaction Design for a Large Scale Urban Simulation. In *Proceedings of the 9th Annual International Conference on Digital Government Research*, Montréal, Canada, May 2008.

Friedman, B., Kahn, P., & Borning, A. (2006a). Value Sensitive Design and information systems. In Zhang, P., & Galletta, D. (Eds.), *Human-Computer Interaction and Management Information Systems: Foundations* (pp. 348–372). New York: M.E. Sharpe.

Friedman, B., Kahn, P., Hagman, J., Severson, R., & Gill, B. (2006b). The watcher and the watched: Social judgments about privacy in a public place. *Human-Computer Interaction, 21*(2), 233–269. doi:10.1207/s15327051hci2102_3

Gallopín, G. (1997). Indicators and their use: Information for decision-making. In Moldan, B., & Billharz, S. (Eds.), *Sustainability Indicators: Report of the Project on Indicators of Sustainable Development* (pp. 13–27). New York: Wiley.

Ghose, R. (2001). Transforming Geographic Information Systems into Community Information Systems. In *Transactions in GIS* (Vol. 5, pp. 141–163). Use of Information Technology for Community Empowerment.

Habermas, J. (1979). *Communication and the Evolution of Society* (McCarthy, T., Trans.). Boston: Beacon Press.

Habermas, J. (1984). *The Theory of Communicative Action* (Vol. 1). (McCarthy, T., Trans.). Boston: Beacon Press.

Haklay, M. (2002) *Public Environmental Information Systems: Challenges and Perspectives*. PhD thesis, Department of Geography, UCL, University of London.

Haklay, M. E., & Tobon, C. (2003). Usability Evaluation and PPGIS: Towards a User-centered Design Approach. In *Geographical* []. New York: Taylor & Francis Ltd.]. *Information Science, 17*, 577–592.

Hamilton A., Burns, M., Arayici, Y., Gamito, P., Marambio, A. E., Abajo, B., Pérez, J., & Rodríguez-Maribona, I. A. (2005 September). *Building Data Integration System*. Final Project Deliverable in the Intelligent Cities (IntelCities) Integrated Project, IST – 2002-507860, Deliverable 5.4c. University of Salford, UK.

Hamilton, A., Curwell, S., & Davies, T. (1998). A Simulation of the Urban Environment in Salford. In *Proceedings of CIB World Building Congress*, Gåvle, Sweden, 7-12 June 1998 (pp. 1847-1855).

Hirschheim, R. (1992). Information Systems Epistemology: An Historical Perspective. In Galliers, R. (Ed.), *Information Systems Research: Issues, Methods and Practical Guidelines* (pp. 28–60). Oxford: Blackwell.

Hirschheim, R., & Klein, H. (1989). Four Paradigms of Information Systems Development. *Communications of the ACM, 32*(10), 1199–1215. doi:10.1145/67933.67937

Kahn, P., Friedman, B., Perez-Granados, D., & Freier, N. (2006). Robotic pets in the lives of preschool children. *Interaction Studies: Social Behaviour and Communication in Biological and Artificial Systems, 7*(3), 405–436. doi:10.1075/is.7.3.13kah

Knapp, S., Chen, Y., Hamilton, A., & Coors, V. (2009). An ePlanning Case Study in Stuttgart Using OPPA 3D. In Reddick, C. G. (Ed.), *Handbook of Research on Strategies for Local E-Government Adoption and Implementation: Comparative Studies*. Hershey, PA: IGI Global.

Landorf, C. (2009). Social inclusion and sustainable urban environments: an analysis of the urban and regional planning literature. In *Proceedings of Second International Conference on Whole Life Urban Sustainability and its Assessment*, Loughborough, UK, 22-24 April 2009 (pp. 861-878).

Laurini, R. (2001). *Information Systems for Urban Planning: A Hypermedia Co-operative Approach*. New York: Taylor & Francis.

Lindgaard, G. (1994). *Usability Testing and System Evaluation: A guide for designing useful computer systems*. Technical Communications (Publishing) Ltd.

Miller, J., Friedman, B., Jancke, G., & Gill, B. (2007). Value tensions in design: The value sensitive design, development, and appropriation of a corporation's groupware system. In *Proc. of GROUP 2007* (pp. 281-290).

Mugumbu, W. (2000). *GIS into Planning*. Retrieved from http://www.hbp.usm.my/thesis/heritageGIS/master/research/GIS%20into%20Planning.htm

ODPM (Office of the Deputy Prime Minister). (2004). *Planning Website Survey 2004: Survey of Planning Websites in England and Wales*. Retrieved from http://www.pendleton-assoc.com/planningsurvey2004.html

Onsrud, H. J., & Craglia, M. (2003). Introduction to the Special issues on Access and participatory Approaches in Using Geographic Information. *URISA Journal, 15*, 5–7.

Rowe, D., McGibbon, S., & Bell, O. (2005). Shared Source and Open Solutions for e-Government. In Cunningham, P., & Cunningham, M. (Eds.), *Innovation and the Knowledge Economy: Issues, Applications, Case Studies* (*Vol. 2*, pp. 375–381). Amsterdam: IOS press.

Sengers, P., Boehner, K., David, S., & Kaye, J. (2005). Reflective design. In Proc. 4th Decennial Conference on Critical Computing: Between Sense and Sensibility (Aarhus, Denmark, Aug 2005) ACM Press, 49-58.

Sester, M., Bernard, L., & Paelke, V. (2009). *Advances in Giscience*. Berlin: Springer.

Song, Y., Bogdahn, J., Hamilton, A., & Wang, H. (2009). Integrating BIM with Urban Spatial Applications: A VEPS perspective. In *Handbook of Research on Building Information Modeling and Construction Informatics: Concepts and Technologies*. Hershey, PA: IGI Global.

The Planning Service. (2004). *Planning Service Website*. Retrieved from http://www.planningni.gov.uk/

VEPs. (2006). *VEPs Project Home Page*. Retrieved from http://www.veps3d.org

Vircavs, I. (2006). Development of e-services in Urban Development Issues in Riga City. In *Proceeding of the 11th International Conference on Urban Planning and Regional Development in the Information Society*, 13 – 16 February 2006, Vienna, Austria.

von Ranke, F., Puhler, M., Wolf, P., & Krcmar, H. (2005). Software Engineering for e-Government Applications. In Cunningham, P., & Cunningham, M. (Eds.), *Innovation and the Knowledge Economy: Issues, Applications, Case Studies* (*Vol. 2*, pp. 397–404). Amsterdam: IOS press.

Waddell, P., & Borning, A. (2004, February). A case study in digital government: Developing and applying UrbanSim, a system for simulating urban land use, transportation, and environmental impacts. *Social Science Computer Review, 22*(1), 37–51. doi:10.1177/0894439303259882

Wang, H., & Hamilton, A. (2009). Extending BIM into Urban Scale by Integrating Building Information with Geospatial Information. In *Handbook of Research on Building Information Modeling and Construction Informatics: Concepts and Technologies*. Hershey, PA: IGI Global.

ADDITIONAL READING

Avison, D. E., & Fitzgerald, G. (2006). *Information System Development: Methodologies, Techniques and Tools* (4th ed.). New York: The McGraw-Hill Companies, Inc.

Booth, P. (1989). *An Introduction to Human-Computer Interaction*. London: Lawrence Erlbaum Associates Publishers.

Buede, D. M. (2009). *The Engineering Design of Systems: Models and Methods* (2nd ed.). Hoboken, NJ: John Wiley and Sons. doi:10.1002/9780470413791

Checkland, P., & Poulter, J. (2006). *Learning for Action: A Short Definitive Account of Soft Systems Methodology and its use for Practitioners, Teachers and Students*. Hoboken, NJ: John Wiley & Sons, Ltd.

Hamilton, A., Curwell, S., & James, P. (2008). The BEQUEST Toolkit. In Curwell, S., Deakin, M., & Symes, M. (Eds.), *Sustainable Urban Development* (Vol. 3). New York: Routledge.

Hamilton, A., Trodd, N., Zhang, X., Fernando, T., & Watson, K. (2001, November). Learning Through Visual Systems To Enhance The Urban Planning Process. *Environment and Planning. B, Planning & Design*, *28*(6), 833–845. doi:10.1068/b2747t

Knapp, S., & Chen, Y. (2007). The Impact of ePlanning Systems on Public Participation in Planning Processes. In *The Proceedings of XXI Aesop Conference*, Napoli, Italy, 11-14 July 2007.

Newman, P., & Thornley, A. (1996). *Urban Planning in Europe: international competition, national systems and planning projects*. London: Routledge. doi:10.4324/9780203427941

Pooley, R., & Wilcox, P. (2003). *Applying UML: Advanced Applications*. Woburn, MA: Butterworth-Heinemann Ltd.

Ranke, F., Puhler, M., Wolf, P., & Krcmar, H. (2005). Software Engineering for e-Government Applications. In Cunningham, P., & Cunningham, M. (Eds.), *Innovation and the Knowledge Economy: Issues, Applications, Case Studies* (Vol. 2, pp. 397–404). IOS press.

Reddick, C. G. (2009). *Handbook of Research on Strategies for Local E-Government Adoption and Implementation: Comparative Studies*. Hershey, PA: Idea Group Inc.

Sharp, H., Rogers, Y., & Preece, J. (2007). *Interaction Design: Beyond Human-computer Interaction* (2nd ed.). New York: Wiley.

Song, Y. (2004). *Development of an Integrated Geo-Spatial System to Improve Accessibility of Urban Planning Information*. PhD thesis, University of Salford.

Waddell, P., & Ulfarsson, G. (2004). Introduction to Urban Simulation: Design and Development of Operational Models. In Stopher, P., Button, K. J., Haynes, K. E., & Hensher, D. A. (Eds.), *Handbooks in Transport: Transport Geography and Spatial Systems* (Vol. 5, pp. 203–236). Oxford, UK: Pergamon Press.

Waddell, P., Wang, L., & Liu, X. (2008). UrbanSim: An Evolving Planning Support System for Evolving Communities. In Brail, R. (Ed.), *Planning Support Systems*. Cambridge, MA: Lincoln Institute for Land Policy.

KEY TERMS AND DEFINITIONS

E-Planning System: A computer-based information system to facilitate Planning activities by enhancing accessibility and to planning related information, guidance and services.

Modelling & Matching (M&M): A methodology for complex information system design, which involves multiple stakeholders and requires multi-disciplinary cooperation, such as e-Planning systems.

Stakeholder Engagement: In the context of information system development, this refers to users participation in the design process, to ensure a human-centred approach. It also means stakeholders engagement/participation in the urban planning process.

Sustainable Regeneration Project (SURegen): a UK Government EPSRC funded project (£2.5 Million, 2008-2012) to recreate an open learning environment and decision-support system (i.e. Regeneration Workbench) for sustainable urban regeneration.

UrbanSim: a simulation system for projecting the impacts of alternative policies and transportation infrastructure projects over periods of 20-30 years.

Value Sensitive Design (VSD): is an approach to the design of technology that accounts for human values in a principled and comprehensive manner throughout the design process.

Virtual Environmental Planning Systems (VEPs): An EU-funded collaborative project (E109, INTEREG 3b, €4.7 million, 2004-2008) to create e-Planning systems for facilitating the multi-way consultation process among citizens and local authorities.

Chapter 4
The Future-Making Assessment Approach as a Tool for E-Planning and Community Development:
The Case of Ubiquitous Helsinki

Liisa Horelli
Helsinki University of Technology, Finland

Sirkku Wallin
Helsinki University of Technology, Finland

ABSTRACT

As e-planning takes place in a complex and dynamic context, consisting of many stakeholders with a diversity of interests, it benefits from an evaluation approach that assists in the monitoring, supporting and provision of feedback. For this purpose, we have created a new approach to e-planning, called the Future-making assessment. It comprises a framework and a set of tools for the contextual analysis, mobilisation and nurturing of partnerships for collective action, in addition to an on-going monitoring and evaluation system. The aim of this chapter is to present and discuss the methodology of the Future-making assessment-approach (FMA) and its application in a case study on e-planning of services in the context of community development, in a Helsinki neighbourhood.

INTRODUCTION

The emergence of network, information and knowledge societies in the last decades of the twentieth century has created great expectations for the revitalisation of cities and its neighbourhoods due to the availability of *information and communication technology (ICT)*. However, the history of the early shaping of urban internet space discloses that the hope and hype of ICT have not been fulfilled (Kasvio & Anttiroiko, 2005). The assumed digital cities in Europe in the 1990s were nothing more than electronic brochures, except for a few holistic digital experiments (Aurigi, 2005).

Nevertheless, from the beginning of the 21st century a digital citizenship has started to emerge.

DOI: 10.4018/978-1-61520-929-3.ch004

According to Karen Mossberg (2008, pp.1-2), "*digital citizenship* is the ability to participate in society online... It represents the capacity, belonging and potential for political and economic engagement in society in the information age". Also the application of urban and community informatics [1] and the appropriation of ubiquitous computing have begun to turn some places into real-time cities in which amateurs become urban planners (Foth, 2009). *Ubiquitous computing* means in its idealised form that ICT is present anywhere and anytime serving people through embedded electronic devices, programmes and sensory networks. It is envisioned that environments become intelligent and cities function online and in real time. The future ubiquitous society also promises to enhance the management of global issues on the local level and vice versa. On the other hand, Mika Mannermaa (2008) alerts that all people, as private citizens, public authority or as entrepreneurs, will continuously monitor and will be monitored. "Some brother" will always oversee, know and will not forget.

Even though a great variety of web-based examples of e-planning currently exists, it is the socio-cultural and political context that conditions and shapes the appropriation of ICT and its eventual benefits. For example, Denmark which scored number one in the United Nations E-Government Readiness Survey (2008), has had a long tradition of participation in most sectors of society. In contrast, Finland, which is technologically well advanced, but culturally lagging in participatory efforts, was placed number 45 in the section of the same survey that tapped citizen participation in the co-production of services. Nevertheless, some Finnish communities tend to be islands of internet-assisted cultures. For example, the website of Helsinki scored third out of 100 cities in the worldwide Digital Governance in Municipalities Survey, by Holzer & Seang-Tae (2008). Seoul is the leading city in the application and appropriation of digital technology. Also the USA has many technology-led experiments, especially with wireless community networks. However, they tend to enhance the so called networked individualism (Foth et al., 2008) instead of collective action, due to the weakness of American public sector in local communities.

Recent developments around the social media or web 2.0 have provided new opportunities for participatory e-planning and the development of local communities. Characteristic of the e-planning experiments is that they take place in a complex context comprising many actors with different interests. The goals and foci of action vary in terms of level, scope, depth and temporal regime. Conflicts often arise between aspirations towards networked individualism and collective action. Also the imbalance between technological determinism versus the social shaping of technology may hinder the progress of the endeavour. In addition, the fuzziness of digital terminology, which is still under construction (Medaglia, 2007), may increase difficulties, such as the confusion between e-planning, e-participation and e-governance.

Consequently, there is a demand for methodologies that can advance the process of e-planning and that can provide some coordination of the fragmented bits and modules. For this purpose, we have created a new approach to e-planning, called the *Future-making assessment (FMA)*. It comprises a framework and a set of tools for the contextual analysis and mobilisation of partnerships for collective action, in addition to an ongoing monitoring and evaluation system. The aim of this chapter is to present and discuss the methodology of the FMA-approach and its application in a case study on e-planning of services in the context of community development.

We argue that e-planning, which usually deals with complex and multilevel issues, benefits from an evaluation approach that assists in the monitoring, supporting and provision of feedback of the multi-dimensional and multilevel activities that take place during the process of planning. The research questions in this chapter concern the ways in which the FMA-approach may enhance e-planning of services and its role in the overall process of community development.

The chapter starts by putting e-planning in the context of other e-activities after which the methodological framework of the FMA-approach and its evaluation characteristics are described. The approach will then be applied in a case study, called Ubiquitous Helsinki. Finally, the methodology and future trends will be discussed.

BACKGROUND

E-Planning Within E-Governance

Given that in many countries, the public has lost trust in the functioning of its governmental institutions, e-government and e-governance have been suggested as a means to improve or at least to complement governmental activities (Tolbert & Mossberger, 2006). The European Union and the United Nations have both recommended a renewal of the interactions between governments and citizens (Chadwick & May, 2003), by increasing opportunities for e-government and changing the orientation from government to governance. Governance refers to the processes and mechanisms that need to be followed to reproduce a successful community or a region (Thesaurus, 2009).

According to the OECD (2002), e-government provides opportunities to develop a new relationship between governments, citizens, service users and businesses, by using ICT for the purpose of service delivery, decision making and accountability. It is closely tied with e-democracy, e-participation, e-governance and even with e-planning, although the latter is not usually included in the e-family (Medaglia, 2007). However, as the research community lacks consensus on the core concepts and definitions of digital activities, we define them here from the perspective of e-planning, which we insert at the centre, between e-government and e-governance (see also Anttiroiko, 2003; Pessala, 2008).

E-government and e-governance are closely related to the digital administrative processes (*e-administration*), the organisation and delivery of services (*e-services*), democratic processes (e-democracy and e-participation), as well as e-planning.

If *e-government* refers to the totality of the political and administrative activities and processes of public bodies, assisted by ICT, *e-governance* is the practical management and development of those processes in the context of the four Ps – public, private, people-partnerships. According to Kickert et al. (1997), the policy networks, through which governance takes place, are stable patterns of social relations between interdependent actors which take shape around policy problems and programs. The same also applies to e-governance in its largest form.

E-democracy is a broad inclusive term that comprises all sectors of democracy (legislature, executive), all democratic institutions (legislative, consultative, civil society) and levels of government (local, regional, national). It also impinges all kinds of participants (West, 2000; Norris, 2001; Vegh, 2003; Buchsbaum, 2008). In practice, it means the linking of citizens to the democratic processes and decision making via traditional and new, digital means. *E-participation*, which promises to lead to a more participatory form of democracy, is applied in e-voting, e-referendums, e-initiatives, e-consultations, e-petitions, and e-party meetings. E-participation can be direct or indirect. Its scope of impact ranges from the reception of information, via consultation to real participation or transaction, such as e-voting, and blogging (McCaughey & Ayers, 2003).

E-planning, especially participatory e-planning, can be an important instrument of both e-democracy and e-governance. E-planning services are increasingly being offered in the UK, Australia, USA and India in order to help the planning system deliver more efficient and accessible services and information. The problem is that the digitalisation of planning processes has so far been based on the traditional top-down approach which hinders the development of new approaches and methods (Kingston, 2008; Wallin & Horelli, in press). Therefore, we are focusing on the participatory and evaluative forms of e-planning in which the internet and other digital means reinforce and enhance the participation of all kinds of stakeholders in planning and community development. We define *participatory e-planning and community development*, as a socio-cultural, ethical, and political practice in which women and men, young and older people take part online and offline in the overlapping phases of the planning and decision-making cycle. It can take place via the internet or other digital and non-digital means (see Horelli 2002; Figure 3). Consequently, e-planning requires the understanding and appropriation of both process and content theories of planning (Taylor, 1998).

THE FRAMEWORK OF THE FUTURE MAKING ASSESSMENT APPROACH

The Future-making assessment-approach (FMA) is a formative and action-oriented type of evaluation. Its *framework* assists in the interpretation of the mechanism of change, by providing concepts for understanding the context and process of planning. The framework, which guides the evaluation design, is based on future research, the co-production of ubiquitous technology and participatory planning.

Future Studies

Future research, foresight or future studies are an interdisciplinary field that tries to open up and expand the number of choices and their meaning in terms of the future. Part of the discipline seeks a systematic understanding of the past and present, and tries to determine the likelihood of future events and trends (Inayatullah, 2007). A recent finding by Mika Aaltonen (2007a) is the recognition of the importance of multi-ontology for decision-making and the consequent choice of methods of implementation and interventions. Multi-ontology means that stakeholders live in different systems of realities or strategic landscapes (linear, disruptive or visionary) which imply different causal assumptions (Figure 1).

Figure 1. A Chronotope in the chronotopic landscape of different systems of causality (© Mika Aaltonen 2007, adapted with permission)

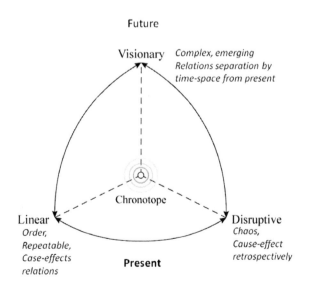

In the linear system cause and effect relationships are discoverable and repeatable, whereas in the chaotic disruptive system causal relations are only found retrospectively or they are not at all coherent. In the visionary system, the relations are separated by time and space from the present moment and they are often quite complex. The strategic landscapes can be assessed by the concept of chronotope (a place in time) that allows a reflection on the complex tradeoffs between the particular time frame (present or future) and the properties of the landscape. For example, moving from orderly towards chaotic circumstances means moving from known strategic landscapes, where the means and resources are more or less fixed, towards more unpredictable landscapes characterised by discontinuity. Also moving from the near future to more distant time horizons means a step from tactical or operational to visionary measures and management. In the linear state where the cause and effect relationships are clearly repeatable, it is possible to apply methods, such as road-mapping and forecasting[2], whereas the visionary landscape might benefit from interactive scenarios, Delphi or weak signals[3]. The disruptive, chaotic state is of course the most difficult one. It can be approached through action research, pattern recognition and risk analysis (Figure 2; Aaltonen, 2007b).

As the future is not an extrapolation of the present, it is important to seek approaches that allow to see and to influence the future by responding to and influencing what is emerging. According to Aaltonen (2007c), the future is story-driven. Good stories are able to express both the causal and temporal sequence of events. The task of evaluation is to analyse the ontological status of the planning and development by assessing the diagnosis of chronotopic landscapes and the choice of tools for sense-making and storytelling (see also Sandercock, 2003).

Co-Production of Ubiquitous Technology

Ubiquitous computing deals with information processing that envision a thorough integration of everyday objects, activities and environments. It means that many computational devices and systems are simultaneously engaged in the course of ordinary activities. Ubiquitous technology, urban and community informatics included can

Figure 2. Examples of methods of sense-making in the chronotopic landscapes (© Mika Aaltonen 2007, adapted with permission)

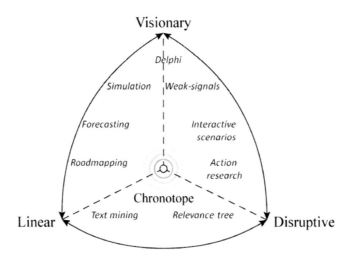

be accessed and distributed via many channels and e-devices, such as sensory networks, radio-frequency identification tags, interactive whiteboards and above all mobile phones and the internet. The co-ordination and interactivity of the devices make the environment intelligent.

On top of the centralised infrastructure of informatics in the 20th century, layers of tools and modular platforms have been juxtaposed which enable material sensing, as well as decentralised participation and cooperation among groups and communities (Townsend, 2009). Urban informatics will also allow users to understand the larger impacts of their everyday decisions. People will be able to appropriate not only the particularities of the local but also connections between cities, and to engage with broader global networks (Williams et al., 2009). Consequently, users will become actors who are embedded in the *glocal networks* of mobile people, goods and information. *Glocal* means the combination of local, regional and global, by using ICT-assisted and non-mediated social networking for shared purpose, such as politics, business or environmental protection.

Social networking is enhanced by the so called *social media* (web 2.0). It refers to the new interactive, social tools that enable shared and user generated content in the internet (Bouton, 2009). Social networking services, such as MySpace and Facebook, have exploded the use of internet. From the perspective of e-planning, we are interested in the kind of social networking that provides e-services for the re-vitalisation of local communities. For example, the US-based, but globally disseminated Craiglist (www.craiglist.com) and Meetup-service (www.meetup.com) provide glocal services and enhance the building of online communities of interest for people who also share the same locality. They can even be accessed in Europe, but they have not been well embedded and appropriated in Helsinki, nor in Stockholm.

Although a great deal of hype exists around ubiquitous computing, the real-time city is already with us in part. The so called mobility tools, such as cars, cycles, public transport, the internet and mobile phones, as well as the simultaneous reduction in travel and communication costs have increased the geographies of social networks and the consequent activity space of people (the geography of locations known to a person; Larsen, Axhausen & Urry, 2006). Internet and mobile technologies also affect the way people make use of urban space in work and leisure time (Jovero & Horelli, 2002; Forlano, 2009). In addition, the mobile phone allows us to map the city-dynamics and to transmit information about air pollution, street repair or cultural events to city administrators. Thus we might say that the city is gradually re-engineered or re-designed through many small-scale relationships (Townsend, 2009).

According to the translation model of Bruno Latour (1987), technology is not a stable and independent entity, but part of the organisation, implementation and used processes. Technology may then be approached as a network of human and non-human elements which are constituted and shaped in network relations. The interaction of humans with technology generates change, which can be regarded as a co-production process of technology and its context. This also means that the transferring of different technologies from one place to another requires the rebuilding of the whole hybrid, namely the technology and its network.

Akrich (1995) and also Parker and Heapy (2006) describe the development of technologies and the services around them by using the concept of *script*. The latter is a scenario that defines, how an innovation or a service should be deployed and organised, as well as what roles should be taken and by whom. The script defines the socio-material network, but the actors perform and co-produce the technology and its organisation (see Koivisto, 2008).

The task of evaluation here is to monitor and evaluate the co-development process; that is the mobilisation and organisation of necessary resources in the form of people, activities, material and money that are needed in a successful application. The results will then be incorporated in the improved script of the technology and its services.

PARTICIPATORY PLANNING

The framework also encompasses a special version of participatory planning, called the *Learning-based network approach to planning (Lena)*. It is both a method and a set of tools to analyse, plan, implement, monitor and evaluate planning and community development. It was originally shaped within participatory projects with young people and women, and later on applied in the context of time policy and time planning. (Figure 3; Horelli, 2002; 2006; Horelli & Wallin, 2006). Lena is based on communicative and post structural planning theories (Booher & Innes, 2002; Hillier, 2008), as well as on the theory of *complex coevolving systems* (Mitleton-Kelly, 2003). The latter implies the parallel existence of tensions, created by order and chaos, the emergence of phenomena and processes, the self-organisation of different stakeholders, and their co-creation of products and systems.

The purpose of planning is to support the communicative transactions of the participants that take place in a specific environmental, organisational, economic, cultural, political and temporal context. Therefore, the transactions are enhanced by a variety of culturally sensitive enabling tools during the overlapping and iterative phases of the planning and development process (Figure 3). The tools are both enabling methods (consensus building instruments and other heuristics, e-techniques included), as well as traditional research methods. An on-going monitoring and self-evaluation provides the participants with feedback on the quality of the change process and its results.

The process of design and planning is iterative and recursive by nature. The designer or planner goes back and forth between the problem definition and its solution, between the material and the

Figure 3. A schema of the methodological approach to participatory e-planning and community development

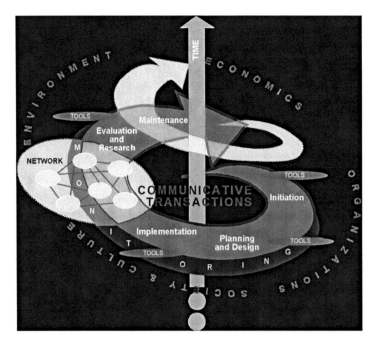

symbolic levels of subjective and socio-cultural awareness building. According to Zeisel (1981), two types of information are used in this creative process. On the one hand, synthetic image information, such as models or designs, provides a general understanding of important issues or the physical ideas pertinent to their resolution. Analytic test information, on the other hand, is necessary for evaluating the good and bad points of a given hypothesis in design. Even if it is an oversimplification to speak about the various phases in planning, it is evident that different kinds of knowledge are required in these stages of the planning process (see Siemens, 2006). The projects usually start with an analysis and sense-making of the context with partners. The initial phase also includes the preliminary visioning of the future, for example imagining the spread of accessible e-services.

Citizen groups tend to see participatory planning and development as a form of empowerment, if it is fairly organized. However, Booher and Innes (2002) indicate that only the network approach to planning provides an authentic situation for participation. As networks cannot be commanded, but only nurtured, they require just a few, core principles or strategies of implementation and embedding. Embedding refers to the collective capacity building, learning and coordination process of the stakeholders and key actors, supported by different techniques (see Siemens 2006).

The strategies of planning within community development might comprise the creation of meaningful events (buzz), participatory networking, capacity building, application of ICT, informing and marketing within all interventions, application of art and creative methods, as well as the co-production of on-going monitoring and evaluation (Horelli, 2003). Thus, the chosen strategies will hopefully mobilise the stakeholders to create and reproduce the nodes and connections of networks that eventually provide the supportive infrastructure of everyday life and eventual social capital (Engeström, 2008).

According to the Finnish experiences (Horelli, 2006; Horelli & Wallin, 2006), gender-, age- and ethnically sensitive coordination is of utmost importance in Lena. It is not about enforcement, but about constant negotiating and interacting with different partners. This presupposes that special attention is paid both to the variety of temporalities (Bryson, 2007) and to the gendered necessities and contingencies of everyday life.

The task of evaluation is to assess and provide feedback of the evolving process and expected or unexpected impact by enhancing the co-creation and integration of a multi-level and multi-dimensional evaluation system in the planning and development.

THE EVALUATION CONTEXT AND FEATURES OF THE FMA-APPROACH

In addition to the evaluation tasks also the policy context of evaluation is important. In general, the evaluation should match the context of its object. The policy context of e-activities and e-planning is complex and dynamic due to the evolving policy systems with new policy goals and diverse forms of multi-level governance. The latter imply divergent modes of policy making with interventions and instruments, which focus on the coordination of activities and knowledge exchange. Also the orchestration and facilitation of interactions within and between networks, as well as the provision of platforms for new practices and capacity building are seminal parts of governance (Valovirta & Stern, 2006).

Systemic and evolutionary evaluations seem to be able to cover the complexities described above, as they do not lean on linear and mechanistic models and causal patterns, but rely on the non-linear shaping and assessment of the context. They tend to recognise both the strategic and operational dimensions of projects and programmes, as well as the variety of stakeholders ranging from communities of practice to single organisations. Key questions are, whether new practices and fora for deliberation have been created in the institutional network context and whether individual and organisational capacity has increased. Interpretations by the stakeholders may play an important role in the development process of such new practice (Valovirta & Stern, 2006).

However, the evaluation of e-activities is so far a contested field. It lacks frameworks and the methodology is still under construction. E-planning is not even mentioned among the evaluation of e-activities (Aicholzer & Kozeluh, 2007; Macintosh & Whyte, 2007).

E-planning, which was earlier defined as a transactive socio-cultural, ethical and political practice, can be classified according to different types. E-planning as an object of evaluation can refer to the: 1) provision and delivery of planning services (building permits etc.); 2) offline planning with e-tools as one technique; 3) co-production and application of e-tools and platforms in community development; 4) planning of virtual objects and spaces with e-tools (for example in Second Life).

E-planning in this chapter refers to the second and third type. Their evaluation requires focusing on both online and offline planning activities. The evaluation process starts by asking the basic questions of what, why, how, as well as by who and for whom to evaluate. Consequently, evaluation implies the identification of the goals of the e-planning project, resources to achieve the goals, aims of the evaluation process and the actors involved. Professionals, citizens and politicians of different backgrounds, sex and age have diverse values and interests that are reflected both in the goals of the project and in the aims of the evaluation.

This diversity affects the choice of the core criteria and measurable qualitative and quantitative indicators concerning the functioning of the tools, evolution of the process and relevance of the outcomes. Evaluation questions operationalise the criteria whose indicators are then monitored or measured by various methods, such as observation, surveys, focus groups, log files and document analysis. The triangulation of theories, actors and methods is advisable.

In practice, the monitoring and evaluation system of e-planning is guided by a multi-dimensional and multi-level evaluation design. It might include elements of the following perspectives with accompanying criteria (see Macintosh & Whyte, 2007; Freschi et al., 2007; Figure 4):

1. *the socio-technical perspective* that focuses on social acceptability (trust, security, relevance), usefulness (accessibility, appeal, content clarity, responsiveness) and usability (navigation, efficiency, error recovery).
2. *the project perspective* that focuses on the inclusiveness of the project (representation, transparency, conflict, consensus), on the relationship, formation of networks and on the results and impact (effectiveness and relevance; better functioning environments and communities; wellness or social capital).
3. *the institutional or systemic perspective* that focuses on the relationship of the project with the system (local, regional, national), power relations included (political equality and community control).

FUTURE-MAKING ASSESSMENT OF THE UBIQUITOUS HELSINKI-PROJECT

The Future-making assessment (FMA) approach was applied in the Ubiquitous Helsinki-project (2007-2009), funded by the Finnish Funding Agency for Technology and Innovation (Tekes). The consortium was a private, public, people-partnership. It comprised several companies, University of Technology (HUT), Technical Research Centre of Finland (VTT) and the Helsinki Neighbourhoods Association (Helka). Their representatives formed the core group of the

Figure 4. The ellipse of the monitoring and evaluation criteria of the Ubiquitous Helsinki project

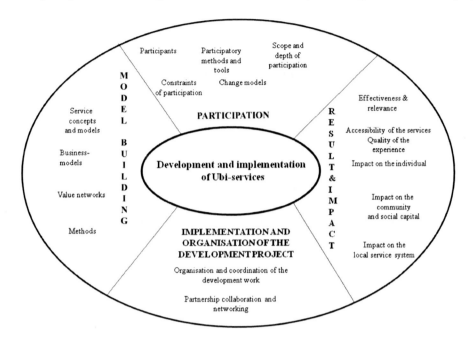

stakeholders. We will describe the main features of the project, the evaluation strategy and the assessment of the process and outcomes.

The Project

The goal of the project was to plan and co-produce ubiquitous services of everyday life and events, in the context of community development of two Helsinki neighbourhoods. The project consisted of research and evaluation, and a living lab test of digitally provided local services. The living lab was a real-life iterative experiment of co-production and e-planning in the neighbourhood.

The resources consisted of the budget and the evolving network. Two thirds of the total budget, 1.250 000 euros, was targeted at the development of two software tools, the "recommendation machine" and the "meeting point" which will be described later. The rest of the budget was used for the service pilot, management and mobilisation of the network. It also financed the purchasing of other programmes and technical applications, such as GIS (geography informatics system) based navigation data, web 2.0 tools and service data.

The network of participants consisted of fifteen different organisations. The number of network members varied between 60 - 100 people, depending on the purpose and intensity of involvement. The network comprised enablers, users and end-users. There were: i) representatives of the enablers who stood for the technology, management and research, as well as the service providers, who were the main contributors to accessible services in practice; ii) providers of local content information and services, the local forum group, the community house, libraries, child care-centres, local enterprises and business to business (b-to-b) organisations; iii) end-users, such as inhabitants and local workers.

There were as many women involved as men, however, some gender differences were conspicuous. The majority of women participated in the management and service provision, as well as in the building of the portal while men co-created most of the technical applications. This horizontal gender gap is mirrored generally in Finland, where men work with technical and operational tasks, and women in public services and administration. The digital gap is low in Finland, according to international comparisons (Pessala, 2008). The majority of working people has access to broadband internet connections and also know-how to behave in the mobile and PC environments.

The implementation of the project meant constant iteration between the developers and users. The Helsinki Neighbourhoods Association (Helka) implemented the living lab test. It worked to enhance the collective capacity and social capital of its 56 neighborhoods by developing their local web-sites, called Kotikatu (Home street). The lab test took place by co-creating a locally based service and partnership platform that could be accessed by the cellular phone, PC and TV screens in public space. Helka steered not only the objectives of the ICT-tools, but also formulated the technical requirements of web and mobile applications, which is unusual among non-ICT-related NGOs. The HUT-research group (which included the authors) supported Helka both financially and with practical know-how. In addition, the VTT research group, which was responsible for the management of the project, provided mobile and semantic web tools that were co-produced with ICT-enterprises.

Monitoring and evaluation covered the whole life cycle of the project. The HUT research group was responsible for the evaluation design, though the data was gathered in conjunction with the stakeholders.

Evaluation Strategy

The aim of the evaluation was to monitor and evaluate the process and outcomes of the project, as well as to enhance the networking, deliberation and capacity building of the network members. Thus, the evaluation was formative and empowering. It also sought to serve the implementation of the co-evolutionary project. In addition, it aimed to create new interpretations and knowledge in order to open up the future development. Accountability was also on the agenda, but to a lesser degree.

The users of the evaluation were the participants themselves, but included town planning and service professionals, citizens and the eventual external evaluators of the funding programme.

The choice of criteria and measurable indicators were negotiated in the core group. The criteria were influenced by the aim, object and context of the evaluation, as well as by the framework, described at the beginning of this chapter. The monitoring and evaluation criteria tapped the nature and characteristics of participation, model building, the implementation of the project and its outcomes (Figure 4 and Horelli et al. 2007). Thus, they covered parts of the criteria belonging to the socio-technical, project and system perspectives.

The evaluation questions focused on several operational areas. The main question explored the role of ubiquitous technology in the planning of services within community development.

The evaluation design steered the building of the multilevel monitoring and evaluation system (m&e system). It consisted of the following set of intertwined assessment activities: i) assessment of the chronotopic diagnosis (ontological perspective); ii) monitoring of the operational level (socio-technical perspective); iii) collective appraisals of the cycles of planning and development (project perspective); iv) thematic evaluations (institutional or systemic perspective).

The methods and instruments that were applied, included questionnaires and interviews, analysis of documents, usability studies and, above all, timely dialogues in different focus groups, meetings, seminars and workshops.

Evaluation of the Process

Assessment of the chronotopic diagnosis of the project"s strategic landscape was conducted by the researchers with core stakeholders (Figure 1; Aaltonen, 2007a). It was an assessment from the ontological perspective that took place alongside the contextual analysis of the initial conditions and policy context of the project. On the basis of this assessment, appropriate visionary techniques and road mapping were applied (Figure 2). Inspired by the Metaverse scenarios (2007) the guiding vision was constructed on the basis of two continua: the spectrum of technologies and applications ranging from augmentation to simulation and the spectrum ranging from external, world-focused to a shared one in the virtual world. The continua bring forth four different types of future scenarios, which are already emerging (Figure 5). In "Augmented Reality", the use of location-aware systems and interfaces enhance the external physical world for the individual, for example tags in the hotel lobbies or packaged goods that can be read using the mobile phone. "Life Logging" describes the capture, storage and distribution of everyday experiences and information. It serves as a means of providing useful historical, as well as current status information that can be shared with others by sending on-line messages via the web or mobile phone, linked to pictures and personal experiences, as in social networking sites, such as Facebook. "Mirror Worlds" are informationally enhanced virtual models or maps of the world around us that can be annotated, such as Google Earth. They help people orient in their environment. "Virtual Worlds" augment the economic and social life of physical world communities. A key component of this Virtual World scenario is the avatar. It is the user"s personification in the Virtual World, which can learn new skills and competencies, for example in Second Life.

The vision was discussed by the core group who partly operationalised it into a roadmap that showed the partners what their goals were in terms of time, space and outcomes. The road-map formed the preliminary version of the script (Akrich, 1995) that defined what should be developed and how the services should be deployed and organized. In practice, it implied the co-creation of new software for mobile, internet and urban TV screen transmission, as well as appropriate service contents for the multi-channel environment, such as the local calendar and Help Desk services which will be described in the next paragraphs.

Figure 5. The vision of the Ubiquitous Helsinki-project (adapted from Metaverse, 2007)

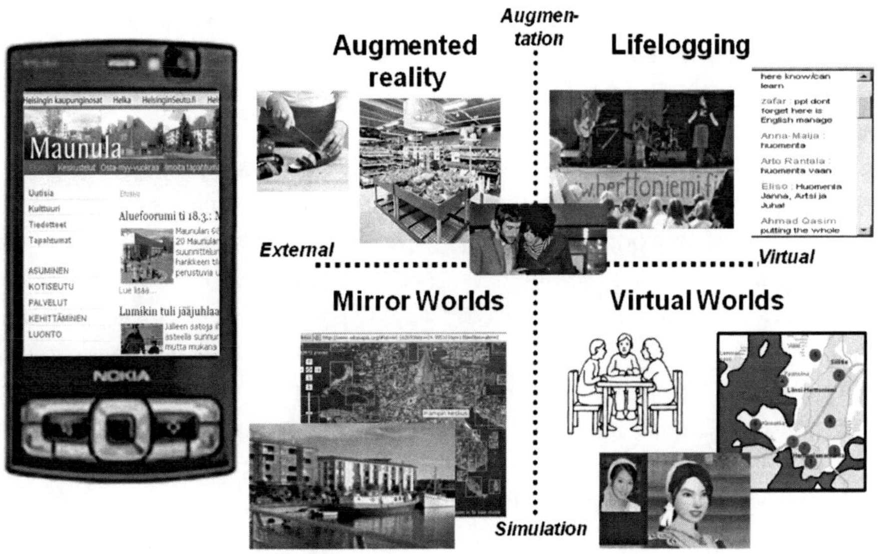

The script took the form of a state of the art-report (Horelli et al., 2007) which described the possibilities of living lab testing with new technologies, service contents and e-planning. The road-map and the report were part of the *thematic evaluations* which were conducted from the systemic or institutional perspective. The report positioned the partners in the network and enabled them to further elaborate their own goals in relation to the project.

From the end of 2006 to the end of 2008, the project progressed from the initial visions to the first co-produced software applications. *The monitoring of the operational level*, (the socio-technical perspective), took place among others by testing the applications in a festival of seven events. Later on, the applications were tested in contexts where the local inhabitants could use the software in order to get public, private and NGO-services in their neighbourhoods. The usability of the applications, as well as the contents, was co-developed according to the feedback and individual wiki-corrections during the testing.

The consortium partners met in a large assembly every four months. In addition, smaller task meetings or "timely dialogues" were organised several times a month. The possible outcomes were sketched meeting by meeting, negotiation after negotiation. Gradually, the interpretations of the goals started to change from purely technological ones towards socially-oriented aspirations. The dialogues were monitored and facilitated by the HUT- research group. HUT also arranged special workshops in which partners could give feedback and asses the total value of networking. *The collective appraisals of the cycles of planning and development* (the project perspective) took place in these sessions.

The planning and development process was time-consuming and at times quite strenuous. The biggest problem was the lack of socially sensitive management. The responsible organisation for the latter was mainly focusing on the technological goals, ignoring the embedding and appropriation of the artefacts. It took one year before a common language was found during which some business partners quit the project. This had a negative impact on the formation of value networks and new business models. Gradually the rest of the partners learned how to collaborate smoothly, at least in part, due to the FMA-approach.

Evaluation of the Outcomes

The outcomes of the project were assessed according to the chosen criteria, such as effectiveness and relevance (Figure 4). Several of the concrete goals were achieved, such as developing product innovations like the *Recommendation engine.* The latter is a simple semantic web tool that selects and combines information from different web pages[4]. Another innovation was the new semantic application, the *Meeting point-tool,* which can be deployed via the internet and the mobile phone. It enables people to invite their friends to a certain event, fix the time and place of the meeting and follow the journey of their friends in the city region.

The service innovations, such as the web-based calendar programmes, RSS-feeds and other sophisticated links, made it possible to create digital service and community portals or shared web spaces, where people can communicate and link the sites with real-life places. The expansion of the former service help desk into *Everyday life-services* allowed people to reach a variety of local services, make orders and monitor the accessibility and availability of services, such as child care, cleaning, repairing etc.

These tools and services have transformed the local neighborhood web-site into *a local service and partnership platform.* The latter refers to the technical and social co-construction of the web and mobile devices and services by the stakeholders. This platform enables local interest groups and individuals to share their knowledge about events, services or local news. This way a larger audience is reached than through individual web sites. Consequently, the platform allows local inhabitants to see, what is going on in the area. They can call their friends and neighbours to join the events. Some context aware content, such as weather, transportation time tables, event locations and details, are also available as a mobile service. Thus, three of the scenarios in the initial vision (Figure 5), the Augmented Reality, Mirror World and Life Logging, have partly been realized, although not yet appropriated by large groups of people.

Evaluation from the socio-technical perspective revealed that the end-users found the chosen software applications useful. According to the test group interviews and usability studies, people wanted to have new tools and contents which would enable a smoother and faster use of the web.

The digital information and services were not the only amenities in the portal. The service and partnership platform also enables informal capacity building, as people get acquainted with planning and development issues and take action in local politics, or co-produce new services. The estimated users for these services in Helsinki range from a few dozens to over 500 000 people depending on the service. It is likely that the new applications will improve the accessibility of the current services and provide some new ones. This might strengthen the infrastructure of everyday life of the neighbourhoods. In addition, the public, private, people-partnership has strengthened the informal e-governance, as the local work groups and the local forum have been intensively involved in the co-planning of the web-site.

The traditional innovation process was turned around in the project. The users were drawn to the same operative level with enterprises and enablers. The applied FMA-approach had an important broker role in the networking, mobilisation and capacity building of the project stakeholders (Kao, 2009). The local network of NGOs and SMEs were connected to the city administration and big enterprises, to funding bodies and to technology policy, which in turn provided resources to the local level. In conventional development work such networking would have been difficult.

The co-production of the local service and partnership platform has gradually begun to transform the pilot neighbourhoods to collective digital spaces. At this stage, it is not possible to assess the impact of the project, as it will take more time to see the effect on the networking of the stakeholders and the consequences for individuals and the community.

To answer the evaluation question about the role of ubiquitous technology in the enhancement of services within community development, one has first to define what ubiquitous technology signified in this project. In practice, it meant a set of new mash-ups and software applications that are suitable for multi-channel delivery through PC and mobile environments, as well as through urban screens. When coordinated, they make a hybrid infrastructure for the real-time city (Saad-Sulonen, 2005; Aurigi & de Cindio, 2008). Like ICT in general, ubiquitous technology is not deterministic, but it seems to have a catalytic role that inspires and engages stakeholders to take action for the community (see also Rettie, 2008).

In sum, the FMA-approach assisted the core participants in the choice of both visionary and operational objectives. It also enhanced the making of connections between the three Ps in the co-production of services. Above all, it provided multi-layer feedback on the conditions, structure and content of desirable digital services. However, the FMA-approach could not compensate for the poor management of the project and its negative consequences. The approach is also quite time-consuming and demands a great deal of resources.

CONCLUSION

We have argued in this chapter that, as e-planning takes place in a complex and dynamic context, consisting of many stakeholders with diverse interests, it benefits from a new approach, called the Future-making assessment (FMA). The key elements of this action research-oriented evaluation (Roininen, 2009) are: i) the diagnostic assessment of the chronotopic landscape of the project and its policy context; ii) the monitoring and evaluation of the mobilisation and application of the necessary resources (people, activities, money); and iii) the provision of continuous feedback on the evolving process and expected outcomes by constructing a multi-level monitoring and evaluation system.

The FMA-approach requires a transdisciplinary framework. In this particular example it consisted of future studies, the co-production of ubiquitous computing and participatory planning. The framework can also comprise other disciplines depending on the context. The FMA-approach was applied in the Ubiquitous Helsinki-project, which dealt with the e-planning of services in the context of community development.

We will now discuss the ways in which the FMA-approach may enhance the e-planning of services and its role in the overall process of community development.

FMA-Approach as Speeding the Embedding of Outcomes

E-planning was defined in this chapter as a socio-cultural, ethical, and political practice in which women and men, young and older people take part online and offline in the overlapping phases of the planning and decision-making cycle. Participation can take place via the internet or other digital and non-digital means. E-planning implied in our case study the planning and co-production of web-based tools, as well as a service and partnership platform in the local web-site. The FMA-approach assisted in the co-creation and embedding of the platform, which was at least partly appropriated by a network of stakeholders, inside and outside the community. The stakeholders represented women and men in diverse roles, such as professionals, politicians, enablers, service providers and end-users. Thus, the animated and embedded local web-site became the outcome of the development process and eventually a real instrument of urban planning.

When the web-tools of the case study are examined closer and compared to the glocal Meetup-service and the Craig list, the new Meeting point and Everyday life-services, co-created in the project, were better tailored to and embedded in the local context than the American exemplars. This happened at the cost of losing the global dimension. This is a serious problem, as the aim of the Ubiquitous Helsinki was to spread the tools and the service and partnership platform in the capital city area and beyond to other cities. According to Bruno Latour (1987), technology cannot be transferred without taking into consideration its context; i.e. the network in which it has been co-produced, implemented and embedded. Hopefully, the FMA-framework, especially the Learning-based network approach to planning and the social construction of technology will assist in the future efforts to transfer the results. The glocal dimension can be further enhanced through both networked individualism and collective community action (Foth et al., 2008; Horelli & Wallin, 2009) around the local web site with themes, such as climate change, environmental protection, etc.

The FMA-Approach as Co-Creation and Script Enactment in Context

The FMA-approach has been influenced by empowerment evaluation (Fetterman, 2001) and evolutionary evaluation (Valovirta & Stern, 2006), although its transdisciplinary framework differs considerably from both. As a genre of evaluation, FMA is process-oriented and closely intertwined with the planning and development process (Horelli, 2009). It is not goal-bound, nor goal-free, but characterised by sensitivity to the diverse and evolving goals of the stakeholders. For example, in the Ubiquitous Helsinki-project the scope and level of goals extended from tools and platforms to the local web-site with the potential to become an instrument of urban planning.

In evaluation, the outcomes usually have to be explained by a change theory or a change mechanism. According to Aaltonen (2007c), change can be explained by examining the interplay of three forces: sensitivity to initial conditions, the final cause and circular cause. In complex contexts, the challenge is to identify and influence the system"s initial conditions, as they are emerging. *Sensitivity to initial conditions* took place in this case study by encouraging the stakeholders to conduct a contextual and chronotopic analysis. It resulted in the strategic envisioning of the future and an operational roadmap with concrete goals. Thus, it set the scene for *the final cause*; i.e. the goals. It also meant the beginning of the collective script-writing that was enacted by the stakeholders during the project and even after it. *Circular causality* works in loops and circles. Activities on the micro-level may give rise to effects that can be identified on the macro-level and vice versa. Emergence can be explained by the great amount of small causes interacting all the time. Therefore, it is important to carefully nurture the emerging network of participants and its nodes and relations (Horelli, 2009). The FMA-approach was closely integrated with the management of the project, but still separate from it in terms of power relations. With hindsight this relationship should have been made more transparent than was the case in our project.

As connectivity and feedback influence the evolution of the time-consuming development process, the FMA applied a whole set of orchestration tools: spaces for deliberation and negotiation, networking and knotworking (Engeström, 2008), capacity building, and interpretation of terminology and statements. It also implied a self-critical and reflexive approach in which sensitivity to gender, age and culture are seminal. The recognition of multiplex causality and the application of appropriate intervention and assessment methods seem to be important for the balancing of the desired degree of chaos and control, during the various cycles of e-planning. This was not sufficiently well done in our case, as some important business partners left the project. According to Carden (2008), it is the evaluator"s role to become the integrator of suitable approaches and methods in terms of the causal landscape. However, the FMA-evaluators could not, and, it can be argued, should not, compensate for the deficiencies in management.

In practice, the FMA-approach seems to have the potential to enhance e-planning, first, by assisting in the choice and clarification of both visionary and operational objectives leading to the script of the planning and development. Second, by connecting the three Ps, private, public, people, to the co-production and embedding of services. Thirdly, by providing multi-layer monitoring and assessment of the conditions, structure and content of desirable outcomes. In addition, the approach can also play a mediator role by enhancing the connections between the operational, strategic and policy-making levels. Thus, it makes the necessary feed-back loops between and within levels shorter. This is important, since e-planning seems to deal with a design that requires the creation of rapid feed-back loops. Issues at stake also presuppose a short path between policy making and the day to day activities.

FUTURE RESEARCH DIRECTIONS

The planning, implementation and embedding of the digital infrastructure are vital components of community life. We agree with Harvey (2007) that e-planning and ICT-assisted community development do not just take place by designing web-sites but by providing the necessary infrastructure and resources through the co-creation, appropriation and evaluation of community portals. This kind of process will drive foot traffic to local businesses, neighbourhood organisations and community events. Eventually, it might revitalise the community. The future trends will imply the strengthening of community informatics, meaning that ICT will be increasingly applied for the empowerment of communities. In this trend, the FMA-approach will have an important role to play.

The side effect of the co-production of digital devices, and e-planning in general, is that the context of local communities will gradually be transformed to digital urban or even rural space. Digitisation of the environment means that the whole set of urban and community informatics, such as interactive screens and wireless networks, becomes part of the e-planning network. As the FMA- approach relies on the networking and negotiations of a great number of diverse stakeholders, it provides capacity to anticipate the positive and negative impacts of the digitised future on the experience and behaviour of citizens. Thus, it is important to continue to study the question whether e-planning, with the increasing palette of mobility, social and digital tools, will enhance the opportunities to master everyday life in the glocal context.

REFERENCES

Aaltonen, M. (2007a). *The Third Lens. Multi-ontology Sense-making and Strategic Decision-making*. Aldershot, UK: Ashgate.

Aaltonen, M. (2007b). Chronotope Space – Managing Time and the Properties of Strategic Landscape. *Foresight*, *9*(4), 58–62. doi:10.1108/14636680710773830

Aaltonen, M. (2007c). The Return to Multi-Causality. *Journal of Futures Studies*, *12*(1), 81–86.

Aaltonen, M., & Sanders, T.I. (2006). Identifying Systems" New Initial Conditions as Influence Points for Future. *Foresight: Journal of futures studies, strategic thinking and policy, 8*(3), 28-35.

Aichholzer, G., & Kozeluh, U. (2007). eParticipation and Democracy: Evaluation dimensions and approaches. In B. Lippa (Ed.), DEMO-net: Research workshop report – Frameworks and methods for evaluating e-participation (pp. 31-43). Bremen, Germany: IST Network of Excellence Project, Akrich, M. (1995). The de-scription of technical objects. In W. E. Bijker & J. Law (Eds.), Shaping Technology/Building Society: Studies in Sociotechnical Change (pp. 205-224). Cambridge, MA: MIT Press.

Anttiroiko, A.-V. (2003). *eGovernment. eGovernment-alan tutkimuksen ja opetuksen kehittäminen Tampereen yliopistossa* [Development of the research and teaching of eGovernment at the University of Tampere]. Retrieved December 3, 2008, from http://www.uta.fi/laitokset/ISI/julkaisut/eGovernment-raportti_1-2002.html#Luku1

Aurigi, A. (2005). *Making the Digital City. The Early Shaping of Urban Internet Space*. Aldershot, UK: Ashgate.

Aurigi, A., & de Cindio, F. (Eds.). (2008). *Augmented Urban Spaces. Articulating the Physical and Electronic City*. Aldershot, UK: Ashgate.

Booher, D., & Innes, J. E. (2002). Network Power in Collaborative Planning. *Journal of Planning Education and Research*, *21*(3), 221–236. doi:10.1177/0739456X0202100301

Bouton, N. (2009). *Designing for the Community Experience*. Retrieved May 28, 2009 from http://www.slideshare.net/nickbouton/designing-for-the-community-experience-vanue-may- 2609-1497974

Bryson, V. (2007). *Gender and the Politics of Time. Feminist theory and contemporary debates*. Bristol, UK: Policy Press.

Buchsbaum, T. (2008). *E-democracy and E-Parliament. Thoughts and Standard-setting of the Council of Europe*. Retrieved December 3, 2008 from http://www.bmeia.gv.at/fileadmin/user_upload/bmeia/media/AOes/e-Democracy/CAHDE_2008/Sofia_IPAIT_-_Buchsbaum_080610_final.pdf

Carden, F. (2008, October). *Ordinary Word, Extraordinary Confusion: Cause in development evaluation*. Paper presented at the Conference of the European Evaluation Society, Lisbon.

Chadwick, A., & May, C. (2003). Interaction between States and Citizens in the Age of the Internet: E-Government in the United States, Britain, and the European Union. *Governance, 16*(2), 271-300. Retrieved December 3, 2008 from http://www.bmeia.gv.at/fileadmin/user_upload/bmeia/media/AOes/e-Democracy/CAHDE_2008/Sofia_IPAIT_-_Buchsbaum_080610_final.pdf

Engeström, Y. (2008). *From Teams to Knots. Activity-theoretical studies of collaboration and learning at work*. New York: Cambridge University Press. doi:10.1017/CBO9780511619847

Fetterman, D. (2001). *Foundations of Empowerment Evaluation*. London: Sage.

Forlano, L. (2009). Codespaces: Community wireless networks and the reconfiguration of cities. In Foth, M. (Ed.), *Handbook of Research on Urban Informatics: The Practice and Promise of the Real-Time City* (pp. 292–309). Hershey, PA: IGI Global.

Foth, M. (Ed.). (2009). *Handbook of Research on Urban Informatics: The Practice and Promise of the Real-Time City*. Hershey, PA: IGI Global.

Foth, M., Choi, J. H., Bilandzic, M., & Satchell, C. (2008, October). *Collective and Network Sociality in an Urban Village*. Paper presented at the MindTrek-conference, Tampere, Finland.

Freschi, A. C., Raffini, L., Balocchi, M., & Tizzi, G. (2007). White paper. In Lippa, B. (Ed.), *DEMO-net: Research workshop report – Frameworks and methods for evaluating e-participation* (pp. 76–85). Bremen, Germany: IST Network of Excellence Project.

Glenn, J. C., & Gordon, T. J. (2003). *Futures Research Methodology – V2.0*. AC/UNU Millennium Project.

Gurstein, M. (2007). What is community informatics (and Why Does It Matter)? Milano, Italia: Polimetrica.

Harvey, L. A. (2007). Digging the Digital Well. *Urban, 10*(1), 24–26.

Hillier, J. (2008, September). *Are we there yet?* Key note speech at the 40th Anniversary Conference of the Centre for Urban and Regional Studies, Helsinki University of Technology.

Holzer, M., & Seang-Tae, K. (2008). *Digital Governance in Municipalities worldwide (2007). A longitudinal assessment of municipal websites throughout the world*. Newark, NJ: Rutgers University.

Horelli, L. (2002). A methodology of participatory planning. In Bechtel, R., & Churchman, A. (Eds.), *Handbook of Environmental Psychology* (pp. 607–628). New York: John Wiley & Sons.

Horelli, L. (2003). *Valittajista tekijöiksi (From complainers to agents, Adolescents on the arenas of empowerment)*. Espoo, Finland: Helsinki University of Technology.

Horelli, L. (2006). A Learning-based network approach to urban planning with young people. In Spencer, C., & Blades, M. (Eds.), *Children and Their Environments: Learning, Using and Designing Spaces* (pp. 238–255). Cambridge, UK: Cambridge University Press. doi:10.1017/CBO9780511521232.015

Horelli, L. (2009). Network Evaluation from the Everyday Life Perspective. A Tool for Capacity-Building and Voice. *Evaluation, 15*(2), 205–223. doi:10.1177/1356389008101969

Horelli, L., & Wallin, S. (2006). Arjen ajan hallintaa, kokemuksia suomalaisesta aikasuunnittelusta [Mastering of everyday life, Experiences from time planning in Finland]. Helsinki, Finland: Helsingin kaupungin tietokeskus.

Horelli, L., & Wallin, S. (2009, June). *Gendered community informatics for sustaining a glocal everyday life*. Paper presented at the City Futures 09 Conference, Madrid.

Horelli, L., Wallin, S., Innamaa, I., & Jarenko, K. (2007). *Introduction to the user-sensitive Ubi-Helsinki – a report on the preconditions for the user-sensitive design and its evaluation in the context of commercial, public and community services*. Espoo, Finland: Helsinki University of Technology, Centre for Urban and Regional Studies.

Inayatullah, S. (2007). *Questioning the Future: methods and tools for organizational and societal change*. Tamsui, ROC: Tamkang University.

Jovero, S., & Horelli, L. (2002). *Nuoret ja paikallisuuden merkitys? Nuorten ympäristösuhteen tarkastelua Vantaan Koivukylä-Havukosken alueella (Young people and the meaning of locality)*. Espoo: TKK/YTK Julkaisuja.

Kao, J. (2009). Tapping the World's Innovation Hot Spots. *Harvard Business Review, 87*(3), 109–114.

Kasvio, A., & Anttiroiko, A.-V. (Eds.). (2005). e-City. Analysing efforts to Generate Local Dynamism in the City of Tampere. Tampere, Finland: Tampere University Press.

Kickert, W. J. M. Klijn. E-H., & Koppenjan, J. F. M. (Eds.). (1997). Managing Complex Networks: Strategies for the Public Sector. London: Sage.

Kingston, R. (2008). *The role of participatory e-Planning in the new English Local Planning System*. Retrieved December 3, 2008, from http://www.ppgis.man.ac.uk/

Koivisto, J. (2008, October). *Relational evaluation: A Step forward or a lapse in evaluation practice?* Paper presented at the Conference of the European Evaluation Society, Lisbon.

Larsen, J., Axhausen, K., & Urry, J. (2006). Geographies of Social Networks: Meetings, Travel and Communications. *Mobilities, 1*(2), 261–283. doi:10.1080/17450100600726654

Latour, B. (1987). *Science in action: How to follow scientists and engineers through society*. Cambridge, MA: Harvard University Press.

Macintosh, A., & Whyte, A. (2007). Towards an evaluation framework for e-participation. In Lippa, B. (Ed.), *DEMO-net: Research workshop report – Frameworks and methods for evaluating e-participation* (pp. 43–57). Bremen, Germany: IST Network of Excellence Project.

Mannermaa, M. (2008). Jokuveli. Elämä ja vaikuttaminen ubiikkiyhteiskunnassa [Some Brother, Life and Participation in the Ubiquitous Society]. Helsinki: WSOYpro.

McCaughey, M., & Ayers, M. D. (Eds.). (2003). *Cyberactivism. Online Activism in Theory and Practice*. New York: Routledge.

Medaglia, R. (2007). The challenged identity of a field: The state of the art of eParticipation research. *Information Polity, 12*(3), 169–181.

Metaverse. (2007). *Metaverse roadmap foresight framework*. Retrieved October 15, 2008, from http://www.metaverseroadmap.org/inputs.html

Mitleton-Kelly, E. (2003). Complexity Research – Approaches and Methods: The LSE Complexity Group Integrated Methodology. In A. Keskinen, M. Aaltonen, & E. Mitleton-Kelly (Eds.), Organisational Complexity (pp. 55-74). Turku: Finland Futures Research Centre, Turku School of Economics and Business Administration, Mossberger, Karen (2008). Digital citizenship: the internet, society, and participation. Cambridge, MA: MIT Press.

Norris, P. (2001). *Digital Divide? Civic Engagement, Information Poverty and the Internet Worldwide*. Cambridge, UK: Cambridge University Press.

OECD. (2002). *Public Governance and Management. Definitions and Concepts: E-government*. Retrieved December 3, 2008 from http://www.oecd.org/EN/aboutfurther_page/0,EN-about_further_page-300-nodirectorate-no-no--11-no-no-1,FF.html

Parker, S., & Heapy, J. (2006). *The Journey to the Interface. How public service design can connect users to reform*. London: Demos.

Pessala, H. (2008). Sähköisiä kohtaamisia: Suomalaisten yhteiskunnallinen osallistuminen internetissä Electrical meetings [Societal participation in the internet by Finns]. Helsinki: Helsingin yliopisto, Viestinnän tutkimuskeskus CRC.

Rettie, R. (2008). Mobile Phones as Network Capital: Facilitating Connections. *Mobilities*, 3(2), 291–311. doi:10.1080/17450100802095346

Roininen, J. (2009). *Alue- ja yhdyskuntasuunnittelun arvioinnin fragmentoitunut luonne ja eheyttäminen* [Fragmentation and defragmentation of evaluation in regional and urban planning]. Unpublished doctoral dissertation. Espoo, Finland: Teknillinen korkeakoulu.

Saad-Sulonen, J. (2005). *Mediaattori – Urban Mediator: a hybrid infrastructure for neighborhoods*. Master of Arts Thesis in New Media, University of Art and Design, Helsinki. Retrieved January 15, 2009 from http://www2.uiah.fi/~jsaadsu/thesis.html

Sandercock, L. (2003). Out of the closet: The importance of stories and storytelling in planning practice. *Planning Theory & Practice*, 4(1), 11–28. doi:10.1080/1464935032000057209

Siemens, G. (2006). *Knowing knowledge*. Retrieved June1, 2008 from http://www.knowingknowledge.com

Taylor, N. (1998). *Urban Planning Theory since 1945*. London: Sage Publications.

Thesaurus. (2009). *Governance*. Retrieved April 7, 2009 from http://www.answers.com/governance&r=67#Thesaurus

Tolbert, C. J., & Mossberger, K. (2006). The Effects of E-Government on Trust and Confidence in Government. *Public Administration Review*, 66(3), 354–369. doi:10.1111/j.1540-6210.2006.00594.x

Townsend, A. (2009). Foreword. In Foth, M. (Ed.), *Handbook of Research on Urban Informatics: The Practice and Promise of the Real-Time City* (pp. xxii–xxvi). Hershey, PA: IGI Global.

UN E-Government Readiness Survey. (2008). Retrieved November 30, 2008, from http://unpan1.un.org/intradoc/groups/public/documents/UN/UNPAN028607.pdf

Valovirta, V., & Stern, E. (2006, October). *The evaluation of new policy instruments. Complexity and governance*. Paper presented at the Conference of the European Evaluation Society, London, UK.

Vegh, S. (2003). Classifying Forms of Online Activism. The case of Cyberprotests against the World Bank. In McCaughey, M., & Ayers, M. D. (Eds.), *Cyberactivism. Online Activism in Theory and Practice*. New York: Routledge.

Wallin, S., & Horelli, L. (in press). Methodology of a user-sensitive service design within urban planning. *Environment and Planning, B*.

West, D. M. (2000). *Assessing E-Government: The Internet, Democracy, and Service Delivery by State and Federal Governments*. Washington, DC: World Bank.

Williams, A., Robles, E., & Dourish, P. (2009). Urbaning the City: Examining and refining the assumptions behind urban informatics. In Foth, M. (Ed.), *Handbook of Research on Urban Informatics: The Practice and Promise of the Real-Time City* (pp. 1–20). Hershey, PA: IGI Global.

Zeisel, J. (1981). *Inquiry by Design. Tools for Environment-Behavior Research*. Monterey, CA: Brooks/Cole.

WEBSITE

Craiglist. (n.d.). Retrieved from http://www.craiglist.com

KEY TERMS AND DEFINITIONS

Chronotope: The unification of space and time. Chronotopic analysis enables to reflect upon the ontological nature of different systems or strategic landscapes so that the sense-making and consequent decision-making concerning interventions can become more appropriate.

Everyday Life: The self-evident subjective experience of everyday, in contrast to the structures or systems made of institutions, financial flows etc. Scientifically everyday life can be approached as a process in which people shape in their homes, at work or in the living environment the structural conditions into lived life. The mastering of everyday life means then the coordination of those multi-dimensional processes with which people shape the conditions.

Future-Making Assessment Approach (FMA): A special form of evaluation that is integrated with e-planning. It comprises a framework and a set of tools for the contextual analysis and mobilisation of partnerships for collective action, in addition to an on-going monitoring and evaluation system.

Learning-Based Network Approach to Participatory Planning and Community Development (Lena): Comprises a method and a set of tools to analyse, plan, implement and monitor development processes in an iterative way. As an action research strategy, Lena provides possibilities to develop social, spatial and temporal structures that provide the basis for accessible services in (real) living lab sites.

Local Partnership And Service Platform: The stakeholders' technical and social co-construction of web and mobile devices and services in a way that transforms the neighbourhood website into an environment enabling local interest groups and individuals to share their knowledge about events, services or local news.

Participatory E-Planning And Community Development: A socio-cultural, ethical, and political practice in which women and men, young and older people take part online and offline in the overlapping phases of the planning and decision-making cycle. It can take place via the internet or other digital and non-digital means.

PPP-Partnership: The organisational model for co-producing, co- planning and co-development of initiatives or services. The private or commercial services that are provided by enterprises, serve business purposes. The public services, which serve the common good, are provided, for example by cities and districts. The community services, which are created and provided by individual users, either by themselves or by the community of actors, serve the common good of certain groups of interests.

Script: A scenario that defines how a technology, a plan, an innovation or a service should be deployed and organized, as well as what roles should be taken and by whom. The script defines the socio-material network, but the actors perform and co-produce the technology and its organisation.

Ubiquitous Computing: In its idealised form that ICT is present anywhere and any time serving people through embedded electronic devices, programmes and sensory networks. Thus, the environments may turn intelligent and cities function online and in real time.

ENDNOTES

[1] Community informatics (CI) means the application of information and communications technology (ICT) to enable and empower community processes (Gurstein, 2007, p.11).

[2] Forecasting is the process of estimation in unknown situations based on time series, cross-sectional or longitudinal data (Glenn & Gordon, 2003).

[3] A weak signal is a factor for change hardly perceptible at present, but which will constitute a strong trend in the future.

[4] The Recommendation engine (RE) was introduced to all sorts of local web pages, such as the sites of public institutions, private enterprises and individual persons. It gathers and combines information according to chosen attributes, for example, from certain genres of art, local typologies or schedules of events and services. The RE requires that cookies are allowed in the computer. The cookies tell the RE which pages have been browsed on the basis of which it makes the recommendations. The cookies enable the tool to be used without sign-in protocols and profilations of the user. They also ease the privacy preserving issues both for the users and the web pages that provide the information for the RE.

Chapter 5
Local Internet Forums:
Interactive Land Use Planning and Urban Development in Neighbourhoods

Aija Staffans
Helsinki University of Technology, Finland

Heli Rantanen
Helsinki University of Technology, Finland

Pilvi Nummi
Helsinki University of Technology, Finland

ABSTRACT

The Internet is shaking up the expertise and production of knowledge in the planning institution. Digital citizens are searching for information from different places, combining formal and informal sources without apology, and are debating and speaking out on matters. Public planning organisations will be fully stretched to adapt their practices and services to meet these demands. This chapter will present the research results of a project that embarked on gathering and combining local information and knowledge on urban planning on Internet forums. Interactive applications were also developed for these forums to support public participation in ongoing land use planning and development projects in the City of Espoo, Finland. The research results demonstrate how fragmented local, place-based knowledge is, how difficult it is to combine informal and formal information in urban planning, and how inaccessible public data systems still are.

INTRODUCTION

Urban planning involves major social interests. For example, in Finland, the preparation of and decisions on land use plans are municipal monopolies, so the connection of land use to governance and political decision-making is strong.

The connection of land use planning to policy-making and governance is also reflected in planning theory, which over the years has focused specifically on the political nature of planning and power relations. The most important theorists in the field have stressed the social and institutional nature of planning. They have emphasised institutional design (Healey, 1997) and the position of the individual planner as a central actor in the urban planning institution (Forester 1989).

One of the general features in the debate on planning over recent years has been the promotion of communication and collaboration. There is broad support for increased interaction, but there has not been much debate, however, on what effects the increasing use of information and communication technology will have on the content of knowledge and on expertise. The connection between the exercise of power and information has been highlighted (e.g. Flyvbjerg, 1998), but not whether the strong expert institution would really be willing to open itself up to genuine public debate on what types of cities and environments should be planned and constructed.

The Internet is the most important knowledge building environment in today's world. "Digital citizenship" includes the idea of the ability of citizens to effectively participate in social activities in real time via data networks (Mossberger et al, 2008). Participation in the production of knowledge in online environments is determined through its members own capacities, interests and objectives (Wenger et al, 2005).

Digital citizens, or at least the "born digital" generation, digi-natives (digital natives, Prensky, 2001), expect the same kind of high-quality usability, flexibility and reliability from electronic services provided by public administration as they do from commercial services. Applications like Wikipedia and Facebook have spawned a generation that is not content simply to read articles by others, but which wants to comment on and add to the knowledge itself, both as members of a community or network and as individuals (Foth et al, 2008).

The expansion in expertise and knowledge building is challenging the monopoly position of expert organisations in urban planning as producers of urban knowledge. Planners have to consider their own ways of working and the methods, through which planning information is created, distributed, processed and used (Goodspeed, 2008). The use of the Internet in planning projects has also raised questions concerning the utilization of formal and informal knowledge that has been generated in public online environments (Rantanen & Nummi, 2009).

In this chapter, we will examine urban planning, not from the perspective of the institution, but from the local perspective of urban areas and neighbourhoods and the people who live there. The aim has been to study the use of the Internet in interactive land use planning processes. We ask, with the aid of a few Internet applications that have been developed and implemented, how online environments are shaking up practices in urban planning.

BACKGROUND

Place-Based Urban Planning

Only a decade ago, the placelessness of the information society was hailed with enthusiasm. But as Castells (1996) states, people still live in places. In fact, the opposite has happened to what was assumed: the importance of places has further increased. Information technology frees people to choose the places where they live, work and spend their free time. According to Madanipour (2001), the metropolis of the future will be an integrated region that has various attractive places and space for different cultures. Planning a metropolis requires knowledge based on a "soft" understanding of places and the more detailed incorporation of hard, physical facts (Madanipour, 2001).

A place-based approach builds a bridge between the practice by municipal managers of looking at urban areas and neighbourhoods from departmental silos, and the practice of people living in these areas of looking at cities in terms of experiences. Places are no longer viewed in nostalgic terms of traditional, homogeneous communities, nor as mere locations on a map. They are conceptualized as dynamic locales – with their own diversity and power relations – where the larger forces and flows that structure daily life are contested and given meaning (Bradford, 2005). Places and the local knowledge linked to them are a way of analysing and compiling knowledge in a new way. This local knowledge has three interconnected aspects, as it relates to the new challenges of urban policy-making and planning (Bradford, 2005):

- Knowledge *of* communities: input from the "policy clients" themselves based on their lived experience and intimate familiarity with conditions "on the ground and in the streets" of their place.
- Knowledge *about* communities: statistical data disaggregated to the local scale, tracking trends in the city or community that provide a profile of the place.
- Knowledge *for* changing communities: theoretical models that articulate plausible links between reform strategies and outcomes.

The emphasis of this chapter is in the first category of local knowledge, knowledge of communities.

Planning over the Internet

The Internet activates people to participate: the use of the Internet promotes both economic well-being and democratic participation (Mossberger et al, 2008), and, therefore, is no longer only a kind of tool but an important action environment, the development of which should be the primary task of public administration.

Urban planning processes can no longer be taken forward without the professional use of Internet-based methods. This requires knowledge on the technical and functional aspects of the laws underpinning these tools. The challenge in the design of applications is to control the considerable and heterogeneous mass of planning data and similarly to create interfaces that are easy for citizens to use. The background information on planning areas should be easy to find and it should be of a high quality and reliable. What makes information reliable is that it is produced openly, critically and through collaboration (Eräsaari, 2006).

Planning practices and especially political decision-making do not yet give adequate consideration to methods required for the production of information in online environments (Rubinstein-Montano, 2003; Foth, 2008; Goodspeed, 2008). Interactivity requires the planning organisation to have online media skills: the ability and willingness to take part in debate by highlighting its own expert knowledge or by otherwise expressing a personal opinion on issues. Online debate also raises many questions for public administration: what is formal and what is informal information or knowledge? It is not clear what types of information can be presented in what kinds of environments – or if it can be presented at all. It can also be asked whether planners and civil servants are active "digital citizens" in the same way as others, or whether their roles in the online planning process are something completely different.

Learning-Based Urban Planning, OPUS

The research presented in this chapter is part of an extensive research project called Learning-based urban planning, OPUS (the acronym comes from the project's name in Finnish), which was carried out by Helsinki University of Technology in 2005–2008 (Staffans & Väyrynen, 2009). Learning-based urban planning is a comprehensive approach to local development stretching the institutional processes of land use planning towards long-term collaboration and partnerships between the public, private and residential (civic) parties. Learning-based urban planning refers to integrating practices that are used to build a bridge between three different urban perspectives: a democratic self-governing city, a competitive metropolis and a local village town (Figure 1). The key concepts in developing the

Figure 1. Learning-based urban planning

methods have been local knowledge, perceived environmental quality, shared processes and 4P partnership, which refers to cooperation between the public sector, private sector and citizens (4P, public-private-people partnership).

Several methods utilizing the Internet were developed in the OPUS project and their suitability for interactive urban planning processes was studied. The local Internet forums presented below are part of these methods. Model forums were established for the City of Espoo.

LOCAL INTERNET FORUMS

Defining OPUS Forums

An OPUS forum is an Internet-based concept where local place-based knowledge, information and data are compiled, processed and shared. It is a knowledge building platform where knowledge is linked to local land use planning and development projects. OPUS forums act as meeting places for formal and informal information. They offer local service providers, developers and planners a platform for partnership and collaborative projects. The forums compile many types of information from various sources and the information accumulates in such a way that a locally important databank is created.

The forums do not operate by themselves simply with the participation of users, but require a local maintenance group. The forums also need information facilitators who analyse and condense the information accumulated from the various sources and raise issues. Users, on the other hand, produce news and comments. The facilitators can be members of the maintenance group, researchers or civil servants.

From the perspective of the planning organisation, an OPUS forum is a locally focused communications tool that can assist in reaching the residents through more flexible tools than official websites (Figure 2). The forums have no direct connection to municipal decision-making in land use planning, which referring to Arnstein means mostly public consultancy (Arnstein, 1969).

Theoretical Starting Point of the Forums

Local Knowledge

Local knowledge is defined here as empirical knowledge of residents and other local actors, such as NGOs, which can be, for example, every-

Figure 2. The concept design of the OPUS forums

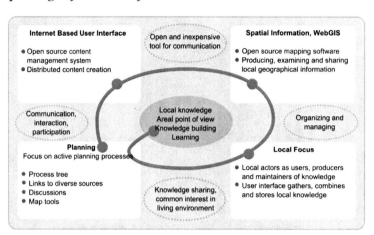

day experiences of the lifeworld and memories of places. It can be the impressions of an area's reputation, safety etc. that are stored in the community's memory or observed in daily life. This local knowledge is often defined as personal and experiential place-based knowledge (Coburn, 2003; Fischer, 2000; Van Herzele, 2004). When local knowledge is shared within a residential community, it evolves through interaction with other people who share the same local environment. Defined like this, local knowledge can be called *community knowledge* (Bradford 2005).

As local knowledge is based on first-hand experiences, it lacks the verification and status of expert knowledge (Coburn, 2003). This distinction is not, however, clear as the planners also have experiential knowledge of their own, and laymen may represent many types of expertise and professions. In fact, the stakeholders may today hold expertise and knowledge in many fields that is superior to that of the planner. The concept of local knowledge is thus attached not only to places (one's living environment) but also to knowledge producers' other roles and competences. Local knowledge can also be expert knowledge of an area that the planner has accumulated over a period of years.

It has been argued that the issue of the interrelationship between expert and experiential knowledge has become crucial in communicative planning (Khakee, Barbanente & Borri, 2000). According to communicative planning theories, in an ideal planning process the planners, residents and other stakeholders try to find the best solutions together by communicating with each other. The exchange of knowledge and the development of ideas through communication during a planning process are seen vital for its successful outcome (Van Herzele, 2004).

The differences between planners' and residents' types of knowledge are frequently based on impressions of technical expertise and its objectivity and accuracy, and, on the other hand, impressions of citizens as laymen and the vagueness and unreliability of the knowledge they produce. From the perspective of *OPUS forums*, the information or knowledge produced is equal in value but is considered from different directions, of which no single one forms a "correct" overall view of the matter.

Knowledge Building via the Internet

Knowledge building can be defined as an activity by communities to improve and promote various conceptual structures such as plans, strategies, theories, ideas and models (Scardamalia & Bereiter, 2002). Knowledge building aims at developing the knowledge and capacities of a community. The essential element in successful knowledge building is interaction between people. Likewise, the ability to put one's own knowledge into proportion and incorporate it into the knowledge of others is essential (Parviainen, 2006). Reference is often made in the organisational sciences to Nonaka's and Takeuchi's spiral model (1995). In this model, knowledge is created in a dynamic process in which explicit knowledge is created from individuals' tacit knowledge through social interaction in the community, which the individual re-adopts through internalization and which becomes new tacit knowledge again. Knowledge building requires networking, and it produces a wider perspective on the issues being debated for those included in the network. The objective of knowledge building is not consensus but a synthesis in which the participants can compare their own ideas (Scardamalia & Bereiter, 2003).

Through knowledge building, the community and individual learn about the planning or other problems under consideration, but also how the "system works". The local development process appears as a continuum in which some arguments and practices prove to be more successful than others. Those involved in developing the process take part in creating new knowledge through action, whereby a deepening in understanding and change in social practices takes place simultaneously (Kurki, 2005). Knowledge building can be seen as a *collective activity* that aims to solve a certain problem and in which those involved have a common goal or objective (Parviainen, 2006). If the problems become more complicated, an extensive knowledge base is required as well as the crossing of boundaries between sectors and experts.

OPUS forums can be considered as *virtual knowledge building environments* in a participative and interactive urban planning process. The forums are designed for use by local communities and experts. Local knowledge appears alongside expert knowledge, giving the issues an experiential framework and historical perspective. The starting point is that issues relating to the built environment are of interest to the residents and that the forums can bring added value, especially in processes relating to land use planning and construction by making local knowledge visible. The forums support a model of progressive debate where the augmenting databank supports the continuity of the region's development and planning.

The natural place for knowledge produced by the municipality itself is its official website. However, these usually lack the tools that could be used to engage in debate on topical issues in a flexible way with residents, service providers and other actors. The information relating to projects is also easily lost after the active stage. OPUS forums aim to improve these shortcomings in public administration data systems, which limit open knowledge building.

Volunteered Geographical Information and Public Participation

Open source WebGIS tools have only recently made it possible to use GIS applications on the Internet, which previously were only used by professionals. Consequently there has been an explosion of interest in using the Web to create, assemble, and disseminate geographic information provided voluntarily by individuals. The most important value of this volunteered geographical information (VGI) is seen to lie in what it can tell about local activities in various geographic locations that go unnoticed by the world's media, and about life at a local level (Goodchild, 2007).

OPUS forums combine VGI with other types of social media and with formal information, as well, thus being more a comprehensive knowledge creation system than an institutional and formal geo-information system. As the forums are closely connected to public planning they can also be considered to be so-called public participative GIS applications (Carver et al, 2001). Participation in this case means mostly public consultancy as there is no direct connection to political decision-making in land use planning (Arnstein, 1969).

Forums for Different Stages of Urban Development

The pilot forums of the OPUS project were developed and implemented in the City of Espoo for three neighbourhoods in different stages of urban development. Referring to these experiments, the OPUS forums can be divided into three categories: inventory, planning and development forums (Figure 3).

Inventory Forum – Before a Planning Process

An inventory forum is suitable for assessing the public opinion of residents in an existing residential area. The tools in an inventory forum include commentary map, a message board and a local knowledge map on which residents can mark their place-based comments. Information about important places, routes used by residents, development areas and potential sites for infill building can be marked and commented on the map. The active involvement of the planner is important. In this way, formal expert knowledge can be combined with the experiences of the residents.

An inventory forum was set up for the Lillhemt residential area of houses, where the City of Espoo was initiating a town plan process that will improve the area. Experts carried out an inventory of the area's building stock and these evaluations were placed on the forum so that the area's residents could make comments and additions to them. Similarly, the residents' own points of view on the area's merits were also assessed. (www.lillhemt.fi, Figure 4)

Planning Forum – Planning On-Line

The inventory forum becomes a planning forum once the environmental assessment and inventory stage moves to the active planning stage. The planning forum follows up the planning process and the core content of it includes presentations of the plans in everyday language. Interaction takes places through debate and commentary, which can be linked to published articles and places on a map. The forum's most important user groups are the current residents in the area and possible future residents. Reaching new residents is challenging and requires the active marketing of the forum. The planning forum is a tool especially for planners and designers, but the construction companies as well as the other consultants concerned with the project also benefit from it.

An example of a planning forum was carried out in the Hista area, where the City of Espoo is planning a new district for around 20,000 residents. Hista is predominantly an area of agriculture and forestry, and has at present around 2000 residents. The progress of the planning and construction of the area are being followed via the forum. (www.hista.fi)

Figure 3. Forums for different stages of urban development

Figure 4. The inventory forum for the Lillhemt residential area

Development Forum – A Different Kind of Local Portal

Development forum is an Internet-based knowledge building environment which serves the local developer community and residents. It is a platform for interactive planning, and more clearly focused on land use planning and development projects than a traditional areal web portal. Development forum is a local medium that is connected to the area's development efforts (Kurki, 2005; Staffans, 2004).

Local residents, as the users of the site, comment, debate and write news articles. The maintenance group is responsible for the updating and editing of the site; it decides on the local policies and content of the development forum. The people involved work voluntarily, which is motivated by the opportunity to participate in developing the area. In terms of the city, the key actors are the information officers and officials responsible for planning. Municipal managers' support to the forum is important because it is in this way that the city's representatives gain a mandate to operate through this "external", unofficial forum. In practice, there is a need for an intermediary party who gets the project going, brings the actors together and markets the forum.

An example of a development forum was carried out for the Centre of Espoo, which is the district centre for 30,000 residents in Espoo. In spite of the fact that also the administrative centre of the City of Espoo is located in the area, the area's reputation is rather bad, mainly due to the poor construction practices in the 1970s. For this reason, several development projects aim to improve the situation, and the development forum is supporting and pulling together these projects. (www.espoonkeskus.fi).

Tools and Content

Although OPUS forums (inventory forums, planning forums and development forums) serve different types of urban planning situations, their design maintains structural consistency. All the forums can have the following tools and content, albeit with varying emphasis (Figure 5):

1. Local content
   a. News, events and service
   b. Editorial content: history, stories, pictures
2. Interactive tools
   a. Local knowledge map (on top of which the commentary map forms one level)
   b. A message board
   c. Commentary on articles (blogs)
3. Content relating to the urban development projects and land use planning
   a. Process tree
   b. Presentations of the plans

Points 2 and 3, i.e. interactive tools and websites presenting planning projects, are common to all the forums. Point 1, the nature and extent of local content, varies significantly on the different forums. Inventory and planning forums only have information for the planning projects in question, but on development forums, the provision of information, event monitoring and service offering may be quite extensive.

Next we take a closer look at the components of the OPUS forums.

Figure 5. The tools and contents of the OPUS forums

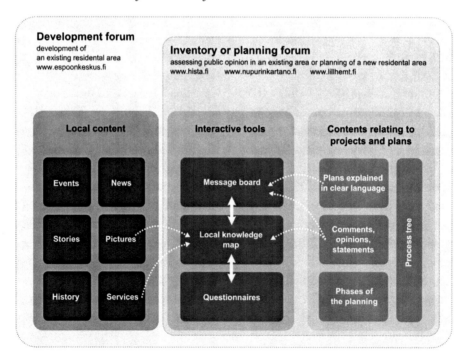

Local Content

Local content like news and list of events form a comprehensive insight to the area but it also requires a certain amount of commitment of local editors and distributed updating. The creation of news items and events has been made easy through a separate news editor, so anybody can send articles to be published. These news items then become material used later for the process tree. Editorial content, such as the local history, personal recollections of residents and pictorial material, is also put on the website, forming a local database of background information and documents.

OPUS forums utilize browser-based content creation software and an open source map application. The content management system used by the OPUS forums is Joomla 1.0 (http://www.joomla.org/) and the map functionality was created using the MapServer (http://www.mapserver.org/) development platform.

Using these open source software instead of commercial products was reasonable because of the low costs, many suitable features and also because there was already some experience of Joomla among the research group. A link between the map application and the content management system was designed. A moderator application was designed to manage and publish data and comments on the map. All the data can be retrieved from the database, for example, onto an Excel file and can also be transferred to a GIS application for further processing.

Interactive Tools

Message Board
The message board is the forum's most important interactive tool, which encourages local actors to create the content for the site. The topics are optional and can also be introduced to the debate at the request of planners. Debate can be carried out in the form of a blog by linking it to the articles and news items. The conversations are moderated afterwards, if necessary; moderating in advance effectively kills the discussion. Users could also be required to register, but this inevitably reduces the number of debaters. Although registering does not require the user to reveal his or her identity, it contributes to keeping the debate civil. Officials generally appear on message boards under their own names.

Local Knowledge Map
A local knowledge map is an interactive map application for examining place-related information (Figure 6). This information could be, for example, formal information concerning projects and plans, or informal local knowledge, such as user experiences, statements from regional associations, or news in other media. What is essential is that the person searching for information finds it on the basis of its geographical position and not, for example, through a hierarchical organizational structure.

Users can also leave their own positive and negative comments on the map, as well as development proposals. The comments are public and anonymous. A link is created from the comments to the message board so contributors can engage further debate. Background information on the respondents can be collected (e.g. age group, gender, neighbourhood). The comments can be studied using a simple analysis tool on the basis of the date, type of comment (positive/ negative/ development comment) or information about the respondent. In this way, the information accumulated on the map can be shared and examined by the entire community.

Further improvements on the forums will focus in particular on the usability of the map application. GoogleMaps has already created new standards for map interfaces. Open source solutions are offering excellent solutions for implementing map interfaces. The use of open source also meets the needs of local NGO's.

Figure 6. The local knowledge map

Content Relating To Projects and Plans

Planning projects are presented in an informal way on the forums. In addition to the official planning documents (e.g. maps and reports) illustrations and other visualizations can be published. Starting points, goals and the planning drafts and proposals are described in plain Finnish.

Process Tree

The process tree has proven to be a good tool for visualizing and archiving the various stages of a given project in chronological order from three different perspectives. The process tree comprises a table that grows upwards, on the left side of which is the formal process, in the middle the general history or other "neutral" frame, and on the right the informal process comprising information in other media, opinions of associations and parties, and other themes relating to the matter. The process tree brings several types of information together in one place, thus creating a general picture of the matter at hand. It also brings together the disparate news and static content within the forum into sensible themes. Planners have used the process tree, for example, at public events to illustrate their presentation material and the history of the project.

Experiences of the Forums

So far the OPUS forums have not been under external evaluation. Two doctoral theses will focus on the forums, one on the usability of the forums and one on knowledge management. The following remarks about the forums are based on monitoring the number of visitors to the forums and the feedback from them, on a survey arranged by the research group to the users of the forums and on researchers' overall experineces of the intensive development phase.

User Feedback

Monitoring user opinions and the number of visitors to the OPUS forums has been an integral part of the introduction of the concept and evaluation of its success. Only a sufficiently large mass of visitors can help to achieve two important objectives: the municipal organisation gets the benefit of this additional communication channel, and the forum users are enabled to build common knowledge within the context of urban development. The number of visitors is also linked to the legitimacy of the participation, since the more people that visit the site and take part in the discussions the more representative the interaction can be considered. It would also be an unwelcome development if several local parties remain outside the forum.

Number of Visitors

The development forum for the Centre of Espoo was opened in October 2006. The statistics show the slow start of the forum from October 2006 until the following summer. Only in autumn 2007 did the number of visitors to the site start to show a clear increase. A local traffic survey mobilised a large number of visitors to the site due to the conflict surrounding some of the issues, and was reflected as a spike in the statistics. In 2008, the website attracted around 4,000–7,500 visitors per month. There is still considerable room for growth in the number of visitors.

Compared to the development forum, the number of visitors to the planning and inventory forums has been smaller as a result of the smaller target group for the pilot websites. However, those who visit these forums have spent a longer time on them. The number of visitors to the Lillhemt forum has been over 700 hits per month at its peak. This is a high figure for a residential area of around 600 people. The corresponding figure for the Hista forum has been slightly over 600 hits per month at its peak.

User Survey

User surveys of the forums are currently being drawn up. Preliminary results are only available for the survey targeted at users of the development forum for the Centre of Espoo.

In the preliminary survey, the development forum was seen, above all, as the residents' and other actors' own unofficial website. According to the respondents, the website for the Centre of Espoo was considered to be a success, especially with respect to the contents and discussions concerning ongoing construction and planning projects. News items, information about locally important events and the message board were closely followed. The website also reinforced the feeling of home about the area. The accumulation of knowledge and compilation of historical knowledge was considered to be important. One user comment stated that the regional website acts as a "knowledge databank, as the preserver of a collective memory."

In terms of its usability, the site was considered to be good – only the map application had clear usability problems. Three out of four respondents believed that the development forum helps residents have a say in matters. However, the residents hope that officials would have a greater presence on the website.

Understanding the Users and the User Context

Forum for planners: The adoption of OPUS forums as a tool by planners and other officials is only in its infancy. Inventory and planning forums are closely linked to formal planning processes, and therefore they are easier to integrate into institutional practices. The development forum, on the other hand, requires extensive a real partnership and cooperation models. Establishing a development forum in an area is therefore a long process, but can result in the most useful benefits being achieved.

From the perspective of planners, the forum is above all an information management tool. The planner or development manager acts as the knowledge manager in the process. This, however, requires an active use for the forum and the adoption of more communicative working practices in the entire organisation.

The usability of the forums must be further developed. The production of knowledge must not require the planner to have special ICT skills. In the future, the planner should be able to easily produce the content relating to the different stages of the project over the Internet alongside his or her routine work. This would demand a certain degree of integration between the city's own data systems and the forums.

Forum for inhabitants: From the perspective of residents, the forum works best when information on planning is provided openly and alternative solutions are debated at an early stage. The experiences from participative processes demonstrate that residents appreciate the option of giving feedback early on in the process. Accordingly, the forums could act as proactive platforms for producing new ideas and development needs or for making these visible. This would enable the residents to shape the agenda for the area's future development.

For local NGO's, a development forum may be a new tool that needs to be learned. Establishing new communication practices takes its own time – this process is still ongoing in the Centre of Espoo. However, once the actors commit themselves to common local objectives, awareness of the forum will increase. Furthermore, over the course of time, local actors will also become more familiar with social media tools in general.

It is still difficult to say what impact OPUS forums have had on residents in the area. It is evident that officials should participate in discussions more visibly as this would be an important signal for the residents with respect to the implementation of transparent governance. In Espoo, the manager of the development project has been involved in the discussions and produced material for the site. He has highlighted the official point of view, explained future projects and quashed incorrect rumours. The planners, on the other hand, have not been so active in the discussions. The researchers' own work as producers of material and activators of debate has also been important throughout the whole process.

Forum for local service providers: From the business perspective, the forums are a regional communications tool that can act as a joint tool for sharing and publishing information for local partners and cooperation. The involvement of business partners is desirable, but also fairly challenging. It can be difficult for companies to see the direct benefit of participating in the forum's activities. Therefore, the cooperation models should be easy to implement from the point of view of businesses. However, this cooperation could still be important for the forum and its image.

No adverts or material produced by sponsors have been published on the OPUS forums. However, it is possible that a development forum can, as it increasingly takes the form of a local portal, start publishing sponsored content linked to the local economy.

Forum for politicians: For decision-makers, OPUS forums offer a place where they can participate in local debate. Decision-makers are usually well-informed and have influence over many local matters. Politicians often have the kind of "inside information" that others do not necessarily have, and they can share this information on the forum. They could, if needed, promote tools such as the OPUS forums within municipal organisations. The forums highlight the common problems associated with citizen participation in decision making: what significance does local discussion have on the planning process, what weight should be given to opinions? Decision-makers can also get to find out about all kinds of "tacit knowledge", which would not necessarily otherwise be communicated to the decision-making process.

The input from a couple of active councillors on the development forum for the Centre of Espoo has been significant. However, decision-makers always take a risk: their comments can be publicly criticised. The readers of their messages are also voters.

It can be asked whether the OPUS forums promote *direct democracy*. In the current crisis of representative democracy – voter turnout in elections is currently at record lows in Finland – this question is especially relevant. Basically, the OPUS forums promote representative democracy indirectly as communal knowledge building increases understanding of planning issues. Thus, voters become increasingly aware of factors that affect the development of an area, which, on the other hand, activates people to participate directly. This assumption is supported by the research, referred to in the introduction, on the impact of the Internet on the willingness of "digital citizens" to participate (Mossberger et al., 2008).

Multichannel Marketing

The usual problem with Internet-based research projects is the inadequate marketing of the website that is the object of the research. Several means were employed to increase awareness of the OPUS forums. Various informative events were organised, and visits were undertaken to events organized by local stakeholders, as well. Writing articles in local newspapers and placing advertisements on shop notice boards are useful but laborious ways of promoting the forums in a district of over 30,000 residents. Alongside these means, it has also been important to consider search engines and networking via links to other websites.

The primary way of searching for information by Internet users is nowadays by using search engines such as Google. People are directed to the forums once they have searched for information on particular projects, events or services. For the search engine to be effective, it is important to acquire a good and unambiguous domain name for the website.

The website of the City of Espoo also has several links to the OPUS forums. The statistics from the server shows that a relatively large number of visitors are directed to the OPUS forums specifically from City's website. The forums are also accessed by the City's intranet, i.e. from the workstations of officials.

The OPUS project has cooperated with neighbourhood associations by cross-linking the forums and association's websites. In addition to this, the maintainers of these websites publish news items on the OPUS forum websites, and the maintainers of the forum pick out news items from the associations' websites.

In-Between the "Formal" And the "Informal"

The maintainers of the OPUS forums have continuously had to consider what the relationship between the forums and the official websites of the municipalities is, and how the content of the cities' websites can be treated on the forums (Figure 7). Another issue related to this is that it has been important to clearly highlight on the forums where the original information is located, i.e. whether is it a formal or informal source. There is a need to specify the position of the OPUS forums as being in between the formal and informal online environments.

Evaluating the degree of formality of the information is often based on which organisation produced the information. Information produced

Figure 7. Formal and informal web sites in Espoo

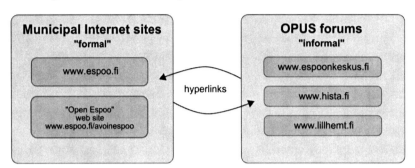

by public administration mainly comprises various types of reports, press releases, statements and decision documents. They are drawn up in accordance with a certain set of rules, under particular terms, and fulfilling technical and legal requirements.

Official decision and planning documents do not necessarily create an adequate knowledge base for an overall picture of the area. Residents may feel that official websites do not present enough locally important knowledge – informal, experiential knowledge is needed to make up for this shortcoming (Rantanen & Kahila, 2008).

From the administrative perspective, OPUS forums are informal online environments outside the municipal data systems. However, from the perspective of residents, the forums are at least semi-formal. The status of this kind of 'in-between' environment should be clarified in relation to the municipal organisation. From a maintenance point of view, it has been important that the OPUS forums have not been owned or published by the city, but that the city is an actor among others. The intermediary party, i.e. the university, has been considered as being neutral and thus as a suitable actor in between the residents and the city administration.

Many levels of structural, organisational, and technical problems as well as problems relating to operating practices emerge when integrating informal and formal information and knowledge. There are no agreed practices on how local websites, such as the OPUS forums, should be monitored by municipal departments, or how the information added to the map, for example, should be used in the municipal GIS system. The closed intranets within public administration, incompatible data systems and deficient interfaces are also frequent problems. There are no structures that could be used to import the information compiled on the forums to formal organisations.

Analysing Comments – The Difficulty in Interpreting Open Discussions

The interactive maps and the message boards generate qualitative data, the use of which is not unequivocal. The map comments are processed by content analysis, i.e. by reading through the comments and classifying them into relevant categories. Local knowledge is needed to avoid misinterpreting the discussions and map comments (Rantanen & Nummi, 2009). Opinions can in general be ranked according to their quantity (how many are of the same opinion) or content (interesting proposal). It is often considered that an opinion is credible in the eyes of the authorities only if it is presented by a sufficient number of citizens. However, a good argument is not necessarily formed as the result of a vote.

Online discussions are still fairly unconnected to decision-making and planning processes. Innovative ideas from residents are easily left unexploited because they are presented in the wrong place and at the wrong time. In the OPUS project, the aim has been to connect online discussions on forums to actual planning and development processes in terms of their timing and locality. However, there is no clear evidence of the effectiveness of the discussions on planning and decision-making. At best, the information on message boards is shared and processed so that understanding of an issue by all the parties improves and better decisions are formulated.

Online discussions are held, in any case, all over the Internet. Discussions on interesting subjects, such as a person's own residential area, are nowadays conducted to an increasing extent on local websites and in informal web communities. Why should the residents use a particular forum organized by the administration for a debate? Should the city's representatives, on the other hand, monitor these forums and participate in all these informal conversations?

From Knowledge Gathering To Knowledge Management

Knowledge management refers to processes that are employed to manage the creation, dissemination and utilisation of knowledge (Gupta et al., 2003). The aim of knowledge management is to find the means as to how knowledge work, such as urban planning, is managed within organisations (Tuomi, 1999). The task of expert planners is increasingly to distil the information and ideas produced by various parties and to mediate the interests of various groups of people (Eräsaari, 2006). Planners are nowadays increasingly knowledge managers working in collaboration than isolated virtuosi.

Knowledge management requires that the organisation's structures are developed to support knowledge processes (Gupta et al., 2003). Knowledge management is not possible without the data systems and tools supporting it. Organisations have traditionally used various groupware and online learning environments. Nowadays, organisations are learning to use social media applications such as wikis and blogs, as well as networking tools such as Facebook and Linked-In.

The data systems for urban planning include GIS systems, various registers, databases, and databanks. The greatest challenge of knowledge management in urban planning is, however, the fact that many other parties produce knowledge in addition to municipal organisations. The linking of these parties to the urban planning knowledge management process as fully equal partners is the starting point of the OPUS forums. Information technology offers means through which the knowledge gathered by many actors can be imported for use by the planning process.

Web 2.0 incorporates the idea of an integrated Internet-based operating environment in which people's work and leisure time become interconnected, information is shared openly and people's roles as producers, users and upgraders of knowledge is closely connected to the (virtual) community they belong to (http://en.wikipedia.org/wiki/Web_2.0). The concept of one-way communication seems increasingly old-fashioned: the publication of information is only one step in the processes in which the information is further processed and adapted in various ways.

The Knowledge Management Model On the Development Forum

The challenges in knowledge management are emphasised on forums, such as the development forum, which have an extensive and diverse number of actors and quantity of material. The functioning of the development forum is loosely based on the knowledge management organisation model of Gupta et al. (2003). It consists of three functional levels and an additional intermediary level that is represented by the OPUS project's research group (Figure 7). The knowledge management level on the planning and inventory forums lacks an actual maintenance group; therefore, the intermediary level also carries out this function.

1. **Data source level**. The information producers can be the municipality, other authorities, the parish, associations, individual residents, political parties, housing companies, businesses, etc. At the data source level, a common knowledge databank is compiled from these various sources of information. These sources all produce electronic information, i.e. documents, Internet sites, GIS data. A databank is created when the existing sources of information are linked, new syntheses are created or previously unpublished digital content is published.
2. **Intermediary level**. The mediators of the forums can be researchers, consultants, voluntary local residents or hired employees. The people operating at the intermediary level launch the activities, put the maintenance team together, and choose different subjects to be discussed by the maintenance group as well as receive the information that has been compiled. In practice, they also produce the majority of the content – depending on the interests.
3. **Knowledge management level**. At the knowledge management level, the information is actually gathered, combined and arranged. The knowledge managers are usually members of the forum's maintenance group. They publish the articles and comments on the map as well as monitor the discussions.

Figure 8. Knowledge management model of the development forum

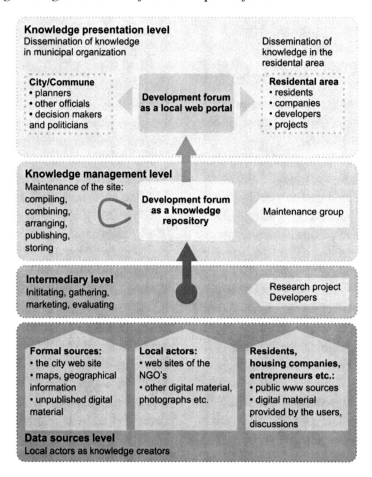

4. **Knowledge presentation level.** Information is shared over the Internet at the knowledge presentation level. The forum takes the form of a public local portal. Information is mutually shared between the planning institution and the local actors, as well as between the residents and the municipality.

The knowledge management model tested on the development forum also requires, in practice, changes in operating practices within planning organisations and local associations (NGOs). This will allow more effective and systemic dissemination of knowledge. The key role of the intermediary level should be reduced in future. However, this requires not only an improvement in the awareness of the forums but a new operating culture at the local and institutional level.

FUTURE RESEARCH DIRECTIONS

Integrating Urban Planning Through Knowledge Management

Urban planning data systems are usually not very effective at simultaneously examining and managing heterogeneous information and ongoing areal processes. Therefore, making knowledge management and data management systems more integrated is a matter of great importance.

In the future, data systems must meet the needs and expectations of various user groups. The challenge is to make the systems and content designed for experts publicly available, and to establish informal Internet applications (e.g. message boards) as serious tools within expert organizations. Particular attention should be paid to the openness, flexibility and usability of the systems and their content.

Urban planning data systems should be developed on two levels. On the one hand, a system of compatible data systems should be created, allowing information from various municipal sources to be searched in an easy way and in which new information, irrespective of whether it is produced by the planning institution itself or by other actors, can easily be incorporated. This kind of *urban information system* would create a comprehensive planning databank and a platform where knowledge is preserved and individual projects would be considered from a longer-term perspective.

On the other hand, urban planning data systems are being increasingly challenged by open source, mash-up-type platforms that combine information from several different sources and which have to a great extent been adopted by the business world and online communities. Open interfaces and the ability to combine databanks are a prerequisite of social media and they are specifically often lacking from the data systems of municipalities. The dialogue between these systems will demand not only a change in attitudes but also a redefinition of officials' jobs. Time and resources will have to be allocated to online interaction.

Paradigm Shift in Knowledge Production

In the future – or even today – social media tools in knowledge production will change public administration. However, municipal strategies and planning objectives will not be put together on Wikipedia for some time yet, even though the technology would make this possible. The communal adaptation and processing of wiki-type information would require a coherent and internalized perspective of at what stage, how and who could participate in producing a common knowledge base. Once the tools are available to all and the platform is open, the actors in the planning process can become active knowledge builders. There should be a clear definition of the roles, rights and responsibilities within knowledge production. Planners and development managers would in this case increasingly be knowledge managers and facilitators in urban planning processes.

The formal nature and validity of the information will be further considered by public administration, but at certain stages it will be submitted for evaluation and supplementation to the public. Expert knowledge is basically no more important or cogent than experiential knowledge of laymen. Utilising experiential knowledge produced by the public will be as important as using any other type of formal information such as GIS or register data. Knowledge of locally significant problems or possibilities for solving them will be moulded through open debate and the sharing of information.

Planning knowledge is shaped over time through dynamic processes and within various historical contexts. The linking of individual planning projects to this perspective requires that the knowledge produced during these processes is preserved. In this way, it can be used for evaluation and background information should the need for development arise again after a few years.

The paradigm shift in knowledge production will require planning organisations to model practices through which they use information gathered from citizens: an ever increasing amount of the information collected will be published and disseminated to citizens for further processing.

CONCLUSION

The Internet forums presented in this chapter are part of methodological framework that can be used to develop practices for learning-based urban planning. The key concepts are local knowledge and interactive knowledge building in which social media and volunteered citizen sensoring of the daily environment are in focal roles. The following conclusions can be drawn from the research on the internet-based OPUS forums: '

a. The forums create the conditions for place-based, locally focused policies by bringing together and fostering debate on areal knowledge;
b. The forums expand the urban planning knowledge base by integrating formal and informal information that comes from various sources of information. From the perspective of the planning institution, this means challenges in terms of data management and the need to develop information systems.

The methodology of the learning-based urban planning includes user-centered development of the online planning environment. The most important aspects to the overall usability are:

a. the usability of the tools;
b. the information quality, which means illustrative, accessible and local content;
c. the competence and willingness of the municipal organisations to work interactively;
d. the impact and meaningfulness of participation on urban planning from inhabitants point of view.

Planning institutions cannot become enclosed in their own information systems any longer. Internet-based participative applications will operate to an increasing extent on the principle that the information is gathered from databanks produced by several different actors and organisations. The basic challenge is to genuinely link planning-related local forums to the actual decision-making process. Technical possibilities for this already exist. The planners will get eventually familiar with online interaction with the citizens: municipal interaction in local and other forums will be understood as part of the openness and transparency of governance as declared in the municipal strategies.

Surprising little consideration has been given in planning theory to the impact of social media on urban planning expertise and the future of the planning institution. The experiences of the OPUS forums only hint at how far current planning practices are from the future knowledge society. The experiences concern not only the technical systems and operating practices but also the attitudes of expert institutions towards new knowledge producers.

REFERENCES

Arnstein, S. R. (1969, July). A Ladder of Citizen Participation. *Journal of the American Planning Association. American Planning Association, 35*(4), 216–224. doi:10.1080/01944366908977225

Bradford, N. (2005). *Place-based Public Policy: Towards a New Urban and Community Agenda for Canada.* Research Report F/51. Family Network.

Carver, S., Evans, E., Kingston, R., & Turton, I. (2001). Public participation, GIS, and cyber-democracy: evaluating on-line spatial decision support systems. *Environment and Planning. B, Planning & Design, 28*, 907–921. doi:10.1068/b2751t

Castells, M. (1996). *Rise of the Network Society: The Information Age: Economy, Society and Culture.* Cambridge, MA: Blackwell Publishers, Inc.

Coburn, J. (2003). Bringing Local Knowledge into Environmental Decision Making: Improving Urban Planning for Communities at Risk. *Journal of Planning Education and Research, 22*(4), 420–433. doi:10.1177/0739456X03022004008

Eräsaari, R. (2006). Objektiivisuus, asiantuntijat ja instituutiot. In J. Parviainen (Eds.), Kollektiivinen asiantuntijuus (pp. 19-54). Tampere, Finland: Tampereen yliopistopaino.

Fischer, F. (2000). *Citizens, Experts, and the Environment. The Politics of Local Knowledge.* London: Duke University Press.

Flyvbjerg, B. (1998). *Rationality and Power.* Chicago, IL: The University of Chicago Press.

Forester, J. (1989). *Planning in the Face of Power.* Berkeley, CA: University of California Press.

Foth, M., Choi, J. H., Bilandzic, M., & Satchell, C. (2008). Collective and Network Sociality in an Urban Village. In Mäyrä, F., Lietsala, K., Franssila, H., & Lugmayr, A. (Eds.), *Proceedings MindTrek: 12th international conference on entertainment and media in the ubiquitous era* (pp. 179–183). Tampere, Finland.

Goodchild, M. (2007). *Citizens as Sensors: The World of Volunteered Geography.* Retrieved May 10, 2009, from http://www.ncgia.ucsb.edu/projects/vgi/docs/position/Goodchild_VGI2007.pdf

Goodspeed, R. (2008). *Citizen Participation and the Internet in Urban Planning.* Unpublished master's thesis. Retrieved in January 1, 2009, from http://goodspeedupdate.com/wp-content/uploads/2008/11/goodspeed-internetparticipation.pdf

Gupta, J. N. D., Sharma, K. S., & Hsu, J. (2003). An Overview of Knowledge management. In Gupta, J. N. D., & Sushil, K. S. (Eds.), *Creating Knowledge Based Organizations.* Hershey, PA: Idea Group Publishing.

Healey, P. (1997). *Collaborative planning. Shaping places in fragmented societies.* London: MacMillan.

Khakee, A., Barbanente, A., & Borri, D. (2000). Expert and experiential knowledge in planning. *The Journal of the Operational Research Society, 51*, 776–788.

Kurki, H. (2005). Lähitalouden ymmärtäminen ja asuinalueiden kehitys. In Neloskierrettä kaupunginosiin. Kumppanuudet ja roolit alueiden kehittämisessä (pp. 69-80). Kulttuuriasiainkeskus, Helsingin kaupunki.

Madanipour, A. (2001). Multiple Meanings of Space and the Need for a Dynamic Perspective. In Madanipour, A., Hull, A., & Healey, P. (Eds.), *The Governance of Place. Space and planning processes* (pp. 154–168). Aldershot, UK: Ashgate.

Mossberger, K., Tolbert, C. J., & McNeal, R. S. (2008). *Digital Citizenship. The internet, society and participation*. Cambridge, MA: The MIT Press.

Nonaka, I., & Takeuchi, H. (1995). *The Knowledge-Creating Company. How Japanese Companies Create the Dynamics of Innovation*. New York: Oxford University Press.

Parviainen, J. (2006). Kollektiivinen tiedonrakentaminen asiantuntijatyössä. In J. Parviainen (Ed.), Kollektiivinen asiantuntijuus [Collective expertise] (pp. 155-187). Tampere, Finland: Tampereen yliopistopaino.

Prensky, M. (2001, October). Digital Natives, Digital Immigrants. *Horizon, 9*(5).

Rantanen, H. (2004). *Paikallisyhteisöt internetissä. Julkaisujärjestelmät ja kolmas sektori. Sitran raportteja 44*. Helsinki, Finland: Edita Prima Oy.

Rantanen, H. & Kahila, M. (2008). The SoftGIS approach to local knowledge. *Journal of Environmental Management, 90*(2009), 1981-1990.

Rantanen, H., & Nummi, P. (2009). Alueella on tietoa. In A. Staffans & E. Väyrynen (Eds.), Oppiva kaupunkisuunnittelu [Learning-based urban planning] (pp. 29-78). Helsinki University of Technology, Department of Architecture, Research Publication 2009/98.

Rubinstein-Montano, B. (2003). Virtual communities as role models for organizational knowledge management. In Gupta, J. N. D., & Sushil, K. S. (Eds.), *Creating Knowledge Based Organizations*. Hershey, PA: Idea Group Publishing.

Scardamalia, M., & Bereiter, C. (2003). Knowledge Building. In Guthrie, J. W. (Ed.), *Encyclopedia of Education* (2nd ed., pp. 1370–1373). New York: Macmillan Reference.

Staffans, A. (2004). Vaikuttavat asukkaat. Vuorovaikutus ja paikallinen tieto kaupunkisuunnittelun haasteina [Influencial inhabitants. Local knowledge and interaction challenging urban planning, in Finnish]. Helsinki University of Technology, Centre for Urban and Regional Studies, Publication A 29. Helsinki, Finland: Yliopistopaino.

Staffans, A., & Väyrynen, E. (Eds.). (2009). Oppiva kaupunkisuunnittelu (Learning-based Urban Planning, in Finnish). Helsinki University of Technology, Department of Architecture, Research Publication 2009/98. Espoo, Finland: Painotalo Casper Oy.

Tuomi, I. (1999). *Corporate knowledge. Theory and Practice of Intelligent Organizations*. Helsinki, Finland: Metaxis.

Van Herzele, A. (2004). Local Knowledge in Action. Valuing Nonprofessional Reasoning in the Planning Process. *Journal of Planning Education and Research, 24*, 197–212. doi:10.1177/0739456X04267723

Wenger, E., White, N., Smith, D., & Rowe, K. (2005). *Technology for Communities*. CHEFRIO Book Chapter v. 5.2. Retrieved from http://technologyforcommunities.com/CEFRIO_Book_Chapter_v_5.2.pdf

KEY TERMS AND DEFINITIONS

Formal Knowledge: Knowledge and information created and shared by formal institutions and their representatives like municipal governance, civil servants and other authorities. Typically, it is based on professional expertise, formal decisions and official documents.

Knowledge Management: Knowledge management can be defined as a process in which knowledge is created, validated, presented, distributed and applied. It helps many kinds of organizations and communities to find, select, disseminate and transfer important information and expertise that is needed in problem solving, learning, strategic planning and decision making. The internet offers many usable tools for managing knowledge.

Learning-Based Urban Planning: Learning-based urban planning is a comprehensive approach to local development stretching the institutional processes of land use planning towards long-term collaboration and partnerships between the public, private and residential (civic) parties. Dialogue and interaction between the stakeholders is supported by various tools, e.g. the local, internet-based forums.

Local Internet Forum: Local internet forum is a web site that is focused on publishing and discussing on many kinds of local issues and is maintained by various local actors. In an ideal case, a local internet forum is managed by a local community or a maintenance group that constitutes of actors from third, private and public sectors, as well as of active inhabitants and researchers.

Local Knowledge: Local knowledge is considered as the knowledge of inhabitants and other local actors in the neighbourhood. This knowledge is often based on people's every day experiences, and it is formulated in many ways within the community and individuals. Thus, local knowledge is attached to the physical places where people live, work and act. In a planning process local knowledge is often valued as common or informal knowledge, which is hard to specify and translate into formal or technical language.

Local Knowledge Map: A local knowledge map is an interactive map application for examining place-related information. This information could be, for example, formal information concerning projects and plans, or informal local knowledge such as user experiences and comments of the daily environment.

OPUS: OPUS is an acronym derived from Finnish words "Oppiva kaupunkisuunnittelu" which refers to learning-based urban planning. OPUS was a name for a research project in the field of urban planning conducted in the Helsinki University of Technology in 2005-2008.

OPUS Forum: OPUS forum is the name of a neighbourhood specified and Internet-based local forum for urban planning, developed in the OPUS-project. It is a knowledge building platform where local place-based knowledge, information and data are compiled, processed and shared and, where knowledge is linked to local land use planning and development projects. OPUS forums act as meeting places for formal and informal information. There are three different applications of OPUS forums: inventory forum, planning forum and development forum.

Place-Based: Place-based refers to a general planning approach, which emphasizes the characteristics and meaning of places as a fundamental starting point for planning and development. Place-based knowledge or information has a geographical position.

Process Tree: A process tree is a tool for visualizing and archiving the various stages of a given planning and development project in chronological order. It brings several types of information together in one place, thus creating a general picture of the matter at hand.

Usability: Usability is related to interactive tools and their users. Knowing the users, their goals and the context they use the tool in are the main principles of usability. Efficiency of use, learnability, memorability and subjective satisfaction can be considered as dimensions of usability, but also other criteria can be defined. Usability research explores the relationship between the user, his goals or tasks and the use context. The objective in usability design is to find harmony between these elements.

4P: 4P refers to public-private-people partnership in planning and developing neighbourhoods and local communities. It is an integrative practice of planning and implementation emphasizing the role of users and local knowledge in developing the living environment.

Chapter 6
Does Computer Game Experience Influence Visual Scenario Assessment of Urban Recreational Paths?
A Case Study Using 3-D Computer Animation

Arne Arnberger
University of Natural Resources and Applied Life Sciences, Austria

Thomas Reichhart
University of Natural Resources and Applied Life Sciences, Austria

ABSTRACT

During the past decades, computer visualizations have been frequently used in urban e-Planning and research. The question arises of whether the degree of experience with the computer during leisure time can have an influence on the assessment of computer-visualized outdoor environment scenarios using visualizations comparable to computer games. We used a computer-animated choice model to investigate the influence of computer game experience on respondents' preferences for an urban recreational trail. Static and animated representations of use scenarios were produced with 3-D computer animation techniques. Three social factors were investigated: number of trail users, user composition, and direction of movement. The scenarios were presented to respondents (N = 149), segmented into groups with different computer game experience. The results indicate that the individual experience with computer gaming and the presentation mode influences the evaluation of trail scenarios. Animated trail scenarios seem to be more useful than static ones.

INTRODUCTION

Research on preferences for urban planning issues, such as the design of public green spaces, has become an important scientific field driven by the rapid changes in urban and social structures of recent decades. At the same time, the systems of governance for planning and managing urban public space have changed, with a shift towards governance

DOI: 10.4018/978-1-61520-929-3.ch006

at the local and regional level (Silva & Syrett, 2006) and increased public participation. Several tools for participatory planning have been developed. The newest approaches use 3-D computer graphics or 3-D computer animations for presenting urban and rural developments and recreational scenarios (Karjalainen & Tyrväinen, 2002; Lange, 2001; Reichhart, Arnberger, & Muhar, 2007; Rohrmann & Bishop, 2002; Vallerie, Park, Hallo, Stanfield, & Manning, 2006). In particular, the capability of these methodologies to assess the acceptance of hypothetical and future scenarios is of great value for urban planners. Therefore, computer simulated environments are now widely used in environmental planning (Lange, 2005; Chen, Bishop, & Hamid, 2002).

Nowadays, Western urban societies are often familiar with computer generated graphics. This familiarity can be gained by work with the computer and/or video games. For example, 76% of Austrian households have at least one computer (Statistik Austria, 2008), and a high share of the population play computer games more or less regularly. Only 24% of the 11-18 year olds do not play computer games (Großegger & Zentner, 2008). The question of whether this familiarity with computer generated graphics can have an impact on the individual assessment of virtual environments that are increasingly used in e-Planning of urban environments arises. This topic is receiving added attention because research on video games has found that there are interconnections between real and virtual environments (Fischer, Kubitzki, Guter, & Frey, 2007).

BACKGROUND

Methods Used In Perception Research on Outdoor Environments

Environmental perception research predominantly relies on the human perception-based approach which derives from the psychophysical tradition in psychology (Daniel & Meitner, 2001; Zube, Sell, & Taylor, 1982). Most of the approaches used the concept of preferences (Aoki, 1999). Early methods applied in environment preference research, such as text-based descriptions or on-site visits, were limited in assessing the complex dependences and interactions between a wide range of social and physical factors (Tahvanainen, Tyrväinen, Ihalainen, Vuorela, & Kolehmainen, 2001; Manning & Freimund, 2004). Thus photographs were used as a visual stimulus and have a long history in the context of preference research on natural and man-made environments (Aoki, 1983; Daniel & Meitner, 2001; Nasar, 1983). In particular, the aesthetic evaluation of natural and urban landscapes in the field of environmental psychology has relied on photographs (Daniel & Meitner, 2001; Van den Berg, Vlek, & Coeterier, 1998). These studies have shown that humans prefer natural environments to built environments.

During the last two decades, digital photomontages have become relevant. Photomontages were often used for the assessment of hypothetical planning scenarios and landscape changes, and most research findings indicate that these computer supported approaches brought more consolidated results. These studies investigated a range of topics such as recreational scenarios (Manning, 2007; Manning & Freimund, 2004; Needham & Rollins, 2005), impact of recreation use on the natural resource (Manning, 2007), landscape changes (Arnberger, Eder, Brandenburg, & Reichhart, 2007), forest management (Tahvanainen et al., 2001), and water-bodies (Junker & Buchecker, 2008). For example, studies in urban and suburban environments using manipulated photos investigated recreational trail scenarios (Arnberger & Haider, 2007) and the design of urban park settings (Jorgensen, Hitchmough, & Calvert, 2002).

However, photographs and photomontages have several limitations, specifically when presenting motion related elements such as water flow or walking or bicycling behavior on multi-use trails

(Hetherington, Daniel, & Brown, 1994; Manning & Freimund, 2004; Reichhart, Arnberger, & Muhar, 2007). For example, Heft and Nasar (2000) found that differences in the preference results appear when comparing static and dynamic visual stimuli. Additionally, non-visual environmental elements – in particular, sound and smell – could have an impact on the validity of photographs (Newman, Manning, Pilcher, Trevino, & Savidge, 2006). Therefore, Daniel and Meitner (2001, p. 63) argue that photographs should be used for relatively passive environmental experiences such as sightseeing, driving for pleasure, walking and picnicking.

Recent advances in computer simulation allow for digital 3D-visualizations of natural and built environments using CAD- and GIS-based approaches (Lange, 2001). Simulations of urban and peri-urban environments are used to elicit city dwellers' preferences for design variations in public natural and built spaces and alternative development scenarios (Laing, Davies, & Scott, 2005; Lange & Hehl-Lange, 2005; Loiterton & Bishop, 2005). Additional tools enable the animation of landscapes such as the growth of plants (Muhar, 2001). These tools are useful for scenario development, presenting different stages for example of forest growth.

Real-time rendering provides the opportunity for observers to move dynamically through virtual environments. Thus, computer visualizations offer the ability to view current and hypothetical environments from different perspectives (Bishop, Ye, & Karadaglis, 2001; Loiterton & Bishop, 2005). The possibility of computer animations to be interactive, and their capacity to control all visible features precisely, opens this technique up to a wide field of research applications. Consequently, such approaches have become more and more important in the field of research on public preferences for urban and suburban environments (Rohrmann & Bishop, 2002; Lange, 2001).

Technologies and software developed by the computer game and cartoon film industry now make it possible to produce interactive and animated environments (Herwig, Kretzler, & Paar, 2005). These computer animations can show moving elements such as walking and cycling behavior in detail (Reichhart, Arnberger, & Muhar, 2007). Such approaches are particularly useful for presenting highly dynamic environments such as heavily-used shared recreational trails.

Research on preferences for outdoor environments has found that the results depend heavily on the quality of the stimuli used (Daniel & Meitner, 2001; Manning & Freimund, 2004). When using visual stimuli, in the form of photos, films, static or animated computer visualization, one has to ask whether the respondents' stated preferences are valid. Modern 3-D computer graphics achieve a high grade of realism, but such simulated environments remain abstractions of the real world (Bishop & Rohrmann, 2002; Daniel & Meitner, 2001). A range of studies using visual presentations of outdoor environments in the form of photographs or computer-manipulated photographs (Manning & Freimund, 2004; Arnberger & Haider, 2007; Daniel & Meitner, 2001) concludes that the results are suitable for the estimation of user preferences in a reliable way. Nevertheless, several studies indicate that there are serious limitations regarding the validity of visualizations – in particular, when the graphic quality is not close to photorealism (Daniel & Meitner, 2001; Rohrmann & Bishop, 2002).

Research on the Influence of Computer Experience

In a study on the representational validity of various stimuli representing a building, Oh (1994) found that people with experience in computer simulation and visual modeling differ in their evaluation from subjects without experience. The latter group gave lower confidence ratings to the video textured and digitized photo representations.

A completely different research field focuses on the impact of computer games on human behavior.

These studies document that experiences with video games have an influence on the respondents' behavior insofar as violent video games lead to aggressive real world behavior (Anderson, 2004; Fischer et al., 2007). The fact that video games can influence reality is of interest for e-Planning. If sensation patterns learned in virtual environments have an impact on real-world sensations, it could probably be the case for other virtual scenarios. One has to ask whether and how experience with video games has an impact on the validity of computer graphic based preference research.

3-D visualizations are being increasingly used for preference research and public participation processes. The more computer graphic approaches are applied, the more important the critical assessment of its validity becomes. So far, research on the validity of computer graphics as surrogates for the real world has focused either on the human perceptions of virtual environments (Bishop, Ye, & Karadaglis, 2001; Rohrmann & Bishop, 2002) or indirectly on the impact of violent media content on actual human behavior (Fischer et al., 2007). Whether the degree of experience with computer graphics in computer gaming has an impact on the assessment of semi-virtual recreational environments is – to the best of our knowledge – not sufficiently known, although an increasing number of people play these games. This question addresses the validity of animated 3-D-visualizations which can be comparable to computer game environments.

The Survey

This empirical study explored the influence of computer game experience on the assessment of trail preferences for social conditions in a standardized manner using static and animated computer visualizations. Investigations on trail preferences have been a longstanding issue in outdoor recreation research (Manning, 2007). Trails provide access to, and within, recreational areas and are important infrastructure components of green spaces. Multiple and intense uses characterize many recreational trails, particularly in urban und suburban settings (Arnberger & Haider, 2007; Gobster, 1995). In order to avoid conflicts and improve the recreational quality of urban pathways, trail managers and recreation planers need information about users' preferences and the influence of use levels in relation to user conflicts and other the trail environment determining factors.

Trail research found that the visitor preferences depend on many social factors such as visitor numbers, visitor activities, visitor behavior, group size and visitor characteristics. In particular, user conflicts (Cessford, 2003; Jacob & Schreyer, 1980) and crowding (Arnberger & Haider, 2007; Manning, 2007; Needham & Rollins, 2005) received much attention, while little research has so far addressed the role of the direction of movement, motion, the relative importance of various social trail factors and the interactions between these factors.

While photomontages have been widely used in trail preference research (Manning, 2007), few attempts exist for the urban context and the use of 3-D animations. Reichhart, Arnberger and Muhar (2007), for example, compared use displacement intentions from a recreation trail elicited by static visualizations with those gathered from animated visualizations. They concluded that animated representations make a more accurate assessment of the influence of trail factors possible than static images.

This study investigated whether respondents who play computer games have different preferences for social trail scenes compared to non-gamers, and whether the presentation mode has an influence. The research dealt with the following questions:

- Are animated trail use scenarios more valid than static ones?
- Do computer gamers express different trail preferences for social conditions compared

to non-gamers because of the similarity of the animated trail visualizations to computer game environments?
- Are computer gamers less influenced by the presentation mode than non-gamers because of their familiarity with computer games?

Methodology

This study relies on the human perception-based approach which is frequently used in environmental perception research (Daniel & Meitner, 2001). In this study, 149 participants evaluated digitally calibrated still-rendering sets and 3-D animation sets of trail scenarios. Each computer-generated image and 3-D animation contained exactly the same systematic representations of the following three attributes, with three attribute levels each: (1) trail users (4, 8, or 16 trail users), (2) user type (shares of bikers to walkers: 25/75%, 50/50% or 75/25%), (3) and the direction of movement (shares of trail users moving towards/away from the observer: 25/75%, 50/50% or 75/25%). The number of trail users was defined as the number of persons in the observer's field of vision. These factors were chosen because of their strong connection to motion, which was one of our major interests.

The still renderings and 3-D computer animations were created in 3-D Studio Max (Figure 1). A marginally retouched digital photo of a shared trail of a larger urban recreation area in Vienna, called the Prater, was used for the background of the trail environment. This trail is heavily used and user conflicts frequently occur – predominantly between walkers and bicyclists.

The pedestrians and cyclists were modeled with 3-D polygon meshes and mapped with textures (faces, clothes). For the dynamic animations, all the persons were rigged and the character animation was made via footstep animation for walkers and via free form animation, hierarchical motion connections, and path follow constrains for the bikers. Avoiding collisions was solved by establishing pre-defined tracks, which the characters

Figure 1. A virtual trail scenario with 3-D wire models

followed. For the still renderings, a typical moment was extracted from the 20 second animation strip and saved as a static slide. Each animated filmstrip had 400 frames. In order to ensure that only the investigated factors changed controlled from animation to animation all the unchanged parts where copied from clip to clip.

A full factorial design was used to compile the trail scenarios, and 27 still renderings and 27 animations were produced. The trail scenarios, and the way the scenarios were combined, were exactly the same for the animated and the static sets. All the choice sets where shown in a standardized manner to landscape architecture students of the University for Natural Resources and Applied Life Sciences, Vienna. Most students were about 20 years old and mainly in the second semester of their landscape planning study. About 60% were female and most of the students have had first experience with CAD and GIS which are frequently used in landscape planning.

Each student evaluated eight dynamic 3-D animation sets and eight still-rendering sets. Each set of the choice model presented two trail use scenarios. The order of presentation was changed to avoid any starting point biases. Each pair of trail scenarios, both static and animated, was shown for exactly 60 seconds. Therefore, the 20-second animation strips were looped three times. During this period, the students had to choose their most pleasant trail scenario out of each visualized set.

After the presentations of the trail scenarios, the respondents had to provide the following information: The respondents' feeling about the presentation methods, their computer experience and their socio-demographic background, as well as the importance of motives for visiting urban green spaces, using 12 items and the frequency of park visits during the last year.

To establish their degree of computer experience, the respondents were asked to specify the average number of hours they play computer games per week. The experience with computer games indicated experience with computer graphics. However, we have not differentiated between different types of computer games. An additional question asked them about the average number of hours they work on the computer. The respondents' feelings about the presentation methods were evaluated using a Likert-scale ranging from 1 to 5 for the following questions: How realistic was the presentation of the trail users' behavior? How easily could you put yourself into the scene?

Binary logit regression (Hosmer & Lemeshow, 2000) was used for the analyses and the factors were indicator coded. The regression produced a parameter estimate, standard error and the Wald-statistic for each attribute level, and the interactions between all attributes. Two segments of respondents with different computer game experience were defined and separate analyses of their preferences were undertaken and tested on group differences. A significance level of $p < .05$ was chosen. The identified interactions improved the models compared to the main-effects-only models at least at the $p < .01$ level using the log likelihood ratio test (Louviere, Hensher, & Swait, 2000; p. 53). As multiple responses were given by each respondent, the unobserved error component in the logit model will be correlated across each respondent's observations. Other analysis methods, such as latent class modeling, could account for response heterogeneity.

Results

Respondents' Computer Experience

On average, the respondents used the computer 1.6 hours per week for playing games ($Mdn = 0$) and 9.5 hours per week for work ($Mdn = 6$). Women spent less time playing computer games ($M = 0.8$ hrs. per week) than men ($M = 3.0$ hrs. per week) but no differences were found for the hours spent working. The hours playing computer games and working on the computer were not correlated. No relationships were found between the frequency of park visits and the hours spent working on the computer or playing games.

Two groups of computer experience were formed. The respondents were divided into one group that did not play computer games ($n = 99$) and one that played at least one hour per week ($n = 41$). Nine students who play computer games for less than one hour were dropped from the analysis. The decision to form these groups was driven by the wish to contrast groups with either much or no computer game experience as well as to have a reasonable sample size.

Respondents' Preferences for Trail Use

The models correctly predicted about 70% of all cases and explained between 24 and 30% of variance (Table 1). All investigated attributes influenced respondents' choices. Students disliked a high number of visitors, a high share of bicyclists and – mainly oncoming – other visitors. The number of trail users was highly significant in all models, no matter what computer experience the respondents had and which presentation mode was applied, and the most important predictor. The factorial design allowed for analyzing 2-way interactions. These indicated that, under specific

Table 1. Regression results for the computer game experience

Factor levels	animated		static	
	0 hrs.	>=1 hrs.	0 hrs.	>=1 hrs.
Number of trail users				
4	0	0	0	0
8	***-1.131	***-.995	***-1.213	***-1.443
16	x***-1.763	***-2.070	x***-2.218	***-2.239
Direction of movement				
25% oncoming/75% going away	0	0	0	0
50% oncoming/50% going away	.081	-.307	.281	-.095
75% oncoming/25% going away	***-.685	-.287	**-.425	-.104
Composition of users				
25% cycling/75% walking	0	0	0	0
50% cycling/50% walking	a, y***-.818	a-.209	y*-.282	-.344
75% cycling/25% walking	-.361	**-.686	***-.749	**-.674
Interactions				
(8 trail users) x (50% cycling)	***.959	n.s.	n.s.	n.s.
(16 trail users) x (75% cycling)	***-1.387	*-.836	n.s.	*-1.057
(50% oncoming) x (75% cycling)	*-.639	n.s.	*-.662	n.s.
(75% oncoming) x (50% cycling)	n.s.	**-1.393	n.s.	n.s.
Intercept	***1.671	***1.798	***1.631	***1.779
n	99	41	99	41
Log Likelihood	-1801.41	-737.19	-1851.55	-752.69
Correctly predicted cases	69.7%	70.6%	68.5%	70.7%
Nagelkerkes R-Quadrat	.269	.301	.243	.283

a Significant differences at the $p < .05$ level between the segments
x Significant differences at the $p < .05$ level; y at the $p < .01$ level between the presentation modes; n.s. = not significant

circumstances, cumulative effects, such as the combination of many bicyclists and many trail users, which increase additionally the negative evaluation of trail scenarios, can exist. On the other hand, significant interactions and insignificant main effects indicated that the dislike of one specific social factor was only given if it appeared in combination with one other social factor. No significant interactions were found between use levels and direction of movement.

Differences between the Presentation Modes

Overall, the results gained by both approaches pointed in the same direction and the variance explained was similar, although variance explanation was higher for the animated approach (Table 1). Few differences emerged for the presentation modes regarding the magnitude and significance of factors and interactions. More significant interactions were observed in the animated mode between user composition and use levels, and between the direction of movement and use levels or user composition. For the gamers, no differences between the presentation modes appeared. For participants without game experience, however, two factor levels were significantly different. Differences appeared for the main effects 16 people on the trail and the mixed user composition. In the animated approach high numbers of trail users received fewer negative evaluations, while for the mixed user composition their negative evaluation increased.

The Influence of Computer Game Experience

Several differences were found between the trail preferences for social conditions of gamers and non-gamers (Table 1). The models of gamers explained more of the variance. The direction of movement played no role for the gamers' trail preferences, neither as main effect nor as interaction, while non-gamers disfavored higher shares of oncoming trail users. Both groups disliked higher shares of bicyclists. For the gamers, however, this dislike increased particularly under crowded trail conditions. For the non-gamers the interaction between the direction of movement and user composition indicated that a mixed movement direction in combination with many bicyclists scored negatively on their preferences.

The animated approach showed that more interactions played a role for both groups. However, only one interaction was relevant for both. The interactions between user numbers and composition indicate that, under crowded conditions, many bicyclists were disfavored. For the non-gamers, two additional interactions between user composition and trail user numbers, as well as direction of movement, contributed to their preferences: A mixed movement direction in combination with many bicyclists was disliked, while a mix of trail user groups in combination with eight visitors contributed positively to preferences. A difference was found for the main effect describing user composition. Non-gamers scored more negatively on a mix of user groups. For the gamers, the direction of movement was only significant as an interaction between the approaching direction of movement and mixed user groups.

Respondents' Perception of Digital Visual Stimuli

For both approaches, more than half of the students could put themselves easily or very easily into the shown scenarios (Table 2). They stated that this was more the case when the trail scenarios were presented in an animated way ($t = -2.345, p = .020$). However, respondents did not perceive the image and animation based presentations differently based on their personal computer game experience.

The extent of realism perceived for the animated approach was additionally asked (Table 2). Although the graphic style of the animations was

Table 2. Stated degree of realism and being able to put oneself into the scenes (N = 149)

Answer scale	Realism	Images	Animations
bad	1.4%	2.1%	0.7%
not so good	15.9%	6.9%	5.5%
acceptable	15.9%	35.2%	34.5%
good	61.4%	49.0%	43.4%
very good	5.5%	6.9%	15.9%
Mean	3.54	3.52	3.68

somewhat poor, two-thirds of the students evaluated the animations of the trail users as realistic or very realistic. Again, we could not find any differences between the segments based on their computer experience. A positive correlation was found between the perceived degree of realism and the perception of putting oneself into the presented static scenarios ($r_p = .301; p = .000$) and animated scenarios ($r_p = .456; p = .000$).

FUTURE RESEARCH DIRECTIONS

This study investigated the influence of computer game experience on the assessment of non-interactive semi-virtual outdoor environments. So far, research on the validity of computer graphics as surrogates for the real world has rather neglected the question of whether familiarity with computer gaming has an impact on the assessment of virtual environments. Because of the increasing use of computer visualizations in research and planning (Daniel & Meitner, 2001; Rohrmann & Bishop, 2002) and, as a result of the increasing familiarity with computer graphics in Western societies, this question has gained in importance.

Therefore, we analyzed whether computer game experience can have an influence on preferences for computer visualized trail use scenarios in a standardized manner using digitally calibrated films and still renderings. We investigated whether this influence differs between the presentation modes – static and animated – and analyzed the influence of computer game experience on the perceptions of the visualizations (Table 1). We used a fairly homogenous group, students in the same field of study and mostly in the same semester.

The results of the analysis of these images containing three social variables controlled by an experimental design document that all variables affected the respondents' trail preferences. Consequently, trail managers should take all these into consideration for recreational trail planning and management. Both modes of presentation indicated that respondents preferred lower trail user numbers and less bicycling use. Specific trail conditions, such as those with high shares of bicyclists in combination with high numbers of trail users, are increasingly disfavored. Use levels played a much more prominent role for the preferences than user composition and direction of movement. These results are in line with previous trail use research about user conflicts (Cessford, 2003; Jacob & Schreyer, 1980) and crowding (Manning, 2007; Needham & Rollins, 2005). The perception of recreational conflict (Jacob & Schreyer, 1980) is often associated with people's safety concerns, particularly on crowded shared trails with fast moving bicyclists.

Results indicate that the direction of movement plays a role for trail preferences. So far, this factor has not been researched extensively in recreation research (Arnberger & Haider, 2007; Fredman & Hörnsten, 2001), and the evaluation of it as a function of user composition brought new insights. The finding that trail users are preferred walking away from the observer than oncoming can be interpreted as a coping strategy avoiding higher levels of social stimulation. Arnberger and Haider (2007) assumed that respondents seem to dislike to be confronted with many fleeting social contacts, and prefer to walk behind others. Arnberger and Haider (2007) also found in their study an interaction between direction of movement and use levels. In this study an interaction was given between user composition in form of bicyclist-rich trail situations and direction of movement, particularly in the animated approach. In this presentation mode, the direction of movement became relevant for gamers only in combination with the mixed user composition.

The presentation mode of the visual stimuli, static versus animated, delivered different preference results. The inclusion of motion in the design resulted not only in better regression models, but also in more interactions, predominantly between user composition and other factors (Table 1). The animation seems to increase the capability of respondents to evaluate more precisely the influence of different user groups on trail preferences and the presence of interactions allowed investigating public preferences for shared trail environments in greater detail.

Most of the students found both modes of presentation easy to understand and evaluate. Over 98% of all choice possibilities were answered validly, and most respondents expressed that the computer visualized scenarios could be seen as surrogates for the real world (Table 2). This evaluation was not influenced by computer game experience. In addition, main study results are in line with findings from previous studies using static images. Furthermore, the existence of significant interactions reflects the high quality of the responses to these digitally calibrated visual stimuli. These findings may support the assumption made by several authors (Daniel & Meitner, 2001; Reichhart, Arnberger, & Muhar, 2007; Rohrmann & Bishop, 2002) that results received from computer visualization approaches are usable for the estimation of real world behavior and animated approaches are more valid in presenting dynamic environments than static ones, particularly when presenting shared trail environments.

Study findings indicate that computer game experience has an influence on stated trail preferences (Table 1). Differences were not only found in the relative importance of factor levels, but also in the existence of various factor interactions. One study assumption was that gamers may be less influenced than non-gamers by the animation because of their familiarity with such visualizations. Oh (1994) found that people with experience in computer simulation and visual modeling gave higher confidence ratings to the video textured and digitized photo representations of a building. In this study gamers' evaluations were more stable across both approaches, while animations had more influence on non-gamers' trail use preferences. This may indicate that the presentation mode can have an influence on trail preferences depending on the familiarity with computer visualizations.

Another assumption was that, because of the similarity of the animated trail visualizations to 3-D computer game environments, computer gamers may express different trail preferences compared to non-gamers. While game research is concerned with the influence of video games on real world behavior and cognitions (Fischer et al., 2007), this study indicates that there are relationships between the frequency of gaming and trail preferences. Our findings indicated that gamers focused on bicyclists and user numbers and specifically disliked a crowded bicyclist-rich situation, while they were hardly affected by the direction of movement. This result could be interpreted as a specific dislike of user conflicts.

More factors and factor levels influenced the trail preferences of non-gamers. They disliked higher shares of oncoming trail users, a mix in the user composition and mixed movement directions in combination with many bicyclists. The disfavor of mixes in social situations and the direction of movement by non-gamers could be interpreted as a dislike of crowded and heterogeneous situations. This dislike was more relevant in the animated approach. On the other hand, a mix of trail user groups in combination with eight visitors contributed positively to preferences.

The question of why these different preference patterns emerged arises, although there were no differences in the perception of both visual stimuli between gamers and non-gamers (Table 2). Research on video games has found that there are interconnections between real and virtual environments (Fischer et al., 2007). A hypothesis could be that respondents who play computer games might have equated the presented trail scenes with computer game scenes (Großegger & Zentner, 2008); our 3-D visualizations were possibly similar to photographically highly realistic game environments. This could mean that computer gamers did not transform the scenarios into real recreational scenarios, and their preferences could not be seen as absolutely valid. One can also speculate that non-gamers may have paid more attention to the trail scenarios because of the higher number of significant factors and interactions. What these data definitely demonstrate is that there are differences in trail preferences for social conditions depending on computer game experience. Further research, therefore, should investigate whether computer gamers evaluate virtual scenarios similar to real word scenarios, and whether people who like complex scenarios are more likely to take up computer games. The fact that most computer game players were males also has to be taken into account (Großegger & Zentner, 2008). For example, Tlauka, Brolese, Pomeroy, and Hobbs (2005) found significant differences in the sexes' preference for virtual shopping centers.

The presented results provide only a limited insight into the topic because of the small sample size. On the other hand, the homogenous structure of the sample reduced the influence of other individual factors such as age and allowed studying differences more precisely. We collected computer game experience data using only one variable and addressed only one specific segment of the population. Further research might specify experience with computer game graphics and include other population groups.

There is also room for improvement in the graphic quality of the presented virtual scenarios. The somewhat robotic motion of the trail users should be improved. In the real world, people also experience other sensations such as smell, taste and noise (Newman et al., 2006; Manning & Freimund, 2004). Only three social factors were included in this explorative study investigating the relative importance of these on social trail preferences. However, animated choice experiments could include many more social, as well as physical, attributes to investigate trail preferences in general (Arnberger & Haider, 2007) and the influence of computer game experience in particular. Future studies may add factors describing vegetation, trail characteristics, and larger variations in use levels and user composition. Animating environmental factors and how the respondent would really move through a virtual scenario (Bishop, Ye, & Karadaglis, 2001) are future aspects to analyze whether this type of interactive behavior depends on computer game experience. One specific issue using 3D-animations for future shared trail research is the exploration of speed. Different users, such as bicyclists and walkers, move at different speed levels, which may result in user conflicts. In order to avoid user conflicts, trail managers need information about the influence of speed in relation to the other trail environment determining factors and how they can reduce the impact of speed without excluding some user groups.

CONCLUSION

This explorative study found that the degree of computer game experience, as well as the presentation mode, resulted in different preferences for trail scenarios. This research suggests that researchers and planners in the field of e-Planning have to be aware of the potential influence of computer game experience and presentation mode. Although the found differences do not substantially impact the overall results, data on computer game experience should be regularly collected in trail preference surveys to identify its potential influence, particularly because of the increasing proportion of the population who plays computer games (Großegger & Zentner, 2008).

Visualization technologies support the sustainable and cost-efficient planning and design of urban environments and are already widely used in planning processes. Additionally, researchers and planners gain greater understanding of human perception and behavior in the urban environment (Loiterton & Bishop, 2005). New planning and research tools which rely on 3D-visualisation are becoming more and more available and the application of animated and interactive visualization tools will increase. Because of the rapid development of visualization technologies, such as real-time render- and sound engines, and the trend towards photo-realism of computer graphics, preference research in complete virtual realities will soon be a standard approach. 3D-visualizations in combination with stated choice surveys will allow collecting preference data for a wide range of urban planning related issues. Web 2.0 applications integrating such approaches could be a powerful tool for participatory planning of future urban projects

However, there are reasoned concerns regarding the validity of visualizations. Planners should be aware that the medium is not the message but the contents which are conveyed by applying visualizations (Lange, 2005). Further research is needed in order to understand comprehensibly the interactions between human beings and the virtual world, and to prove the validity of computer animations for environmental planning.

REFERENCES

Anderson, C. A. (2004). An update on the effects of playing violent video games. *Journal of Adolescence, 27*, 113–122. doi:10.1016/j.adolescence.2003.10.009

Aoki, Y. (1983). An empirical study on the appraisals of landscape types by residential groups – Tsukuba Science City. *Landscape Planning, 10*, 109–130. doi:10.1016/0304-3924(83)90055-2

Aoki, Y. (1999). Trends in the study of the psychological evaluation of landscape. *Landscape Research, 24*(1), 85–94. doi:10.1080/01426399908706552

Arnberger, A., Eder, R., Brandenburg, C., & Reichhart, T. (2007). Assessing landscape preferences of urban population for terraced areas. In Faculty of Forestry (Ed.), Landscape Assessment - From Theory to Practice: Applications in Planning and Design (pp. 111-119). Beograd, Srbija: Planeta Print.

Arnberger, A., & Haider, W. (2007). Would you displace? It depends! A multivariate visual approach to intended displacement from an urban forest trail. *Journal of Leisure Research, 39*(2), 345–365.

Bishop, I. D., Ye, W., & Kardaglis, C. (2001). Experiential approaches to perception response in virtual worlds. *Landscape and Urban Planning, 54*(1-4), 117–125. doi:10.1016/S0169-2046(01)00130-X

Cessford, G. R. (2003). Perception and reality of conflict: Walkers and mountain bikes on the Queen Charlotte Track in New Zealand. *Journal for Nature Conservation, 11*(4), 310–316. doi:10.1078/1617-1381-00062

Chen, X., Bishop, I. D., & Abdul Hamid, A. R. (2002). Community exploration of changing landscape values: the role of the virtual environment. In D. Suter & A. Bab-Habiashar (Eds.), *Proceedings of the sixth digital image computing - techniques and applications conference - DICTA2002* (pp. 273-278). Melbourne, Australia: Australian Pattern Recognition Society.

Daniel, T. C., & Meitner, M. M. (2001). Representational validity of landscape visualizations: The effects of graphical realism on perceived scenic beauty of forest vistas. *Journal of Environmental Psychology, 21*, 61–72. doi:10.1006/jevp.2000.0182

Fischer, P., Kubitzki, J., Guter, S., & Frey, D. (2007). Virtual driving and risk taking: Do racing games increase risk-taking cognitions, affect, and behaviors? *Journal of Experimental Psychology, 13*(1), 22–31.

Fredman, P., & Hörnsten, L. (2001). *Perceived crowding, visitor satisfaction and trail design in Fulufjäll National Park Sweden. Report.* Östersund, Sweden: European Tourism Research Institute.

Gobster, P. H. (1995). Perception and use of a metropolitan greenway system for recreation. *Landscape and Urban Planning, 33*, 401–413. doi:10.1016/0169-2046(94)02031-A

Großegger, B., & Zentner, M. (2008). *Computerspiele im Alltag Jugendlicher. Gamer-Segmente und Gamer-Kulturen in der Altersgruppe der 11- bis 18-Jährigen.* Wien, Austria: Institut für Jugendforschung.

Heft, H., & Nasar, J. N. (2000). Evaluating environmental scenes using dynamic versus static displays. *Environment and Behavior, 32*(3), 301–322.

Herwig, A., Kretzler, E., & Paar, P. (2005). Using games software for interactive landscape visualization. In Bishop, I. D., & Lange, E. (Eds.), *Visualization in Landscape and Environmental Planning – Technology and Applications* (pp. 62–67). New York: Taylor & Francis.

Hetherington, J., Daniel, T. C., & Brown, T. C. (2004). Is motion more important than it sounds? The medium of presentation in environmental perception research. *Journal of Environmental Psychology, 13*, 283–291. doi:10.1016/S0272-4944(05)80251-8

Hosmer, D. W., & Lemeshow, S. (2000). *Applied Logistic Regression.* New York: Wiley. doi:10.1002/0471722146

Jacob, G.-R., & Schreyer, R. (1980). Conflict in outdoor recreation: a theoretical perspective. *Journal of Leisure Research, 12*(4), 368–380.

Jorgensen, A., Hitchmough, J., & Calvert, T. (2002). Woodland spaces and edges: their impact on perception of safety and preference. *Landscape and Urban Planning, 59*, 1–11.

Junker, B., & Buchecker, M. (2008). Aesthetic preferences versus ecological objectives in river restorations. *Landscape and Urban Planning, 85*(3-4), 141–154.

Karjalainen, E., & Tyrväinen, L. (2002). Visualization in forest landscape preference research: A Finnish perspective. *Landscape and Urban Planning, 59*, 13–28. doi:10.1016/S0169-2046(01)00244-4

Laing, R., Davies, A.-M., & Scott, S. (2005). Combining visualization with choice experimentation in the built environment. In Bishop, I. D., & Lange, E. (Eds.), *Visualization in Landscape and Environmental Planning – Technology and Applications* (pp. 212–219). New York: Taylor & Francis.

Lange, E. (2001). The limits of realism: perceptions of virtual landscapes. *Landscape and Urban Planning, 54*, 163–182. doi:10.1016/S0169-2046(01)00134-7

Lange, E. (2005). Issues and questions for research in communicating with the public through visualisations. In Buhmann, E., Paar, P., Bishop, I., & Lange, E. (Eds.), *Trends in Real-Time Landscape Visualization and Participation* (pp. 16–26). Heidelberg, Germany: Herbert Wichmann Verlag.

Lange, E., & Hehl-Lange, S. (2005). Future scenarios of peri-urban green space. In Bishop, I. D., & Lange, E. (Eds.), *Visualization in Landscape and Environmental Planning – Technology and Applications* (pp. 195–202). New York: Taylor & Francis.

Loiterton, D., & Bishop, I. (2005). Virtual environments and location-based questioning for understanding visitor movement in urban parks and gardens. In Buhmann, E., Paar, P., Bishop, I., & Lange, E. (Eds.), *Trends in Real-Time Landscape Visualization and Participation* (pp. 144–154). Heidelberg, Germany: Herbert Wichmann Verlag.

Louviere, J. J., Hensher, D. A., & Swait, J. D. (2000). *Stated choice methods – Analysis and application*. Cambridge, UK: University Press.

Manning, R. (2007). *Parks and carrying capacity*. Washington, DC: Island Press.

Manning, R. E., & Freimund, W. A. (2004). Use of visual research methods to measure standards of quality for parks and outdoor recreation. *Leisure Sciences*, *36*(4), 557–579.

Muhar, A. (2001). Three-dimensional modelling and visualisation of vegetation for landscape simulation. *Landscape and Urban Planning*, *54*(1-4), 5–19. doi:10.1016/S0169-2046(01)00122-0

Nasar, J. L. (1983). Adult viewers' preferences in residential scenes. A study of the relationships of environmental attributes to preference. *Environment and Behavior*, *15*(5), 589–614. doi:10.1177/0013916583155003

Needham, M. D., & Rollins, R. B. (2005). Interest group standards for recreation and tourism impacts at ski areas in the summer. *Tourism Management*, *26*, 1–13. doi:10.1016/j.tourman.2003.08.015

Newman, P., Manning, R. E., Pilcher, E., Trevino, K., & Savidge, M. (2006). Understanding and managing soundscapes in national parks: Part 1- indicators of quality. In D. Siegrist, C. Clivaz, M. Hunziker & S. Iten (Eds.), *Exploring the Nature of Management. Proceedings of the Third International Conference on Monitoring and Management of Visitor Flows in Recreational and Protected Areas* (pp. 193-195). Rapperswil, Switzerland: University of Applied Sciences.

Oh, K. (1994). A perceptual evaluation of computer-based landscape simulations. *Landscape and Urban Planning*, *28*, 201–216. doi:10.1016/0169-2046(94)90008-6

Reichhart, T., Arnberger, A., & Muhar, A. (2007). A comparison of still images and 3-D animations for assessing social trail use conditions. *Forest Snow and Landscape Research*, *81*, 77–88.

Rohrmann, B., & Bishop, I. (2002). Subjective responses to computer simulations of urban environments. *Journal of Environmental Psychology*, *22*, 319–331. doi:10.1006/jevp.2001.0206

Silva, C. N., & Syrett, S. (2006). Governing Lisbon: Evolving forms of city governance. *International Journal of Urban and Regional Research*, *30*(1), 98–119. doi:10.1111/j.1468-2427.2006.00646.x

Statistik Austria. (2008). *Haushalte mit Computer, Internetzugang und Breitbandverbindung 2004-2008*. Retrieved December 3, 2008, from http://www.statistik.at/web_de/static/ergebnisse_im_ueberblick_haushalte_mit_computer_internetzugang_und_breitba_022206.pdf

Tahvanainen, L., Tyrväinen, L., Ihalainen, M., Vuorela, N., & Kolehmainen, O. (2001). Forest management and public perceptions – visual versus verbal information. *Landscape and Urban Planning*, *53*, 53–70. doi:10.1016/S0169-2046(00)00137-7

Tlauka, M., Brolese, A., Pomeroy, D., & Hobbs, W. (2005). Gender differences in spatial knowledge acquired through simulated exploration of a virtual shopping centre. *Journal of Environmental Psychology, 25*, 111–118. doi:10.1016/j.jenvp.2004.12.002

Vallerie, W. A., Park, L. O. B., Hallo, J. C., Stanfield, R. E., & Manning, R. E. (2006). *Enhancing visual research with computer animation: A study of crowding-related standards of quality for the loop road at Acadia National Park.* Retrieved December 4, 2008, from http://www.issrm2006.rem.sfu.ca/abstractdsip_popup.php?id=629

Van den Berg, A. E., Vlek, C. A. J., & Coeterier, F. J. (1998). Group differences in the aesthetic evaluation of nature development plans: A multilevel approach. *Journal of Environmental Psychology, 18*, 141–157. doi:10.1006/jevp.1998.0080

Zube, E. H., Sell, J. L., & Taylor, J. G. (1982). Landscape perception: Research, application and theory. *Landscape Planning, 9*, 1–33. doi:10.1016/0304-3924(82)90009-0

ADDITIONAL READING

Arnberger, A. (2006). Recreation use of urban forests: An inter-area comparison. *Urban Forestry & Urban Greening, 4*, 135–144. doi:10.1016/j.ufug.2006.01.004

Arnberger, A., Brandenburg, C., & Muhar, A. (Eds.). (2002). *Monitoring and Management of Visitor Flows in Recreational and Protected Areas.* Proceedings of the First International Conference on Monitoring and Management of Visitor Flows in Recreational and Protected Areas. Vienna: Institute for Landscape Architecture and Landscape Management.

Arnberger, A., & Eder, R. (2007). Monitoring recreational activities in urban forests using long-term video observation. *Forestry, 80*(1), 1–15. doi:10.1093/forestry/cpl043

Arnberger, A., & Eder, R. (2008). Assessing user interactions on shared recreational trails by long-term video monitoring. *Managing Leisure, 13*(1), 36–51. doi:10.1080/13606710701751385

Arnberger, A., & Haider, W. (2005). Social effects on crowding preferences of urban forest visitors. *Urban Forestry & Urban Greening, 3*(3-4), 125–136. doi:10.1016/j.ufug.2005.04.002

Arnberger, A., & Haider, W. (2007). A comparison of global and actual measures of perceived crowding of urban forest visitors. *Journal of Leisure Research, 39*(4), 668–685.

Bishop, I. D., & Lange, E. (Eds.). (2005). *Visualization in Landscape and Environmental Planning – Technology and Applications.* NY: Taylor & Francis.

Buhmann, E., Paar, P., Bishop, I., & Lange, E. (Eds.). (2005). *Trends in Real-Time Landscape Visualization and Participation.* Heidelberg: Herbert Wichmann Verlag.

Carothers, P., Vaske, J. J., & Donnelly, M. P. (2001). Social values versus interpersonal conflict among hikers and mountain bikers. *Leisure Sciences, 23*(1), 47–61. doi:10.1080/01490400150502243

Faculty of Forestry (Ed.). (2007). *Landscape Assessment - From Theory to Practice: Applications in Planning and Design.* Beograd: Planeta Print.

Gatersleben, B., & Uzzell, D. (2007). Affective appraisals of the daily commute – Comparing perceptions of drivers, cyclists, walkers, and users of public transport. *Environment and Behavior, 39*(3), 416–431. doi:10.1177/0013916506294032

Graefe, A. R., Vaske, J. J., & Kuss, F. R. (1984). Social carrying capacity. An integration and synthesis of twenty years of research. *Leisure Sciences*, *6*(4), 395–431. doi:10.1080/01490408409513046

Gramann, J. H. (1982). Toward a behavioral theory of crowding in outdoor recreation: An evaluation and synthesis of research. *Leisure Sciences*, *5*, 109–126. doi:10.1080/01490408209512996

Hall, T., & Shelby, B. (2000). Temporal and spatial displacement: Evidence from a high-use reservoir and alternative sites. *Journal of Leisure Research*, *32*(4), 435–456.

Hammitt, W. E. (2002). Urban forests and parks as privacy refuges. *Journal of Arboriculture*, *28*(1), 19–26.

Kaplan, S., & Kaplan, R. (1989). *The Experience of Nature: A Psychological Perspective*. Cambridge, UK: University Press.

Lieber, S. R., & Fesenmaier, D. R. (1984). Modelling recreation choice: A case study of management alternatives in Chicago. *Regional Studies*, *18*(1), 31–43. doi:10.1080/09595238400185031

Manning, R. E. (1999). *Studies in Outdoor Recreation, Search and Research for Satisfaction*. Oregon: State University Press.

Shelby, B., & Heberlein, T. (1986). *Carrying Capacity in Recreation Settings*. Corvallis, OR: Oregon State University Press.

Shelby, B., Vaske, J. J., & Heberlein, T. A. (1989). Comparative analysis of crowding in multiple locations: Results from fifteen years of research. *Leisure Sciences*, *11*, 269–291. doi:10.1080/01490408909512227

Siegrist, D., Clivaz, C., Hunziker, M., & Iten, S. (Eds.). (2006). *Exploring the Nature of Management*. Proceedings of the Third International Conference on Monitoring and Management of Visitor Flows in Recreational and Protected Areas. Rapperswil, Switzerland: University of Applied Sciences.

Stokols, D., & Altman, I. (Eds.). (1991). *Handbook of Environmental Psychology* (2nd ed., *Vol. 1*). Malabar, FL: Krieger Publishing.

Taczanowska, K., Muhar, A., & Arnberger, A. (2008). Exploring Spatial Behavior of Individual Visitors as Background for Agent-Based Simulation. In Gimblett, R., & Skov-Petersen, H. (Eds.), *Monitoring, Simulation, and Management of Visitor Landscapes* (pp. 159–174). Tucson: The University of Arizona Press.

Watson, A. E., Williams, D. R., & Daigle, J. J. (1991). Sources of conflict between hikers and mountain bike riders in the Rattlesnake NRA. *Journal of Parks and Recreation Administration*, *9*(3), 59–71.

Westover, T. N., & Collins, J. R. (1987). Perceived crowding in recreation settings: An urban case study. *Leisure Sciences*, *9*, 87–99. doi:10.1080/01490408709512149

Whyte, W. H. (1980). *The Social Life of Small Urban Spaces*. Washington, D.C.: The Conservation Foundation, Wiberg-Carlson, D., & Schroeder, H. (1992). *Modeling and Mapping Urban Bicyclists' Preferences for Trail Environments*. Research Paper NC-303. St. Paul, MN: U.S. Dept. of Agriculture, Forest Service, North Central Forest Experiment Station.

KEY TERMS AND DEFINITIONS

3-D Computer Animation: Trail users' motion behaviour, i.e. walking, bicycling, is animated using 3-D computer animation.

Computer Game: A computer or video game is a 3D electronic game that involves interaction with game players.

Photomontages: Images which has been, mostly digitally, manipulated to presented variations in, for example, trail conditions, i.e. few trail users, or many trail users, without changing other trail environment settings.

Semi-Virtual Environments: These environments combine computer-generated elements, for example 3D animated trail users, with a photo of a real trail setting.

Shared-Use Recreational Trail: The trail is used by several user groups such as pedestrians and bicyclists for recreational purposes.

Social Trail Preference: Preference of respondents for trail conditions which are characterized by social stimuli such as trail user numbers, user activities.

Trail Scenario: A trail scenario consists of a specific combination of three social attributes.

Chapter 7
Communicating Geoinformation Effectively With Virtual 3D City Models

Markus Jobst
Hasso-Plattner-Institute at University of Potsdam, Germany

Jürgen Döllner
Hasso-Plattner-Institute at University of Potsdam, Germany

Olaf Lubanski
Jobstmedia Präsentation Verlag, Austria

ABSTRACT

Planning situations are commonly managed by intensive discussions between all stakeholders. Virtual 3D city models enhance these communication procedures with additional visualization possibilities (in opposite to physical models), which support spatial knowledge structuring and human learning mechanisms. This chapter discusses key aspects of virtual 3D city creation, main components of virtual environments and the framework for an efficient communication. It also explores future research for the creation of virtual 3D environments.

INTRODUCTION

Virtual 3D city models include important tools for urban planning and management. Their power to support naïve geography is used by the public and results in easy understandable geospatial presentations. In contrast to traditional 2D maps with highly abstract contents, elements in 3D maps follow some natural and naive coding that can easily be accessed even by layman in map-reading. For instance, Gaerling and Golledge (1993), Keller (1993) or Robinson (1976) see map-reading as subject to education processes: the reconstruction of coded information must be learned for specific cases of application (e.g., politics, topography, among other cases).

The effective use of virtual 3D city models as public communication tools requires the correct handling of the media, which means that the content on specific media must be unmistakably perceivable. This handling of transmitting media will have to bear in mind resolution, extension and interaction

DOI: 10.4018/978-1-61520-929-3.ch007

issues. In most cases transmitting media will be in form of computer displays, touchable points of interest (touch-screens) or even multi-touch large displays. Resolution issues mainly follow rules of expressiveness, where the content has to be adapted to the transmitting characteristics of the media. Issues of transmitting media extension underlie expressiveness as well as effectiveness (Mackinlay 1986), especially when a large display also enhances resolution and thus supports deeper information contents. This aspect expands field of vision as important depth cue parameter at the same time. Interaction issues are mainly subject to effectiveness, when the impact of contents can be enhanced via multiple coding and mouse over events. All cases are focusing at appropriate use of transmitting values. Picture elements are generally restricted and thus have to be specifically used. Each interface-pixel is able to transport one information value. Therefore, the concentration of several information values to one single pixel leads to perceptive problems that will cause undifferentiated pixel areas. These areas can be called "dead values" of the presentation interface. Especially, virtual 3D city models easily contain dead value areas due to lacking generalization algorithms for 3D building aggregation, simplification or appropriate highlighting. At least the aim is to reduce dead value areas in virtual 3D city models by view port variations and simple cartography-oriented design. Furthermore, these specific preparations lead to a manipulation of mental images.

This chapter presents state-of-the-art virtual 3D city modelling, which considers actual creation processes of virtual 3D city models, geomedia techniques as framework for efficient communication as well as the 3D user interface. Future research directions for the work with virtual 3D environments discuss "smart approaches", explain the enhancing of geospatial communication in terms of expressiveness and effectiveness and give a short insight into studies on the usage of virtual 3D city models.

BACKGROUND

State-of-the-Art Virtual 3D City Modelling

Planning procedures make use of models that show historic and present situations and communicate planned situations (Marcinkowski, 2007). Up to now, maps and physical 3D models delivered graphics for this geospatial communication. Nowadays, digital 3D presentations provide a more immersive interaction that may support understanding, cognitive processing and expanding of individual spatial knowledge bases. This main characteristic of 3D delivers important potential for an innovative geospatial communication in terms of linguistics, naïve geography, learning as well as cognitive, affective and conative communication aspects. The variations within virtual 3D city models can support this spatial communication more effectively in comparison to physical models. This situation leads to considerations of state-of-the art virtual 3D city creation, communication frameworks and actual user interfaces for virtual 3D city models.

Actual Aspects of Virtual 3D City Model Creation and Visualization

Nowadays easiness for the creation of virtual 3D city models and its visualization makes this technique more common. The following points will give an overview of the state-of-the-art model creation and visualization. Techniques for surface model creation, the control of light and aspects of vegetation in virtual city environments will be examined.

Surface Models and Virtual 3D Cities

The procedures to create virtual 3D cities vary from precise photogrammetry to block model extrusion and smart city algorithms. Each procedure results in a specific data model that allows for cartographic information preparation.

Main production processes for virtual 3D city models span various methods of airborne surveying and photogrammetry, terrestrial surveying with ground plan extrusion or smart city algorithms. Airborne surveying and photogrammetry make use of high detailed remote sensing, aerial photographs and laser scanning (LIDAR). Detailed remote sensing, if the resolution of satellite pictures is high enough, and aerial photography produce top view pictures that form a starting material for the analysis of ground plan. In addition stereoscopic methods allow extracting heights and therefore detailed structures of building roofs (Kraus 2007). Because a manual evaluation with its point wise recording is highly time consuming, semi-automated procedures were developed and can support the modelling of buildings and roofs (Kraus 2007). As a result, a detailed and precise structure of buildings including roofs is offered, which is sufficient for fly trough's in very large scales. For its cartographic usage these data mostly have to be simplified, aggregated and their most important characteristics highlighted for the bigger part of transmitting media. This has to be done because this high detailed information cannot be successfully transmitted in most of the used scales in combination with low media resolution (displays). Airborne laser scanning delivers data with similar precision. But the general result, without any post-processing, is a point cloud of single measurements in form of the regular raster of the laser scanner (Kraus 2007). A first post-processing leads to a covering hull of the recorded surface that does not include any object wise information than the height. At least filtering and further post-processing lead to a covering hull of single buildings that can be enhanced with further (meta) information. Similar to the results of aerial photogrammetry, laser scanning also requires generalization for most of transmitting media because of its detail richness.

Terrestrial surveying with ground plan extrusion uses building plots and the average height of buildings to reconstruct a building block by extrusion of its outline. This procedure cannot reconstruct the roofs. Therefore the virtual 3D city model resulting from this technique is a simple block model, which may be sufficient for many applications and analysis of districts at a medium scale level. Depending on the outline detail of buildings, simplifications have to be done to remove imperceptible details (Kada 2005). The extrusion process is done automatically by delivered data (outline, height). Additionally, for reason of aesthetics synthetic roofs can be added by algorithms. These synthetic roofs will visualize main characteristics but not represent a reconstruction of real world objects.

Besides building reconstruction at various scale levels, the modelling of terrain is another part in topographic reconstruction that has to be considered. Data structures for terrain models span from regular grids to triangulated irregular networks (TIN) and regular grid with break lines (Pfeifer 2002). The modelling of terrain is lead by reduction of data size and modelling precision. In terms of visualization it is important that the virtual 3D city model and the virtual terrain model do not perceptibly diverge.

In addition to geodetic recording methods and their precise results, further automatic approaches lead to useful results. The aim to adapt information contents to transmission media leads to automated building construction and generalized city models. Smart X approaches are able to satisfy these needs and offer new possibilities in managing complex virtual 3D city scenes.

Following the model creation, the control and modelling of light is a further step to make virtual environments more attractive. Especially lighting has enormous impact on perception of the environment and its impact.

The Control of Light

Lighting and its consequence within the environment is a central factor for the understanding of spatial representation. Shadowing is a key for

Figure 1. An urban 3D model with a lightmap as texture. This situation emulates physical model characteristics

referencing objects relatively to a ground plane and to other objects. But if the graphical impact of shadowing mechanisms is too strong, a perception of object silhouettes may be disturbed. For this reason more smooth solutions, like ambient occlusion, offer more natural lighting situations that imitate a natural behaviour of smooth light (light as it occurs with a cloudy sky). The problem for computer graphics arises with the need of dynamic applications. Lighting situations are for the most cases too complex to be rendered in real-time. Therefore various basic mechanisms, like texture baking and multitexturing, have to be used to improve dynamic lighting.

The procedure of texture baking (unwrapping) describes a technique that allows for projecting planes of a 3D object to a 2D plane. It is used to map texture coordinates of whole objects to a single texture area/file. This is helpful for an easier texturing and modification of these textures in an image. On the other hand this method is used to render all the material characteristics of an object in a single file. These characteristics span from diffuse behaviour, transparency to light and shadow. Thus the complete light situation can be pre-rendered and stored as texture for all the objects in a scene. The advantage of texture baking lies in the possibility to render a light situation before a dynamic use of the environment and in this way incorporate complex light situations in virtual real-time environments. One disadvantage is the static characteristic of baking: a baked texture presents one specific moment of lighting situation and does not dynamically adapt. So this method cannot be used for moving light sources. Instead, this method has to be combined with dynamic shadowing methods.

The techniques of multitexturing and lightmapping allow for simultaneous usage of several textures, whereas each texture becomes mapped with the help of its UV coordinate set. Whereas previous GPU's needed two rendering passes, several textures can be rendered in one single pass nowadays. The combination of textures uses various methods of calculation, which span from adding to multiplication. One of the most prominent examples of multitexturing is lightmapping, which combines an object texture with the lightmap of that object. The advantage of this technique is its combination of multiresolution textures: the resolution of the lightmap can be very low, because the depiction of light does not need higher qualities. Due to the usage of UV coordinates the textures of various resolutions can be combined.

An embedding of global illumination effects dark model areas by brightening these areas

Figure 2. Example for a hybrid tree model, where billboards are used for branches and leaves

up with light diffusion, which leads to lighting that reminds of physical model characteristics. This kind of lighting is very neutral, because the light sources are placed in any direction. There is no dominant lighting direction, which makes it easier to copy instances of objects and their baked lightmaps, like it is done with trees or street lights.

City furnishing, especially vegetation, forms another challenge for virtual 3D city environments. In that process, resources are confronted with visualization details. The following section presents some advanced techniques for the visualization of city furnishing.

A Glance on City Furnishing – Vegetation

The furnishing of a virtual city is a further aspect that calls for attention, especially those parts that use a huge amount of processing resources. In particular virtual models of vegetation request smart ideas of modelling. From a mathematical point of view the creation of a plant follows some basic principles and recurring patterns. Additionally further parameters like feasibility or randomization have to be considered. Then, all these parameters can be used in a description model to automatically create specific virtual 3D plants (specific kind of trees, etc.) (Paar 2005).

One of the most important description models is called L-system: originally this model was used for the development of multi-cellular organisms by Aristid Lindenmayer (Prusinkiewicz et. al. 1990). L-systems recursively replace an initial structure with a generation law similar to fractal models.

A L-grammar describes how and when parts of the existing model should be replaced. With this technique high detailed trees and plants can be created. Their detail, with hundred-thousands of polygons, blasts a useful level (in terms of real-time applications) of complexity. Therefore, alternative ways for vegetation models, like billboard-based models, simple geometries or hybrid tree models (billboards and simple geometries) were developed (Colditz et. al. 2005, S97ff).

Billboards are a very popular method for vegetation depiction, but it is not convincing. Flat vertical polygons become textured with images. The front side of the texture orients to the camera, so that the whole texture is visible all the time. The main advantage is its resource-saving mechanism: only one or two polygons are needed for texturing. A further step of development is the hybrid tree model. This uses simple structures of a L-system (small number of polygons) and combines this 3D model with billboards for the leaves or even branches.

The question arises if symbols for trees should be introduced for smaller scales (e.g. a scale of 1:500 to 1:2000 in the physical model). This higher abstraction then simulates classical (physical) model-building, while small spheres and specific distortions of these spheres represent broad-leafed trees. Further abstractions seem to be essential with decreasing scales, like 1:5000 and downwards. In these cases, physical models use areas of dash to mark vegetation. In virtual 3D environments the depiction of single trees is not useful either. Instead, vegetation areas can be extruded by their polygons and simulate wooden areas. In this case straight edges should be avoided.

As a matter of fact an abstraction of plants and vegetation areas follows the idea of cartography-oriented design. Depending on the scale and the transmission media used, the presentation form has to be changed. Therefore, the transition of an object's representation, from one scale to another, gains importance, especially because virtual 3D environments are generally characterized by a combination of an infinite number of scales in one central perspective view.

Particularly lighting and vegetation can have enormous impact on the understanding of virtual space. These are the key factors for a comprehensive spatial transmission. Due to the fact that physical models are still widely used, the actual impact of virtual 3D models in planning situations is still limited.

The following section focuses on the modern framework for spatial communication (e.g., geomedia techniques), which enhances naïve understanding and knowledge acquisition.

Geomedia Techniques as Framework for Communication

Virtual 3D environments offer new potential for an innovative communication by combining immersive and interactive geospatial transmissions. On one hand this potential is based on the naïve understanding of space. On the other hand human learn types and geospatial knowledge structuring provoke affective and conative aspects of communication.

Naïve Understanding

The naïve understanding of space is based on human sensual modes and their processing of stimuli, which originate from space, spatial related objects and their representation. The results of this treatment are organisational and system-oriented structures (whereas the system describes the individual). Zimbardo (1997) and many others in the field of learning theories are convinced that individual knowledge structures become more efficient when cognitive processed information follows spatial structures. In general, these structures use similarities, sizes, appearance and temporal behaviour. All of these characteristics are accordingly incorporated in cartographic coding, where geometric, semantic and temporal classes as well as spatial hierarchies will be used (Buziek 2001). In this connection virtual 3D environments deliver further dimensions to establish similarities, sizes, appearance, etc. especially by their natural combination of footprints and upright projections. As long as the transmitting media (visual user interface) provides enough resolution for a visual differentiation of spatial information contents, these parts can be incorporated into a new knowledge structure. In doing so, an existing knowledge structure works as a filter for the new knowledge acquisition (Neisser 1996). In the case of geospatial contents, knowledge creation refers back to the psychological development of dealing with space. The spatial environment forms the main fundament for any spatial reference of the human individual. According to Piaget's theory of mental development, where he identifies three main stages of psychological development in space, a fully developed psychology is needed to respond to any received spatial stimulus (Piaget 2003, S. 53 ff). This means that the usage of virtual 3D environments assumes the existence of a completed development of spatial psychology. Otherwise the recipient will not be able to understand the content, deal with occlusion, suggest spatial relations or come to a logical conclusion, neither in real world nor in a virtual one. At least the success of this naïve geospatial communication with virtual 3D environments can be measured by cognitive maps. Cognitive mapping describes the conscious and motivated coding/drawing of spatial knowledge in a way that it can be used for solving questions like "Where am I?", "Where are special coded objects in space?" or "How do I reach place A?". Cognitive mapping forms an initial point for individual spatial planning and communicating space to others (Golledge 1999). We can conclude that virtual 3D environments offer an efficient geospatial presentation technique due to its congruence with real world characteristics. The expansion of spatial knowledge by virtual space will be done similar to the naïve exploration most recipients/users are used to do in their daily life. But this task of exploration underlies some characteristics of learn-types. Not all users are confident with a holistic and active exploration environment, which most of virtual 3D environments represent.

Learning Issues in Space

Generally, learning processes for the creation of mental representations of spatial information occur in two steps: in a first step new information becomes absorbed (Jonassen et al 1993); the second step covers information processing (Mayer 2000). Learning applications have to consider both steps (Brunner-Friedrich 2004), which have their impact on the classification of recipients. The step of information absorption mainly divides recipients in active and passive (Mayer 2000) as well as in holistic and serial (Jonassen et al 1993). An active user independently looks for specific information in order to solve a problem, like "I am looking for a specific shop!". The passive user requests all the relevant information. Additionally to these classes of activity, a variation of information detail will support the holistic or serial type. The holistic type requires an overview before detailed information can be absorbed. The serial type starts with detailed information and successively completes an overall view. The second step of information processing classifies recipients by sensual modes (Blumenstengel 1998), like acoustic or visual, and by an abstract- / graphical-oriented user group (Paivio 1978). Of course, these learning types are extreme cases and will not be distinctively observable. But a spatial application with the demand to efficiently support exploration will have to consider existing learning types. This circumstance exceeds previous description of naïve virtual environments and calls for highly interactive virtual 3D environments, which means that photo-realistic rendering and camera motion is not enough to support all learn types. On the contrary, non-photorealistic rendering, interactive objects and the option for predefined fly-through within the 3D environment should expand a useful exploration application. These factors easily provoke the affective and conative aspect of spatial communication. The affective aspect refers to the experience of feeling or emotion at the recipient side, which is an emotional starting point for an interaction with stimuli. As soon as virtual objects affect a recipient, a reaction can be expected. The conative aspect of communication covers the motivation for reaction. This component is responsible for the reaction after a stimulus.

Knowledge Structuring

The main potential for spatial communication lies in the consideration of knowledge structuring and human learning types. Therefore, innovative approaches focus on non-photorealistic rendering, highly interactive contents and naïve user interaction for virtual 3D environments. In the case of geospatial information transmission, learning theories and aspects of human psychology illustrate the support of virtual 3D environments for the transmission process (Edwardes, 2007). In fact, virtual 3D environments show new aspects in the context of learning: virtual 3D environments allow for imitation and evaluation of learned knowledge sequences directly inside the virtual environment.

In the end, all aspects of innovative approaches highly depend on user interface characteristics and their resolution. Thus, the role of user interfaces for a visual information transmission is the focus of the next section.

The 3D User Interface

User interfaces are the most important part within the human computer interaction. These input- and output-interfaces enable virtual "real-world" movements in terms of psychological perception and form the technical fundament for information transmission. Besides specific interfaces that support naïve input and movement, the visual transmission of information is a key factor for cartography-based 3D environments. In this section the focus of user interfaces simply lies on visual information transmission.

The visualization of virtual 3D environments becomes most effective on digital transmission media, like displays or projectors. Although the rendering of a 3D scene on an analogue media (paper) may offer new possibilities for planning situations, especially when planned situations can be shown in photo-realistic expression, an interactive virtual 3D environment can deliver much more impact in terms of knowledge acquisition, as it was discussed in the previous section. Whereas the dynamic and interactive visualization forms the main advantage of digital transmission media, its relatively low resolution is a main disadvantage compared to "paper". Analogue media and their reproduction techniques offer a resolution (e.g. printing granularity) that is much higher than the human eye can dissolve. Therefore information that should be unmistakably transmitted has to follow minimum values for perception. The "minimum value" describes a critical size of graphical signs, which barely allows for differentiation of form and dimension (Malic 1998). According to theoretic considerations these critical values correspond to the aligning power of the eye. Hence the technical limitations of transmission media and physiological limitations of the eye create a perceptive barrier (Neudeck 2001). The perceptive barrier is a value in which graphics can just be perceived and interpreted. This barrier becomes maintained if critical values of cartographic elements and minimal distances between elements do not go below the limit (the aligning power of the eye). This circumstance of perceptive barrier and its critical values lead to a cartographic dilemma (where the cartography-based aspect describes the unmistakable transmission of geospatial information): on one hand the resolution of transmission media has to be lower than the aligning power of the eye in order to enable homogeneous areas. On the other hand graphical information must not go below the critical value. These facts are not the dilemma itself. The dilemma starts with the fact that the resolution of actual display technologies lie within the aligning power of the eye (of course depending from the viewing distance) (Ibrahim 2007). Actual displays use a pitch of 0,19mm to 0,24mm within their picture mask. The aligning power of the eye lies around 0,2mm in a viewing distance of about 30cm (Spiess et al 2002). This fact leads to the dilemma that there is few potential left to build up clear perceivable signs. For example, 10 pixels are needed to draw a homogeneous circle in a contrast picture (Neudeck 2001).

Geospatial information within virtual environments mostly results in graphics on digital displays, which leads to the need of high resolution. For digital reproduction an amount of 1200dpi is requested for high quality results. In some cases 300dpi are sufficient (depending on the crispiness and contrast of graphics). In opposite to this demand, a display delivers about 96dpi, depending from its pitch, size of the screen and overall screen resolution. Due to the contrast graphic of geospatial contents, anti-aliasing and soft-focus effects, as these are used in photographic depictions, can hardly be adapted. From this point of view information depth, which is the amount of differentiable elements, has to be adapted to the display resolution. This circumstance and the request for expressive and effective geospatial information transmission lead to modern ways to support cartographic communication concepts in 3D.

This section described actual aspects of virtual 3D city creation, shortly focused on the framework for spatial communication and the importance of user interfaces for virtual 3D environments. The next section expands these reflections to topics of actual research and future research directions in virtual 3D city creation and enhancement of spatial communication.

FUTURE RESEARCH DIRECTIONS

Virtual 3D City Models

Future trends in virtual 3D city models cover advanced creation methods, enhancements for effective and expressive communication and empirical studies on the usage of virtual 3D city models. Advanced creation methods describe automated procedures for virtual 3D city creation and their graphical abstraction techniques (Jiang 2005). Enhancements for effective and expressive communication evolve new approaches for the usage of user interfaces (more focused: geomedia techniques). Empirical studies on the usage of virtual 3D city models focus on the acceptance of virtual 3D environments, in opposition to physical 3D models.

Smart Approaches – Advanced Virtual 3D City Creation

Smart approaches describe various kinds of concepts that focus on an incremental development of virtual 3D city objects, like buildings, vegetation or furniture. The main question arises how virtual 3D city models can be maintained and how the process of authoring and customization can be defined. For example urban planners want to incorporate a new planning situation for the development of a district into an existing virtual 3D city model. Therefore they have to remodel, rebuild and refine parts of the model. The most efficient way to do so would be an incremental and automated development of all estimable components.

Smart building approaches provide a concept for continuous level-of-detail modelling of building models and target at their incremental development. Due to their per-floor concept, they can be perfectly used by direct-manipulation interfaces, providing an intuitive tool for building refinement. The main use case of project-driven and event-driven customization and re-engineering of city model components drives this approach. It can be observed that approximation of complex building models in a timely efficient way can be done with smart buildings in a variety of use cases. Smart buildings do not intend to substitute CAD models. But this approach provides a graphics-centred, application-centred modelling scheme. It is also suited for large-scale 3D city models that can be easily mapped to an internal graphics representation that allows for real-time photorealistic and non-photorealistic rendering (Döllner et al 2005). One possible application that would benefit of these approaches is a decision-support system in urban planning. A usage of smart buildings or smart cities enables interactively performed changes within the geovirtual 3D environment in a way that the effect of the modification can be evaluated and discussed immediately. Another application is concerned with managing interactive 3D location maps. A smart-building editor may then create and maintain building models, which exhibit characteristic exterior and interior features.

Smart city algorithms use cell-based clustering to generate generalized 3D city models. The clustering bases on the infrastructure network and decomposes city areas into clusters. Therefore, the infrastructure network uses an implicit hierarchy of network elements that is used to create generalizations at various scales (Glander and Döllner 2007). As input data, the infrastructure network (street hierarchy, areas, etc.) is used for the clustering process. The height of clusters is then calculated by the heights of a 3D city model, which may be the result of laser scanning, aerial photography or similar. The infrastructure network uses weighting of its vectors to build up categories and thus follow

Figure 3. An example for a generalized virtual 3D city, which was generated by smart city algorithms

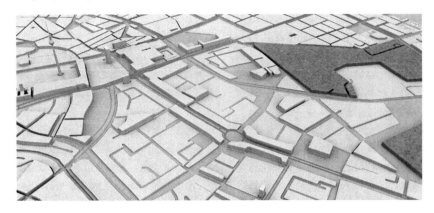

a desired degree of generalization. According to a chosen weight and its selected polylines a cell structure can be computed (Glander and Döllner 2007). By cartographic means this resulting cell structure of a virtual city is applicable for small scale applications. On account of the underlying 3D city model, which is used for the heights of clusters, and the hierarchical structure of infrastructure network various levels of generalization can be visualized. This characteristic of the smart city method seems to be most promising for cartographic applications that call for transmitting media adapted scales.

A smart terrain model uses base vectors to generate virtual 3D objects. It is organized in important classes, like Ground Area, Water Area, Stair, Wall, Kerb and Barrier. Most terrain elements correspond to exactly one base vector-object. The attributes of a base vector-object determine the specific class of a corresponding terrain element. Depending on the respective class, the required attributes for 3D shape and appearance of a terrain element are also extracted out of the attribute table of the base vector-object. The primary design goal was to simplify the refinement of pure 2D base vector-data to full smart surface-model specification for typical cases. These cases can be intuitively described via 2D polygons or polylines, which support optional refinement of individual objects by external 3D tools. For this

Figure 4. Snapshot of the automatically created urban terrain model as redevelopment of a downtown area in Potsdam

reason terrain elements and their automatically generated 3D models can be exported, externally refined and finally referenced by the terrain element again. This procedure helps to keep the 3D model's relation to the underlying base data and avoids the need for fitting an externally created 3D model in the scenery again.

The presented smart approaches simplify and enhance the construction, manipulation and usage of complex urban terrain models. Its major advantages include the rule-based and heuristic-based automated model generation, the inherent functionality and "smartness" of virtual 3D city- and terrain objects. The procedures seamlessly integrate basic 2D geodata from a GIS into the 3D modelling process. The ability of smart X models to maintain semantic and thematic information provides a technical basis for smart 3D geovisualization tools.

Smart approaches enable new and more efficient procedures for the creation of virtual 3D city models. On one hand, modern structures of geobasis data form the basement for smart approaches. On the other hand, the question for the main usage of these virtual city models arises. Therefore new approaches for spatial communication have to focus on the enhancement of expressiveness and effectiveness within geospatial communication.

Enhancing Expressive And Effective Communication

Since cartography was involved in spatial information transmission, usable graphical variables were a key factor for the graphical coding of information. By traditional means eight core variables can be identified. These are color, size, brightness, form, pattern, orientation and position (x, y) of the element (Bertin 1982). Bertin splitted for x and y the variable of position in two parts. In combination with the graphical elements point, line, area, diagram and font, all imaginable coding can be realized. The extension of this traditional 2D semiotic structure with design mechanisms of 3D does not only lead to an extension, but to an extended semiotic structure for 3D (see Figure 1). An extension of Bertin's list, like the one done by MacEachren (1995) with the variables crispiness, resolution and transparency, is not sufficient for the field of 3D cartography and virtual 3D environ-

Figure 5. Extended semiotic structure for designing virtual 3D environments. The mutual impact of identified classes leads to unclear geospatial information transmission that has to be avoided in cartography-based design. Therefore this structure has to be considered in design processes of virtual 3D applications

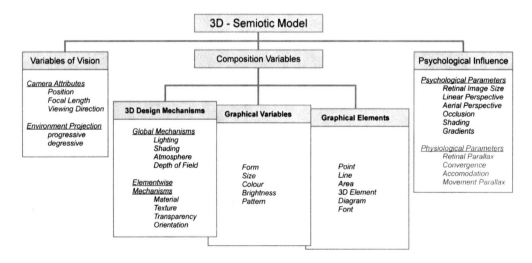

ments. An extension for 3D cartography means that all added design mechanisms of 3D massively influence the coding with graphical variables and elements. Therefore a semiotic structure for 3D includes variables of vision, composition and the psychological influences. Composition variables consist of global and element-wise 3D design mechanisms, graphical variables and graphical elements. These components will have mutual impact on each other.

The following aspects discuss new semiotic considerations, non-photorealistic rendering and exemplary describe semiotic enhancements, which include viewport variations with the aim to reduce dead pixel zones.

New Semiotic Considerations

The semiotic structure for 3D results in a confrontation of graphical variables and design mechanisms. As the core elements of graphical composition in 2D lead to a visual clear result on the transmitting media and their mutual impact in 2D (Bertin 1982, Brunner 2001, Hake and Grünreich 1994, Spiess et al 2002), a dynamic 3D environment with its demand of intuitive geospatial information transfer, expands the semiotic model and shows up new graphical conflicts, which should be described by Table 1.

The critical interplay of graphical variables and design mechanisms show that composition variables have to be precisely considered for cartographic 3D applications. Then it becomes possible to find an appropriate graphical expression and an optimal information transmission. However, composition variables in 3D offer additional design mechanisms. The rendering of the virtual 3D environment provides relevant tools to incorporate and manipulate most graphical variables and design mechanisms. The most important rendering techniques for enhancing expressiveness in virtual 3D worlds are called non-photorealistic rendering (NPR), which reach beyond the establishment of virtual reality (VR).

Reaching Beyond VR

The field of non-photorealistic rendering within computer graphics deals with abstraction and stylization of virtual 3D elements. In opposite to the photorealistic paradigm of virtual reality, non-photorealism aims at supporting human perception by enhancing a rendered depiction with stylistic elements such as shape, structure, colour, light, shading and shadowing. Most researchers agree that the term "non-photorealistic" is not satisfying because neither the notion of realism itself nor its complement, the non-photorealism, is

Table 1. Graphical conflicts of 3D design mechanisms with graphical variables

	A.Form	B.Size	C.Colour	D.Brightness	E.Pattern
Global Mechanisms:					
1.Lighting			C1	D1	
2.Shading	A2		C2	D2	E2
3.Atmosphere			C3	D3	
4.Depth of field		B4			
Element wise mechanisms:					
5.Material	A5	B5	C5	D5	E5
6.Texture					E6
7.Transparency			C7	D7	E7
8.Orientation	A8				

Figure 6. Example for simple scenery as non-photorealistic rendering. According to the needs of expression, stylistic variables, as these are defined in a semiotic structure, are used as toolbox for graphical design

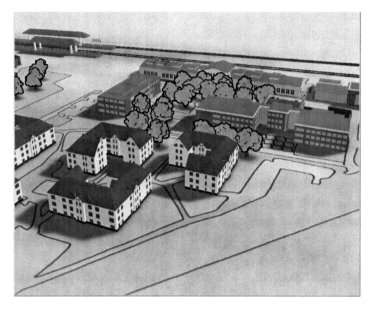

clearly defined. Nevertheless, "non-photorealism" (NPR) has established itself as a key category and discipline in computer graphics starting around 1990. An introduction to concepts and algorithms of non-photorealistic computer graphics can be found in Strothotte and Schlechtweg (2002) as well as in Gooch and Gooch (2001).

The visualization concepts of virtual 3D city presentations use photorealistic and non-photorealistic methods. Photorealistic methods try to reconstruct reality in virtual space. Therefore several graphical variables are used for this simulation procedure and few potential is left for additional information coding. Additionally the high resolution and detailed nuances are often not perceptible and have to be aggregated anyhow. In opposition to that, non-photorealistic rendering, like cartoon rendering, offers precise control over colours, sizes, brightness, forms, etc. for appropriate information coding at various levels. The following points show up the main limitations of photorealistic methods for virtual 3D city models.

Photorealistic visualization implies a number of limitations with respect to virtual 3D city models:

- To produce convincing photorealistic depictions, complete data in high quality have to be processed, for example an exactly matching of façade textures and high-resolution aerial photography. The larger the virtual 3D city model, the higher the costs for data acquisition. In most cases, required data will not be available for a whole 3D city model. As a consequence, the images are faced by a breach of graphics style.
- To incorporate thematic information (e.g., state of repair, average rental fees) into photorealistic depictions, the information needs to be visually combined with the virtual 3D city model. This turns out to be difficult, because textured façades, roofs, and road systems dominate the image space.
- To visualize complex information, photorealistic details increasingly interfere with a growing number of information layers.
- To express objects of city models in different states (e.g., existing, removed, and planned buildings), photorealism does not

offer a broad range of graphics styles to communicate these variations such as by sketchy and outlined drawings.
- To generate compact depictions for displays with minimal capacities (e.g., on mobile devices), photorealism frequently fails due to the visual complexity inherent to photorealistic images. For example, a scaled-down version of a digital photography has a drastically lower information level compared to a scaled-down version of a hand-drawn sketch of the same scenery.

These drawbacks and limitations are the main arguments for non-photorealistic rendering, which needs specific requirements of the virtual model. On one hand, geometry should follow a clear structure/outline for a most expressive cartoon rendering and on the other hand an object-oriented data structure splits up the element in smaller parts that are individually render and combinable. The aspects beyond VR, namely non-photorealistic rendering, provide graphical methods to enhance geospatial visualization. The following section describes semiotic enhancements by examples, which help to maximize the amount of information pixels on user interfaces and support viewport variations.

Exemplary Semiotic Enhancements

The variation of rendering techniques provides a powerful toolbox to adapt the information contents to specific needs. Additionally the influence of transmitting media's resolution leads to problematic areas on the interface that does not expressively support information transfer. These problematic areas occur by variation of viewport attributes.

The depiction of a virtual 3D environment on the transmitting interface/media is rendered according to viewport attributes. These attributes define a camera orientation, distance and field of view (FoV) in case of perspective views, and camera orientation and orthographic height in case of parallel projections. The amount of dead values changes according to these camera's attributes and to the transmitting interface's resolution. Dead values describe information pixels that are not appropriately used for information transfer. This means that the information of a pixel cannot be related to a specific element or the overall visualization contents. Due to a restricted amount of pixels for an information transfer, a most efficient use of these pixels is the aim.

Central perspective views combine linear perspective and multiple scales in one view, which leads on one hand to geometric distortions of the elements and on the other hand to a high variance of element sizes (on the presentation plane/viewing plane). Geometric distortions of elements assume that more than one element of the same sort exists in the view in order to retrieve geometric relations (e.g., size, primitives). The high variance of scales enables the direct comparison of large and very small scales in one directed view, but also leads to dead values if elements fall below the resolution of transmitting media. If this uniqueness-relation between transmitting media and content element is larger than one, a single picture element of transmitting media has to represent several pixels of the content.

Figure 7. The area of dead values in a central perspective view is marked red in the right picture. The picture on the left shows this "undefined" pixel area in detail

Parallel perspective views offer one single scale throughout the view, which allows a direct comparison of element sizes and orientation. Due to the isometric character of this view, the scene seems to arch upward in the background. Actually the geometric correct characters of parallel perspectives create discrepancies with the human cognitive system. Thus the main disadvantage is its disturbance of naïve perception. The main advantage lies in a scale dependent illustration.

In the case of standard perspectives we can conclude that their use of transmitting media's viewplane for virtual 3D environments is not perfect. Due to dead value areas for various camera parameters the requirements of cartography-oriented design are mainly not met. In order to reduce these areas, viewport variations show one possible approach.

Viewport Variations for Reducing Dead Pixel Values

A variation of viewport perspectives follows the aim to reduce dead value areas and thus improve the use of transmitting media's viewplane. This approach is guided by syntactic considerations in terms of expressive geospatial communication. Hence the following collection of perspective variations is evaluated by their impact on dead value areas, communication of spatial relations and its syntactic dimension.

The progressive approach intensifies perspective impression by forcing ground-view zones nearby the camera and front-view zones for far elements. This means that the perspective view becomes arranged depending on the distance to the camera. The grouping to "view angle" zones enables the specific element enhancement in that zone. For example, a ground-view generally delivers footprints that are combined with very restricted front-view information. In this situation it is important to keep the footprint information by its semantics, which will lead to highlighting and aggregation for specific scales. On the other hand, in a front-view zone the footprint almost disappears. The main transferred information is build up by the upright projection. Therefore, the front-view and its semantic/outline has to be highlighted for specific, orientation enabling elements.

The Progressive Central Perspective makes use of "view angle" zoning that allows for element adjustment according to the view angle of viewport camera. In addition its perspective impression becomes intensified. The enhancement of the ground-view in the foreground presents an overview around the current camera position, which mainly relies on buildings footprints and topographic relations. The force for front-views in the background enhances visual landmarks by their front-view. Highlighting important and well known elements directly supports the usage of landmarks for orientation within the virtual 3D environment. Spatial relations between the foreground's overview and the background's landmarks are strengthened, especially because standard perspectives will not provide the extension (on the transmitting interface) to show both.

Progressive Parallel Perspective uses the same constellation of "view angle" zoning and results in the same argumentation for using ground-view

Figure 8. The conceptual view of a progressive central perspective and progressive parallel perspective easily shows the importance of depth cues that are heavily supported in central perspectives. The front-view zones in the background enhance the depth cue of linear perspective

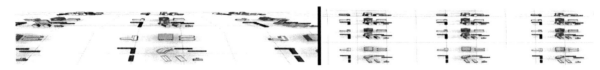

and front-view situations. Additionally its metric constitution, which disturbs a naïve perception, gets some "perspective" character. Thus the parallel perspective can use its metric and simulates perspective characteristics.

In contrast, the degressive approach with a Degressive Central Perspective generally destroys any naïve perspective perception by its use of front-view zones in the foreground and ground-view zones in the back. Its aim is to simulate user's perspective at the camera's position in order to generate high identification in the specific area and deliver an overview of the prospective tour at the same time. The request in this case is a high detailed model in the front and a scale-dependent map design in the back. The rendering style definitely has to change in order to reduce confusion when a map is used instead of clouds and sky.

All viewport variations have in common that dead values on the transmission interface become minimized. The reason is that "view-angle" zones reduce the infinite number of scales to several groups and therefore allow for specific element generalization. The information transmitting area theoretically becomes more appropriate used according to its purpose (generating an overall overview with naïve geography or supporting navigation in virtual 3D environments). As result of this reduction of dead values and a more extensive use of transmission interface, spatial relations are enhanced as well (because more information pixels are available for overview generation).

This section focused on enhancing expressiveness and effectiveness of user interfaces in geospatial communication. New semiotic considerations for virtual 3D environments showed up possible graphical conflicts that originate with the characteristics in 3D (light, perspective, …). Non-photorealistic rendering techniques described graphical accentuation, which goes beyond the simple reconstruction of reality (VR) in terms of communication. At least the exemplary appliance of all these new techniques lead to semiotic enhancements, which cause a reduction of dead pixel zones.

The main argument for the implementation of semiotic complexity lies in a better impact and usage of virtual 3D city models (in comparison to physical models). For this reason the next section shortly describes studies on the employment and usage of virtual 3D city models.

The Impact of Virtual 3D City Models

3D environments/models were established as additional decision support for planning purposes in the past. Whenever spatial decisions have to be discussed within a team of experts or the public, photorealistic renderings, movies, physical models or even virtual 3D environments translate an abstracted knowledge/information of a map to easier understandable presentation forms. But the usage of virtual 3D environments, which enables highly interactive contents and easy dissemination, is still demoted as side product of the planning process. One main reason is requirements on hard-, software and interaction, which often do not exist at business client's equipment. There is a strong demand on easy handling, intuitive interaction and usability on standard office computers. In addition to spatial communication, cartography-based design and the use of virtual 3D models, latest technological developments were a motivation to perform user studies on the employment of virtual models.

Studies on Employment of Virtual 3D City Models

In the course of rapid developments in computer-graphics-processing since the eighties, several studies and comparisons on the employment of virtual 3D environments, especially in the field of architecture were done. Several results were published during the second architecture conference in 1995 in Vienna: for example a direct comparison of CAD and physical models for urban development by courtesy of empirical studies was published by Keul (Keul et al 1995). This study based on slides, recorded data by semantic differential and included experts as well as laymen. It showed that the opinion about the CAD´s or physical model's usefulness depended on which slide had been shown at first. In one case the CAD model was preferred, in the other case the physical model's slides lead to preference. From this effect, the authors of this study concluded that the choice of communication media has impact on the result of the study.

In another study an objective comparison of an architectural endoscope model and CAD model was done by evaluating interviews (Siitonen 1995). Since both methods use a digital display as transmission media, these methods were comparable. Thus a comparison list according to various criteria could be established. The main criteria of this list covered expense of modelling, staging, lighting, animation, interface, image quality, dissemination, possibility of modification, overall costs, and architectural sensation. At least the architectural endoscope model won with 27 points instead of 20 points for the CAD model, although the authors remarked that CAD is catching up due to increasingly powerful personal computers.

For any judgement of previous studies and comparisons it has to be seen that enormous performance increases took place in terms of soft- and hardware. Completely new possibilities for real-time visualization were created. Although interactive models were used in these previous studies, their visual appearance was "reduced" and raw. These simple architectural models were not able to represent reality. Nowadays the requirements of landscape architecture with their detailed vegetation models can be considered and embedded in real-time applications, even though the complexity of vegetation calls for smart technologies. Many studies use depictions with various abstraction levels and compare these with pictures of reality (Lange 1999). Thus simple two-dimensional pictures are compared. Some studies focus on dynamic visualizations: for example video simulations were used to explore the impact of various abstraction levels of virtual 3D environments on the reaction of recipients compared to videos of physical models (Boerwinkel et al 1994).

In the following section a comparison between physical models and their virtual pendant was the focus of a study in order to quantify the practical suitability of real-time 3D environments for geospatial communication within a group of experts.

Comparing Virtual and Physical 3D Models

The following study focused on the practical suitability of real-time virtual 3D environments as comprehensive geospatial presentation method. Therefore two different environments were built. Both models were oriented towards their existing physical models with the main aim to allow for direct comparison within groups of experts.

The two different environments were initiated by the central problematic of scale: physical models precisely follow a specified scale. In opposition to this general valid scale for the whole physical model, virtual 3D environments generally do not follow one specific scale, but incorporate an infinite number of scales due to linear perspective

Figure 9. Example for a larger scale virtual building model with detailed objects and varying light situation

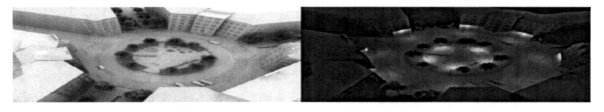

and camera movement. For this reason the first virtual 3D environment was designed according to the physical model in a scale from 1:100 to 1:200 and the second according to the physical model in a scale from 1:1000 to 1:2000. The first environment with the bigger scale allows for incorporating precise design elements, like trees, that are important for the planning process. Additionally, elements like human beings or cars were embedded to demonstrate sizes and proportions. A reference to a location could be supported by detailed building facades. The smaller scale comes along with draft ideas, abstract vegetation and buildings. Objects in this scale mainly serve for the transmission of masses and proportions. Models of persons or cars were generally omitted.

The impact of the virtual 3D environment on a group of experts was the main aim of this study, which used a normalized questionnaire. Participating experts came from the fields of regional planning, landscape planning, urban planning, architecture and civil engineering. The main parameters of the questionnaire were aiming at practical suitability of virtual 3D environments and its acceptance for participating experts. Thus interaction/navigation within the virtual 3D environment was one key factor in this survey. In the end a profile of polarity (according to Frohmann 1997) helped to interpret the results.

During the procedure of this study both models, the physical- and the virtual 3D environment, were presented simultaneously, whereas the virtual model became projected to a wall. After an explanation of main characteristics and differences, the navigation interface within the virtual environment was explained. Then the participants explored the model and filled out the questionnaire. A final discussion helped to collect further comments and ideas: 13% of participants stated their novice experience, which means that they have never used a virtual 3D application before; 57% of the participants stated their active usage of virtual 3D environments for their planning issues.

Results of the Virtual- And Physical Model Comparison

The results of this study concerned immersion, aesthetical experience, navigation and additional information yield. These factors are still main arguments against virtual 3D environments, whereas production cost and model flexibility are main advantages. For this section most of the results, which were explored for the two environments at various scales with simultaneous comparison between virtual and physical model, were subsumed to one single statement.

The grade of immersion was very high for about 48% of participants. For the smaller scale the effect of immersion was higher than for the bigger scale. Immersion describes a transmission of reality in terms of naïve experience and the reference of virtual graphics to the real-world. Men seem to imbibe this effect more than women. They rate the grade of immersion higher.

The aesthetical experience followed the results of immersion for some extend. The aesthetical impact of the smaller scale environment was higher, whereas the experience within the bigger scale was higher (~54%). Aesthetical experience

Figure 10. A comparison between reality and detailed virtual 3D city model, which is using non-photorealistic rendering, lightmaps and hybrid trees

describes the urgent impression of a planning situation in order to understand the design of lighting, infrastructure and landscape/buildings. Again this parameter showed that female participants had problems to get an immersive spatial impression of the model. Hence a usage of virtual 3D environments instead of physical models was imaginable for only 50% of the participants. The rest would use virtual 3D environments as additional presentation. No one was against the usage of virtual 3D environments.

Navigation within the virtual 3D environment was a key factor for this study. Navigation issues have the potential to destroy an immersive spatial impression and aesthetical experience by steering participants cognitive processing mainly to the interface for navigation. Therefore the navigation within these two virtual 3D environments of the study was kept as simple as possible for standard interaction interfaces, the mouse and keyboard. The mouse handled all the rotations of the camera, like the movement of the head. As it could be expected for various learning-types, not all the participants were confident with an interactive camera movement. 79% of all the participants preferred an additional automatic camera movement along a path (like a movie). Only 17% were fully satisfied with the interactive camera movement.

At least there remains the question for additional information yield. Is the virtual 3D environment possible to enhance knowledge transfer? Only 13% of participants experienced no new impressions by means of the virtual 3D application. 87% were able to gain new knowledge and further impressions by virtual pedestrian perspectives. The most important of these additional impressions were the relation of dimensions, scales, spatial depth, visual axes, relations of sight, lighting situations (shadowing of trees or buildings).

The study on virtual 3D environments enlightened on important factors for the usage of virtual 3D environments. The demographic structure of participants was not considered due to its possible empirical influence on the key factors for the usage of virtual 3D environments. If computer scientists are very confident with the navigation of a virtual 3D environment, these stakeholders often forget that most of the others are not used to deal with these interaction mechanisms. Similarly, the importance of cartography-oriented design may be overvalued. Maybe other parameters than graphical design aspects are more important.

Figure 11. A screenshot of the smaller scale 3d city model with abstracted trees and ambient occlusion light map

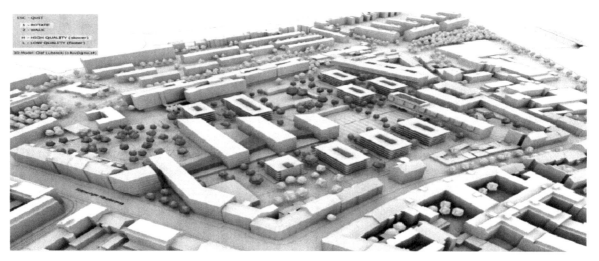

Further studies will have a closer look on user communities and try to embed their needs. This section easily showed that use and usability within virtual 3D environments is indeed a key factor for an effective dissemination.

It could also have been shown that virtual 3D environments provide substantial potential for innovative geospatial communication. The main assumption to develop this potential is appropriate use of the semiotic toolbox as well as optimising the usage of the view plane. The semiotic toolbox bases on the semiotic structure for 3D, which highlights mutual impact of design components. The optimal usage of the view plane focuses on a limited pixel amount with digital displays, which can be used for information transmission. At least every pixel on the view plane plays its role for information transmission. These aspects are concentrating on geomedia techniques, which is almost independent from data models and virtual 3D city creation.

CONCLUSION

The aim of this chapter was to show the importance of virtual 3D city models by focusing on interaction, contents coding/description, user-interfaces for geospatial communication, production of virtual 3D city models, and on the needs of spatial e-planning. The success of geospatial applications depends on the mutual reaction of interaction, content description and appropriate use of user-interfaces. Therefore, cartographic communication concepts, which focus on expressive and effective transmission of spatial information, should be incorporated in geovisualization technologies, as the structures of human learn-types and complexity of geospatial information can show. These approaches allow for specific enhancements within the depiction that particularly support mental imagery and unambiguous decoding of embedded information.

Additionally the acceptance of virtual 3D city models had to be discussed and studied, especially because physical 3D city models have been established in planning processes. At least all the theoretical and technological investigations lead to an overall impact of virtual 3D city models, which is a prospective issue to mental imagery research in cartography. For instance, the precise needs of planning experts in terms of lighting, relations of masses or sight, may be too detailed for a wide public. Sometimes very few, but expressive information is needed for the spatial communication process. Further investigations will help to gain more insight in an effective use of this interdisciplinary combination.

REFERENCES

Bertin, J. (1982). *Graphische Darstellungen – Graphische Verarbeitung von Informationen, übersetzt und bearbeitet von Wolfgang Scharfe*. Berlin: Walter de Gruyter.

Blumstengel, A. (1998). *Entwicklung hypermedialer Lernsysteme* (Dissertation). Berlin: Wissenschaftlicher Verlag.

Boerwinkel, H. W., & Jansen, W. J. A. (1994). CAD-Modelle on line bewertet. In: Garten + Landschaft 10/1994

Brunner, K. (2001). Kartengestaltung für elektronische Bildanzeigen. *Kartographische Bausteine, 19*.

Brunner-Friedrich, B. (2004). *InMuKIS – Konzept eines benutzergruppenangepassten interaktiven multimedialen kartographischen Informationssystems für die Schule zur Präsentation raumbezogener Informationen*. Dissertation; Technische Universität Wien.

Buzin, R. (2001). *Multimedia-Kartographie – Eine Untersuchung zur Nutzer-Orientierung kartomedialer Atlanten*. Diss., TU Dresden, Der Andere Verlag, Osnabrück.

Colditz, C., Coconu, L., Deussen, O., & Hege, H. C. (2005). Real-Time Rendering of Complex Photorealistic Landscapes Using Hybrid Level-of-Detail Approaches. In Trends in Real-Time Landscape Visualisation and Participation – Proceedings at Anhalt University of Applied Sciences. Heidelberg, Germany: Herbert Wichmann Verlag.

Döllner, J., Buchholz, H., Nienhaus, M., & Kirsch, K. (2005). Illustrative Visualization of 3D City Models. In *Proceedings of SPIE – Visualization and Data Analysis 2005 (VDA)*, San Jose, CA, USA (pp. 42-51).

Edwardes, A. J., & Purves, R. S. (2007). A theoretical grounding for semantic descriptions of place. M. Ware & G. Taylor (Eds.), *Proceedings of 7th Intl. Workshop on Web and Wireless GIS, W2GIS 2007* (pp. 106-120). Berlin: Springer.

Frohmann, E. (1997). *Gestaltqualitäten in Landschaft und Freiraum*. Wien, Austria: Österreichischer Kunst und Kulturverlag.

Gaerling, T., Golledge, R. G., et al. (Eds.). (1993). Behavior and Environment: Psychological and Geographical Approaches (Advances in Psychology Vol. 96). Amsterdam: North-Holland.

Glander, T., & Döllner, J. (2007). Cell-Based Generalization of 3D Building Groups with Outlier Management. In ACMGIS 2007. New York: ACM.

Gooch, A., & Gooch, B. (2001). *Non-Photorealistic Rendering*. Natick, MA: AK Peters Ltd.

Hake, G., Grünreich, D., & Meng, L. (2002). *Kartographie – Visualisierung raum-zeitlicher Information*. Berlin: Walter de Gruyter.

(1999). Human Wayfinding and Cognitive Maps. In Golledge, R. (Ed.), *Wayfinding Behavior* (pp. 5–46). Baltimore: The John Hopkins University Press.

Ibrahim, K. F. (2007). *Newnes Guide to Television and Video Technology, The Guide for the Digital Age – from HDTV, DVD and flatscreen technologies to Multimedia Broadcasting, Mobile TV and Blue Ray* (4th ed.). Burlington, MA: Newnes.

Jiang, B., & Li, Z. (2005). Geovisualization: Design, Enhanced Visual Tools and Applications. *The Cartographic Journal, 42*(1), 3–4. doi:10.1179/000870405X52702

Jonassen, D. H., & Grabowski, B. L. (1993). *Handbook of Individual Differences; Learning and Instruction*. Hillsdale, NY: Lawrence Erlbaum Associates.

Keller, P. R., & Keller, M. M. (1993). *Visual Cues: Practical Data Visualization*. Los Alamitos, CA: IEEE Computer Society Press.

Keul, A. G., & Martens, B. (1995). Simulation – How Does it Shape the Message? In *The Future of Endoscopy – Proceedings of the 2nd European Architectural Endoscopy Association Conference*. Wien, Austria: Technische Universität Wien, ISIS-ISIS Publications.

Kraus, K. (2007). *Photogrammetry - Geometry from Images and Laser Scans* (2nd ed.). Berlin: Walter de Gruyter.

Lange, E. (1999). *Realität und computergestützte visuelle Simulation*. Vdf Zürich.

MacEachren, A. M. (1995). *How Maps Work: Representation, Visualization and Design*. New York: Guilford Press.

Mackinlay, J. (1986). Automating the Design of Graphical Presentations of Relational Information. *ACM Transactions on Graphics, 5*(2), 111–141. doi:10.1145/22949.22950

Malić, B. (1998). *Physiologische und technische Aspekte kartographischer Bildschirmvisualisierung*. Dissertation; Schriftenreihe des Instituts für Kartographie und Topographie der Rheinischen Friedrich-Wilhelms-Universität Bonn.

Marcinkowski, F. (2007). Media system and political communication. In Klöti, U., Knoepfel, P., Kriesi, H., Linder, W., Papadopoulos, Y., & Sciarini, P. (Eds.), *Handbook of Swiss politics* (2nd ed., pp. 381–402). Zürich, Deutschland: NZZ Publishing.

Mayer, H. (2000). *Einführung in die Wahrnehmungs-, Lern- und Werbe-Psychologie*. Oldenburg, Deutschland: Oldenburg Wissenschaftsverlag.

Neisser, U. (1996). *Kognition und Wirklichkeit*. Stuttgart, Deutschland: Klett-Cotta.

Neudeck, S. (2001). Zur Gestaltung topografischer Karten für die Bildschirmvisualisierung; Dissertation; Fakultät für Bauingenieur- und Vermessungswesen. In *Schriftenreihe des Studiengangs Geodäsie und Geoinformation* (Vol. 74). Neubiberg: Universität der Bundeswehr München.

Paar, P., & Rekittke, J. (2005). Lenné3D - Walkthrough Visualization of Planned Landscapes. *Visualization in landscape and environmental planning*, 152-162.

Paivio, A. (1986). *Mental representations: a dual coding approach*. Oxford, UK: Oxford University Press.

Pfeifer, N. (2002). 3D Terrain Models on the Basis of a Triangulation. In Geowissenschaftliche Mitteilungen (Vol. 65).

Piaget, J. (2003). *Meine Theorie der geistigen Entwicklung* (Fatke, R., Ed.). Berlin: Beltz Verlag.

Prusinkiewicz, P., & Lindenmayer, A. (1990). *The Algorithmic Beauty of Plants*. New York: Springer Verlag.

Robinson, A., & Petchenik, B. (1976). *The Nature of Maps*. Chicago: The University of Chicago Press.

Siitonen, P. (1995). Future of Endoscopy. In *The Future of Endoscopy – Proceedings of the 2nd European Architectural Endoscopy Association Conference*, Technische Universität Wien, ISIS-ISIS Publications.

Spiess, E., Baumgartner, U., Arn, S. & Vez, C. (2002). Topographsiche Karten, Kartengraphik und Generalisierung. *Schweizerische Gesellschaft für Kartographie, Kartographische Publikationsreihe*, 16.

Strothotte, T., & Schlechtweg, S. (2002). *Non-Photorealistic Computer Graphics: Modeling, Rendering and Animation*. San Francisco: Morgan Kaufman.

Zimbardo, P. G. (1997). *Psychologie (5. Aufl.)*. Heidelberg, Germany: Springer Verlag.

ADDITIONAL READING

Dykes, J., MacEachren, A. M., & Kraak, M.-J. (Eds.). (2005). *Exploring Geovisualization*. Amsterdam: Elsevier.

Kraak, M.-J., & Ormeling, F. (2002). *Cartography: Visualization of Spatial Data*. Upper Saddle River, NJ: Prentice Hall.

MacEachren, A. M. (1994). *Some Truth with Maps: A Primer on Symbolization & Design*. University Park. PA: The Pennsylvania State University.

Meister, D. (1999). *The History of Human Factors and Ergonomics*. Mahwah, NJ: Lawrence Erlbaum Associates.

Norman, D. A. (2002). *The Design of Everyday Things*. New York: Basic Books.

Peterson, M. P. (1995). *Interactive and Animated Cartography*. Upper Saddle River, NJ: Prentice Hall.

Shneiderman, B. (1980). *Software Psychology*. New York: Little, Brown and Co.

Slocum, T. (2003). *Thematic Cartography and Geographic Visualization*. Upper Saddle River, NJ: Prentice Hall.

Taylor, D. R. F. (Ed.). (2005). *Cybercartography: Theory and Practice*. Amsterdam: Elsevier Science.

Wickens, C. D., Lee, J. D., Liu, Y., & Gorden-Becker, S. E. (1997). *An Introduction to Human Factors Engineering* (2nd ed.). Upper Saddle River, NJ: Prentice Hall.

Wood, D. (1992). *The Power of Maps*. London: The Guilford Press.

KEY TERMS AND DEFINITIONS

Geomedia Techniques: Techniques that take account of transmitting media and user interface characteristics in order to adapt information processing and enhance the communication process.

Geospatial Communication: describes the transmission of geospatial knowledge from one person to another.

Geovisualization: stands short for Geographic Visualization. The notion refers to a set of tools and techniques supporting geospatial communication and information analysis through the use of interactive maps.

Semiotics: focuses on sign processes in order to understand the impact and usage of signs and symbols.

Smart Approaches: Automation of modelling and creation processes that result in a virtual 3D city model.

Virtual 3D City Models: Digitally reconstructed cities in virtual space, which enable virtual movement and analysis in three dimensions.

Chapter 8
Political Power, Governance, and E-Planning

Kheir Al-Kodmany
University of Illinois at Chicago, USA

ABSTRACT

Spatial information is a crucial cornerstone of e-planning. This paper explains the process of constructing a mega geospatial database for the Hajj, the annual Muslim Pilgrimage to Makkah, Saudi Arabia. It discusses the complex process and influence of top-down political power on the comprehensive planning process for the Hajj. It specifically examines how the process led by a politically powerful agency impacted the creation and adoption of a mega geospatial database. The chapter provides transferable and useful lessons on GIS misconceptions and solutions, as well as insight on how and when political power may help in advancing the planning process.

INTRODUCTION

Since the beginning of this millennium, there has been an ever-increasing volume of literature on e-planning. An e-planning model presented by Budthimedhe, et. al, 2002, may help in contextualizing this case study. The presented e-planning model divides activities into three categories: debates, technologies, and applications. It further breaks technologies into three categories: geospatial data, geospatial analysis, and technology interface. Budthimedhe, et. al argues that given that urban planning is essentially a spatial problem, spatial data is vitally important for e-planning. This chapter focuses on the geospatial database needed for planning purposes, specifically for Hajj.

Although there is an extensive research on e-planning and geographic information systems, there is little research that documents the impact of politics and power on the process of creating large geospatial database (Obermeyer and Pinto, 2007). Through a detailed case study, in a storytelling format, this paper attempts to examine how political power can dramatically steer the planning process and profoundly influences the outcomes. It examines the positive and negative impact that

DOI: 10.4018/978-1-61520-929-3.ch008

a politically powerful, government-sanctioned agency can have on leading the implementation of a GIS database, in terms of quality and quantity of data, participation by key stakeholders, and delivery timeframes. It also observes how political leaders, despite their immense resources, had to seek compromised solutions in pursuing GIS. The paper provides insights on common GIS misconceptions and intricacies about data, software, hardware, basemaps, layers, and GIS expertise. It also provides useful lessons on problems and a solution on the way for building GIS. These lessons are particularly applicable to projects concerned with building large geospatial database.

The case study presented in this chapter is unique for several reasons. First, it deals with a very complex planning problem. Hajj is a multi-faceted event that requires considerable planning by more than 20 government agencies to safely accommodate and transport three million pilgrims during a week's time. It encompasses a large region (68,000 km2) with multiple activities happening at multiple sites at a given time. Hajj involves a series of rituals conducted at various Holy Sites on different days and specific times with a slightly varied order. Therefore, planning for Hajj must take into account a complex spatial-temporal-ritual phenomenon that involves moving of large masses of people and their services at multiple sites at different times.

Second, the case study deals with an intricate political context. There are more than 20 agencies that are closely or remotely involved in planning, preparing, and running Hajj. The audience of Hajj is immense as it concerns almost the entire Muslim population for Hajj is obligatory on every financially and physical capable Muslim. Collectively, this would make the process of planning far more complex than a project with fewer stakeholders. The involved agencies have varying levels of political "weight" that must be observed and respected to ensure the smoothness of the process.

Third, Hajj is of paramount importance to the Saudi government for religious, political, and economic reasons. The government of Saudi Arabia takes great pride in its role as the sole host of the Hajj. Playing host to all Muslims from around the globe confers certain legitimacy upon the government. In response, the government makes every possible effort to ensure the ease of Hajj and the safety of pilgrims. The Custodian of the Two Holy Mosques, HM[1] King Abdullah bin Abdulaziz, and his brother, the former king, HM King Fahd bin Abdulaziz, have taken a keen interest in the affairs of Hajj and matters that affect the two holiest cities of Islam, Makkah al Mukarramah (Makkah the dignified) and Madinah al Munawwarah (Madinah the illuminated). In recent years, the government has invested heavily in research and planning to solve Hajj problems. Finally, the research presented in this chapter is based on personal experience. The author of this paper worked on the project for two years in Saudi Arabia (2004-2006).

This chapter first sets the overall context of Hajj and its planning activities. It explains the newly formed organization charged in planning for Hajj and speaks to the initial planning stages that involve creating a mega geospatial database. Then, it details out the process of building the database. The paper dwells on the obstacles on the road and the role of political power in easing the problems. The discussion section attempts to extract meaningful lessons. The conclusion summarizes the case study and provides a future direction for research.

BACKGROUND

Hajj and Planning

Hajj is the fifth pillar of Islam. The pilgrimage to Makkah is an obligation that must be carried out at least once in the lifetime of every physically-able,

Figure 1. Hajj prime geographic area (Source: Google Earth)

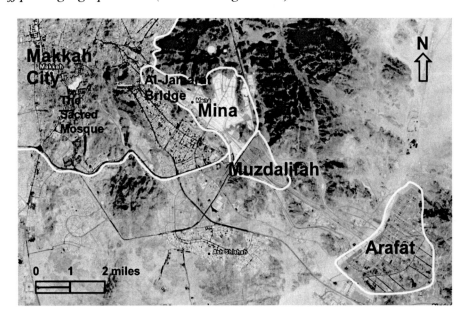

financially-capable Muslim. It is the largest annual pilgrimage in the world, with an estimated 3 million people participating in 2008. The geographic area involved in the Hajj includes the Makkah region (Makkah City and three Holy Sites: Mina, Muzdalifa, and Arafat) and Madina, as well as numerous entry points throughout Saudi Arabia (Al-Qahtaani, 2002; Long, 1981).

Hajj follows the lunar calendar and happens on the twelfth month, known as *Zil-Hijjah*. The actual days of Hajj extend from the 8th through the 13th. Although there are slight variations in the rituals based on the sub-type of Hajj, the common rituals are as follows:

- Day One: Pilgrims visit the Sacred Mosque in Makkah and then go to Mina, about 3 miles southeast of Makkah, and stay overnight.
- Day Two: In the morning, the pilgrims depart for Arafat, about 7 miles southeast of Mina. At sunset, they make their way back and stay overnight in Muzdalifah, about 2 miles southeast of Mina.
- Day Three: In the morning, the pilgrims go to the Stoning of the Devil Site, conduct rituals, and then rest in Mina.
- Days Four through Six: The pilgrims revisit the Stoning of the Devil Site. They conclude Hajj by paying a farewell visit to the Sacred Mosque in Makkah (Al-Qahtaani, 2002), (Figure 1).

In the last half century the number of pilgrims has increased exponentially, primarily due to the availability and affordability of transportation to Makkah. The number of pilgrims has increased from about 200,000 in the early 1950s to more than three million in recent years. Such an increase has caused extreme overcrowding, which has resulted in traffic congestion, shortages of services, and tragic incidents, such as deadly stampedes in 2004 and 2006 that left hundreds dead and hundreds more injured (Still, 2008).

Such tragedies have compelled the Saudi government to undertake massive planning efforts. H.M. King Fahd bin Abdul Aziz has implemented several planning initiatives, including the establishment of a new oversight planning organization, named the "Development Commission of Makkah, Madinah, and Mashair," (DCMMM). The DCMMM's role is to facilitate collaborative planning among all governmental organizations that are involved in Hajj. The prime mission, as stated by the former King Fahd, is to ensure ease in performing Hajj, safety, and security of all pilgrims. The King authorized the agency to supervise all Hajj activities. In addition, the agency has the responsibility of setting strategic plans (long-term and short-term) for cities involved in Hajj (Makkah and Madina) and to ensure the well-being and prosperity of these cities. The agency enjoyed authority, power, and discretion in forming planning teams and in getting national and international consultations.

The budget of this project was anticipated as "unlimited" for several reasons. The project is very important to the country and to the Muslim world and the government has been extremely generous on all Hajj related activities. The organization was formed by an order from the king and its leaders have direct connection to top officials of the country and enjoy tremendous power. However, there were various incidences as presented in the following sections that explain situations where budget considerations were of major concern in making decisions and contracting work.

The leader of the newly formed DCMMM is a political appointee, handpicked by HM King Fahd. This individual, who has the title of "His Excellency" (H.E.), is well respected and highly admired throughout Saudi Arabia. He had previously served in several important government positions including Minister of Transportation, earning him a reputation as an effective administrator. Moreover, he has an engaging and charming personality, helping him to get things done in spite of political obstacles. Perhaps most importantly, the new head of the DCMMM has the full confidence and political backing of HM King Fahd.

For the aforementioned reasons, it was expected that H.E. would be able to effectively manage the input of all of the participating organizations to quickly produce the desired outcome that is easing Hajj problems. Hajj problems have been immense and planning for Hajj has been challenging for it deals with a wide spectrum of interconnected issues. It involves transporting, housing, feeding, and ensuring the safety, security, and health of millions of participants. What is more, in the past two decades, there have been series of tragic incidences that resulted in the death of hundreds and the injuries of thousands individuals. Such tragic incidences have marred Hajj and the newly formed organization is looked at as the problem-solver for Hajj.

The head of the DCMMM quickly gathered his scientific advisory cabinet and assembled an international planning team. The team is comprised of planners of various specialities including urban design, community development, economic development, environment, housing, transportation energy and natural resources, and engineers specializing in transportation, traffic, and infrastructure (electricity, water, sewer, etc). It also has remote sensing and GIS experts with various sub-specializations in spatial analysis, modelling, geospatial data management, and data collection. Finally, it consists of computer scientists, mathematicians, physicists, religious scholars, and lawyers. Consultants and experts for specific advice on technical issues frequently assist this interdisciplinary team.

The team quickly reviewed e-planning tools. As a team member, the author of this chapter presented his work on e-planning tools. He has developed collaborative tools that facilitated e-planning between the University of Illinois at Chicago, and neighboring communities such as the Pilsen and North Lawndale communities. He also used GIS, virtual reality, linear and nodal movies, and still and animated photographs to engage participants in the planning process. He also utilized the Web and argued that it is an efficient medium for organizing and displaying immense visual images, maps, and textual information (Al-Kodmany, 2000a, 2000b, 2001, and 2002)

After extensive reviews of e-planning tools, the head of the DDMMM completely believed that GIS is the proper tool to plan for Hajj. GIS effectively integrates vast amounts of planning data such as transportation, housing, economic, geology, infrastructure, and the like. Much research and literature suggest that GIS is a powerful planning and decision-making tool. In addition, GIS has powerful visualization capabilities that would effectively present massive information in relatively easy to understand graphical formats. Furthermore, other e-tools and applications such as virtual reality can be integrated with GIS databases.

The planning team also agreed that GIS is a powerful e-planning tool. They viewed that geography is central to planning and e-planning makes geographic data available to all participants through a computer and the Internet. The team viewed that building a GIS is a prerequisite for e-planning. It is particularly an appropriate vehicle since there were more than twenty organizations involved in planning. An online GIS would be a suitable planning tool. Online GIS is increasingly possible and supported by several commercial software such as ArcIMS and ArcSDE.

New technological features provided by online GIS technology are making it a further attractive e-planning tool. For example, new features allow the planning team to create integrated documents, i.e. integrating text with intelligent and interactive maps and map layers. Interactive documents link specific sections of text to specific features on maps and allow users to click on map features to view relevant text. Participants can also select specific document text to identify the associated features and can comment on these documents online. Additional features would allow conducting comment analysis to sort, track, and prepare responses on draft plans. Collectively, as envisioned by the planning team and the head of the organization, e-planning will establish a new mechanism for planning the Hajj that would allow an openly participative and collaborative environment. The first step for e-planning is to create a comprehensive geospatial database of the Hajj geography.

Building the Geospatial Database

There is no clear road map to build the geospatial database for several reasons. First, the organization is newly formed and has no previous experience in building geospatial database. Second, the project is large and complex; it is one of the largest geospatial database undertaken in the country. Third, there is little precedent and examples that could help in guiding the vision and the process because Hajj is unique and different from other large scale events. For example, the Maha Kumbh Mela in Allahabad, which happens every 12 years, has a different agenda of activities and rituals. Literature shows that there is little GIS work done to help in this event. Furthermore, the Olympics have a different set of activities and it takes places in various geographic locations. As such, the Olympics experiences were viewed of little relevance. Overall, the pressure on the organization contributed to rushing the process and not taking the time to do adequate ground research and planning for constructing the geospatial database.

Overview

The geospatial database had several prime components including:

A. The three cities that are involved in Hajj are Makkah, Madina, and Jeddah. This component had to reflect the status of these cities (in terms of demand on city services) at each season of 1) Hajj, 2) Ramadan, and 3) Umrah. It also needed to reflect these cities beyond these seasons. Therefore, it needed to represent the status of these cities at four distinct times, in three religious-intensive

seasons and out of these seasons. This component had extensive attribute data about the cities' infrastructure and services.
B. Three Holy Sites: Mina, Muzdalifah, and Arafat. These sites are not used beyond a few specific days during Hajj.
C. Entry-points to Saudi Arabia through land, air, and sea with their road networks connecting to Hajj geographic region.

The database also needed to contain a spatial-temporal component of the Hajj detailing the sequence of Hajj activities in relation to their geographic areas. Core activities are spread over a week, while ancillary activities are spread over two weeks. By relating spaces and activities to their specific times, there was a need to devise a Spatial Temporal Model that fit into the overall geospatial database scheme. Furthermore, topology of the database had to consider the various applications, e.g., for the transportation applications, the link and node structure had to be accommodated. Also, the database had to allow for continuous updates and annual archiving of Hajj.

The Process of Building the Geospatial Database

More than 20 organizations are closely or remotely involved in planning and overseeing the Hajj each year, with some significant overlap in tasks and responsibilities. The formation of the politically powerful DCMMM was an attempt by the Saudi government to coordinate these disparate agencies in planning this significant event, with the first task being the implementation of a GIS.

One of the first challenges that emerged in the implementation of the GIS was dealing with the considerable disparities in knowledge of GIS. While some organizations were more technically sophisticated and possessed invaluable GIS knowledge and data, others were in the beginning stages. Still others had no clue about GIS and had no intention of adopting the system. Organizations had to agree on who would be responsible to update the database, which would have rights to modify the database, and who would have rights to view it and to retrieve data.

In addition, there was a lack of consensus on the type of GIS that was to be adopted. The process went through several interrelated stages for finding a GIS solution: "Departmental GIS," "Enterprise GIS," "Quick and Dirty GIS," "The Best GIS," and "Best Possible GIS."

1. Departmental GIS

In the first stage, decision makers wanted to create a gigantic centralized departmental GIS to be hosted in a central location. The DCMMM pursued this goal with the intention of serving as the host for database and sharing it with its departments. However, the GIS data was fragmented and scattered across various agencies and institutions with some level of undefined redundancies. The DCMMM had to request data from the various agencies for building its geospatial database.

The DCMMM sent letters requesting all agencies involved in Hajj to provide all types of data they possessed about Hajj. The letters explained the new role of the DCMMM, the intended and symbiotic nature of the project, and the importance of supplying the data. The request was ambitious in terms of asking for all data of different types and formats including maps, attribute and descriptive data, both analogue and digital. The request also pointed out to the need to provide the data quickly for the time sensitivity of the project and its importance to the government of Saudi Arabia.

The responses varied tremendously in terms of speed and supplied materials. While some agencies responded immediately, others lagged behind, and others did not respond at all. Follow-up letters were sent, resulting in some additional responses. Some agencies still did not respond after repeated requests. The delivered materials and data were good but had some problems. Data arrived in different formats, and some were outdated. The lack of consistency could be blamed to some extent on the request letter, as it was not specific enough regarding the type of data requested. The letter used generic language and was open for discretionary interpretation.

The DCMMM realized through this process that the data was of great value to these agencies, and they were reluctant to give it up. Therefore requesting data through letters needed a follow up for agencies considered data as sources of great power and importance. Agencies considered data as "privately" owned as they spent much time and effort to create them. Data are crucial to conduct their research, so data are the backbone of the results of their research. Giving out the data would expose any errors and raise questions about the validity of the agencies' research. If the data were complete and accurate, then giving such data for "free" was inconceivable. For these reasons, many of the delivered data were not in the best shape; some were disorganized, meaningless, or outdated. Therefore, the DCMMM started to realize gradually that their political power was no match for the power inherent in the data. Some concluded that the data was the *"real power."*

Consequently, this direction, building a mega centralized GIS database, received considerable resistance from all of the other organizations. They viewed such a project as a threat to their power in the Hajj planning process. Giving data as requested would empower the new and rival agency, the DCMMM. Since some agencies did not send data, and others sent outdated data or disordered data, the outcome was disappointing. It became clear that building a mega-centralized GIS database was infeasible.

2. Enterprise GIS

Decision makers then resorted to a decentralized solution. A decentralized enterprise GIS solution was practical since data resided in various organizations, departments, and locations. It was also an attractive solution to other agencies since they would be partners in this project. In contrast to the first approach where agencies were to "give" data, under the second approach agencies were to "share" data. Each agency would benefit from the project while maintaining its sovereignty.

In order to avoid the lack of specificity about the requested data that happened in the first stage, the DCMMM designed a detailed questionnaire for data acquisition. The lengthy questionnaire, about 60 pages long, was intended to be clear in terms of the needed data. It also addressed the technical issues of the data in great depth. For example, it asked agencies to provide metadata for each GIS file they possessed. Such metadata included dates of production and modifications, names of producer, nature of modifications, map projections, topology, number and names of layers, and the like.

Overall, responses to the questionnaire were slow for several reasons. First, the questionnaire was too long. Second, it contained technical sections that only specialized staff could understand and intelligently answer. Finally, knowing that the data would be shared with many other organizations, security issues and concerns arose. For example, the Saudi Geological Survey was reluctant to provide their GIS data about the Hajj region for they considered them to be sensitive information.

Therefore the decentralized enterprise solution also faced obstacles and serious technical issues. In addition, the lingering mistrust from the first phase resulted in a lack of cooperation in the second phase. Altogether, these issues impeded the implementation of a decentralized GIS. Participation was sometimes tense. Agencies were sceptical about the real intention of the project, resulting in a reluctance to actually share their data. Schisms deepened and the process slowed down.

3. Quick and Dirty GIS

In the third stage, decision makers set about creating a stand alone "quick and dirty" GIS that would be hosted, shared, and utilized only by the DCMMM. The rationale was shaped by the need to speed the process and use the GIS for planning for the next event. Decision makers realized that obtaining a comprehensive GIS would take far longer time than was expected. Therefore, the desperate nature of the project made a "quick and dirty solution attractive and fitting to the situation.

However, the planning team soon advised that a "quick and dirty" GIS solution would be a disservice. It contradicted with the propagated purpose of the GIS as an ideal collaborative planning tool for Hajj. Since Hajj is an extremely complicated planning problem, a weak database would hurt rather than serve the planning process. In e-planning, a strong database is the backbone of all planning. A thin database would not help produce trustworthy planning applications. Furthermore, producing a poor GIS product would hurt the DCMMM's image and credibility, given that other Hajj organizations had built sophisticated GIS databases.

4. Best GIS

As a reaction to the "quick and dirty" solution, and as a result of the DCMMM becoming bigger and stronger, there was a notion to build the "best GIS" and house it in the DCMMM for good. This solution matched the high expectations of the agency as being equipped with "state-of-the-art" knowledge and expertise.

However, the data contained at the DCMMM, which were given by other agencies, were outdated. For example, many of the basemaps were built based on outdated aerial photographs. In other cases, basemaps were built based on low-resolution satellite imageries. In still other cases, basemaps were built based on low-resolution and outdated imageries.

In response to these problems, the DCMMM decided to build up-to-date basemaps that would cover the entire region. Consultant agencies were contacted to bid on a project that involved flying over the region and then building high resolution and detailed GIS basemaps of topography, vegetation, geology, urban infrastructure and utilities, transportation, and parcels. Unfortunately, the bids received were too expensive. Providing such maps at high resolution was extremely laborious and would require expensive technology.

Therefore, the "best GIS" solution needed operational and technical definitions. The definitions should address questions such as: how precise do we want our maps to be versus how much can we afford? How much precision do we *really* need? When is high precision a must and when it is unnecessary? What regions need to be at what best resolution? For example, it was determined that urban regions should have a higher resolution. The question then became: what is the best urban resolution? What is the affordable resolution? The same applied to the city core where density is higher, and the same applied to rural areas where density is lower.

Consequently, the "best GIS" solution was theoretically a great idea, but when it came to implementation there were serious hurdles. The "best GIS" solution was prohibitively costly and lacked clear operational and technical definitions. This phase of the project paved the way to a final phase, explained in the next section of the chapter, a "best possible" solution that resided in between the "quick and dirty" and the "best GIS."

Best Possible GIS

This approach manifested a negotiation approach despite the strong political will and power in pursuing the best GIS. The following section details out the negotiation of technical and political issues in the process of building the geospatial database.

A. *Complete Data Sets.* Planners often advocate the need for considering all scenarios in planning. They look for data completeness particularly in modern planning, a data-driven planning (Hoch, 2007). In the case of Hajj, learning about the importance of data to GIS, decision makers requested acquiring complete data sets about all planning topics and subtopics. Being fascinated with the power of GIS and the importance of data to GIS, decision makers wanted to obtain complete data.

To boldly express their enthusiasm on gathering all data, they often repeated the following phrase: "we want every data concerning every aspect of Hajj and the cities involved; above ground and underground."

In reality, it was impossible to obtain complete data for the five cities involved (Makkah, Madinah, Mina, Muzdalifah, and Arafat), and the data of Hajj. In addition, getting too concerned with data would shift the organization's focus from planning to data gathering, or becoming a data warehouse organization. Furthermore, too much data would be difficult to handle at technical, managerial, and financial levels. It could become more of a burden than power. The process ended by taking a practical approach of seeking what was possible to acquire.

A complete GIS data set is difficult to acquire even in the most advanced countries. This is true particularly when it comes to data concerning details. For example, in the U.S. forest management agencies are likely to have extensive data regarding growth rates and distribution of commercial trees species. They are less likely to have readily available data regarding distribution and growth of non-commercial species such as shrubs, grasses and bracken, which can also form important elements of the visual landscape (Williams et al., 2007).

Ideally, comprehensive planning means complete knowledge that would help illustrate causes and consequences of uncertainty, knowledge that can be used in plans to reduce uncertainty. Instead of using comprehensive to means complete, the process resorted to a practical sense to describe a richer and more meaningful grasp of unfamiliar relationships in terms of more familiar ones. So when considering comprehensive plans, it was not necessary to insist on including a complete understanding of all elements of the region, but rather to describe important but unfamiliar and meaningful relationships. Therefore, the process had to seek a practical approach and to compromise a complete and perfect database for one that was feasible.

B. *Data for Specific Applications*. At some point in the discourse, a few specific GIS applications were proposed including: 1) Establishing an Operational GIS, 2) Developing Environmental Applications; 3) Developing Location-Based Applications; and 4) Developing Real-Time Navigation and Vehicle Tracking.

Tailoring the data gathering for specific applications was ideal for application developers. However, it was too early to know what specific questions the applications should answer. The following table shows the need for different data to answer different questions. The questions are organized in a progressive order, from simple to complex. Table 1 illustrates how a slight shift in the question may result in a considerable addition of required data.

Table 1.

Question	The Needed Data
Where are the hospitals in Mina City?	One Geocoded layer of data with symbols that represent locations of hospitals
Where are the hospitals that have treatment for X problem?	Plus, tabular information about the treatment available in each hospital in Mina
What is the shortest path (in time) to the hospital from a specific location for a particular treatment?	Plus, transportation and network data such as average daily traffic, average travel time of each road, speed limit, road conditions, etc.
During Hajj: What is the shortest path (in time) to the hospital for a particular treatment taking in considerations the issue of pedestrian intrusion on streets?	Plus, real-time spatial-temporal transportation data and density of pedestrian's intrusion on streets

The fourth question embodies one of the most important uses of GIS and IT in addressing Hajj problems. However, in reality, it is difficult to obtain the needed data to answer this question. The practical need sounds right; however, technical capacity could be limiting. Real-time spatial-temporal data would require "freezing" one or more commercial remote sensing satellites (depending on available swath and resolution) at a Low Earth Orbit (LEO) and positioning them straight over the Hajj scene. If this were achieved, then two-way communication and real-time data via satellite imageries would tell the location of street closures and pedestrian intrusion on transportation routes (Figures 2 and 3). It would also tell the location of traffic congestion. This is

Figure 2. Squatting and pedestrians' encroachment on automobile roadway

Figure 3. Pedestrians' encroachment on automobile roadway

difficult to attain since satellites are in constant motion orbiting around the earth in a high speed, about 8 km/second.

The process took on additional practical approach. Rather than gathering data to answer specific research concerns and questions and to serve specific applications, the process moved toward gathering generic but essential data. The aim became to create a "container of GIS data" that could be utilized for multiple purposes, not to any specific research inquiry, or application (Figure 4). The "container" could grow in the future to assist in building specific application and planning decision support systems (PDSS).

For the purpose of identifying the contents of the geospatial database, the Development Commission held several workshops and conducted national and international consultations. These extensive discussions resulted in the identification of eight areas of required data for planning Hajj: 1) Natural resources; 2) Transportation; 3) Socioeconomic; 4) Environmental; 5) Hajj Services; 6) Local Municipalities and Land Use; 7) Urban Design, Zoning and Building Codes; and 8) Religious Affairs. Collecting data for these topics led to bidding and selection process on eight separate projects.

C. *Data Accuracy*. Another example that illustrates how the strong political power resorted to compromise in building the geospatial database is related to data accuracy. Politically powerful decision makers were adamant to acquire the most precise and accurate data. This was partially a result of finding significant errors when maps

Figure 4. Creating a "GIS container": Makkah, Madinah, and Holy Places (MM&Holy Places) geospatial database scheme

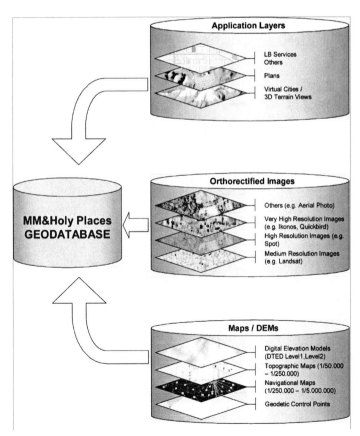

were superimposed on recent and accurate satellite imageries. However, they had to compromise accuracy based on several factors including costs, application need, geographic scale, and required time of production. For example, maps and images of urban areas needed to be of higher resolution than that of rural areas to incorporate greater details about buildings, infrastructure, and street networks. A hierarchy of map accuracy needed to respect geographic scales between regional, rural, semi-urban, and urban areas (Figures 5, 6, 7, 8, 9, and 10).

Topography is more crucial in urban areas. For these reasons, for compiling precise Digital Elevation Models and Building Heights, the process distinguished among three levels of required accuracy:

1. 15m grid spacing DEMs would be compiled for the entire project area, 68,000 km2.
2. 10m grid spacing DEMs would be compiled for Makkah, Madinah, and Holy Places and their immediate surroundings, amounting to about 20,000 km2.

Figure 5. The project area against 1/250,00 maps

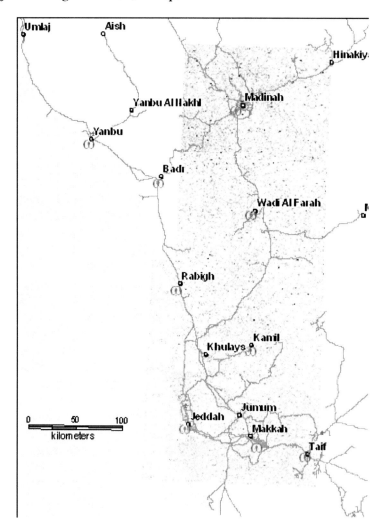

3. 5m grid spacing DEMs would be produced for Makkah-Madinah urban areas and Holy Places, amounting to about 2,000 km2.

Similarly, the process of compiling precise Geodetic Control Points distinguished among two levels of accuracy:

1. 50km spacing with meters accuracy in the entire project area, 68,000 km2
2. 5km spacing with dm-accuracy in Makkah, Madinah, and Holy Places, amounting to about 2000 km2.

In the same way, for developing imagebase maps, the process distinguished between two levels of accuracy:

1. Medium resolution. Orthorectified LANDSAT-7, in the entire project area (68,000 km2) Rs. 15m panchromatic and 30m color, Accuracy = 1.5 pixels, 22.5m.
2. Very High resolution. Orthorectified QuickBird in Makkah – Madinah urban areas and Holy Places (2000km2), Rs. 0.6m panchromatic, 2.4m color, Accuracy = 1.5 pixels, 1.0m (Figures 11, 12 and 13).

Figure 6. The project area against LANDSAT-7

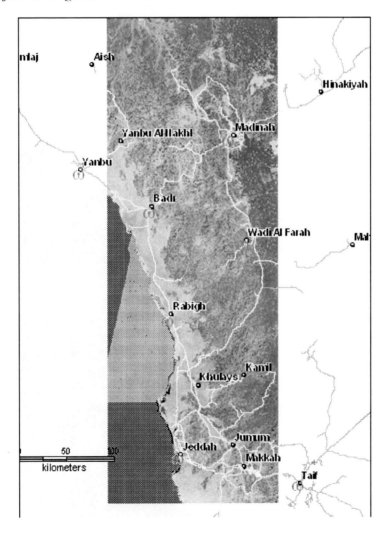

Figure 7. 1/250,000 coverage of the project area

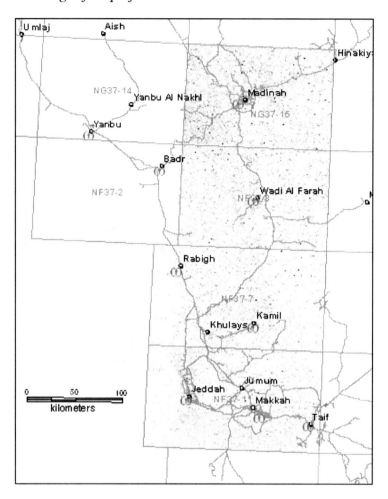

Prices of these maps varied tremendously (Figure 14). Therefore, as a guiding principle in considering the multiple variables related to accuracy (costs, production time, and purpose), it was later realized that only accuracy of data that would make a difference should be discussed and debated. If it would not make a difference, then there was no need to consider it. This led to an in-depth examination of the practical consequences of considering various levels of data accuracy. Satisfactory or sufficient accuracy was based on its practical application and use. It was a surprise to decision makers that even by deploying immense resources absolute accuracy of GIS data is unattainable. It is common that boundaries have some errors; for example, geological boundaries can never be accurately mapped in GIS (Sprinkel and Brown, 2008). However, decision makers observed that accuracy can always be enhanced by quality control, validation, and verification of raw data.

D. Basemap. Another example of the "best possible" or practical approach is related to the establishment of the GIS basemap for the project. There have been misconceptions about obtaining an ideal GIS basemap, as found in common planning practices (Craig, et al., 1998). However, upon examining various theoretical definitions as well as examining the GIS databases, it became apparent that there was no consensus on the definition

Figure 8. 1/50,000 coverage of the project area

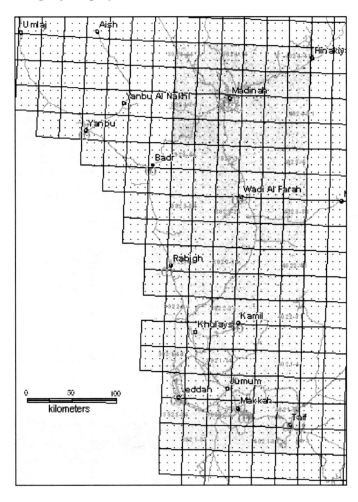

or elements of a GIS basemap. In fact, definition and contents varied tremendously from one field to another. For example, in environmental planning, elements of a basemap are different from that of an urban design basemap. An environmental basemap may include basic natural elements such as water body, greenery, soil types, and topography (Arendt, 1996), while an urban design basemap may include building footprints, dimensions and heights, detailed street networks, and sidewalks (Esnard, 2007). Decision makers had to give up the idea that there is an ideal basemap.

E. *GIS Experts.* Another migration from best to best possible was that decision makers had to compromise on bringing to the table experts in all fields of GIS. There are numerous specialization and sub-specializations within GIS, and there are experts in various areas of GIS. The following is not inclusive, but an illustrative list of some experts in GIS: GIS image analyst, GIS remote sensing analyst, GIS developer, GIS technician, GIS software engineer, GIS photogrammetric technician, GIS cartographer technician, GIS programmer, GIS Internet developer, GIS mapping specialist, GIS database specialist, GIS precision measurement vehicle operator, GIS program manager, Enterprise GIS manager, etc.

Although the resources allocated to the DC-MMM were immense, having one expert in each area of GIS would be extremely costly and difficult

Figure 9. 1/10,000 map coverage

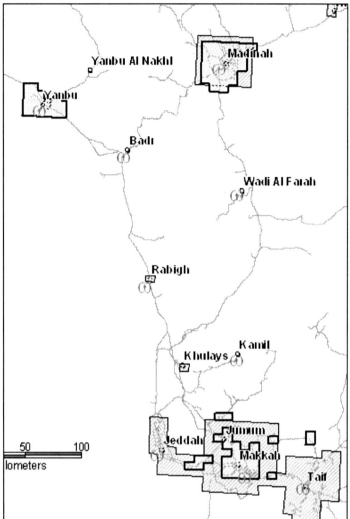

to manage. On the other hand, it became apparent that it would be impossible to have one GIS expert who could span all of the required GIS expertise. Therefore, decision makers resorted to a middle ground, a practical solution that resulted in hiring individuals with expertise in more than just one GIS area. Fewer experts who could cover multiple areas of expertise seemed to be a practical solution.

F. *Software/hardware.* As would be expected, obtaining the "best" software and hardware was desirable by powerful decision makers. After lengthy exploration and examination of different operating systems, computer science experts insisted that UNIX was most appropriate for the project since it could handle a large geospatial database for five cities combined. Also, UNIX is superior for ensuring security of data and protection from computer viruses.

In examining GIS software, the ESRI product seemed to be the most practical for several reasons. ESRI had already made a strong presence in the Middle East including Saudi Arabia, and several participating organizations used the ESRI product. Also, expertise on the ESRI product was readily

Figure 10. 1/1,000 map coverage

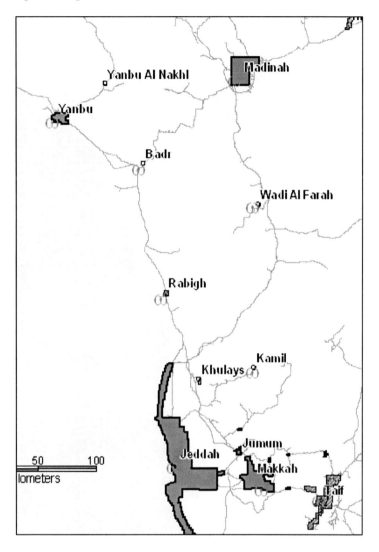

available. Nevertheless, there were roadblocks. Planners discovered that the ESRI software did not support the UNIX platform. The discussion reverted to discussing operating system options. The process ended by giving up on UNIX and adopting MS Windows. Although Windows has much less power for handling large data sets, it does handle ESRI software effectively. Selecting software and operating systems for the project illustrates another example of negotiating technical issues throughout the process.

G. Contracting the work. After coming to terms with the magnitude of data that needed to be collected, the DCMMM determined that it could not build the geospatial database by itself. The Commission decided to contract the overall GIS project through competitive bidding as well. The selected GIS organization would need to coordinate with the eight agencies that won the data collection contracts and would dictate the technical specifications for the data. The Development Commission embarked on the lengthy process

Figure 11. LANDSAT-7 image of mina area

of writing the Terms of References (TORs) for each of the eight data collection projects and for the GIS project. Bids were submitted, evaluated, and selected.

However, some of the data collection companies were slow in collecting data and failed to meet their deadlines. Since completing the overall GIS project hinged on getting all the data in a timely manner and in the correct format, the project was delayed.

FUTURE RESEARCH DIRECTIONS

This case study presents several lessons on the process of building a geospatial database. The planning process led by the DCMMM for Hajj provides useful lessons and discussions on top-down processes, particularly as they relate to GIS. The top-down approach used by the DCMMM illustrates how political power can both challenge and benefit a complex planning process and its outcomes. The political influence of the DCMMM impacted the process in terms of broad-based planning participation, technical decision-making, accomplishment of project objectives, and effectiveness of the completed GIS as a planning tool.

Overall, political power was overwhelming. In the early stages, participating agencies were afraid of the new rival organization, the DCMMM, and were hesitant to provide services to it. Later on, the perceived problem of "taking

Political Power, Governance, and E-Planning

Figure 12. High resolution QuickBird image of central Makkah (The Sacred Mosque at the center) is used to check on topographic accuracy

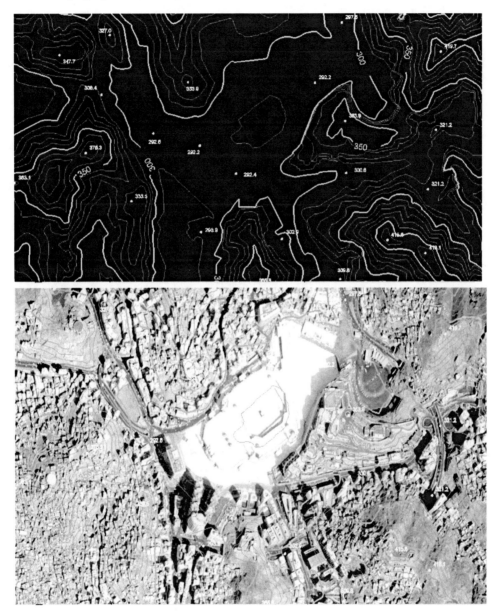

advantage of" the agencies was resolved by offering them profitable contracts. These agencies received a double benefit. First, they already had a great portion of the data that needed to be collected and they were able to "sell" it, rather than give it away for free. Thus, they were able to make a considerable profit. Second, they were able to use part of their fees to enhance their existing geospatial databases by collecting missing data and improving the database structure and format.

Interestingly, the presence of the politically powerful DCMMM served to orchestrate the meetings and mediate negotiations. Political assertiveness helped to sustain the project in the face of competing and conflicting interests among

Figure 13. GTC GTVS package for terrain visualization

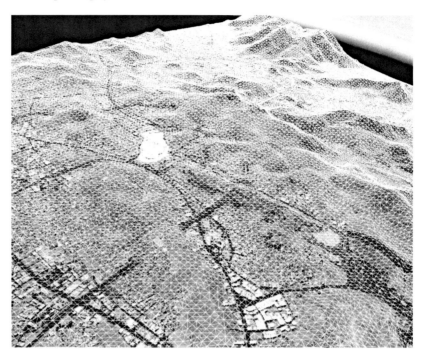

Figure 14. Geodatabase prices for an area of 250 Km2

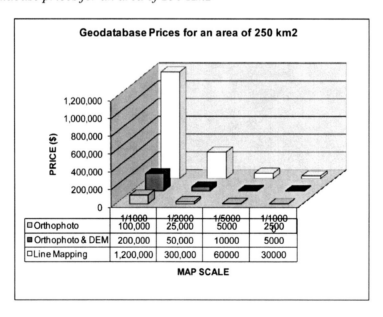

stakeholders. Political power that championed GIS kept the project alive in the face of insurmountable organizational and technical obstacles. The power element contributed a venerated atmosphere that gained the respect of all participants throughout the process.

One interesting lesson to the politically powerful decision makers was that GIS data represented real power and was tough to obtain, despite their supremacy and political influence. Pressuring others through power did not result in obtaining satisfactory GIS data. Another important lesson is that power must partner with sound GIS technical and organizational knowledge since the early stages of the planning process. Politically powerful organizations and decision makers need to set a plan from the start on how to adopt technology by allocating adequate time and resources. This will help to explore technical options and to find the best solutions. Technical issues in mega projects are vague and complex, as it is the case in the Hajj project, and rushing the process could be counterproductive or risk the delivered quality.

The technical challenges of creating the mega geospatial database were expected by the technical experts, not by political decision makers. Technical issues may appear easy, for example producing GIS maps could be viewed as a result of pushing view buttons. However, decision makers may not know the necessary data needed to produce the map. Politicians have agendas and visions and unless technical issues are laid out from the set go, there could be clashes between the technical requirements and visions.

Another observation is that the GIS visual representation of data is real power. GIS visual capabilities helped participants to focus discussions and to collectively analyze, diagnose, and prescribe. In this sense, GIS served the participating agencies to collectively see the presented problems and solutions. Although they were not involved in manipulating the geospatial database, the visual representation helped everyone to see and follow the discussion. It helped to place everyone on the same page.

Interestingly, in rare incidents, when political leaders found that the GIS data and visual outputs were not reinforcing their agenda, they removed GIS altogether and resorted to verbal discussions. Therefore, there are situations when GIS can expose planning mistakes. Despite the ability to manipulate input and output of GIS, there are situations where GIS visual representations will reveal mistakes in a vivid way.

Decision makers were able to speed the process of constructing large structures but were unable to speed building GIS. For example, through DD-MMM, the government successfully replaced an outdated, gigantic pedestrian bridge with a larger one over one mile long. However, they were not expecting that building a large geospatial database to be time-consuming, and they had the assumption that it would be easy to speed it up. Upon viewing GIS maps, decision makers may not realize that the production of these maps could require vast amount of time.

Furthermore, vendors tend to propagate the notion that there is a clear roadmap to create GIS and it is easy and fast to create it. Building GIS at a large scale takes a meandering rather than a straight path. It may follow unclear rather than a straightforward map. Therefore, one lesson that came out of this case study was that building a GIS could be more challenging than constructing large-scale engineering projects. The GIS process was intricate and unique and political will could have a limit in speeding the process. Certain intricacies must be observed in order to optimize the interplay between power and technical knowledge.

Planning profession is political in nature and planning and politics do intertwine. (Carson, 2002). However, this chapter illustrates a new layer of politics, which concerns decisions for building GIS for planning purposes. It suggests that political power and technical knowledge do intertwine as well. Politicians should be aware of this new dimension and should deal more adequately with the technical realities. Also, politicians need to recognize that there is much in the technical world that would benefit planning and there is a whole art and science in dealing with technical issues. Given the diverse and constant change of technology and technical knowledge, this means that revisions and changes of the initially envisioned roadmap could be needed. Politicians may need to learn how to manage these changes so they ensure a sound outcome (Herzele and Woerkum, 2008).

CONCLUSION

Although politics is an integral part of the planning profession, research shows that planners tend to shy away from it. Richard Carson explained "Most planners disdain playing politics. Yet political astuteness was, and continues to be, a predominant characteristic of the field's pioneers and a prerequisite for achieving change" (Carson, 2002). With the proliferation of e-planning, planners are facing a new kind of politics that deal with technical issues. Planners need to be aware of the interplay between politics and technical knowledge and they need to learn how to build consensus and negotiate technical differences.

The topic of political power and GIS in particular has received little observation and analysis. GIS has been employed in developed countries where democracy prevails. This chapter provides a detailed account of a top-down GIS adoption. The narrated story in this chapter illustrates the immense intricacies of the process of adopting a GIS. Unexpectedly and contrary to vendors' claims that it is easy to build a GIS, we find that in the practical world the processes are not straightforward. There is not a clear roadmap to follow; certain political situations may require different roadmaps.

Therefore, the project experience concludes that in GIS planning application projects, decision makers need awareness of the complex dynamics that power discrepancies pose. There are times when exercising political power helps to stimulate discussion, mobilize forces, and optimize resources, and there are times when a power imbalance is counterproductive. There are times when power helps to gain allies and strengthen ties and relationships and there are times when power ruptures them. It is crucial that decision makers are aware of these intricacies and deal with them tactfully.

Since numerous governmental and community-based organizations are on the way to adopting GIS, they may find lessons of this project useful and insightful. Regardless of whether GIS is being adopted in a top-down or bottom-up approach, there are useful lessons on the issue of power and GIS. It is important to fully realize the intricacies of the interplay between power and GIS. Organizations need to acquire a nuanced understanding of the role of power in the various stages of adopting GIS situations. They should explore the wide range of situations when power can serve or harm the process. Education on these issues is a prerequisite for effective utilizations of GIS.

The story is incomplete as the geospatial database building process is unfinished and has not been fully tested on planning applications. It would be interesting to learn about the adequacy of the data collected around the eight planning arenas. Follow-up research will be required to assess the efficacy of the geospatial database for creating strategic plans and the necessary refinements over the long term. Researchers may be interested in examining the usefulness of gathered information for planning purposes and assessing the need to gather additional data.

REFERENCES

Al-Kodmany, K. (2000a). Extending geographic information systems to meet neighborhood planning needs: The case of three Chicago communities. *Journal of the Urban and Regional Information Systems Association, 12*(3), 19–37.

Al-Kodmany, K. (2000b). GIS in the Urban Landscape: reconfiguring neighborhood planning and design processes. *Landscape Research, 25*(1), 5–28. doi:10.1080/014263900113145

Al-Kodmany, K. (2001). Supporting imageability on the World Wide Web: Lynch's five elements of the city in community planning. *Environment and Planning. B, Planning & Design, 28*(6), 805–832. doi:10.1068/b2746t

Al-Kodmany, K. (2002). GIS and the artist: shaping the image of a neighbourhood in participatory environmental design. In Weiner, D., Harris, T. M., & Craig, W. J. (Eds.), *Community participation and geographic information systems* (pp. 320–329). London: Taylor and Francis. doi:10.1201/9780203469484.ch24

Al-Qahtaani, A. (2002). *A Manual on the Rites of Hajj* (2nd ed.). London: Invitation to Islam Press.

Arendt, R. G. (1996). *Conservation Design for Subdivisions: A Practical Guide for Creating Open Space Networks*. Washington, DC: Island Press.

Budthimedhee, K., Li, J., & George, V. (2002). E-planning: A Snapshot of the Literature on Using the World Wide Web in Urban Planning. *Journal of Planning Literature, 17*, 227–246. doi:10.1177/088541202762475964

Carson, R. (2002, October 21). The Art of Planning and Politics. *The Planetizen*. Retrieved July 20, 2009, from http://www.planetizen.com/node/67

Craig, W., Harris, T., & Weiner, D. (1998, October 15-17). *Empowerment, Marginalization and Public Participation GIS*. Report of Varenius Workshop, Santa Barbara, California. Retrieved July 20, 2009 from http://www.ncgia.ucsb.edu/Publications/Varenius_Reports/PPGIS98.pdf

Esnard, A. (2007, January). Interoperable Three-Dimensional GIS: Urban Modeling with ArcGIS 3D Analyst and SketchUp. *ArcUser*.

Herzele, A., & Woerkum, C. (2008). Local Knowledge in Visually Mediated Practice. *Journal of Planning Education and Research, 27*, 444–455. doi:10.1177/0739456X08315890

Hoch, C. (2007). Pragmatic Communicative Action Theory. *Journal of Planning Education and Research, 26*(3), 272–284. doi:10.1177/0739456X06295029

Obermeyer, N., & Pinto, J. (2007). *Managing Geographic Information Systems* (2nd ed.). New York: Guilford Press.

Sprinkel, D. A., & Brown, K. D. (2008). Using Digital Technology in the Field. *UTA Geological Survey. Survey Notes, 40*(1), 1–2.

Still, K. (2008). *Jamarat Bridge –Accidents*. Retrieved July 20, 2009, from http://www.crowddynamics.com/Disasters/jamarat_bridge.htm

Williams, K., Forda, R., Bishop, I., Loiterton, D., & Hickey, J. (2007). Realism and selectivity in data-driven visualisations: A process for developing viewer-oriented landscape surrogates. *Landscape and Urban Planning, 81*(3), 213–224. doi:10.1016/j.landurbplan.2006.11.008

ADDITIONAL READING

Aitken, S. (2002). Public Participation, Technological Discourses, and the Scale of GIS. In Craig, W., Harris, T., & Weiner, D. (Eds.), *Community Participation and Geographic cal Information Systems*. London: Taylor & Francis. doi:10.1201/9780203469484.ch27

Arias, E. (1996). Bottom-up Neighborhood Revitalisation: A Language Approach for Participatory Decision Support. *Urban Studies (Edinburgh, Scotland), 33*(10), 1831–1848. doi:10.1080/0042098966402

Carney, D. (1998). Michigan Communities Launch Largest Regional GIS. *Newsletter of Regional Geographic Information Systems*. An Agency of Grand Valley Metropolitan Council. Retrieved July 20, 2009, from http://www.gvmc-regis.org/news/news4.html

Elwood, S., & Leitner, H. (1998). GIS and Community-based Planning: Exploring the Diversity of Neighbourhood Perspectives and Needs. *Cartography and Geographic Information Systems, 25*(2), 77–88. doi:10.1559/152304098782594553

Esnard, A. (2007, January). Interoperable Three-Dimensional GIS: Urban Modeling with ArcGIS 3D Analyst and SketchUp. *ArcUser*. Retrieved July 20, 2009 from http://www.esri.com/ news/arcuser/0207/urban.html

Hanzl, M. (2007). Information Technology as a Tool for Public Participation in Urban Planning: a Review of Experiments and Potentials. *Design Studies, 28*(3), 289–307. doi:10.1016/j.destud.2007.02.003

Hopkins, L. D., Twidale, M., & Pallathucheril, V. G. (2004, July). *Interface Devices and Public Participation*. Paper presented at the Public Participation GIS Conference, Madison, WS.

James, W. (1908). *Pragmatism: A New Name for Some Old Ways of Thinking: Popular Lectures on Philosophy*. New York: Longman. Retrieved July 20, 2009, from Project Gutenberg: http://www.gutenberg.org/dirs/etext04/prgmt10.txt

Kwan, M. (2002). Feminist Visualization: Re-envisioning GIS as a Method in Feminist Geographic Research. *Annals of the Association of American Geographers. Association of American Geographers, 92*(4), 645–661. doi:10.1111/1467-8306.00309

Li, S., Guo, X., Ma, X., & Chang, Z. (2007). Towards GIS-Enabled Virtual Public Meeting Space for Public Participation. *Photogrammetric Engineering and Remote Sensing, 73*(6), 641–649.

Maxcy, S. (2003). Pragmatic Threads in Mixed Methods Research in the Social Sciences: The Search for Multiple Modes of Inquiry and the End of the Philosophy of Formalism. In Tashakkori, A., & Teddlie, C. (Eds.), *Handbook of Mixed Methods in Social Q4 and Behavioral Research*. Thousand Oaks, CA: Sage.

Nawwab, I. (1992 July). The Journey of a Lifetime. *Saudi Aramco World*, 24–35. Retrieved from http://www.saudiaramcoworld.com/issue/199204/the.journey.of.a.life time.htm

Obermeyer, N., & Pinto, J. (2007). *Managing Geographic Information Systems* (2nd ed.). New York: Guilford Press.

Scott, D., & Oelofse, C. (2005). Social and Environmental Justice in South African Cities: Including Invisible Stakeholders in Environmental Assessment Procedures. *Journal of Environmental Planning and Management, 48*(3), 445–467. doi:10.1080/09640560500067582

Siebenhüner, B., & Barth, V. (2004). The Role of Computer Modelling in Participatory Integrated Assessment. *Environmental Impact Assessment Review, 25*, 367–389. doi:10.1016/j.eiar.2004.10.002

KEY TERMS AND DEFINITIONS

Database: Primarily imageries, textual, and numeric data; no audio-video data is involved.

E-Planning: An online setup for facilitating planning activities among government officials and associates.

Geospatial: Geographic data at both regional/national scale (geo) and urban scale (spatial).

Governance: The administrative infrastructure that governs decision-making process.

Hajj: The annual Muslim's pilgrimage to Makkah that happens on the 12th month of the lunar Islamic calendar.

Political Power: Totality in influence, pressure, and decision making.

Top-Down: An organizational hierarchy where decisions are made by a few individuals at the top.

ENDNOTE

[1] HM stands for His Majesty

Section 2
Citizen Participation in E-Planning

Chapter 9
The Potential of E-Participation in Urban Planning:
A European Perspective

Herbert Kubicek
University of Bremen, Germany

ABSTRACT

Because urban planning affects the living conditions of its inhabitants, most countries, at least western democracies, require some kind of citizen participation by law. The rise of the World Wide Web has led to recommendations to offer participation via the Internet (eParticipation) in various forms. However, many eParticipation applications are not well accepted and fall short of the expectations associated with them. This chapter argues that the electronic mode of participation per se does not change much. Rather, electronic forms of participation have to be embedded in the context of the respective planning processes and participation procedures. If citizens are not interested in participating in an urban planning process, they will not do so just because they could do it via the Internet. Therefore, an analysis of the barriers and deficits of eParticipation has to start with a critical review of traditional offers of participation. Against this background, the forms and methods of electronic participation are described and assessed in regard to expectations and barriers associated with them. It becomes apparent that eParticipation research has still not provided solid knowledge about the reasons for low acceptance of eParticipation tools. This research is largely based on case studies dealing with quite different subject areas. There is also high agreement that electronic tools will not substitute traditional devices for a long time. Instead, they will only complement them. Therefore, online and traditional forms of participation have to be designed as a multi-channel communication system and need to be analyzed against each context together. Accordingly, this chapter starts with summarizing both the institutional context of urban planning and traditional modes of citizen participation and the development and use of technical tools as two backgrounds. Recognizing a certain degree of disappointment with the low use of eParticipation, future eParticipation research should focus on fitting electronic tools better into their context and apply more comprehensive and rigorous evaluation.

DOI: 10.4018/978-1-61520-929-3.ch009

Copyright © 2010, IGI Global. Copying or distributing in print or electronic forms without written permission of IGI Global is prohibited.

INTRODUCTION

The outcome of urban planning affects the living conditions and wellbeing of individual inhabitants in many ways as well as the social, economic and ecological welfare of the population concerned. In most cases, different parts of the population are affected in different ways and prefer different outcomes of a planning process because of different interests. Formal legitimization of urban planning by elected bodies was criticized as insufficient by environmental groups and initiatives in the 1970s, supported by democracy movements in Europe and the United States. In order to overcome resistance and to improve the quality of the outcome of urban and regional planning, the respective laws were revised, requiring planning bodies to publish plans for public consultation and to provide for appellation procedures. However, quite often these kinds of formal citizen participation did not achieve their objectives. Many observers argued that the procedures did not provide for real participation, and were more conducive to manufacturing dissent instead of consent.

As a consequence, experiments with different forms of group-related participation were conducted, such as round tables, focus groups, and consensus conferences. In some cases, these informal methods of participation reached their objectives. But they did not succeed on a broader scale. In the 1990s, citizens' engagement in public affairs, voter turnout and trust in political decision-makers declined even more. At the same time, the Internet made electronic information and communication services such as the World Wide Web, e-Mail and electronic forums available to local governments, business and citizens. The new technological possibilities gave rise to far-reaching expectations of entering a new stage of democracy with more direct participation of citizens (see Hagen 2000). Under the heading of eDemocracy or Digital Democracy or eParticipation, experiments were started in many western countries offering some kind of participation via the Internet, often supported by national funding programs.

However, although a lot of money has been spent for these projects, not much has changed in general. Mirroring the fate of expectations for new economic growth by a New Economy, some kind of disillusion has replaced optimism in regard to the potential of electronic tools for improving democracy and the quality of decision-making, including urban planning. Today there is high agreement in the research community that citizens' attitudes and behavior, such as trust in political institutions and the readiness to engage in public affairs, cannot be changed simply by introducing new technical tools in existing procedures and cultures. In a report for the Council of Europe, Lawrence Pratchett summarizes this position quite well:

"New technologies, in whatever form, are socially and politically neutral devices and have no inevitable consequences for democracy, participation or political engagement. However, the way in which such technologies are used and the purposes to which they are put can have radical consequences for the practice of democracy. The design of particular tools and their association with existing democratic practices (and other aspects of governance) shapes their value and impact, as does the way in which citizens and intermediary bodies (such as the news media, political parties and so on) adopt and use the technologies." (Pratchett 2006, p. 3)

Pratchett proposes to look at new electronic devices in their relationship to institutions, actors, procedures and outcomes of democratic processes.

This has seldom been done so far. And despite the many projects applying some kind of technology to provide for participation in some kind of urban planning, there is a lack of rigorous assessment of outcome and impact as well as a lack of systematic analysis of barriers and success factors to exploit the potential of the technology. A recent OECD report speaks of an "evaluation gap":

"As noted in the 2001 OECD report, Citizens as Partners: Information, Consultation and Public Participation in Policy Making, there is a striking imbalance between the amount of time, money and energy that governments in OECD countries invest in engaging citizens and civil society in public decision making and the amount of attention they pay to evaluating the effectiveness and impact of such efforts. That a significant "evaluation gap" exists is hardly surprising. If public engagement in policy making is a recent phenomenon and evaluation is itself a relatively young discipline, then it may safely be said that the evaluation of public participation is still very much in its infancy". (OECD 2005, p. 9-10)

This evaluation gap applies to traditional participation as well as to its support by electronic means. Only recently the research and grant programmes have begun to address the issue of appropriate concepts and methods for evaluating eParticipation. It is now obvious that closing this gap requires inter- and multidisciplinary approaches. It is not sufficient to assess the usability of the electronic tools with concepts and methods of information systems and human computer interaction research or to assess the outcome by counting the number of users or contributions. Rather the application of the technical tools has to be analyzed in the context of the different participatory procedures and the respective institutional contexts. This requires contributions from political science research on participation and, with regard to urban planning, additional input from planning research, environmental research and others. The underlying premise is that the potential of electronic tools can only be exploited if they are properly embedded in their procedural and institutional context. Therefore eParticipation research should no longer focus primarily on the technical tools, but should start with the analysis of their respective contexts, as urban planning in this case. Figure 1 tries to illustrate this approach:

Accordingly, this chapter requires reference to two overlapping backgrounds, i.e. participation in urban planning in general and options for electronic support of these participation processes. Future trends emerge from the identification of barriers and success factors and concern the organizational, cultural, legal, economic and technological context on the side of governments and citizens, the complex issue of trust building as well as the more comprehensive and rigorous evaluation of the tools' fit into these contexts.

BACKGROUND

Public Participation in Urban Planning

The Domain of Urban Planning

Urban planning is an established area of political decision-making at the local level with established procedures and an established profession. It covers official plans and other activities by public authorities concerning land use, traffic and transport, energy and water provision, waste management, any measures with environmental effects, but also social infrastructure, e.g., schools, fighting poverty and promoting social integration of minorities as well as local financial budgets. This broad range of topics has been summarized under the heading of Sustainable Development by the Rio Declaration, and municipalities have been asked to elaborate their own sustainable development program called Local Agenda 21 back in 1996.

Public Participation in Urban Planning

Almost all urban issues and plans concern a variety of different stakeholders with often conflicting

Figure 1. E-Participation in the context of urban planning

interests. There is an obvious competition for different options of using certain pieces of land. Therefore most laws on regional and urban planning require some kind of consultation. First of all, other administrative units or public entities responsible for aspects such as nature conservation or protection of historical monuments have to be consulted. In addition, responsible bodies of public concern, i.e. NGOs, often have to be consulted as well. They have to be formally invited to formulate their concerns or objections, and the planning offices are obliged to consider these arguments and also discuss them in public meetings. If decisions are taken without these steps of participation, they may be reversed by courts. Accordingly, there are formal rights of appeal.

The obligation to also consult ordinary citizens as well became generally accepted after the formulation of the so-called Aarhus Convention only ten years ago. The full title "UNECE Convention on Access to Information, Public Participation in Decision-Making and Access to Justice in Environmental Matters" enumerates the three "pillars" on which the Convention is based (http://www.unece.org/env/pp/).

The European Commission is party to the Convention and has launched two directives implementing the Convention in 2003 (Directive 2003/4/EC on public access to environmental information and Directive 2003/35/EC providing for public participation in respect of the drawing up of certain plans and programs relating to the environment).

The reasons for demanding public participation considered in the Aarhus Convention are:

"Recognizing that, in the field of the environment, improved access to information and public participation in decision-making

- *enhance the quality and the implementation of decisions,*
- *contribute to public awareness of environmental issues,*
- *give the public the opportunity to express its concerns,*
- *and enable public authorities to take due account of such concerns,*

Aiming thereby to further the accountability of and transparency in decision-making and to strengthen public support for decisions on the environment..." (UNECE 1998, p. 1)

Already in the early 1990s, the Congress of Local and Regional Authorities of the Council of Europe had passed resolutions aiming to reinforce local democracy. With Resolution 91 (2000) the Congress approved the "Guidelines for a policy of responsible civic participation in municipal and regional life", which among others recommend to "strengthen citizen participation at local and regional level":

The participation of citizens in local politics must be guaranteed at all political and administrative levels. The Congress of Local and Regional Authorities of Europe therefore agreed with Agenda 21 that "one of the fundamental principles for the achievement of sustainable development is broad public participation in decision-making".

In particular, this presupposes:

- *the implementation of the subsidiarity principle, according to which public business should only be carried out by the administrative authorities if this cannot be successfully accomplished by the citizens and their voluntary associations;*
- *the right of citizens to be informed and heard on every major plan or project before decisions are taken (transparency encourages a sense of acting for the public good!);*
- *the use of the new information technologies to provide citizens with comprehensive information and establish needs and problems;*
- *the creation of a system of co-operation based on mutual trust between specialised staff on the one hand and the citizens and the representatives on the other;*
- *the greatest possible involvement in political life of all inhabitants, whether or not they are nationals of the country. (Congress of Local and Regional Authorities 2000, sect. II)*

Since then several recommendations and resolutions have been passed by the Congress of Local and Regional Authorities as well as the Council of Ministers (see http://www.coe.int) touching every aspect and factor relevant for increasing well-reflected deliberative opinion building and real influence of citizens on local plans and decisions. These all apply to urban planning as well.

Similarly, since 2001 the Organization for Economic Cooperation and Development advocates "Engaging Citizens in Policy Making" and recommends considering "Citizens as Partners". (OECD 2001a, b)

According to Innes and Booher (2004), most claims to justify public participation in planning can be captured by five purposes:

1. Through participation, decision-makers can find out what the public's preferences are and consider them in their decisions.
2. Decisions can be improved by incorporating citizens' local knowledge.
3. Public participation can advance fairness and justice.
4. Public participation helps getting legitimacy for public decisions.
5. Participation is offered by planners and public officials because the law requires it.

Recently, another very important reason has been identified in the context of measures against climate change. Under the heading of "Environmental Democracy", the Green Mountain Institute for Environmental Democracy (2005) argues that climate change is not only caused by sources which are subject to regulated decisions but also by the behavior of individuals and the choices they make every day. As behaviors cannot be changed by imposing restrictions and obligations, new forms of engagement and self-commitment, which can be initiated via participation in planning procedures and by making citizens partners of planning, are necessary as well to achieve the goals of environmental policies.

The first four propositions mentioned by Innes and Booher, however, are not completely agreed upon and are not proven by empirical evidence.

Objections against Citizen Participation in Urban Planning

There are good arguments questioning the propositions of quality improvement as well as those of fairness and legitimacy (Creasy et al. 2007).

Many professional urban planners claim that they have the expertise to develop plans in the public interest. The public may be consulted, but because they lack the expertise, they should have no influence on the final decision. Political obligations to give the public more influence are perceived as undermining their expertise and professional status.

A second argument questions that participation advances fairness and justice because certain groups of people who have more time and engage more have their voices heard, while others do not have the time and their concerns are not entered into the decision process. Therefore decision-makers cannot find out what "the" public's preferences are, but only the preferences of certain groups.

And a further argument comes from elected representatives who appreciate a light amendment of representative democracy but fear that this extension may not be kept under control and will lead to eroding representative democracy in the end.

But with regard to the legitimacy proposition, much depends on how participation is set up. If contributions are asked for and it does not become transparent how they are handled, this may enforce distrust and not create legitimization at all.

Limits of Existing Formal Participation

Empirical evidence shows that the offered kind of participation by publishing plans and inviting for written comments or public hearings is poorly accepted. There are at least two main arguments why these traditional forms of consultation do not meet their objectives.

First, to many citizens they are not attractive, because it is not transparent how their contributions will be handled and what degree of influence they may exercise. In the planning literature some authors maintain that what is called participation does not deserve this name. For example, Arnstein (1971) defines a ladder of participation with eight steps of increasing citizen influence. Similarly, Wiedemann and Femers (1993) propose a slightly different "public participation ladder" (Table 1).

If participation offers are to be accepted and if public engagement and trust shall be regained, "real" participation has to be established.

A second argument relates to the kind of communication within the formal procedures. Innes and Booher (2004) argue that the formal participation in planning processes does not achieve its objectives because, for example, public hearings put citizens against each other, and force them to speak in polarizing terms. Each individual has to argue his or her point. Often there are many heterogeneous single points which are not related to each other. No attempts are made to look for compromise. In fact, compromise can only be achieved when there is a mutual understanding of interests and arguments. These, in turn, need some time to grow. Therefore a longer process of exchange of arguments is necessary. In the planning and in the democracy literature, this is called deliberation or deliberative consultation (see Coleman and Goetze 2001).

Deliberative consultation is a necessary step to well-reflected votes. According to Coleman and Goetze (2001, p. 6), methods are needed which "encourage citizens to scrutinize, discuss and weigh up competing values and policy options.

Table 1. Different degrees of citizens' participation

Arnstein's Ladder of Participation	Public Participation Ladder by Wiedemann and Femers
Nonparticipation 1. Manipulation 2. Therapy Degrees of Tokenism 3. Informing 4. Consultation 5. Placation Degrees of Citizen Power 6. Partnership 7. Delegated Power 8. Citizen Control	1. Public right to know 2. Informing the public 3. Public right to object 4. Public participation in defining interests, actors and determining agenda 5. Public participation in assessing risks and recommending solutions 6. Public participation in final decision

Such methods encourage preference formation rather than simply preference assertion."

Deliberation does not only mean a dialogue between individual citizens and public authorities, but also a triangle situation where citizens can discuss the arguments of their fellow citizens. This could overcome the critique of Innes and Booher that public hearings do not lead to a public opinion in the sense of collectively shared views or preferences, but rather to a number of contradictory individual statements.

Methods and Devices for Citizens' Participation

Out of the critique of the established formal methods of citizen participation, experiments with more group-oriented and deliberative forms have been started in the 1980s. Several countries have revised their legislation. And at the local level in particular, in order to involve more people, a variety of new forms of participation including focus groups, citizen panels or boards, public surveys and many more have been introduced and tried within urban planning and the Agenda 21 programs.

There are different approaches to classify this broad range of methods and devices. The Organization for Economic Cooperation and Development (OECD 2001a, p. 23) uses a broad classification with three main forms of citizen participation:

- information
- communication
- cooperation

For each of the main categories a variety of devices is identified (Table 2).

Preconditions and criteria for successful participation in urban planning

From the above observations, two preconditions for successful participation of citizens in urban planning can be derived:

1. They must offer some real influence of the public;
2. They must provide space for deliberation to get to some common preferences.

These theoretical conclusions are in line with the Recommendation of the Council of Ministers of the Council of Europe regarding the participation of citizens in local life (Rec (2001)19) pointing to the need to

"give citizens more influence over local planning and budgetary and financial planning; ... to ensure that direct participation has a real impact on the decision-making process, that citizens are well informed about the impact of their participation and that they see tangible results. Participation

Table 2. Devices for citizen participation (compilation by the author)

Information	Consultation	Cooperation
• Written official plans with maps • Leaflets • Booklets • Visualization • Games • Newsletters • Oral lectures/ presentations • Road shows • Hotline • Site visits • Street stalls	• Surveys and polls • Complaint forms • Appeal services • Citizens expertise • Ideas competition • Award scheme • Simulation • Face to face /door to door • Consultation • Townhall meetings • Community planning forum • Citizens request sessions • Invitation of NGOs to council meetings • Complaint hotlines • Action planning event • Experimentation	• Neighborhood planning office • Development trusts (i. e. by independent organizations) • Round tables • Focus groups • Workshops • Neighborhood committees • Consensus conferences Advocacy planning • Mediation procedures

that is purely symbolic or used to simply grant legitimacy to pre-ordained decisions is unlikely to win public support. However, local authorities must be honest with the public about the limitations of the forms of direct participation on offer, and avoid arousing exaggerated expectations about the possibility of accommodating the various interests involved, particularly when decisions are made between conflicting interests or about rationing resources." (Council of Ministers 2001, sect. C5)

The Council of Ministers also recommends "deliberative consultation":

"3. Make full use, in particular, of... more deliberative forms of decision-making, i.e. involving the exchange of information and opinions, for example: public meetings of citizens; citizens' juries and various types of forums, groups, public committees whose function is to advise or make proposals; round tables, opinion polls, user surveys, etc." (Council of Ministers 2001, ibid.)

But this may not be regarded as true participation and has to be accompanied by other provisions. Accordingly, the Council recommends:

"Introduce or, where necessary, improve the legislation/regulations which enable:

1. *petitions/motions, proposals and complaints filed by citizens with the local council or local authorities;*
2. *popular initiatives, calling on elected bodies to deal with the matters raised in the initiative in order to provide citizens with a response or initiate the referendum procedure;*
3. *consultative or decision-making referendums on matters of local concern, called by local authorities on their own initiative or at the request of the local community;*
4. *devices for co-opting citizens to decision-making bodies, including representative bodies;*
5. *devices for involving citizens in management (user committees, partnership boards, direct management of services by citizens, etc.)." (Council of Ministers 2001, sect. C4)*

From the theories mentioned above and the literature on criteria for evaluating processes of public participation (e.g. Rowe & Frewer 2000, Macintosh & Whyte 2006, RTPI 2007), four additional requirements for successful participation can be derived:

- accessibility,
- transparency of the consultation and decision-making process,

- responsiveness,
- accountability of public institutions.

These requirements also apply to any participation process, which is supported by electronic means, i.e. to eParticipation.

Electronic Participation and its Potential for Urban Planning

Definitions and Forms of eParticipation

Following the classification of three forms of eParticipation, OECD (2001b) differentiates eParticipation into

- *"Information (eEnabling) - a one-way relation in which government produces and delivers information for use by citizens.*
- *Consultation (eEngaging) - a two-way relationship in which citizens provide feedback to government, based on the prior definition by government of the issue on which citizens' views are being sought.*
- *Active participation (eEmpowerment) - a relation based on partnership with government, in which citizens actively engage in the policy-making process. It acknowledges a role for citizens in proposing policy options and shaping the policy dialogue, although the responsibility for the final decision or policy formulation rests with government."*

Electronic Tools and Web 2.0

For each form of citizen involvement, several electronic devices or tools may be employed. DEMO-net, following Macintosh et al. (2005), put together a comprehensive list of eParticipation tools with short definitions (see Table 3).

This list includes tools such as Wikis and Blogs, which are considered as prototypes of a new generation of Internet services called Web 2.0 or "participatory web" (OECD 2007) and which raise new hopes for increasing citizen engagement and participation. Another term for the new tools and new modes of using existing tools is "user-generated content" (UGC). Many of these tools are not really new because the early newsgroups on the net, long before the World Wide Web, consisted only of user-generated content. However, it was plain text, black and white (or green or amber on black, referring to the monochrome monitors widely in use at that time) and produced by a few thousand people only. Today, user-generated content is text, audio and video, accessible by billions of people in portals, tagged and ranked by other people. It covers webcasts, RSS feeds, podcasts, photos and video portals, i.e., features and platforms for distributing self-produced audio and video files, which can be employed in urban planning processes as well. But there are other participative dimensions of the Web 2.0 as shown in Figure 2.

Of particular relevance for participation are Wikis, which allow for collaborative writing of plans or other documents instead of single messages and comments. Via a Wiki, a statement of public opinion, a shared view on a subject can evolve.

But not many citizens are willing to write. So far they could only read, and their voice was not heard. Now rating features allow for a form of comment appropriate to this group. Just by clicking one to five stars or "agree"/"disagree" they can make their own voice heard.

Another relevant Web 2.0 feature is social tagging. People attach keywords to elements of plans, which they choose themselves. The tools can link one person to others who have used the same keyword, and so shared views across existing social groupings may emerge.

Consultation-Related Tools

Frequently, online consultation consists of offering an e-mail address or a response form or in providing

Table 3. eParticipation tools (Source: DEMO-net 2007)

eParticipation Chat Rooms	Web applications where a chat session takes place in real time especially launched for eParticipation purposes
eParticipation Discussion forum/board	Web applications for online discussion groups where users, usually with common interests, can exchange open messages on specific eParticipation issues. Users can pick a topic, see a "thread" of messages, reply and post their own message
Decision-making Games	These typically allow users to view and interact with animations that describe, illustrate or simulate relevant aspects of an issue; here with the specific scope of policy decision-making
Virtual Communities	Web applications in which users with a shared interest can meet in virtual space to communicate and build relationships; the shared interest being within eParticipation contexts
ePanels	Web applications where a 'recruited' set, as opposed to a self-selected set, of participants give their views on a variety of issues at specific intervals over a period of time
ePetitioning	Web applications that host online petitions and allow citizens to sign in for a petition by adding their name and address online
eDeliberative Polling	Web applications which combine deliberation in small group discussions with random sampling to facilitate public engagement on specific issues
eConsultation	Web applications designed for consultations which allow a stakeholder to provide information on an issue and others to answer specific questions and/or submit open comments
eVoting	Remote Internet enabled voting or voting via mobile phone, providing a secure environment for casting a vote and tallying of the votes
Suggestion Tools for (formal) Planning Procedures	Web applications supporting participation in formal planning procedures where citizens' comments are expected to official documents within a restricted period
Webcasts	Real time recordings of meetings transmitted over the Internet
Podcasts	Publishing multimedia files (audio and video) over the Internet where the content can be downloaded automatically using software capable of reading RSS feeds
Wikis	Web applications that allow users to add and edit content collectively
Blogs	Frequently modified web pages that look like a diary as dated entries are listed in reverse chronological order
Quick polls	Web-based instant survey
Surveys	Web-based, self-administered questionnaires, where the website shows a list of questions which users answer and submit their responses online
GIS-tools	Web applications that enable the users to have a look at maps underlying planning issues and to use them in various ways
Search Engines	Web applications to support users find and retrieve relevant information typically using keyword searching
Alert services	One-way communication alerts to inform people of a news item or an event, e.g. email Alerts and RSS Feeds
Online newsletters	One-way communication tools to inform a general audience or a pre-registered audience of specific news items and events
Frequently asked questions (FAQ)	A 'tree' of questions and answers that can be searched using keywords or by inputting a question or statement
Web Portals	Websites providing a gateway to a set of specific information and applications
Groupware tools	Tool environment to support computer-based group works
LIST SERVS	Tool for information provision and two-way interaction that can be used for Citizen2Citizen, Citizen2Administration, Citizen2Politicians etc

an online questionnaire. While the questionnaire does not allow for deliberation and can only be a component within a broader procedure, individual messages can be handled in small pilots, but not when thousand or ten thousand citizens or more are expected to participate.

Figure 2. Web 2.0 formats

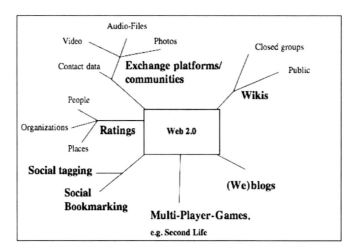

Honor Fagan et al. (2006) in their report on eConsultation in Ireland emphasize four stages of an online consultation and discuss different technical devices for each stage.

"a. Defining the problem: In order to find out what shall be discussed in a consultation and how a problem shall be defined from the point of view of different stakeholders and interested parties, electronic forms, issue forums or story-telling blogs are recommended.
b. Exploring the problem, i.e. discovering different sub-problems and views, understanding reasons for concerns etc. can be supported by online chats, video conferences, uploads of photos and videos. Furthermore there is a need for and there are tools for mapping different arguments (e.g. mindmaps).
c. Choosing and developing solutions: When a few solutions emerge, they can be commented and ranked. Or different people can change and add a draft, if it is offered as a Wiki, a collaborative writing tool.
d. Managing the consultation process covers issues of registration as well as structuring contributions and producing reports."

As urban planning to a large extent consists of information with some spatial relation, it is important to link and integrate the participation tools with geographical information systems (GIS) providing the reference points for the consultation.

Monitoring

Transparency has been emphasized as being important for motivating participation and building trust. There are several ways to employ electronic tools to improve transparency, either by public authorities themselves or by third parties.

Transparency about the voting behavior of elected representatives is provided by a website run by a non-government organization http://www.theyworkforyou.com in the UK.

Many city portals offer forms for notifying local problems regarding pavement, street lighting etc. Via the portal http://www.fixmystreet.com, citizens can enter a problem together with postal code and details. The portal sends it to the respective council, tracks replies and produces listings of responsiveness, i.e. how long it takes to get a confirmation and a report about fixing the problem.

Tracking and tracing features can be provided by public authorities themselves regarding complaints or appeals in the context of urban planning. In the UK the planning department provides information about open consultation on a web portal (http://www.planningportal.gov.uk). Citizens can enter appeals online, look for other appeals which have been accepted as valid and after having received a reference number view the status of their request and the final decision. And citizens can also comment on other people's appeals.

ePetitions

Petitions and referenda are means to give the public a real say and therefore were recommended by the Committee of Ministers.

The Scottish Parliament was the first, after a five-year pilot phase, to allow online petitions. The system allows

- to propose a text for a petition,
- to collect signatures,
- to open a forum for comments on the proposed petition,
- to moderate the forum, and
- to produce reports out of the comments.

Because the petitions may be made public as well as the signatures they have received, any interested party can take a look, who and how many have signed, read comments, and thus can take a better founded decision. The whole process is more transparent and effectively overcomes the physical challenge in the offline world to collect the necessary minimum of signatures.

Meanwhile other national parliaments have adopted the system (in Germany see http://www.bundestag.de) as well as local councils, e.g. Bristol City Council (http://www.bristol.gov.uk/item/epetition.html) or the Royal Borough of Kingston upon Thames.

Expectations Regarding eParticipation and their Validation

There is high agreement about the technical potential of the eParticipation tools with regard to improving information and communication in participation processes (e.g., OECD 2003, p. 33).

- Online offers of participation are easier and cheaper to access than physical meetings and allow for greater flexibility in time, and thereby reach a larger number of people.
- Many people do not like to speak in front of a larger audience and prefer to write comments in a forum.
- Information can be visualized and animated.
- Different levels of aggregation of information can be offered and linked.
- Online offers allow for interactivity, allowing more in-depth consultation and support deliberative debate.
- Individual replies and comments can be published and shared.
- Online offers allow for more transparency of dialogues, are easier to monitor and to evaluate.

There are cases where these effects have been observed. Therefore they bear witness to the potential of eParticipation tools. But as emphasized in the introduction, these are no automatic effects; whether they are achieved or not depends on how these tools are embedded into participatory processes and how they are used.

Based on the now available technical capabilities, there are still much wider hopes. Only recently the Congress of Local and Regional Authorities of the Council of Europe in a resolution on "Electronic democracy and deliberative consultation on urban projects" has declared that:

"Electronic democracy, also known as cyberdemocracy, offers a powerful tool to counteract citizen disengagement and disillusion and to develop the dialogue and trust essential for good governance. Electronic deliberative consultation for sustainable urban spatial development offers a way to revitalise local democracy and citizenship by facilitating people's participation in shaping policies and decisions which impact on their lives and environment." (Congress of Local and Regional Authorities 2008a, p.2)

However, there is no empirical evidence to sustain this hope on a general basis. It is also very unlikely that such positive impacts can be assessed sufficiently. Because trust in politicians and political bodies depends on so many factors, there is no valid method to assign a change in trust or engagement to the employment of technical tools or devices. If people are not interested in a certain subject, they will not participate in a consultation, regardless of whether it is held offline or online. Only if there is an interest in the subject, the online mode may appear easier or more attractive than offline devices.

State of the Art: Surveys and Studies of eParticipation in Urban Planning Projects

Research on eParticipation is largely case-oriented. There are only a few comparative studies on the state of the art of eParticipation within one country or across a few countries. Most of them look at the national level only. None of these surveys and studies shows a serious attempt to evaluate the quality of the processes described. They mainly "count" participatory acts such as comments, participants, votes, etc. and rely to a large extent on data provided by the organizer of participation or on reports of national governments. They are of poor data quality and have almost no relevance for answering the crucial question of impact as well as for the "right way" of using the technical devices.

The overall impression gained from these studies is that the Internet is still used for disseminating information and receiving questions or comments from citizens. Very seldom there is a dialogue, which is a constitutive feature of deliberative consultation.

Implementation of the Aarhus Convention

The only international surveys covering a larger number of countries were conducted in order to assess the implementation of the Aarhus Convention with its obligation to improve access to information, citizen participation and access to justice.

In 2005, the UNCEC tried to assess the implementation of these three pillars, based on national reports by member states covering the national level only. The national reports of 20 member states and the synthesis report are available at http://www.unec.org/env/pp/reports.

Regarding access to environmental information, the synthesis report summarizes that many countries are using ICT for dissemination but that there are some restrictions:

- not all public authorities submit their information,
- in many countries there is only one national centre (to publish information),
- Internet can make access easier, but in many eastern and central European countries Internet penetration is so low that many citizens cannot make use of this new access channel.

Regarding public participation, implementation by legislative measures is less advanced and quite dispersed. Again, the focus of the study here is on regulations and institutions at the national level, and the use of ICT is not mentioned at all.

In 2006, The Access Initiative, a global coalition of public interest groups monitoring the implementation of the Aarhus Convention, published a report on comparative levels of "environmental democracy" among a selected number of mostly eastern European countries, i.e., Bulgaria, Estonia, Hungary, Latvia, Lithuania, Poland, Portugal and Ukraine (Kiss et al. 2006). The summary of this report confirms the general impression that "Access to information is generally satisfactory in practice", in particular with regard to official reports, while "public accessibility of individual facility data is significantly worse". Participation in decision-making exists, "but cannot guarantee that the public is heard". While the report mentions the use of ICT for information provision, there is no information about eParticipation in this context in the assessed countries.

Taking the implementation of the Aarhus Convention as an example, the state of the art does not achieve a satisfactory level of accepted procedures into which electronic devices may be implemented in order to make them more effective or to improve their quality. Rather there are a lot of lacks and deficiencies in the institutional contexts which cannot be healed by technical devices.

Studies at the Local Level

This general finding applies to the local level as well.

(1) Local Governments' Websites

The Local eDemocracy National Project in the UK is probably the largest and best-documented set of experiments with eParticipation at the local level in Europe. In their survey of local government websites, Pratchett et al. (2005) found that only 32% of all council websites offer some kind of online forum, but it is not clear to what extent the discussion is moderated and to what extent they are listened to by the councilors or officers.

In Germany, the Initiative ePartizipation analyzed the websites of all 82 larger cities in regard to offers for citizen participation. In 2005, compared to 2004, there was an increase of the number of cities offering some kind of citizen participation. While information about the government structure and decision-making procedures is evaluated positively, there are only 13 cities offering some kind of informal participation on a few selected topics. Regarding formal participation and consultation processes, 48 out of 82 cities provide information about the offline procedures, only 17 allow for online submission of comments. The report therefore summarizes that citizen participation via the Internet is still an exception, even where it is required and recommended by law (Initiative ePartizipation 2005).

In a comparative study on local eDemocracy initiatives covering Estonia, Hungary, Italy, Spain, Switzerland, the United Kingdom and the United States, Peart and Diaz (2007) also found only a few cases of moderated forums, deliberative interaction or participation by contract, for example. The few good practice cases were not only found in the UK and US, but also in Hungary and Spain.

Torres, Pina and Acerete (2006) conducted a survey of 35 cities with more than 500,000 inhabitants in 12 European countries. In 2003 and 2004, the websites of these cities were surveyed according to 173 criteria items. Items referring to eDemocracy included information about the mayor and council members, minutes and reports, press releases and offers for citizen dialogue such as complaint boxes, forums and other ways of democratic engagement and participation. While on average more than 60% of the cities' websites contained informational items and even 65% a complaint box, only 25% offered a forum and 37% other kinds of engagement or participation. The different items were put together to an index on political information and citizen dialogue, reaching 45% and 30% respectively as mean values.

(2) Online Planning in Germany

Only one larger quantitative study dealing with online consultation in urban planning was found.

In 2006, the German Institute for Urban Studies (Deutsches Institut für Urbanistik) wanted to assess the effects of a new clause in the German Federal Building Act, which allowed the use of electronic media in town planning procedures. Its scope extended to participation procedures for other public authorities and the general public. Out of the sample of 235 responding municipalities, 60% publish their maps and documents on the Internet and use it for consultation of other authorities or the public, but only 9% publish comments they receive (Strauss 2005).

The online offers did not meet their expectations. They did not increase the number of comments nor did they save costs in the majority of cases. The most frequently mentioned obstacles and difficulties concerned hardware and software facilities of citizens and download time (23-60%), and readability of plans (37%).

(3) State of the Art of Geographical Information Systems in Participatory Processes

As mentioned above, eParticipation tools have to be linked with geographical information systems (GIS) in urban planning applications, where plans are generated and administered for internal purposes. Expectations are high, expressed among others by the term Public Participation GIS (PPGIS). The term was coined at meetings of the US National Center for Geographic Information and Analysis (NCGIA) and denotes the possibility of empowering and including marginalized population groups in planning at the local level by employing spatial and visual tools (e.g., Craig et al. 2002; Sieber 2006). Technically there are several concepts and pilot systems for spatial decision support systems or collaborative planning support systems. However, in a comparison of the use of GIS-based websites for spatial planning in Denmark and Italy, Campagna found that these websites offer mainly just one-directional information, whereas interactivity is limited to search functions.

While research papers show promising opportunities for development in this field, "realworld surveys on GIS-based public administration websites in Italy, Austria and Denmark) show as a general trend a weak attitude of local communities in developing online participatory processes so far." (Campagna 2007)

Finally, low acceptance by the public is found for GIS-based participation systems as well. Steinmann et al. compared case studies of experiments and pilot projects of GIS-based citizen participation in the US and Europe. They summarized that technology is well developed but not applied in a useable way for citizens. Therefore, in their empirical case studies, participation could not be increased significantly. The authors call this reaction of citizens "rational ignorance", meaning that the cost of getting the information to participate in a planning procedure, including learning how to use the GIS tools, is greater for many citizens than the gratification caused by this information, and therefore it is rational not to use the systems (Steinmann et al. 2004).

FUTURE RESEARCH DIRECTIONS

Recently, DEMO-net, a network of eParticipation researchers funded by the European Commission, conducted a review of the state of the art of their research area. Macintosh and Coleman state that after 20 years this field of research is still explorative and descriptive. Acknowledging the start of some explanatory and multi-disciplinary research efforts they maintain:

"That a key challenge is to encourage more - and more sophisticated - collaborative multi-disciplinary research in this area with a view to establishing a set of research questions and problems that are specific to eParticipation. Achieving this will ultimately require more dialogue on the part of eParticipation researchers. One initial and broad

suggestion is to focus on the commonalities and links between our shared object of research, by analysing, differentiating and comparing ecologies of eParticipation. This means refusing the research convenience associated with treating specific examples of eParticipation in isolation. It would mean exploring the commonalities and links that exist between different eParticipation activities, in terms of technology, systems, structure, and patterns of use, as analysed by the different research methodologies, expertises and perspectives outlined in this chapter" (Macintosh and Coleman 2006, p. 37).

Similar conclusions were drawn by the international ESF Research Conference on Electronic Democracy, held in November 2007 in Sweden, where leading eDemocracy researchers from more than 10 countries discussed the state of the art and the challenges of eDemocracy research. Present research was assessed as mostly oriented to singular cases, which cannot be compared because subjects, target groups, ways of participation, electronic tools and other communication channels differ. Due to so many variables, no case corresponds to another. Despite the diversity of the relevant aspects and the great number of disciplines that have to join in, there is no trend toward joint theoretical or methodological foundations discernible. Before success factors can be theoretically determined and empirically proven, there must be a common understanding of what a success is and how it can be measured. This has to be done through multidisciplinary collaboration (Kubicek 2008).

There are three future trends, which promise major improvements in eParticipation research in the context of urban planning:

1. Analyzing e-tools in regard to their fit within a multi-dimensional context;
2. More emphasis on trust and confidence building;
3. More rigorous evaluation according to multi-layer frameworks, including a stronger concern for context-related impact.

E-Tools and their Fit within a Multi-Dimensional Context

The general requirement of putting **eParticipation tools** in their procedural and institutional context can be operationalized by referring to organizational, cultural, legal, economic and technical aspects. Participation tools and procedures are not autonomous or isolated entities but offerings for communication between public authorities and citizens, or more general the public, and have to fit into the respective contexts on both sides to the communication (Figure 3).

Organizational Fit

Organizational fit on the supply side of eParticipation means that tasks for developing and applying the technical tools, managing the communication process, and taking care of the output have to be assigned to organizational entities. The relevant workflows have to be amended, adapted or re-engineered. In many cases, at least three parties are involved on the supply side: political decision-makers, one or more offices or departments responsible for the subject at stake (zone planning, traffic, energy) and the ICT department, often with external providers as subcontractors. And often there is no clear agreement on the distribution of tasks and the objective. Politicians may ask for a forum, and a technical tool will be implemented. But it is not clear who shall moderate according to which rules and to which regard contributions shall be summarized by whom and for whom. In the UK review, Pratchett et al. (2005) found that the lack of subject and democracy related knowledge on the side of the providers of the technical tools prohibits clear and common objectives for an eConsultation. So again many officials question the benefits, but indeed have no clear objec-

Figure 3. Five-Dimensional double-sided fit of eParticipation tools

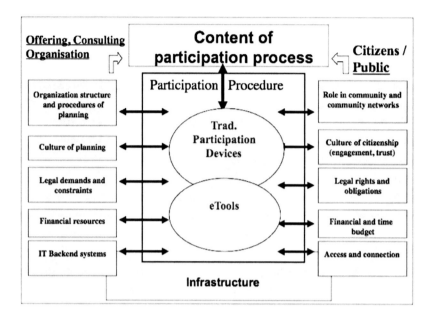

tives and strategies for exploiting the potential. But as with any other large process innovation, eParticipation projects have to be organized and managed well. Otherwise there is a vicious circle: A poorly managed project will deliver poor results and reduce the readiness to put more efforts into this area.

On the side of the citizens or public, it is important to make a difference between individual citizens and communities or networks to be addressed. If communities or networks within the target population exist, their internal organization has to be considered when designing tools and processes.

Cultural Fit

E-Participation is culturally well embedded and successful when the communication via technical devices corresponds to the expectations, values and informal norms of all participating organizations and individuals. In the context of urban projects this includes the professional norms of urban planners, the norms of elected councilors and officers regarding local politics and democracy as well as the attitudes of citizens towards politicians and the ideologies of political or civic society organizations.

In their interviews with local councilors and officials, Pratchett et al. found that a restricted understanding of local democracy is the main barrier for limited offerings of eParticipation. To many councilors, the primary role of local government is not democracy but service delivery. They assume that this is what their customers want (2005, pp. 23f.).

The same is true for urban planners who according to Creasy et al. (2007) perceive their profession as developing plans for the public and to the public, but not with individual citizens. Therefore many planners perceive citizen participation as a threat to the quality of plans and a critique of their professional expertise.

Both findings point to the existence of a vicious circle with self-fulfilling prophesies. Elected political decision-makers and officials assume that citizens primarily expect high quality services from their local government, just like commercial services of banks, or insurance companies. They regard citizens as customers and assume that they do not care for democracy. This assumption is frequently confirmed when offline or online discussions are offered and only a few people take part. But the reason might not be lack of interest, but rather the wrong offer. Surveys show that many citizens are interested in having their voice heard. But from most of the offerings today they do not expect that they are really listened to. In other words: Only when a culture of active local democracy is developed and citizens regain trust in their electorate, they will re-engage.

Legal Fit

As shown in the first chapter, there are several legal requirements for citizen participation in different areas of urban planning. If eParticipation and communication via e-tools is offered in these contexts, it has to conform to the respective legal requirements. As in eGovernment, the authentication of the consulted citizens and/or the integrity of the message may be required by law. For e-devices this means that they have to provide some authentication and encryption features. In Germany, planning offices in some formal procedures accept e-mails as official contributions, while others in other procedures require written and signed letters, which would require digitally signed messages for online appeals. There are also concerns whether maps provided on a web platform meet the integrity requirements of an official document.

Economic Fit

Of course e-consultation requires financial resources for the technical tools and the management of the communication process, i. e. moderation, analysis and summaries, which cost often more than the hard- and software. In the UK survey, limited financial resources appeared "to be particularly problematic". The organizational confusion, who is responsible for what in eParticipation – see above, directly leads to the financial confusion about who is to pay. While e-government activities are expected to provide some cost savings in service provision, eParticipation projects will cause additional expenditure for a long time because they will mainly be offered as an additional communication channel as long as there is a digital divide. As many projects have a general political dimension and a specific subject, different budget positions are affected. Another aspect in regard to the economic dimension is an often prevalent concern that consensus reached in consultations might lead to more expensive solutions.

Economic fit also concerns the side of the addressed constituency. The use of the e-tools has to be affordable to everyone addressed. This means that they cannot be limited to user groups which have broadband access from their home. In order to achieve some degree of equality of chances to participate, public access points and support have to be provided. However, these are very expensive.

As Internet access, even though very widespread now, will remain to be non-universal for quite some time, no formal consultation will be conducted online only. This means that for the foreseeable future no costs for traditional methods can be saved and eParticipation tools will be regarded as additional costs – at least until a culture develops which will demand it.

Technical Fit

Technical fit means that the eParticipation applications offered, for example through a website, have to be integrated into the existing ICT infrastructure and the special software programs and tools used in the offices and departments involved. The empirical evidence from the few examples of electronically supported formal consultation in urban planning showed that exchanging documents between the workflows of consulting party and consulted agencies is a major barrier. Also, maps could only be downloaded effectively if the participating citizens had broadband access. And thirdly, the ease of use when accessing documents from the outside largely depends on the quality of internal document management systems.

Building Trust and Confidence

One of the major objectives political bodies and politicians want to achieve by offering participation is to regain trust and confidence of citizens in the democratic institutions and the decisions they take. However, many case studies show that the eParticipation applications do not contribute to trust building. Instead, trust is already the prerequisite for citizens' engagement in participation processes. If citizens do not trust political bodies, most frequently they will not follow the invitation to participate in a participation process launched by this body. Still, recent research revealed some relevant insight now being looked at by eParticipation practice and research.

Trust Building by an Appropriate Cultural and Legal Environment

Trust can only be gained based on full transparency. Therefore the rules of a consultation or cooperation offer have to be made clear from the beginning. In particular the offering party has to determine how the contributions of the citizens will be handled, who evaluates them, how they may be considered in decisions etc.

Greater motivation can be expected when the decision-making authorities commit themselves to adopt the outcome of a process. In this case, the conditions have to be precisely defined, and it has to be made transparent how these conditions are met. Contracts may be helpful. This was the case in a participation process dealing with the renewal of a public swimming pool in the Free Hanseatic City of Bremen, Germany, where the decision-making bodies committed themselves to adapt the result of the participation process, if this process was fair and neutral (Kubicek/ Westholm 2009).

Balance Between Professional Expertise and Political Fairness

A second crucial aspect identified in recent research on trust in eParticipation is the balance between professional expertise and political fairness. Physical ratios and technical norms in an environmental impact assessment should not be overruled by majority votes of citizens. On the other hand, when there is a choice between different architectural designs for the renewal of a place in town, majority votes might be appropriate.

Therefore the degree of leverage given to the participating citizens should also depend on the subject of participation and the need for professional and scientific knowledge. Because there is distrust in technical norms as well, the best solution is to offer cooperation procedures wherever possible, because here professional planners and experts can work together with citizens and interest groups. Through the course of several meetings, mutual understanding can grow.

There is a certain tension between the professional self-esteem of planners and many other highly educated public servants on the one side and their obligation to consult and adhere to the public. Public servants can limit the potential of citizen participation in many ways. To overcome resistance of the public servants is therefore crucial and can be achieved in two ways (Creasy et al. 2007). To better appreciate what citizens can contribute, practical experience with cooperative procedures is essential. Once a cooperative process has been started, prejudices disappear. And the perception of citizens taking over the job of planners disappears when cooperation is extended to coproduction. In coproduction, the process is not only about improving public plans but also about what citizens can contribute to solve the problem at stake. Therefore a carefully designed trial with citizens' participation can be the first step in a larger change process of participatory urban planning.

Considering Different Levels of Engagement

Public involvement is offered in many public areas. No citizen has the time to engage in several of these at the same time. Therefore broad representation in multiple concurrent participatory processes is not to be expected. Only few people will be ready to engage in cooperative processes.

But the impact of trust building may well go beyond the small number of heavily engaged citizens. There are at least three levels of engagement for which different technical tools can be provided, in particular by Web 2.0 technologies. One can speak of three circles of engagement (cf. Osimo 2008).

- There is an inner circle of heavily engaged citizens who write comments in electronic forums, attend meetings etc. In Web 2.0 terminology, this group produces or generates content.
- Around this inner circle is an intermediary circle of people interested in the subject. For several reasons they do not want to produce content, but are ready to review or comment in an anonymous way. For these groups ratings are an appropriate means of involvement.
- Both circles are watched by a group of people reading the content generated by their fellow-citizens, who also pay attention to how the offering agency responds. They often get interested via a newspaper or local television report and only then look at the respective webpage.

Osimo estimates that out of a total target group of 100% there may be 3% in the inner, 10% in the intermediary and 40% in the outer circle (Osimo 2008, p.18). If this is the case, trust could be built effectively by employing Web 2.0 based ratings and extending an active communication strategy to traditional mass media. To assess this kind of impact, research methods cannot be limited to online surveys of users but have to be complemented by representative telephone surveys.

Legally Binding Procedures and Multi-Channel Applications

One result from the surveys was that there are almost no online offers for legally mandatory consultation procedures or local referenda. They are the form with the highest influence and therefore should be at the end of the deliberative process. But these procedures require equal chances for all entitled to participate. eParticipation applications can make participation possible for people who cannot come to an office to look at a housing plan because of mobility restrictions. But online access is not evenly distributed (digital divide). eConsultation meets new barriers and, as the single channel of communication, can never achieve equal chances. Therefore a multi-channel approach is mandatory. And for the socially and economically disadvantaged additional support has to be organized.

Another reason why there are almost no formal online appeals or referenda is a lack of technical security, which is the same problem as for eVoting. For a legally binding consultation procedure, the same principle of one person, one vote applies. Only the principle of secrecy may not. However, there must be a valid authentication procedure. Digital signatures and electronic identity cards may solve this problem eventually. But for the next ten years, they will not be the ubiquitous means of identification of citizens in most European countries. Therefore other means of authentication have to be provided and probably be recognized by legal directives or amendments to existing legal participation regulation.

Systematic Evaluation

From the point of view of planning authorities and funding agencies, there is an increasing concern for assessing the effectiveness and efficiency of eParticipation.

Multi-Level Frameworks

According to the basic premise of this chapter that **eParticipation tools** have to be analyzed and assessed in their context, a multi-level framework is required for evaluation. The European Research Network DEMO-net has reviewed the state of the art of evaluation research on eParticipation and issued conceptual and methodological recommendations. Based on the work of Macintosh/Whyte (2006), a three-level framework has been proposed (Lippa 2008).

- The first perspective adopts a socio-technical view and looks at the e-tools and their use.
- The second perspective looks at the participation process, which is mostly organized as a project.
- Finally the democratic perspective deals with the long-term impact within the political system including the issue of trust.

So far this concept has not been applied in any comparative assessment. From a methodological point of view the three views do not meet the requirements for valid classifications such as the comprehensives of the classification and the exclusiveness between the classes, With regard to the context related approach proposed in this chapter, the classification seems to focus too much on the e-tools while at the same is disregarding the goal achievements in the subject area of the respective participation processes.

Goal-Related Impact and Subject-Oriented Differentiation

For many participation researchers participation is a goal in itself, and they confine evaluation to the impact on strengthening democracy (i.e. the democratic perspective). However, participation is also a means in the political process to achieve specific goals, such as gathering additional ideas and reaching better informed decisions, to get to plans which are more highly accepted or in the case of environmental issues to reduce emissions

To get hold of these aspects as well and to provide for a more systematic assessment the adoption of a structural model developed for assessment of projecrs within the UN environmental program and recommended by the OECD seems for promising. This model distinguishes five categories of factors to be assessed:

Table 4. The DEMO-net proposal: three perspectives for evaluating eParticipation

Level	Criteria	Method
Tool Level (socio-technical view)	usability, usefulness and usage of eParticipation tools	expert reviews of the websites and usability tests with selected users, content analyses of forum contributions, log-file analyses of use frequency, online questionnaires and offline questionnaires.
Process Level (project view)	different characteristics of participation processes such as transparency, commitment, responsivity, attitudes and behaviour of actors involved, reasons for not participating	stakeholder analysis, expert interviews and focus groups, also online questionnaires and representative surveys (phone interviews)
Political Level (Democratic Perspective)	long-term positive effects of (e-) participation, building of trust in political institutions and politicians and the readiness to political and civic engagement	online surveys and representative phone interviews

The Potential of E-Participation in Urban Planning

- Input = available resources,
- Activities = implementation of resources,
- Output = product or produced service,
- Outcome = immediate and direct effects,
- Impact = conditioned long term and broader effects

To adapt this model for the evaluation of (e)participation processes we may add a category "process" and propose more concrete indicators for each category (see Figure 4). A great advantage of this model is the distinction between the participation offers as the output, the usage of these offers as outcome and the impact of this usage. With regard to the impacts we propose to separate changes in attitudes, e.g. trust in political bodies, from changes in behaviour, e.g. increasing political engagement or participating in elections, but also changing consumer behaviour on the individual level and the overall goal achievement in the subject area as a consequence of these individual changes.

Rose, Lippa and Rios (2009) have compared participatory budgeting at the local level. Pratchett et al recently have submitted the most comprehensive and well founded comparative assessment of more than 150 (e)participation processes, dividing them into six categories: e-participation / e-consultation, participatory budgeting, petitions, redress, asset transfer and citizen (self)-governance (Pratchett et al. 2009). Their comparative analysis, is based on available reports. It is mainly looking for impacts on the individual and community level and for success factors within each category. To compare and assess the impact of different communication channels an international project, under the coordination of the author of this chapter has just started looking at the same form of participation within the same subject area to achieve the highest possible degree of comparability. In Austria, Germany and Spain in three cities each two panels of hundred or more citizens commit themselves to support their city government in reaching the objectives of local CO_2 saving policy goals by changing their consumer behaviour and regularly report data on their energy consumption, mobility and food related behaviour. For this purpose additional tools, in particular a CO_2 calculator have been introduced. Participants submit their data every two month and get a feedback where they are standing in relation to comparable households.

Figure 4. A structural model for evaluating (e)Participation

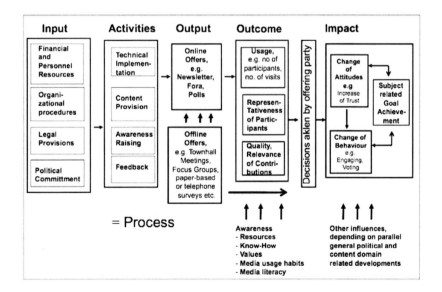

According to Thaler and Sunstein (2008), this kind of feedback creates n special motivation to compete and engage more strongly. In order to assess the overall impact the achievements of the panelists, i.e. the participation effect, their data on energy consumption will be compared with the average of households in comparable neighborhoods, provided by the local utility. The effects of different communication channels will be analyzed for the first time in a large field experiment design: One panel communicates online, enters the data via the Internet into the calculator and immediately gets feedback by statistical figures, receives electronic newsletters and participates in online-fora, while the second panel employs a paper-household book to file their consumption data and reports them via telephone interviews, receives a printed news letter and is invited to physical focus group meetings (for moré details see http://www.e2democry.eu).

CONCLUSION

With the rise of the Internet it was expected that eParticipation could overcome the limits of traditional forms of citizen participation in urban planning. But many case studies show that this did not happen. Often the technical tools created new barriers to access. But most of all, if people are not interested in a subject or do not trust the political bodies, they will not change their mind and engage in a participation process only because it is offered on the Internet. Therefore there is high consensus that eParticipation has to be embedded and analyzed in its procedural and institutional context. Trust and engagement have to be brought about by procedural and institutional innovation. E-tools can support these innovations in several ways.

Web 2.0 technologies in particular do not only allow for more active contributions by citizens and citizen-to-citizen exchange. By offering opportunities for short comments and one-click-ratings they increase the outreach by involving people who are not ready for more extensive engagement.

Regarding eParticipation research there is a trend to go beyond individual case studies and to conduct comparative studies and to explain differences. This requires multi-disciplinary concepts and methods. Recently, proposals have been made for more comprehensive and rigorous evaluation of the effectiveness of eParticipation. In practice, however, there are still several barriers to implement these recommendations. A crucial question arising from these debates concerns the comparability of participation processes dealing with different subjects. The basic idea of putting eParticipation into context in its final consequence requires an evaluation, which needs to add content and context-related to the general democracy-related criteria. eParticipation in local budget planning has to be evaluated by some different criteria than a land-use plan or local climate change policy. Such a differentiation raises new conceptual and methodological issues and might even lead to a paradigm shift in eParticipation research.

REFERENCES

Aichholzer, G., & Westholm, H. (2009). Evaluating eParticipation Projects. Practical Examples and Outline of an Evaluation Framework. *European Journal of ePractice*. Retrieved from http://www.epracticejournal.eu/document/5511

Arnstein, S. R. (1971). A ladder of citizen participation. *Journal of the American Planning Association. American Planning Association*, *35*(4), 216–224. doi:10.1080/01944366908977225

Campagna, M. (2007). *Citizens Participation in Urban Planning*. Summary Note for the ESF-LiU Conference, Electronic Democracy - Achievements and Challenges. Vadstena, Sweden, 21-25 November 2007. Retrieved December 13, 2008, from http://www.docs.ifib.de/esfconference07/conf_programme.html

Coleman, S., & Goetze, J. (2001). *Bowling Together. Online Public Engagement in Policy Deliberation*. London, UK: Hansard Society and BT. Retrieved from http://bowlingtogether.net

Congress of Local and Regional Authorities. (2000). *Guidelines for a policy on citizens' responsible participation in municipal and regional life. Resolution 91*. Strasbourg, France: Council of Europe.

Congress of Local and Regional Authorities. (2008a). *Electronic democracy and deliberative consultation on urban projects. Resolution 267*. Strasbourg, France: Council of Europe.

Congress of Local and Regional Authorities. (2008b). *Electronic democracy and deliberative consultation on urban projects. Recommendation 249*. Strasbourg, France: Council of Europe.

Council of Ministers. (2001). *Recommendation to member states on the participation of citizens in local public life. Recommendation (2001)19*. Strasbourg, France: Council of Europe.

Craig, W. J., Harns, T. M., & Weiner, D. (Eds.). (2002). *Community Participation and Geographic Information Systems*. London: Taylor and Francis.

Creasy, S., Gavelin, K., Fisher, H., Holmes, L., & Desai, M. (2007). *Engage for Change: The Role of Public Engagement in Climate Change Policy. The result of research undertaken for the Sustainable Development Commission*. Retrieved December 11, 2008, from http://www.sd-commission.org.uk/publications.php?id=618

DEMO-net. (2007). Introducing eParticipation. DEMO-net booklet series No. 1. Brussels, Belgium: European Commission European Commission. (2003a). Directive 2003/4/EC. On public access to environmental information.

European Commission. (2003b). *Directive 2003/35/EC*.

Green Mountain Institute for Environmental Democracy. (2005). *Environmental Democracy - What's in it for me?* Montpellier, France. Retrieved December 11, 2008, from http://www.gmied.org/comment.htm

Hagen, M. (2000). Digital Democracy and Political Systems. In Hacker, K., & van Dijk, J. (Eds.), *Digital Democracy: Issues of Theory & Practice* (pp. 54–69). London: Sage.

Honor Fagan, G., Newman, D. R., McCusker, P., & Murray, M. (2006). *E-consultation: evaluating appropriate technologies and processes for citizens' participation in public policy*. E-Consultation Research Project, Final Report, 14 July 2006. Retrieved December 11, 2008, from http://www.e-consultation.org

Initiative eParticipation. (2005). *Elektronische Bürgerbeteiligung in deutschen Großstädten. Zweites Web-Ranking der Initiative eParticipation*. (M. Bräuer & T. Biewendt, Eds.). Retrieved from http://www.initiative-eparticipation.de

Innes, J. E., & Booher, D. E. (2004). Reframing Public Participation: Strategies for the 21st Century. *Planning Theory & Practice*, 5(4), 419–436. doi:10.1080/1464935042000293170

Kiss, C., Poltimae, H., Struminska, M., & Ewing, M. (2006). *Environmental Democracy. An Assessment of Access to Information, Participation in Decision-making and Access to Justice in Environmental Matters in Selected European Countries*. The Access Initiative Europe and The EMLA Association. Retrieved December 11, 2008, from http://www.emla.hu/img_upload/0aa155da39c21509c55c587879f86484/TAI_1.pdf

Kubicek, H. (2008). Electronic Democracy Achievements and Challenges. In *the ESF Research Conference in Vadstena*, Sweden, 21-25 November 2007. Retrieved from http://www.docs.ifib.de/esfconference07

Kubicek, H., Lippa, B., & Westholm, H. (2009). Medienmix in der lokalen Demokratie. Berlin, Germany: Ed. sigma Kubicek, H., & Westholm, W. (20010). Consensus Building by Blended Participation. The Case of the Public Stadium Swimming Pool in Bremen. In S. French & D. R. Insua (Eds.), E-Democracy: A Group Decision and Negotiation Oriented Approach. Berlin: Springer.

Lippa, B., Aichholzer, G., Allhutter, D., et al. (2008). eParticipation. Evaluation and Impact. *DEMO-net Booklet 13.3.* Retrieved from http://www.ifib.de/publikationsdateien/DEMOnet_booklet_13.3_eParticipation_evaluation.pdf

Macintosh, A., & Coleman, S. (2006). Multidisciplinary Roadmap and Report from eParticipation Research. *DEMO-net Deliverable D 4.2.* Retrieved from http://itc.napier.ac.uk/ITC/documents/Demo-net%204_2_multidisciplinary_roadmap.pdf

Macintosh, A., Coleman, S., & Lalljee, M. (2005). *E-Methods for Public Engagement: Helping Local Authorities communicate with citizens.* Bristol City Council for the Local eDemocracy National Project. Retrieved December 11, 2008, from http://itc.napier.ac.uk/ITC/Documents/eMethods_guide2005.pdf

Macintosh, A., & Whyte, A. (2006). Evaluating how e-participation changes local democracy. *eGovernment Workshop '06 (eGOV06)*, 11 September 2006, Brunel University, London.

Macintosh, A., & Whyte, A. (2007). Towards an Evaluation Framework for eParticipation. *Workshop on Frameworks and Methods for Evaluating eParticipation*, Bremen, Germany.

OECD. (2001a). *Citizens as partners: Information, Consultation and Public Participation in Policy-Making.* Paris, France: OECD.

OECD. (2001b). *Engaging Citizens in Policymaking: Information, Consultation and Public Participation. PUMA Policy Brief, 10.* Paris, France: OECD.

OECD. (2003). *Promise and Problems of E-Democracy: Challenges of Online Citizen Engagement.* Paris, France: OECD.

OECD. (2005). *Evaluating Public Participation in Policy Making.* Paris, France: OECD.

OECD. (2007). *Working Party on the Information Economy. Participative Web: User-created Content.* Paris, France: OECD.

Osimo, D. (2008). *Web 2.0 in Government. Why and How. JRC Scientific and Technical Reports.* Brussels, Belgium: Joint Research Centres European Commission.

Peart, M. N., & Ramos Diaz, J. (2007). *Comparative Project on Local e-Democracy. Initiatives in Europe and North America.* Geneva, Switzerland: University of Geneva. Retrieved December 11, 2008, from http://edc.unige.ch/edcadmin/images/ESF%20-%20Local%20E-Democracy.pdf

Pratchett, L. (2006). Understanding e-democracy developments in Europe. Scoping Paper. In *Ad hoc Committee on e-democracy (CAHDE)*, Strasbourg, 18-19 September 2006. Retrieved December 12, 2008, from http://www.coe.int/t/e/integrated_projects/democracy/02-activities/002_e-democracy/CAHDE(2006)E_Scopingpapers.asp

Pratchett, L., et al. (2005). Barriers to e-Democracy. *Local e-Democracy National Project.* Retrieved May 27, 2005, from http://www.e-democracy.gov.uk

Pratchett, L., et al. (2009). Empowering Communities to Influence Local Decisoin_Making. A Systematic Review of the Evidence. Department for Communiuties and Local Government, United Kingdom, June 2009. Retrieved March 12, 2010, from http://www.communities.gov.uk/publications/localgovernment/localdecisionlessons

Rose, J., Lippa, B., & Rios, J. (2009). (accepted). Technology Support for Participatory Budgeting. *International Journal of Electronic Government.*

Rowe, G., & Frewer, L. G. (2000). Public Participation Methods: A Framework for Evaluation. *Journal of Science. Technology & Human Values, 25*(1), 3. doi:10.1177/016224390002500101

Royal Town Planning Institute (RTPI). (2007). Guidelines on Effective Community Involvement and Consultation. *RTPI Good Practice Note 1*. Retrieved December 11, 2008, from http://www.rtpi.org.uk/download/364/RTPI-GPN1-Consultation-v1-2006.pdf

Sieber, R. (2000). Public Participation and Geographic Information Systems. A Literature Review and Framework. *Annals of the American Association of Geographers, 96*(3), 491–507. doi:10.1111/j.1467-8306.2006.00702.x

Steinmann, R., Kerk, A., & Blaschke, T. (2004). Analysis of Online Public Participatory GIS Applications with Respect to the Difference between the US and Europe. In *UDMS - Urban Data Management Symposium*, Chioggia, Italy, 2004. Retrieved from http://www.salzburgresearch.at/research

Stiftung Mitarbeit. (2007). *E-Partizipation*. Bonn, Deutschland: Beteiligungsprojekte im Internet.

Strauss, W.-C. (2006). *Öffentlichkeits- und Trägerbeteiligung in der Bauleitplanung im und über das Internet. Erste Erfahrungen aus den Kommunen. Fachtagung Bauleitplanung und Internet*. Berlin, Germany: Deutsches Institut für Urbanistik.

Thaler, R. H. & Sunstein, C., R. (2008): Nudge. Improving Decisions about Health, Wealth, and Happiness. Yale University Press 2008

Torres, L., Pina, V., & Acerete, B. (2006). E-Governance Developments in EU Cities, Reshaping Government Relation to Citizens. *Governancy, 19*(2), 687–699.

UNECE Convention on Access to Information. (1998, June 25). *Public Participation in Decision-Making and Access to Justice in Environmental Matters*, Aarhus, Denmark. Retrieved December 11, 2008, from http://www.unece.org/env/pp

UNECE Economic Commission for Europe. (2005). *Synthesis Report on the Status of Implementation of the Convention*. Retrieved December 11, 2008, from http://www.unece.org/env/pp/reports%20implementation.htm

Wiedemann, P. M., & Femers, S. (1993). Public Participation in Waste Management Decision-making. *Journal of Hazardous Materials, 33*(3), 355–368. doi:10.1016/0304-3894(93)85085-S

KEY TERMS AND DEFINITIONS

Aarhus Convention: The short term for United Nations Economic Commission for Europe (UNECE) Convention on Access to Information, Public Participation in Decision-Making and Access to Justice in Environmental Matters agreed upon on 25 June 1998 in Aarhus, Denmark, committing signatory states to promote local programs for sustainable development including the obligation to consult ordinary citizens (http://www.unece.org/env/pp/).

Deliberative Consultation: A two-way form of communication within participatory (planning) processes which seek to achieve some kind of consent between the people / parties involved.

Environmental Democracy: The participation of citizens in planning processes with environmental effects and aims at mutual commitment by citizens and public authorities to change their behavior in order to improve sustainable development.

eParticipation: The support of citizen participation in political decision-making supported by information and communication technology, in particular the Internet.

eTools: Applications of World Wide Web functions employed in eParticipation processes such as online consultations. Most prominent tools are electronic fora, polls/surveys, wikis, online newsletters or chat-rooms.

Evaluation of eParticipation: The systematic assessment of the effectiveness and efficiency of eParticipation projects or processes. There are different suggestions for the definition of evaluation criteria, empirical indicators and measurement methods.

Web 2.0: Recent developments of the World Wide Web within the Internet, which in particular allow users to provide and publish their own content (e.g. pictures or videos) and their comments (as text or via rating scales). Synonyms are "participatory web" or "user-generated content".

Chapter 10
Beyond Citizen Participation in Planning:
Multi-Agent Systems for Complex Decision-Making

Domenico Camarda
Technical University of Bari, Italy

ABSTRACT

The new complexity of planning knowledge implies innovation of planning methods, in both substance and procedure. The development of multi-agent cognitive processes, particularly when the agents are diverse and dynamically associated to their interaction arenas, may have manifold implications. In particular, interesting aspects are scale problems of distributed interaction, continuous feedback on problem setting, language and representation (formal, informal, hybrid, etc.) differences among agents (Bousquet, Le Page, 2004). In this concern, an increasing number of experiences on multi-agent interactions are today located within the processes of spatial and environmental planning. Yet, the upcoming presence of different human agents often acting au paire with artificial agents in a social physical environment (see, e.g., with sensors or data-mining routines) often suggests the use of hybrid MAS-based approaches (Al-Kodmany, 2002; Ron, 2005). In this framework, the chapter will scan experiences on the setting up of cooperative multi-agent systems, in order to investigate the potentials of that approach on the interaction of agents in planning processes, beyond participatory planning as such. This investigation will reflect on agent roles, behaviours, actions in planning processes themselves. Also, an attempt will be carried out to put down formal representation of supporting architectures for interaction and decision making.

INTRODUCTION

The increasing orientation of spatial planning toward the so-called 'environmental sustainability' has brought new agents into the process of plan development. Environmental laws and norms, increasingly widespread in different Countries, legitimate a large social participation in planning debates. For example, the diffusion of environmental impact assessment and strategic assessment regulations in Europe, during the 1990s and 2000s, has generated a wide arena of discussion and also an

DOI: 10.4018/978-1-61520-929-3.ch010

important occasion for the development of new technologies.

Different social groups enter both technical and political debates on the two opposite faces of the environmental context. They contribute to the setting up of spatial plans and projects, but meanwhile they are able to delay or speed up them even wrongfully, in the attempt of defending their stakes, considered as absolute principles (Forester, 1999).

A space of mediation and negotiation is continuously created in confrontations and conflicts, and it is managed by mediators or negotiators. They play a role rather similar to the so-called 'intermediate agents' of the multi-agent distributed computer science, who are essential in the coordination of such complex task organizations (Ferber, 1999).

To date, the theory and the practice about these planning 'multi-agencies' is not able to clear up some theoretical and practical problems that look significant. First, the democratic or oligarchic or even aristocratic basis of multi-agencies (commonsense vs. expert agents, for example; Fischer, 2000) is not clear in its process of building and legitimization, and is open to numerous theorems or cases of impossibility (Arrow, 1963; Owen, 1995). Further, the cooperative compromise approach to solutions is 'logically' inappropriate to phenomena that are external to the disputing agents. Finally, the System Theory's postulate on the 'super-sum of parts' seems not always and not clearly affirmable in knowledge domains (Kalman, 1969). These are only few examples of the major problems that planning multi-agencies have to face.

But plans are technical and 'political' exercises per se, in that they are both constructions of action optimization over time and 'social' organizations toward action, therefore requiring coordination and consensus. As a matter of facts, plans can be also individual constructions, as occurring when for example they are drawn out by a business manager. However, even individual plans may involve multiple agents, because they are typically built upon informational materials developed by other agents (Ferber, 1999).

In general, a large complexity comes out in planning knowledge and exercises, implying innovation of planning methods, in both substance and procedure. The development of multi-agent cognitive processes, particularly when the agents are diverse and dynamically associated to their interaction arenas, may have manifold implications. In particular, interesting aspects are scale problems in distributed interaction, continuous feedback on problem setting, language and representation (formal, informal, hybrid, etc.) differences among agents (Bousquet & Le Page, 2004).

In this concern, an increasing number of experiences on multi-agent interactions are today located within processes of spatial and environmental planning. Case studies have not been deliberately set up using formalized and/or predefined multi-agent-system layouts, as described, i.e., by Ferber (1999) or Wooldridge (2002). Basically, the need for the involvement of a number of different agents during the planning process induced issues of gathering and exchanging complex knowledge, representing structured concepts, supporting different formal/informal languages, structuring complex problems, allowing synchronous and/or asynchronous communication. Therefore, a multi-agent approach and supporting system (MAS) emerged in a bottom-up fashion, driven by the needs of the activities being carried out.

Yet, the upcoming presence of different human agents often acting *au pair* with artificial agents in a social physical environment (e.g., with sensors or data-mining routines) often suggests the use of hybrid MAS-based approaches (Al-Kodmany, 2002; Ron, 2005). However, such approaches have not been formalized properly to date, due to both the lack of time allowed by grants associated to projects - mostly target-oriented - and to a research attitude toward the improvement, rather than the formalization of methodologies during that little time available.

In this framework, the chapter will scan some experiences on the setting up of cooperative multi-agent systems, in order to investigate the potentials of that approach on the interaction of agents in planning processes. Considerations and discussions will be carried out not under the perspectives of sociological or policy analysis or methodologies of participation in planning. Rather, arguments will be developed with a computer-science approach, regarding particularly the potentials of system architectures aiming at supporting interaction and participation.

This investigation will reflect on agent roles, behaviors, and actions in planning processes themselves. Also, an attempt will be carried out to put down formal representation of supporting architectures for interaction and decision-making. In this concern, the discussion will be oriented at singling out aspects, features and arguments beyond the traditional and substantial ontology's of participatory planning.

More particularly, the chapter is organized as follows. A first section is devoted to the description of the case studies, with critical considerations, comparisons and possible generalizations. A second section tries to sketch out the different interaction models in the different interaction environments, as basic reference for the subsequent multi-agent research discussion. A third section presents and discusses a possible taxonomy of agents in multi-agent-based processes of environmental planning, with reference to the multi-agent research literature (Ferber, 1999; Wooldridge, 2002). Brief outcomes and remarks are drawn out in the fourth concluding chapter, particularly addressed at outlining the possible learning for urban and environmental planning activities.

BACKGROUND

Multi-Agent Interaction Systems

The key environment for the present work develops on a set of experiences carried out by the Planning Research Group of the Department of Architecture and Urban Planning (DAU) of the Technical University of Bari, Italy. The first set of research activities (1999-2003) has been financed by the European Union, and oriented to building up sustainable development scenarios in the Mediterranean region to support planning decisions. A further analysis is developed on a second set of case-studies carried out in some southern Italian cities (2004-2008), mainly oriented at setting up hybrid human-artificial multi-agent interactions.

In the first set of research activities occurred from 1999 to 2003, the initial activity was concentrated on the Tunis case, and its topic was the interplay between agriculture and urbanization (Khakee *et al.*, 2002a). Then, the activity in Izmir (Turkey) dealt with coastal zone management and the sustainable use of coastal urban areas (Khakee *et al.*, 2002b). The third activity occurred in Rabat (Morocco) and dealt with the impact of market globalization on the actual and potential use of local resources, under the perspective of the emerging Euro-Med free trade zone (Barbanente *et al.*, 2007).

The aim of all three cases was the setting up of a collaborative, interactive, participated community-based process to build future (shared) scenarios to which public managers could orient policies and decision on socio-economic and environmental domains. The scenario-building activity was based on a variant to the *strategic choice* approach by Friend and Hickling (1997), as developed by subsequent evolutions of futures studies. In particular, the Tunis case was based in the well-known future workshop methodology by Jungk and Mullert (1996), by which a sequence of face-to-face brainstorming forums is set up, in order to cooperatively single out strategic lines to build alternative development scenarios. Although paradigmatic in methodological terms, the Tunis case was developed in a very simple context of interaction, with few basic agents and technological support, so raising less interest in terms of MAS architecture.

Figure 1. The future workshop methodology (Khakee et al., 2002a, p.586)

Future Workshops		
PHASE	CONTENTS	EXPECTED RESULTS
1. Preparation	The issue to be analysed is decided and the structure and environment of sessions are prepared.	Summary of contributions.
2. Critique	Clarification of the issue selected, of dissatisfactions and negative experiences in the present situation.	Problematic areas for the following discussion definition.
3. Fantasy	Free idea generation (as an answer to the problems) and of desires, dreams, fantasies, opinions concerning the future. The participants are asked to forget the practical limitations and the obstacles of their present reality.	Indication of a collection of ideas and choice of some solutions and planning guide lines.
4. Implementation	Going back to the present reality, to its power structures and to its real limits, to analyse the actual feasibility of the previous phase solutions and ideas. Identification of obstacles and limits to the plan implementation and definition of possible ways to overcome them.	Creation of strategic lines to be followed in order to fulfil the traced goals. Action plan and implementation proposal drawing.

Basic Process and Methodology

In its original framework, the future-workshop approach starts off with a preparatory phase followed by three operative phases, called critique, fantasy and implementation (Figure 1).

The preparation phase is devoted to the definition of the issue at hand and the practical organization of the brainstorming sessions. When envisioning future, it is commonly assumed that there is a previous general dissatisfaction with aspects of the present and an underlying desire to change the current bases of development (Ziegler, 1991). Therefore, in the phase called 'critique', the participants investigate the selected issues by focusing on problems and negative experiences connected to the present situation. In the phase called 'imagination', the participants are then asked to forget the limitations and the restrictions of the current situation and to generate ideas freely in the form of desires, opinions, dreams and fantasies concerning the future. In the phase called 'implementation', the participants are invited to return to the real world, with its structures of power and constraints and discuss the actual feasibility of the ideas generated in the previous phase. They are also called to find possible ways available to overcome them.

Starting from this framework, the procedure subsequently designed for the Izmir case involved some changes in the future-workshop approach. In particular, the first two phases were extended over a period of four months in order to structure the issues relating to the impact of tourism in the coastal areas of Izmir Province. These issues were wide in scope and were affected by many interrelated factors, so being difficult for scenario analysis (Schnaars, 1987; Huss & Honton, 1987). The choice of using ICT approaches only partially in the future workshop exercise was considered as depending on the fact that the European partners who had the methodological experience had not previously applied computer method in scenario building. It was decided to use the computer tool in two phases only, i.e., in the critique and the implementation phases.

The phase called 'critique' started with a questionnaire survey. The participating agents were asked to describe three major structural changes occurred in the Province of Izmir in the past ten years and three major problems of the present situation. A modified version of *Delphi* method (Rowe & Wright, 2001) was applied in order to acquire a common knowledge base on structural changes and current problems to be used for subsequent phases. The computer software *MeetingWorks*[1] was used to support the participants' focalisation on the major changes and problems, to speed up the processing of the answers to the questionnaires and to improve the

Figure 2. An example of the ITC-based session in Izmir

quantity and quality of the knowledge collected (Figure 2).

The subsequent fantasy and implementation phases were carried out in the form of brainstorming sessions in two consecutive days. The participants were divided into two groups, one of which working in the traditional way, whereas the other one making use of the software method. The ICT approach was not used in the fantasy phase in Izmir, in order to enhance the visioning abilities of a distant future. The scenario time horizon was year 2030: in fact, a crucial feature of future workshop is interactive learning and it was feared that a computer-mediated interaction might inhibit such a process in such a delicate and creative phase of the workshop as fantasy.

Then, the implementation phase was devoted to transform visions into operational strategies. Thus the participants were invited to reflect on the obstacles that would prevent their group's visions from being implemented, to single out measures to remove these obstacles and mobilize necessary resources. For comparison aims, the process followed two parallel paths, with and without the software facility. One day of intensive work was necessary in order to complete this final phase, although in practice a considerable flexibility was necessary to ensure the smooth running of the computer-aided session due to technical difficulties in using the software tool.

Strengths and Weaknesses of A Multi-Agent Architecture

The ICT Base

Because of the participatory orientation of the scenario building architecture, the main features of the software tool needed to be easy access, adaptability and user-friendliness, especially for people who were not familiar with computing. Moreover, there should be no psychological barriers for participants in expressing opinions and visualizing future. On the contrary, the technique should enhance flow of ideas and collaborative mood among stakeholders.

Another important feature of the system architecture was related to the fact that typically automated routines lack the control of process flows, when compared to manual, traditional face-to-face interactions. In a traditional multi-agent interaction, issues come out during a continuous process, attended and controlled by facilitators as well as by other participating agents. In an automated routine process, on the contrary, this is often invisible, with a risk of hidden steps and false heuristics. Therefore, the architecture had to be open enough to allow the control of the interaction process. During the preliminary *Delphi* survey, the architecture had to make the convergence process transparent and controllable while during the implementation phase in scenario building it had to ease the building up and sharing of collective knowledge, without misunderstandings, errors or misallocated answers. In this light, as mentioned, it was decided to choose the *MeetingWorks*™ software, after an examination of several software tools to assist decision-making, build questionnaires and manage interactions in multi-agent-based architectures.

Looking at the planning of the computerized session, a list of all the topics and tasks (called steps) to be dealt with during the interaction was prepared. The steps not only represented the tasks but also the tools necessary to perform those tasks. Steps had to take place in a specified sequence and the results of one step were automatically passed to the following step that used the information to complete its task. If one task produced unexpected results, steps could be added to explore the topic before returning to the agenda in order to proceed to the next planned step.

Following literature, such groupware as *MeetingWorks*™ are supposed to enhance productive interaction (Shakun, 1996). However, the software was conceived as a tool for executive or staff meetings in firms and agencies, where the main goal is to achieve agreement on projects and issues. In the case of participatory scenario building, the major aim is to enhance the sharing of knowledge in the multi-agent system, in order to develop alternative community scenarios and support development strategies. In this concern, the developing of the knowledge base in the critique phase was not easily reachable by the standard *MeetingWorks*™. It required ad-hoc improvement of routines and steps. Given the difficulty in setting up fully automated routines within the schedule of the project, it was decided to replace such routines with semi-automated or manual steps performed by a facilitator during the Delphi sessions. A typical example of this work was the re-formatting of stakeholders' response protocols into manageable worksheets as inputs for subsequent steps and external data manipulation.

Admittedly, such operations using the improvised technical expertise of the facilitator introduced a factor of unpredictability, since the data collected could be affected by a high rate of human-generated errors. It is also able to introduce an element of falsity in the interaction among participants and affect the subsequent activities in scenario building (Sillince & Seedi, 1999). In the Izmir case, errors were few because the small number of participants minimized data exchanges and manual manipulations. However, the human agents' roles proved to be critical in this context and determined the need of building hybrid ICT-human multi-agent systems (Ron, 2005).

Further Features in Selected Stages

As said, the aims of the first two phases of the Future Workshop in Izmir were to constitute the stakeholder group, to get the group aware of the aims, tools and rules of Future Workshop and to acquire a shared knowledge on the problem to be studied. An important task in the first phase was the involvement and selection of the stakeholders. It is not easy to involve a wide range of representatives of local society in envisioning alternative futures in the Mediterranean region because of the dramatic urgency of the current problems that often hinder people in exploring distant futures. Moreover, it is difficult to keep up the interest of local stakeholders for any longer period of time and stimulate them to continue work with future studies even after the European Union project has ended. Moreover, the selection of the stakeholders proved difficult since a major group of them were absentee secondary-home owners who came to Izmir only for vacation, especially during summer. Despite all efforts, it was not possible to involve this group in scenario building. This was only partially compensated by the fact that almost all the municipalities in Izmir region were represented as well as several officials involved in environmental and health care, housing and planning. They were well aware of the problems connected to secondary homes. Moreover, all the participating NGOs were committed to environmental issues, enhancing the significance of the group composition.

In fact, the multi-agent group was dominated by public officials plus some NGOs and business representatives. The average number of participants present at each phase of the Future Workshop was about 30 (a pretty good quantity, especially when compared to previous activities in Tunis; Khakee *et al.*, 2002a), whereas about 60 agents were involved in total.

There was a limited use of software method in the Izmir's Future Workshop. However, a number of general observations can be nevertheless drawn out about the procedure and outcomes of the workshop, focusing on the differences between the traditional and the computer session.

The use of the *MeetingWorks* software proved to be effective in the critique phase of the scenario building. It supported the stakeholders' focalization on major problems and made it easier and quicker to process the answers to the questionnaires. But in the brainstorming the computer session showed a number of limitations when compared with the traditional face-to-face one.

First of all, the software slowed down the idea-generation process because it needed frequent manual operations by the 'chauffeur' in order to arrange intermediate outputs into a format recognizable and usable as an input in the subsequent steps. Moreover, the computer session seemed to be more 'wasteful' than the traditional one because a reduction of proposals was necessary so that the software could manage a limited number of issues and avoid possible "combinatorial explosions". This could occur in the transition from the fantasy to the implementation phases (when the ideas were grouped in issue categories), as well as when the stakeholders were asked to select the most important measures and resources to match the obstacles. Perhaps, the above-mentioned shortcomings have to do with stakeholders' unfamiliarity with the software. That was evident in the steps devoted to the generation of ideas about measures and resources. Here the amount of knowledge produced in the computer session was greater than the traditional one. This is important since a primary aim of Future Workshop is to stimulate creative thinking and to raise as many ideas as possible (Ziegler, 1991).

Anyway, a detailed analysis of the protocols shows redundancy in the greater number of ideas generated in the computer session. The traditional procedure was not free from this problem either, but only a certain number of ideas produced in each step were really taken up in subsequent steps, and repetitions were avoided by the free interaction during collective brainstorming.

A number of problems referred above have to do with the intrinsic rigidity of the computerized procedure. For example, in the traditional session the organization of the knowledge according to issue categories is quite flexible, moving cards from one place to another (Figure 3). In the computer session, the same procedure is quite time-consuming and cumbersome because the policy actions need to be added for every generated idea. Of course, the outcomes can be rearranged in a more compact way. But such an operation minimizes one of the advantages of the computer use, i.e., the speed with which the workshop outcomes are available in a digital format.

Moreover, it seemed that the computer session could not make use of the iterative aspect of the Future Workshop. While the participants in the traditional session could go back to charts displaying the results from the fantasy phase as well as all the previous steps in the implementa-

Figure 3. Moving labeled cards under general categories in face-to-face scenario building

tion phase, this opportunity was not available to the participants in the computer session. In the words of one of the participants of the computer session "we are not as aware of the complexity in the links and interrelationships between future visions and our work in proposing measures and resources to implement these visions".

The intensity of interaction between the participants in the traditional brainstorming session reflected in a way the cultural and social interaction that Mediterranean people are normally accustomed to. While the requirement to work in front of one's own keyboard created a sense of isolation. Moreover it contributed to unwillingness to share ideas with the others in the group. One of the participants in the computer session complained that "the enthusiasm and social interaction of working together was missing while working with the software".

In conclusion, it should be remembered that in the Izmir case all the advantages of the software was not fully exploited. For example, a quick prioritization of ideas through the voting and ranking, the use of the "same-time-same-place" feature that makes the results available to all participants, allowing them to discuss and modify the results face to face. Moreover, the computer could not be used in 'remote' sessions, where stakeholders interact regardless of the geographical distance. Finally the break in the flow of interconnected steps between the fantasy and the implementation phase could be avoided if the software method was used for the entire process.

Anyway, problems and limitations of the use of computer software for the multi-agent system, which emerged from our experimentation, need to be evaluated in the light of the fast growth of the information society. It is well known that technologies support a rapidly increasing range of activities and processes throughout the society. A number of initiatives are being promoted by international and national institutions in order to enhance trust and confidence in computer technologies, and develop tools required by individuals and groups to operate in new organizational environments (Wagenaar & Hajer, 2003).

Constructivists say that computer technology is socially shaped, rather than an autonomously developed by scientific and technological forces. This approach stresses the flexibility of technology, the possibility for different choices, and the basic insight of alternatives (Bijker *et al.*, 1992).

Some Improvements in the MAS Architecture

As an immediate follow up, the specific aim in the Rabat/Casablanca case study was to explore the impacts of globalization on regional social, economic and cultural transformations. In the beginning, this intriguing but intricate issue was supposed to involve only stakeholders from many different private and public areas. This was planned in order to be able to have an intuitive feedback on this subject from different fields of knowledge and experience and to build comprehensive future images. However, because of the complex and difficult theme for the allegedly inadequate knowledge level of local stakeholders, the usefulness of a participatory approach, also in the preparation of knowledge base, was put forward (Bell, 1997). Therefore, it was decided to spend a significant initial time to define the framework within which stakeholders could elaborate their visions and perspectives. This was carried out before the scenario building activity, by interviewing government officials, researchers, business representatives and World Bank experts. The interviews added to the preliminary data gathered from books and public reports that had been collected on the Rabat region to scan the impact of globalization on an urban region in Southern Mediterranean. In fact, owing to economic linkages, as well as to spatial implications of globalization, the study area was extended to the entire Rabat-Casablanca. It gave the opportunity to examine agricultural and industrial transformation in a region consisting of two cities together with several smaller cities and towns and an extensive hinterland that are intertwined in many ways as they struggle to adjust to local and global forces. Within the region, there is substantial migration of rural households from stagnating agricultural areas primarily to the two cities and also to other urban centers, leading to crowd and to the creation of illegal settlements, but also leading to migration to northern Mediterranean countries. The region is also characterized by several environmental issues, such as desertification in rural areas and water shortage in urban areas, pollution and degradation of land.

After the expert interviews, the choice of the appropriate region for futures study and preliminary discussion with the stakeholders about the meaning of globalization, the first scenario-building session was organized in Rabat. Participating agents came mainly from universities, ministries, and also from NGOs. The first part of the scenario-building session was devoted to contextual developments, particularly with respect to globalization. Within the whole process, this step was supposed to determine a common, although complex, awareness of the variety and richness of the issues at hand, as a basis on which to start constructing scenarios.

The process of scenario building broadly followed the 'future workshop' method suggested by Jungk and Mullert (1996), but modified as a result of applying the method in Tunis and Izmir respectively (Khakee et al., 2002a, 2002b) (Figure 1). The Izmir case showed that the process was generally suitable to be managed by automated ICT-based routines, but with some limitations and necessary amendments. Moreover, in the case of Rabat/Casablanca the theme of 'globalization and regional transformation' suggested the intriguing possibility to check if remote web-based brainstorming sessions could encompass the same-time-same-place approach involving participants that were not physically present at the scenario-building sessions. It was therefore decided to implement the scenario-building process by using an ICT-base multi-agent architecture for interaction and knowledge development, and at the same time explore the potentials and drawbacks of the computer-aided approach. However, in order to try to avoid the risks of impersonal interactions, as well as unmanageable black-box results, human hybridization was enhanced as far as possible. An important aspect of such hybridization was the use of scratchpad during brainstorming, as a form of 'chat' or informal communication. This expedient was fairly successful and facilitated scenario-building in many ways.

After this session, it was decided to end up the scenario-building process in subsequent internet-based distant meetings (Figure 4).

In fact, the web-based interaction involved few stakeholders. However, due to several problems related to the network set up, deficient connections, high internet-call costs, the experiment was cancelled. Instead the remaining part of scenario-

Figure 4. Example of MeetingWorks interaction agenda

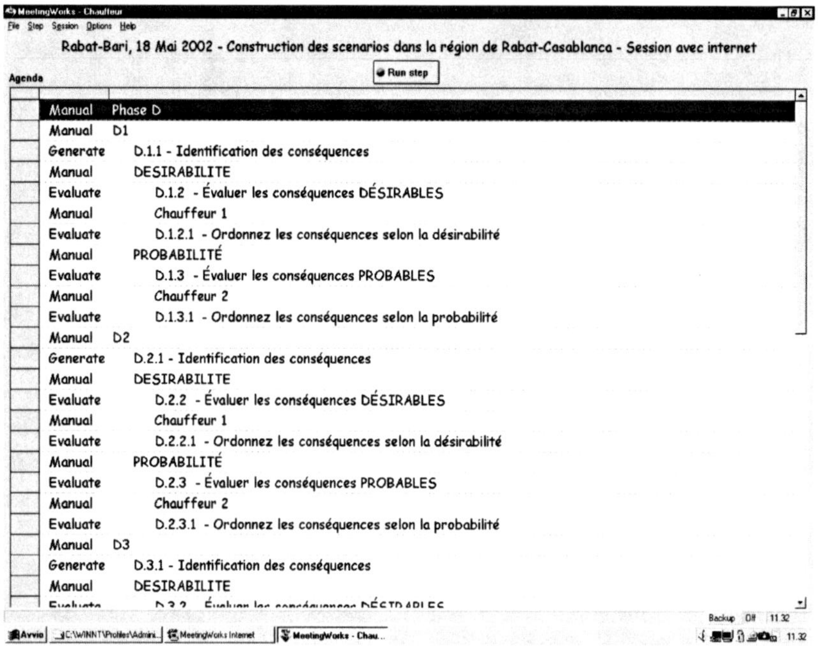

building was carried out in Rabat few months later. The number of stakeholders was increased and included NGO representatives.

MULTI-AGENT ARCHITECTURE FOR DECISION MAKING

The above modeling approaches and methodology have been amended in their sequence and environment, driven by the needs of enlarging the number of interacting agents (that is, the democratic richness of knowledge exchanged) and of making real-time interactions possible and effective with large participation. The initial interacting conditions were then changed with the creation of a more complex ITC-based architecture, and the original process was substantially adapted to those changes. In particular, the Izmir layout was structured in sessions of knowledge agents interacting by pc terminals in a formalized LAN-based indoor (closed-room) environment, whereas the Rabat layout was organized as a two-phase process, i.e., an indoor environment followed by an internet-based remote one (Figure 5).

The original principle for this remote extension was to allow the participation of agents distributed over distant spaces, having interests

Figure 5. The indoor/remote environments

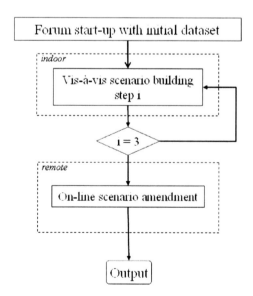

on the context of study. Yet, in the experiences that followed the EU-funded project, a more complex rationale backed the quest for remote interactions. Namely, this was the need of simulating the normal interaction environment of public spaces occurring among agents joining casually and asynchronously (open-room environment), so as to minimize the risks of influencing and forcing knowledge exchanges and maximize the adherence to the complexity of each particular context (Axelrod, 1997).

Such situation occurred in the case study of Foggia (Italy), where an interactive community process was set up to draw the structure plan of its provincial area. Basing on a set of future visions of the province (prepared by the planning office and ranging from environmental to managerial issues), the process architecture aimed at supporting the sharing, enrichment and evaluation of this set of visions and letting decision-makers issue the final plan with related policy programs. In this case, the sequence of iterated face-to-face interactions was just paralleled by a web-based virtual forum (Figure 6), so as to complement a closed-room approach with an open-room one, more coherently with the social and environmental complexity of the real context.

A slight but important amendment of this framework was used in the case of a scenario-building activity for urban regeneration in the small town of Cerignola (Italy). In order to minimize the redundancy of the concepts exchanged during the interaction, and facilitate the delivery of contribution on spatially contextualized issues, the open-room remote interaction was structured so as to allow geo-referenced contributions. Similarly to what described in some literature, this layout proved to be a useful improvement under many viewpoints (Al-Kodmany, 2002; Borri & Camarda, 2006). First, the visualization of a clickable geographic representation of the town area stimulated agents to express their views with an area-based orientation, so focusing issues on micro-scale subjects. Secondly, agents entering the forum in subsequent times were in turn themselves encouraged to deliver new or refined contributions on the same area, following the thread of views previously expressed. Lastly, by generally having a less dispersed contextualization of issues, comments tended to be deepened, so depicting issues with positive and negative aspects.

As a whole, the architecture of interaction processes evolved and changed over years of case studies. From being substantially target-oriented, so as to foster (and force) consensus on issues, it became knowledge-oriented, i.e., aimed at raising the level of involvement and information. The increased number and diversity of interacting agents raised the cognitive aspects of processes that in turn reverberated on a higher complexity of interaction-support systems. Thus, our upcoming awareness is that multiple agent systems

Figure 6. The open/closed room framework

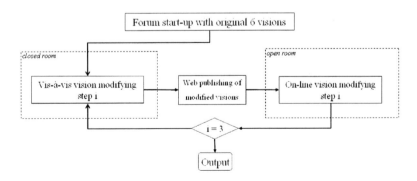

increasingly require architectures coherent with their inherent complexity. These architectures cannot be left to occasional and naive layouts any longer, but need to be studied, articulated and fine-tuned in order to allow informed decision in the social-environmental domain. To the exploration of the implementability of such potential issue is therefore devoted the rest of the chapter.

A Possible Taxonomy

The flow chart of the setting up and carrying out a hybrid knowledge-building process is schematized in Figure 7. Within each phase, numbered A to G, actions are carried out by (single or collective) agents. An in-depth analysis of actions, roles and features of agents can help defining the type of each agent, her cognitive and operational level, so as to provide a raw taxonomy for the building up of the system architecture.

Beyond the layout in Figure 7, agent actions, roles and types are now discussed more in deep. The analysis is carried out in all the stages, basing on compact tables in each stage (Figures 8,9,10,12), as visual guidelines for a better comprehension.

The Preparation Stage

A preliminary phase (00) is needed to conceive and set up the layout of the whole process: this is a strategic activity requiring high-level cognitive and organizational features, as often reported (Anderson & Krathwohl, 2001; Franklin & Graesser, 1997; Van Dyke Parunak *et al.*, 2004). This is an *existing* cognitive community in which human agents carry out activities as *organization* or *coordination agents*, at times supported by *technical agents* in their use of operational tools (Figure 8). Given their higher level of cognitive and organizational features involved, coordination agents can act as technical agents if necessary, by changing level downward, the opposite being hardly possible, of course. However, the level of sector experience and expertise is not equivalent, and this results in occasional, rather than long-time, replacement (as in the case of RFID-based agents) (Ahn & Lee, 2004). Similar level shifts do occur in different phases of the whole process, involving a number of issues that need to be properly analyzed.

In subsequent phase A, the quest for a *new, potential* cognitive community is started off. A preliminary knowledge base is collected by

Figure 7. Flowchart of the closed-open room interaction sessions

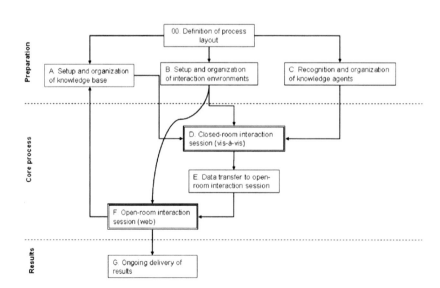

Beyond Citizen Participation in Planning

analysts (*analysis agents*), made up of existing data-sets, maps, raw material, as well as of ad-hoc reports and contributions by sector experts (*knowledge agents*) (Barbanente et al., 2007). Knowledge agents and inference agents draw out and interpret descriptive statistics and trends, aiming at achieving more manageable datasets. Whereas the knowledge agents are essentially the human agents who deliver their formal expert contribution, analysts are often (at least partially) represented by artificial agents, rule-based automatic routines, map/worksheet devices carrying out programmed tasks. In some cases, they are integrated routinary systems (ExSys™, Meetingworks™ etc.), in other cases they are a collection of inter-operational software programmes (search, indexing, inferring, statistical software such as Beagle, Business object™, Gqview, SPSS™, SAS™ etc.). However, especially when dealing with the environmental domain, the overall interaction terrain is too complex to avoid a hybrid human/IT agent involvement. In the end, this informational material is then structured and synthesized into coherent and compact representations, and *communication agents* make the initial knowledge base available and manageable for subsequent process actions.

This preparation stage is aimed also at setting up the interaction environments, and at singling out the agents to be involved in the actual interaction process (phases B and C, Figure 9). These phases play an important role in terms of contents setup, but are critical also under a procedural viewpoint. In fact, although the presence of expert *knowledge agents* is crucial for an adequate definition of initial knowledge, their availability and involvement into activities often need supplementary ad-hoc agents. Such *lobbying agents* are usually provided with a good knowledge of societal roles, organizations, political and/or cultural domains (*intermediate agents*) (Borri et al., 2004), and they are therefore essentially human agents. Very often, only through these agents is it possible to single out, gather and involve experts into the forum arena, in the preliminary as well as in the subsequent phases.

Figure 8. Agent actions, roles and types in the first two stages

	Agent actions in process	Agent role	Actual agent type
	00. Definition of process layout		
1	Defining the layout the specific features of the process	Coordination agents	Human
2	Technical support to layout definition and follow up	Technical agents	Human
	A. Setup and organization of knowledge base		
1	Defining the phase layout	Coordination agents	Human
2	Drawing out basic analyses and reports	Knowledge agents	- if ad-hoc reports: Human - if existing data: Data warehouse (e.g., statistics, libraries, tables etc.)
3	Finding out initial databases	Analysis agents	Rule-based search routine (e.g., websearch)
4	Listing and indexing initial databases	Analysis agents	Rule-based indexing routine (e.g., Beagle, Business object, Gqview etc.)
5	Selecting initial databases	Analysis agents	- Definition of selection criteria: Human (or ad-hoc Expert System) - Actually selecting databases: Rule-based routines
6	Representing initial databases	Analysis agents	Graph & worksheet-based routines (e.g., access, excel etc.)
7	Singling out synthetic indicators	Inference agents	- if needing complex semantics: Human (or ad-hoc Expert System) - if inference from existing data: Rule-based routines (SPSS, SAS etc.)
8	Making descriptive inferences	Inference agents	Rule-based routines (SPSS, SAS etc.)
9	Interpreting inferences	Knowledge agents	Human (or ad-hoc Expert System)
10	Linking and synthesizing inferences	Analysis agents	Rule-based indexing and mapping routines (e.g., Beagle, Business object, Gqview, concept mapping etc.)
11	Representing initial synthesis	Analysis agents	Map, graph & worksheet-based routines (e.g., concept mapping, access, excel etc.)
12	Communicating initial synthesis for general outreach	Communication agents	- if traditional communication: Human - if IT-based communication: Web-publishing routines
13	Making initial synthesis and database accessible for the interaction process	Analysis agents	Rule-based formatting routines
14	Making initial synthesis and database available for the interaction process	Communication agents	- if traditional communication: Human - if IT-based communication: Web-publishing or Web-mailing routines
15	Translating interaction outcomes into feedback data for the initial database	Analysis agents	Rule-based formatting routines
16	Modifying the initial database with process feedback	Analysis agents	Rule-based routines (e.g., access, excel etc.)

Figure 9. Agent actions, roles and types in the third and fourth stages

	Agent actions in process	Agent role	Actual agent type
	B. Setup and organization of interaction environments		
1	Defining the general layout	Coordination agents	Human
2	Defining the closed-room layout	Coordination agents	Human
3	Defining the open-room (web) layout	Coordination agents	Human
4	Supporting the definition of the open-room (web) layout	Technical agent	Human (or ad-hoc self-help-based Expert System)
5	Preparing basic material	Logistic agents	Human
6	Maintenance of the open-room environment	Technical agent	Human (or ad-hoc self-help-based Expert System)
7	Preparing closed-room environment	Logistic agents	Human
	C. Recognition and organization of knowledge agents		
1	Identifying ways of involving agents	Coordination agents	Human
2	Outreach to find out lobbying agents	Communication agents	- if traditional communication: Human - if IT-based communication: Web-publishing or web-mailing routines
3	Lobbying to involve knowledge agents	Lobbying agents	Human
4	Outreach to find out more knowledge agents	Communication agents	- if traditional communication: Human - if IT-based communication: Web-publishing or web-mailing routines

However, as written before, a univocal association between one role/action and one agent is not always realized. Sometimes, an agent who plays an assigned role and develops a typical set of actions decides (or is induced by circumstances) to play even temporarily a role that is typical of another agent, either unconsciously or deliberately. This circumstance, which is typical in some hierarchical-based networks (such as control networks) (Ahn & Lee, 2004), depicts a *multi-level agent*, able to act and/or interact on different levels in the same arena. Nevertheless, the actions of these agents may induce hidden impacts on interactions. In the present interaction process, in particular, if *lobbying agents* are actually knowledge experts as well, the need of decreasing the preliminary period may suggest to include them in the group of *expert agents*. In this case, such agents may use their relational power to boost selected areas of own interest to the detriment of other areas, i.e., to influence interaction. Therefore, it is clear that in many situations *multi-level agents* need to be provided with ad-hoc supporting models to avoid uncontrolled biases and attain effective interactions (Jarupathirun & Zahedi, 2007).

Admittedly, given the reduced size of this phase in the whole process, the building up of the knowledge base is not set up through brainstorming interaction, but rather through dataset and literature analysis, added with series of interviews or *Delphi*-like iterations among the participants (Barbanente et al., 2007). Within such more formalized environment, the risks of influences of the above type are typically minimized. On the contrary, such risks are far more evident in the proper interaction phase of the process.

The Forum Interaction Stage

The proper interaction phase is the core of the knowledge-building process (Figure 10). It occurs under the form of a forum activity, in which each involved social stakeholder is a *knowledge agent*, bringing their knowledge contribution to the arena, changing their or other stakeholders' knowledge contribution or even mind frame (Borri et al., 2004), building shared future visions and scenarios.

The knowledge base collected and synthesized in the previous phase is shown to forum participants as a starting knowledge representation to enrich and build on, by mutually interacting. The interaction process is far more complex than in the preliminary phase, and its layout may vary according to the social/environmental context in which it takes place (Borri et al., 2008). However,

Figure 10. Agent actions, roles and types in the core process

	Agent actions in process	Agent role	Actual agent type
	D. Closed-room interaction session (vis-à-vis)		
1	Supervising the process	Coordination agents	Human
2	Survey of access information on participants	Analysis agents	- If non-IT-based interaction: Human-mediated text-based routines (questionnaires) - If IT-based interaction: sensor-mediated text-based routines (e.g., RFID tags)
3	Cognitive contributions in the interaction	Knowledge agents	Human (or ad-hoc Expert System)
4	Facilitating interactions	Coordination agents	Human
5	Observation and study of interactions for research purposes	Research agents	Human
	E. Data transfer to open-room interaction session		
1	Supervising the process	Coordination agents	Human
2	Synthesizing results of closed-room interaction	Analysis agents	Rule-based indexing routine (e.g., Beagle, Business object, Goview etc.)
3	Transmitting results to web-based open-room	Communication agents	Web-publishing routines
	F. Open-room interaction session (web)		
1	Supervising the process	Coordination agents	Human
2	Survey of access information on participants	Analysis agents	- Without IT devices: Human-mediated text-based routines (questionnaires) - With IT devices: sensor-mediated text-based routines (e.g., RFID tags)
3	Cognitive contributions in the interaction	Knowledge agents	Human
4	Survey of non-written language of participants (complex language)	Analysis agents	Sensor-based routines (e.g., motion-detection cameras)
5	Decrypting complex language	Analysis agents	Rule and ontology-based translation routines
6	Enriching contributions with decrypted complex data	Analysis agents	Rule-based formatting routines
7	Web-publishing complex contributions	Technical agent	Web-publishing routines
8	Technical-logistical support to the process	Technical agent	Human (or ad-hoc self-help-based Expert System)

technical and/or methodological difficulties, as well as cultural diversities may frequently require ad-hoc *coordination agents*, working as *facilitators*. Such agents address a number of different issues, ranging from soft expediting (answering methodological questions, overcoming communication gaps, solving technical troubles), to harder process intromission (fostering dialogue, matching language forms, aggregating scattered concepts, creating consensus) (Jarupathirun & Zahedi, 2007; Warren & Gibson, 2002).

Rather clearly, many of these issues may influence interaction and even affect results remarkably, so challenging the actual effectiveness of the planning exercise. Facilitating agents are either human or artificial, whereas artificial agents are represented by complex supporting systems (groupware interaction programs, WebGISs etc.) or stand-alone software/routines. Although the influence of human mediators may look more evident, also artificial mediators can influence interactions rather strongly, particularly when forum participants must rely on black-box, uncontrolled routines to facilitate interactions (Barbanente *et al.*, 2007).

Also in order to minimize the interposition and influence of mediators, as well as to enhance the randomness and democracy of process participation, an open-room interaction complements the closed-room session. A web-based space of interaction is supposed to create an Agora-like meeting point, where agents strolling throughout the Net leave their contribution in times and circumstances not forced and/or mediated by other agents (Borri & Camarda, 2006)[2]. Knowledge agents are essential to deliver contributions to the virtual forum, but in this phase they cannot be left unsupported. In fact, they need to be integrated and/or helped by different technical agents, devices and sensors that are able to foster a proper, correct, complete, rich and unambiguous language exchange to the participants that make up for non-vis-à-vis contacts. In particular, the so-called *rich language* needs to be introduced into the interaction as much as possible, by complementing typical written sentences with oral, sketched, graphical, gesture languages. Many devices are today regularly used for detecting this rich language (Figure 11) (Veloso *et al.*, 2004; Bravo *et al.*, 2006), that in turn needs ad-hoc agents and software to decrypt

Figure 11. Intelligent motion-detection device (left); RFID-based conference management (right)

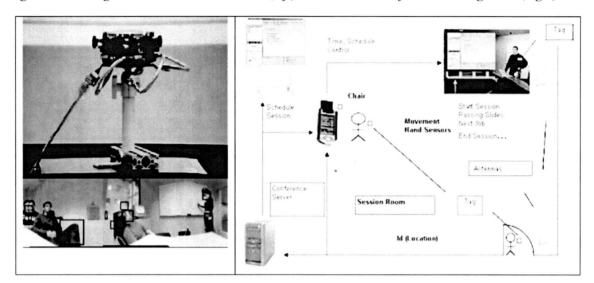

contributions and attach them to written statements so as to complete them. With different features, the agents act as analysts and can be defined and dealt with as *analysis agents*.

The last phase of the process concerns the verification of issues, concepts and data collected and their aggregation and synthesis in proper scenario reports (Figure 12). This is usually considered as necessary in order to avoid both logical inconsistency and data redundancies, as well as to attain more manageable representations for decision-making purposes (Jungk & Mullert, 1996).

The activity of further categories of agents is then involved: they can be called verification agents for the purposes of the process. It is evident that also these agents can, in turn, affect the outcomes of the interaction process, and therefore their activity needs to be deeply understood,

monitored and supported in order to minimize dangerous influences on the process.

An intriguing aspect of this architecture is the identification of the agents involved, their roles, their acting/interacting potentials. In particular, the existence of different possible level of interaction for the same agent is an increasingly frequent issue, due to diverse reasons and contingent opportunities. Chances and reasons for the application of such opportunity are variable, particularly according to the situational contexts of the operation. In the real knowledge-building processes which induced the MAS models dealt with in the present chapter, for example, some issues suggested to reduce and compact the number of human agents. Artificial agents, in turn, are more and more integrated in a complex inter-operational group of routines, or an ad-hoc

Figure 12. Agent actions, roles and types in the last part of the process

	Agent actions in process	Agent role	Actual agent type
	G. Ongoing delivery of results		
1	Verifying data consistency and writing reports	Verification agents	Human
2	Result outreach	Communication agents	- if traditional communication: Human - if IT-based communication: Web-publishing or web-mailing routines
3	Transmitting feedback to the initial dataset	Communication agents	Rule-based routines (e.g. access, excel etc.)

expert system (Ferber, 1999; Wooldridge, 2002). A possible simplified configuration of involved agents, with multiple roles and actions, is reported in Figure 13.

Of course, different levels of action involve a certain ranking of agents according to their lower to higher contextual abilities. That is, agents deputed to supervise or coordinate processes (primary agents) can also act as routinary (secondary) agents upon request; or, similarly, some higher-level agents are deliberately involved in routinary locations of the process because they are expected to perform creative solutions to possible (but expected) problems. This is what usually occurs, e.g., in supply chains (Ahn & Lee, 2004).

FUTURE RESEARCH DIRECTIONS

The theory and practice of multi-agent systems, as applied to decision-making in planning, is particularly addressed at emphasizing roles, behaviors and actions of possibly all agents – living and artificial – involved in knowledge transactions. In this attempt, the spatial environment itself in which transactions take place may represent a key element and play a significant role for the effectiveness of the MAS-based decision support system. However, on this particular side, a diverse scientific context of reference entails diverse considerations of the space in the engineering of intelligent systems, which may in turn strongly affect the architecture and the performance of the system.

For example, on the side of planning and organization studies, space is dealt with either as environment and container of human actions (whose characteristics are considered only from the viewpoint of its adaptability to humanization), or as a relevant, active entity per se (an agent, we could say). Its analysis and modeling reveals parts of human behavior dependent on it (Lynch, 1960; Fischer, 2000; Weyns *et al.*, 2007). On the side of spatial cognition studies space – as abiotic entity – is largely dealt with not per se, but from the point of view of biotic agents who use it to better adapt some of their behaviors that cannot ignore space. An example in elementary actions could be the search for food in a given space, whereas in higher-level actions an example could be the enjoying of a spatial beauty (Ferber, 1999; Tversky, Hard, 2009). There are integrations of the two sides of study, especially where theories and

Figure 13. Knowledge-building process with minimal agent involvement

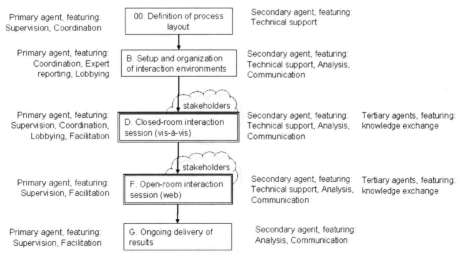

actions of planning and spatial organization – as strongly occurred in recent years – are influenced by knowledge-based approaches (Mockler, 1989; Garling, Evans, 1991).

In the approach that has been presented in this chapter, a 'cognitivist' research perspective has been selected. In it, townscapes and cityscapes (i.e., space) are the prevalent, knowledge-intensive, meaningful spaces and entities humans interact with to carry out their life. However, because of their dynamic complexity, spacescapes seem to offer ill-structured holds to the typical spatial behavior of an agent, and interactions themselves occur in complex, non-obvious and hardly reducible ways.

That is, the space seems to act as an agent itself, and a question arises about the 'fundamentals' of spacescapes from the point of view of the needs of living agents, having to live in and move through them in resilient and intelligent ways. Furthermore, the setting up of a system architecture based on a multi-agent environment in which the environment is an agent per se may put difficult logical, conceptual and management problems.

There have been some experimental attempts to single out features, aspects, elements as proxies able to structure space-environment into the MAS architecture as a proper agent, based on literature on spatial cognition and representation (Danziger, Rafal, 2009; Hartley, 2004, Weyns *et al.*, 2007). For example, the problem connected with the general mechanism of spatial cognition-perception-decision seems interesting, particularly in a human agent who 'navigates' through a given indoor space-environment for the execution of a specific task. In this spatial activity, Goodman (1951) emphasized the importance of the role played by the environment 'structural' components (i.e., 'substances' or 'essences' such as the walls that limit space), and by its 'ornamental' components (i.e., furniture, mobile stuff etc.).

In general, embedding the environment into a MAS-based system represents a hard but challenging issue for an increasing number of research activities. At present, it outlines one of the most interesting study perspectives for the future of participatory planning, both in a logical/ontological and methodological/organizational ways.

CONCLUSION

This chapter has carried out a discussion on the potentials and problems of the setting up of multi-agent systems supporting environmental planning and decision-making processes. Through experiences and case studies some reflections have been developed toward problem solving and setting, but significantly relying also on the support of cognitive and computer sciences. In particular, the today's rising up of an innovative need of non-expert knowledge, beyond the expert one, has determined a new interest toward the cognition aspects of agents' interactions, and toward the ICT architectures able to ease their management. However, in public actions, that awareness has not yet succeeded in changing either the deterministic vision of expert agents or the underestimation of non-expert agents, still intended as mere providers of cognitive contents through a passive delegation to others.

Yet, decision-makers need to rely on distributed expert and non-expert agents to devise and implement planning decisions in socio-environmental domains. In such domains, the habitual cybernetic approach based on determinism and rational process control is now challenged by the participatory and argumentative approach to planning. Here, values become mutable and adaptive, and background structures of thoughts and organizations become flexible themselves. In both human and environmental domains, systems evolve dynamically, at times creating cooperation, but even conflict among multiple agents with multiple cultures. This is then the picture of a weak rationality, not linear nor deterministic, but highly contextual, multi-logic and multi-value, that is destined to replace the typical vision of a strong and absolute rationality.

More than achieving targets, new environmental plans and decisions rely on the building of scenarios and visions related to evolving situations, in which different agents place their stakes, behaviors and cultures while dialectically interacting with other agents and environments. Interaction processes then becomes the key element by which to start off and explore environmental planning paths.

The increasing development of ICT and of computer sciences gives new opportunities to cognitive negotiations, as also witnessed by the described experiences in the Mediterranean context. Negotiations become increasingly multicultural and extended in space and time and such expansion creates in turn inter-cultural and intra-cultural contexts in which negotiations take place.

Meanwhile, however, the roles of the agents involved in the MAS architecture are increasingly critical and less and less obvious than before. Roles change and become differentiated and multifarious, both in their relevant actions and hierarchical location. Even different roles and behaviors are recognizable in the same agent, not necessarily in particular and exceptional processes, but in the normal decision-making contexts of life. Although cognitive patrimony (Anderson & Krathwohl, 2001; Van Dyke Parunak et al., 2004) is a major feature, operational attitudes and representational abilities are rather important as well, even without their own embedded cognitive abilities (Ferber, 1999). The presence of human agents is irreplaceable in many interaction phases, but many other phases do need automatic or semi-automatic routines for real-time, reiterated, influence-free actions. Therefore, a multi-agent system aimed at supporting planning interactions is today more and more intended as a hybrid human/ITC-based architecture (Barbanente et al., 2007; Franklin & Graesser, 1997).

This new hybridist frontier is not free from operational difficulties in system building. In particular, a number of important and critical issues come forth and become evident. For example, we can mention (i) the methods useful to single out agents and roles, (ii) the extent to which they are embedded or context-induced abilities (particularly for human agents), (iii) the impact of a small number of primary agents on complex planning processes, (iv) the impact of compact multi-agent routines on the controllability of data and outcomes (particularly for artificial agents).

Such issues still remain mostly unaddressed. Yet they are very important for both the feasibility and the efficacy of MAS-based architectures, and some of them claim for complex evaluation by different discipline domains. This represents an intriguing research horizon, which deserves further exploration, experimentation and discussion. Therefore, it is not erroneous to predict an increasing attention to such research scenario for MAS studies in a very near future.

ACKNOWLEDGMENT

The author is grateful to A. Barbanente, D. Borri, A. De Liddo, L. Grassini, A. Khakee, M. Puglisi for their cooperation in the research activities that made the present chapter possible. He is particularly grateful to Dino Borri for his precious suggestions in the writing down of the first and the last section of the present chapter.

REFERENCES

Ahn, H. J., & Lee, H. (2004). An agent-based dynamic information network for supply chain management. *BT Technology Journal*, *22*, 18–27. doi:10.1023/B:BTTJ.0000033467.83300.c0

Al-Kodmany, K. (2002). Visualization tools and methods in community planning: From freehand sketches to virtual reality. *Journal of Planning Literature*, *17*, 189–211. doi:10.1177/088541202762475946

Anderson, L. W., & Krathwohl, J. (Eds.). (2001). *A taxonomy for learning, teaching, and assessing: A revision of Bloom's taxonomy of educational objectives*. New York: Longman.

Arrow, K. J. (1963). *Social Choice and Individual Values*. New Haven, CT: Yale University Press.

Axelrod, R. (1997). *The complexity of cooperation: Agent-based models of competition and collaboration*. Princeton, NJ: Princeton University Press.

Barbanente, A., Camarda, D., Grassini, L., & Khakee, A. (2007). Visioning the regional future: Globalisation and regional transformation of Rabat/Casablanca. *Technological Forecasting and Social Change, 74*, 763–778. doi:10.1016/j.techfore.2006.05.019

Borri, D., & Camarda, D. (2006). Visualizing space-based interactions among distributed agents: Environmental planning at the inner-city scale. *Lecture Notes in Computer Science, 4101*, 182–191. doi:10.1007/11863649_23

Borri, D., Camarda, D., & De Liddo, A. (2004). Envisioning environmental futures: Multi-agent knowledge generation, frame problem, cognitive mapping. *Lecture Notes in Computer Science, 3190*, 230–237.

Borri, D., Camarda, D., & De Liddo, A. (2008). Multi-agent environmental planning: A forum-based case-study in Italy. *Planning Practice and Research, 23*, 211–228. doi:10.1080/02697450802327156

Bousquet, F., & Le Page, C. (2004). Multi-agent simulations and ecosystem management: A review. *Ecological Modelling, 176*, 313–332. doi:10.1016/j.ecolmodel.2004.01.011

Bravo, J., Hervás, R., Chavira, G., & Nava, S. (2006). Mosaics of visualization: An approach to embedded interaction through identification process. *Lecture Notes in Computer Science, 4101*, 41–48. doi:10.1007/11863649_6

Danziger, S., & Rafal, R. (2009). The effect of visual signals on spatial decision making. *Cognition, 110*, 182–197. doi:10.1016/j.cognition.2008.11.005

Ferber, J. (1999). *Multi-agent systems*. London: Addison-Wesley.

Fischer, F. (2000). *Citizens, experts, and the environment: The politics of local knowledge*. Durham, NC: Duke University Press.

Forester, J. (1999). *The deliberative practitioner: Encouraging participatory planning processes*. Cambridge, MA: MIT Press.

Franklin, S., & Graesser, A. (1997). Is it an agent, or just a program? A taxonomy for autonomous agents. *Lecture Notes in Computer Science, 1193*, 21–35. doi:10.1007/BFb0013570

Friend, J., & Hickling, A. (1997). *Planning under pressure: The strategic choice approach*. London: Butterworth-Heinemann.

Garling, T., & Evans, G. W. (Eds.). (1991). *Environment, Cognition, and Action: An Integrated Approach*. New York: Oxford University Press.

Goodman, N. (1951). *The Structure of Appearance*. Cambridge, UK: Harvard UP.

Hartley, T., Trinkler, I., & Burgess, N. (2004). Geometric determinants of human spatial memory. *Cognition, 94*, 39–75.

Jarupathirun, S., & Zahedi, F. M. (2007). Dialectic decision support systems: System design and empirical evaluation. *Decision Support Systems, 43*, 1553–1570. doi:10.1016/j.dss.2006.03.002

Jungk, R., & Mullert, N. (1996). *Future workshop: How to create desirable futures*. London: Institute for Social Inventions.

Kalman, R. E. (1969). *Topics in mathematical system theory*. New York: McGraw-Hill.

Khakee, A., Barbanente, A., Camarda, D., & Puglisi, M. (2002b). With or without? Comparative study of preparing participatory scenarios using computer-aided and traditional brainstorming. *Journal of Future Research, 6*, 45–64.

Khakee, A., Barbanente, A., & Puglisi, M. (2002a). Scenario building for Metropolitan Tunis. *Futures, 34*, 583–596. doi:10.1016/S0016-3287(02)00002-2

Lynch, K. (1960). *The Image of the City*. Cambridge, MA: The MIT Press.

Mockler, R. J. (1989). *Knowledge-Based Systems for Management Decisions*. Upper Saddle River, NJ: Prentice-Hall.

Owen, G. (1995). *Game theory*. New York: Academic Press.

Ron, S. (2005). *Cognition and multi-agent interaction: From cognitive modeling to social simulation*. New York: Cambridge University Press.

Rowe, G., & Wright, G. (2001). Expert opinions in forecasting. Role of the Delphi technique. In Armstrong, J. S. (Ed.), *Principles of forecasting: A handbook of researchers and practitioners* (pp. 125–144). Boston, MA: Kluwer Academic Publishers.

Shakun, M. (Ed.). (1996). *Negotiation processes: modeling frameworks and information technology*. Boston, MA: Kluwer.

Tversky, B., & Hard, B. M. (2009). Embodied and disembodied cognition: Spatial perspective-taking. *Cognition, 110*, 124–129. doi:10.1016/j.cognition.2008.10.008

Van Dyke Parunak, H., Brueckner, S., Fleischer, M., & Odell, J. (2004). A design taxonomy of multi-agent interactions. *Lecture Notes in Computer Science, 2935*, 123–137.

Veloso, M. M., Patil, R., Rybski, P. E., & Kanade, T. (2004). People detection and tracking in high resolution panoramic video mosaic. In *IEEE/RSJ International Conference. Intelligent robots and systems (IROS 2004)* (Vol. 2, pp.1323-1328).

Wagenaar, H., & Hajer, M. A. (Eds.). (2003). *Deliberative policy analysis: Governance in the network society*. Cambridge, MA: Cambridge University Press.

Warren, T., & Gibson, E. (2002). The influence of referential processing on sentence complexity. *Cognition, 85*, 79–112. doi:10.1016/S0010-0277(02)00087-2

Weyns, D., Omicini, A., & Odell, J. (2007). Environment as a first class abstraction in multiagent systems. *Autonomous Agents and Multi-Agent Systems, 14*, 5–30. doi:10.1007/s10458-006-0012-0

Wooldridge, M. (2002). *An introduction to multi-agent systems*. London: Wiley.

Ziegler, W. (1991, June). Envisioning the future. *Futures*, 516–552. doi:10.1016/0016-3287(91)90099-N

ADDITIONAL READING

Aboutalib, S., & Veloso, M. (2007, May). *Towards using multiple cues for robust object recognition*. Paper presented at AAMAS'07, Honolulu, Hawaii.

Ågotnes, T., van der Hoek, W., & Wooldridge, M. (2009). Reasoning about coalitional games. *Artificial Intelligence, 173*(1), 45–79. doi:10.1016/j.artint.2008.08.004

Courrieu, P. (2005). Function approximation on non-Euclidean spaces. *Neural Networks, 18*, 91–102. doi:10.1016/j.neunet.2004.09.003

de Hevia, M. D., & Spelke, E. S. (2009). Spontaneous mapping of number and space in adults and young children. *Cognition, 110*, 198–207. doi:10.1016/j.cognition.2008.11.003

Dignum, V. (Ed.). (2009). *Handbook of Research on Multi-agent Systems: Semantics and Dynamics of Organizational Models*. Hershey, PA: IGI Global.

Gillner, S., Weiss, A. M., & Mallot, H. A. (2008). Visual homing in the absence of feature-based landmark information. *Cognition, 109*, 89–104. doi:10.1016/j.cognition.2008.07.018

Kelly, D. M., & Bischof, W. F. (2008). Orienting in virtual environments: How are surface features and environmental geometry weighted in an orientation task? *Cognition, 109*, 89–104. doi:10.1016/j.cognition.2008.07.012

Noriega, P., Vazquez-Salceda, J., Boella, G., Boissier, O., Dignum, V., Fornara, N., & Matson, E. (Eds.). (2007). *Coordination, Organization, Institutions, and Norms in Agent Systems*. Berlin: Springer. doi:10.1007/978-3-540-74459-7

van Harmelen, F., Lifschitz, V., & Porter, B. (Eds.). (2008). *Handbook of Knowledge Representation*. Amsterdam: Elsevier.

Weyns, D., Parunak, H. V. D., & Michel, F. (Eds.). (2007). *Environments for Multi-Agent Systems*. Berlin: Springer. doi:10.1007/978-3-540-71103-2

KEY TERMS AND DEFINITIONS

Agent: Human or artificial (or hybrid) element involved and acting in planning or decision-making exercises.

Cooperative Decision-Making: An organization of agents oriented to issue decisions founded of the exchange of information and/or cognitive contents.

Decision-Support System: A system of elements and actions set up to support complex decision-making.

Environmental Planning: The set up and organization of strategies and actions to achieve mainly social and economic objectives, with particular attention to environmental impacts and issues.

Multi-Agent Interaction: An organization of agents in which exchanges of information and/or general cognitive contents occur.

Multi-Agent System: A system of agents involved in a stable or task-oriented organization.

Scenario Building: Setting up an interacting organization and environment aimed at cooperatively working out future scenarios for given areas and/or situations, particularly in social communities.

Stakeholder: Agent who represents an interest or a stake in a given community.

ENDNOTES

[1] MeetingWorks™ is a groupware product that includes tools for electronic brainstorming, idea organization, ranking, voting, cross impact analysis, and multiple criteria analysis, with reports and graphics. A LAN-based system, MeetingWorks™ provides a Chauffeur station for use by the meeting facilitator in order to create a meeting agenda and run the meeting. The Chauffeur screen is normally projected in the front of the room, for all participants to see. Participants have access to 'participant' stations where they can enter their ideas, votes, comments and other inputs, all anonymously. The participant information is collected and displayed on the Chauffeur screen, allowing the group to view the collective input as well as other information e.g. the level of agreement within the group (Shakun, 1996).

2   Unfortunately, access to Internet is today not yet uniform and somehow elitist, and experience shows that relatively few groups are able to join the process regularly. Therefore, since the very foundations of this virtual-Agora approach are not yet verified, the engineering of the process should be intended as being still explorative, to date.

Chapter 11
The E-Citizen in Planning:
U.S. Municipalities' Views of Who Participates Online

Maria Manta Conroy
The Ohio State University, USA

Jennifer Evans-Cowley
The Ohio State University, USA

ABSTRACT

Municipalities that plan have both a legal obligation and a professional directive to incorporate citizens into the planning process, but garnering sufficient and diverse citizen participation is often a struggle. Online participation tools as a component of e-government provide a potential venue for enhancing the participation process. However, e-government participation raises challenges pertaining to trust, exclusion, and responsiveness. This chapter contributes to the understanding of these issues by analyzing how municipalities in the U.S. view the e-participant. The analysis is based on an ongoing longitudinal study that examines planning department web sites for U.S. cities with 2000 census populations of 50,000 or more. The authors' findings highlight respondents' views of online tools as a means to further efficiency and citizen satisfaction, rather than as a means by which to potentially enhance discussion of community issues.

INTRODUCTION

Local governments engage their citizens to provide them with information and gain support for policy initiatives, to identify unforeseen concerns, and to recognize potential conflicts (Conroy & Berke, 2004; Conroy & Gordon, 2004; Wild & Marshall, 1999). Questions remain, however, on how best to engage citizens in local planning efforts when work, family, and other issues constrain both time and interest (Chess & Purcell, 1999; Day, 1997). While planners are obligated to at least inform citizens of and, preferably, to engage them on land use issues through, for example, the comprehensive planning process, it is often difficult to get input (Conroy & Gordon, 2004). This lack of active participation has created a challenge for city planners, who are responsible for engaging citizens in making decisions about the future of their communities (Brody, Godschalk, & Burby, 2004; Laurian, 2004).

DOI: 10.4018/978-1-61520-929-3.ch011

The Internet has transformed the manner in which people get news and information, shop, find entertainment, and interface with their government (Waldon, 2006). In the 2008 U.S. presidential election campaign, a survey from the Pew Internet & American Life Project found that "46% of all adults are using the internet, email, or phone text messaging for political purposes" (Smith & Rainie, 2008, p. 2). Another work by that same Project noted that "the Internet is the second most popular way for Americans to contact their government" (Dimitrova & Chen, 2006, p. 174). This increased use of the Internet is global, crossing Europe, Africa, and Asia (see, e.g., Wohlers, 2009; Tiamiyu & Ogunsola, 2008; Kalu, 2007; Paul, 2007; and Holliday & Kwok, 2004). For example, according to Wohlers (2009), "In 2000, the federal government in Germany initiated a series of policy initiatives ... to spread the implementation of e-government throughout all levels of government" (p. 112). Local governments in the U.S. and elsewhere have been increasingly adopting applications ranging from simple document delivery to more complex interactive online mapping in order to increase responsiveness and community input and to enhance community renewal (Al-Kodmany, 1999; Conroy & Evans-Cowley, 2006; Kingston, 2007; Lee et al., 2005). Therefore, incorporation of information and communication technology (ICT) into public planning processes represents an area of great promise in which better relationships between government and its citizenry can be built (Lodge, 2003; Weber et al., 2003).

While some studies have suggested that access to the Internet for information and local government services may increase local participation levels, there is little clear empirical evidence one way or the other (Komito, 2007, p. 81). Local governments across the U.S. have been increasing online information and participation opportunities (see, e.g., Evans-Cowley & Conroy, 2004; Evans-Cowley & Conroy, 2009). However, the impact of technology on increasing civic participation with governments remains unclear at best (Komito, 2007). Additionally, there is little insight on how governments themselves perceive the influence of the technology on participation.

The purpose of this chapter is to examine the use of e-government participation applications and insights for public participation in planning processes in municipalities in the United States. Specifically, we build upon previous studies' reviews of online municipal planning offerings by incorporating a survey of planners in certain municipalities to understand their view of the e-participant and the impact online technologies have had on their relationships with citizen participants. The study begins with a review of the citizen participation and technology literature to assess the potential of e-participation and the key models that have been proposed. We then examine planner views of the e-participant with a review of our methodology and a discussion of the survey results. The chapter concludes with a discussion of future trends in e-participation and a summary of our key findings.

BACKGROUND

Citizen Participation and Technology

The goal of citizen participation in planning is often to enhance the outcomes of policy and project decisions in a community. There is a decades-old body of literature that points to improved outcomes, both anecdotally and empirically, when there is successful citizen participation. According to Kingston (2007), "Recent research by the Joseph Rowntree Foundation found that active citizen engagement is likely to improve the effectiveness of neighbourhood regeneration projects both in building personal and community capacity, and in achieving tangible regeneration outcomes, particularly in deprived neighbourhoods (Beresford and Hoban, 2005)" (p. 138). E-government, the use of technology such as the Internet "to enhance access to and delivery of government information and services to citizens, businesses, government employees, and other agencies" (Hernon, Reylea, Dugan & Cheverie, 2002, p. 388 in Jaeger & Thompson, 2004, p. 94), is the broad means by which online participation tools are packaged for citizen engagement. In this section, we first review basic models of e-government and ramifications that they hold for online participation. Then we examine the challenges faced in the pursuit of e-government.

Models of E-Government

Models of e-government capture potential phases of its adoption, implementation, and use. The theoretical underpinnings of each phase—and especially adoption—are often based on the technology acceptance model (TAM). This model, which examines the influence of individual attitudes on intentions to use a specific technology, is based on the long-standing theories of reasoned action (TRA) and planned behavior (TPB) (Horst et al., 2007). According to TAM, use is based on both attitude toward the technology and behavioral intent (ibid). While TAM may capture either the government provider or the citizen user, it does not focus on the interface between the two. As noted by Benyon-Davies (2007), "Most models of contemporary e-government do not satisfactorily include issues of e-democracy" (p. 8). However, some e-government models do focus on the interaction between government and citizen as a means to promote e-governance and democracy. These models are focused on the practice of e-government. Such models are valuable for examining the roles of users and providers and the anticipated benefits of use (Chadwick & May, 2003). The remainder of this section discusses the managerial, consultative, and participatory models of e-government practice and each one's relationship to the goals of this study.

The dawn of the 21st century in the U.S. brought with it a federal push to harness the potential efficiencies and promises of access associated with e-government (OMB, 2007). The E-Government Act of 2002 helped initiate state and local efforts that retained the federal government's views of efficiency and access to information. This managerial model of e-government has remained the dominant model that governments in the U.S. and elsewhere have adopted (Chadwick & May, 2003). It is characterized by the goal of efficient delivery of services to citizens (Bekkers & Homburg, 2007), and it emphasizes a "customer-centric governance, rooted in private sector managerial practices, [which] stems from an approach best captured by Osborne and Gaebler's *Reinventing Government* (1992)" (Dutil et al., 2007, p. 78). The managerial model aligns with Tolbert & Mossberger's (2006) consideration of an entrepreneurial e-government approach. The model is implicit in Asgarkhani's (2007) review of studies that summarize the following local government objectives in adopting e-government: prompt, accurate service; improved quality of service; removing barriers and tackling social exclusion; and providing local access points for citizen inquiries (p. 132). In this regard, while citizens are viewed as users of the system or, essentially, as customers (Dutil et al., 2007), they do not interact with the system to broaden or strengthen a sense of community.

Evans-Cowley and Conroy's (2009) longitudinal review of municipal planning department websites of U.S. cities with year 2000 populations of at least 50,000 highlights the foundational nature of the managerial model. The authors reviewed web-based informational offerings that are typically electronic alternatives to print-based options, such as zoning ordinances, community plans, and meeting agendas or minutes. These offerings represent a more efficient means of distributing information, both in terms of timeliness and cost. All surveyed municipalities increased their information offerings between 2003 and 2007, with 80% or more responding that they provided zoning ordinances, plans, and planning meeting agendas electronically through their websites (Evans-Cowley & Conroy, 2009).

The managerial model evaluates e-government technologies in light of enhancing efficiencies and reducing costs by, for example, making print documents available online for review. Government processes continue largely in the same manner that they did prior to the adoption of these technologies (Chadwick & May, 2003). From a participation viewpoint, the managerial model aligns most closely with an information distribution conceptualization of participation, where participants receive relevant information from their government but have limited ability to affect content (Conroy & Evans-Cowley, 2006). The managerial model is often the introductory level at which local municipalities enter the e-government arena, because "effective and efficient government increases citizen goodwill and sustains a healthy and robust democracy" (Watson & Mundy, 2001, p. 30). It is a level that governments should approach with some caution, however, because, as Dutil et al. (2007, p. 83) highlight in their research, "[b]eing thought of as a client … or customer … ranks far behind and even below the category of taxpayer … (Ekos, 2006)." Such a model may provide "online administration, communal and social services [which] certainly facilitate and improve people's lives" (Netchaeva, 2002, p. 470). However, the managerial model does little to engage citizens in democratic discourse.

A second model of e-government proposed by Chadwick and May (2003) is the consultative model, which provides increasing focus on citizens and the communication of their opinions to government. This occurs when, for example, citizen input is sought through online forms and surveys. It is at the consultative level that governments move away from streamlining information provision and towards offering opportunities for online feedback on policies and projects. While information is reciprocal between citizenry and staff, and it may influence content, it lacks a conversational dynamic that enhances communication both among citizenry and between citizenry and government. However, the consultative model begins a transition towards an e-democracy through expanded citizen opportunities to engage in feedback and dialogue with planners, and governments may employ it during a transition period toward a more participatory engagement strategy (Reddel & Woolcock, 2004). Emails and online fill-in forms are fundamental modes of communication in the consultative model.

The growth of the consultative model in e-government provisions is visible in the Evans-Cowley and Evans-Conroy (2009) study, where 77 percent of the municipalities provided direct email contact to their planning departments. The email addresses were general mailboxes (e.g., planner@mytown.gov) or those of individual staff members. Over the 5-year period of the study, the provisions of staff and planning commission email addresses each grew 9%. Nearly half of the respondents to the survey (44%) noted that staff email was offered in 2007—provision of planning commission email addresses rose 20% from 2003 levels.

The participatory model of e-government, the third model discussed by Chadwick and May, (2003) and the second presented by Tolbert & Mossberger (2006), takes the biggest step toward an electronically based democracy. This model is the basis for discussions of the potential of e-government to transform civic and political participation (Tambini, 1999; Weber et al., 2003). It most notably differs from the managerial and consultative models in that information flow goes beyond the vertical municipality-to-citizen direction. A participatory model "conceives of a more complex, *horizontal*, and *multidirectional* interactivity" (Chadwick & May, 2003, p. 280, emphasis original). It validates citizens as knowledge centers, rather than viewing knowledge as a municipal holding. Citizens are engaged in the process, not simply informed thereof, and they have the ability to affect change in content and decision-making processes. It is through this interaction that citizens can ask questions, get answers, express opinions, and see that they have the capacity to alter policy, which is the essence of a democratic process (Komito, 2007; Netchaeva, 2002). The transition from a user or client of e-government services perspective to an e-democracy view rests in the success of a participatory model.

Interaction opportunities in the participatory model, as described by Evans-Cowley and Conroy (2009), include a mix of consultative and participatory opportunities, depending in part on the manner in which they are employed. Therefore, the email addresses discussed previously not only give citizens the opportunity to ask questions or provide input, they may also serve as a means by which a dialogue can begin between both citizens and staff and among interested citizens. Tools such as wikis, listservs, discussion forums, and chat rooms hold an even greater promise to enhance multidirectional interactions among and between planning staff and citizens. These tools foster an online environment that mimics some of the characteristics of face-to-face interactions, such as turn taking (Tan & Tan, 2006), but they lack some of the nuances inherent in a face-to-face meeting. For example, it can be challenging to interpret intent or subtle meanings behind typed discussions. Use of capitalization, boldface, and italics, as well as emoticons, may help to convey the tenor of a message. Tools such as wikis and listservs may also raise concerns regarding flaming and identity security, in which the perceived safety of the communicative environment is challenged.

Evans-Cowley and Conroy's survey results highlight the lag that exists as communities slowly transition to more consultative and participatory e-government models. While listservs were provided by over half of the respondents (56.1%), discussion forums (5.3%), chat rooms (2.6%), and wikis (0%) each had only a minimal municipal planning online presence (Evans-Cowley & Conroy, 2009, p. 277). Additionally, between 84 to 100 percent of the respondents noted that the latter three are "not planned" in the next 5 years. Therefore, the expectation is that this lag in implementation between managerial or consultative models and participatory models of e-government will continue.

While Dutil et al. (2007) highlight the expressed desires of citizens to be thought of as citizens when using e-government services (as opposed to clients or customers), there are real challenges faced by municipalities in making that transition. Some of the challenges are based on technical or financial resource constraints, while others are more fundamental and raise questions regarding the core role e-government can or should play in municipal governance (Wohlers, 2009; Asgarkhani, 2007).

Challenges of Online Participation

The provision of online information and services provides citizens with enhanced flexibility when interacting with their municipal governments. It has the potential to lower some of the costs of participation in terms of time (e.g., transportation to/from meetings) and resources (e.g., gas to travel to meetings, time from work or family). However, it requires a dedication of resources and expertise on the part of each municipality to provide an open and transparent web offering. This openness is a key factor of advocates' support, in which they assert that such factors will inherently raise the public's ability to voice concerns and demands (e.g., La Port, 2005; Holliday & Kwok, 2004; Tiamiyu & Ogunsola, 2008).

The potential of technology can be contrasted with a variety of obstacles that may be faced as a municipality moves toward e-government. Obstacles may be associated with the technology itself (e.g., system errors and tampering concerns), with access (both infrastructural and individual), or with political and administrative reluctance (Wohlers, 2009; Asgarkhani 2007; Von Haldenwang, 2004). Such limitations may make envisioning an e-democracy future difficult. Access issues are of critical concern, as they underpin the essence of a democratic system.

Therefore, fundamental questions regarding issues of the digital divide and trust in government raise doubts about the potential of democracy-related benefits derived from e-government (see e.g., Asgarkhani, 2007; Tolbert & Mossberger, 2006; Weber et al., 2003).

The concept of the digital divide is based on observed inequities related to the provision of e-government services. Typical disenfranchised groups include minorities and lower income and/or elderly populations (Dimitrova & Chen, 2006; Tolbert & Mossberger, 2006). Full participation through government ICT demands that people are not only able to access the technology, but that they also know how to and want to use it (Netchaeva, 2002). Therefore, there are fundamental issues of technological access, self-efficacy regarding the technology, and trust that the technology is safe for use. The following discussion highlights these factors in the digital divide and trust literature that are relevant to this study.

Prior research that has focused on the demographics of citizens who do use e-government services has coalesced around what may be seen as typical user characteristics. Dimitrova & Chen's (2006) review of the divide literature notes that e-government users (e-citizens) are generally white professionals of higher income and education levels than their non-e-government user counterparts. African Americans, who may be less likely to use the Internet overall, are more likely than other demographics to have visited a local government website, according to an analysis of national survey data by Tolbert & Mossberger (2006). Further, while the Pew Internet Project's May 2008 survey noted that 73% of adults in the U.S. go online and 55% of adults have broadband at home, the "[o]ffline Americans are overwhelmingly over age 70, have less than a high school education, and speak a language other than English" (Fox & Vitak, 2008).

The reasons for youth dominance among e-government users are complex. They include concerns about the physical use of technology, safety, and privacy, as well as challenges to social norms. According to Xie and Jaeger (2008), "…older Americans may choose to not engage in political discussions in order to not disturb social networks with political disagreements" (p. 11). This conceptualization may self-restrict e-government usage to the managerial model rather than in a participatory model, in which there is the risk of engagement. Interestingly, in a January 2009 *Pew Internet Project Data Memo*, a shift had been noticed: "The biggest increase in internet use since 2005 can be seen in the 70-75 year-old age group. While just over one-fourth (26%) of 70-75 year olds were online in 2005, 45% of that age group is currently online" (Jones & Fox, 2009, p. 2).

Other changes are afoot as well. Surveys are showing increased Internet usage by all demographic groups in the U.S. The December 2008 tracking survey from the Pew Internet & American Life Project shows the lowest usage demographic groups as those with less than a high school education (35%) and those over the age of 65 (41%) (Pew, 2009). All other demographic categories had over 50% of the respondents using the Internet. The increases may be due to better access to the technology through publicly available sites, or it may indicate an enhanced comfort level with the technology. This is not to say that the digital divide is a thing of the past, but it does highlight that there has been progress in extending the accessibility of technology.

As alluded to previously, the digital divide deals not only with physical access to technology, but also with the manner in which people can use the technology (Xie & Jaeger, 2008). Impediments due to differences in language and visual, auditory, and/or physical capabilities illustrates that one website will not necessarily fit all. Translation services and web accessibility guidelines, such as that promoted by the Web Accessibility Initiative, are two ways in which municipalities have attempted to extend the reach of their online services (Evans-Cowley & Conroy, 2004; WIA, n.d.).

Improvements to access (physical and interpretive) may be resource dependent, but they are relatively straightforward approaches to help reduce the digital divide. Von Haldenwang (2004) notes that "In a number of countries, public Internet access points (telecentres, kiosks, Internet cafes) ... are already being used as nodal points of civic organisation and debate" (p. 428). However, issues of self-efficacy and trust, each of which is challenging for a municipality to address, are more dependent on the experience and education of users. Dimitrova and Chen's (2006) U.S. study found that "almost one fourth [of respondents] said that any information on the Internet raises privacy issues" (Dimitrova & Chen, 2006, p. 185). National and international news have highlighted cases of identity theft, email scams, and computer viruses, which have rightfully made Internet users cautious. For some users, these threats have been sufficient to resist use completely.

The success of e-government is dependent upon users' confidence in the reliability and security of online offerings. Tolbert & Mossberger (2006, pp. 357-358) present the following six possible benefits of e-government that could lead to increased trust and confidence in government: responsiveness, accessibility, transparency, responsibleness, efficiency and effectiveness, and participatory. These are all user-based benefits that can also increase accessibility (or a sense thereof). The authors reviewed national survey data in the U.S., and their analysis revealed that "visiting a local government Web site led to enhanced trust in local government, controlling for other attitudinal and demographic factors. E-government at the local level was also perceived by citizens as making government accessible and responsive..." (Tolbert & Mossberger, 2006, p. 366). The perceived usefulness of an online service may also enhance trust in e-government (Horst et al., 2007).

The correlation between trust and usage of online tools can also be negative. Research by Dutton and Shepherd (2006) suggests that people who drop-out of Internet use do so based on their perception of usage risk. "Non-users are generally more distrustful of the Internet. ... The risks experienced in using the Internet are most often less than the risks imagined by non-users, who also often underestimate the benefits of the Internet" (Dutton & Shepherd, 2006, p. 448). Dutton and Shepherd also note that related digital divide access issues are linked to trust, resulting in, for example, and lower trust among lower income groups who have less access to technology. The statistical probability of an online security breach is higher for users of a technology than it is for nonusers. Therefore, municipal e-government offerings must promote not only a comfort with technology use, but also a confidence about any information provided to the site by a user.

The digital divide raises critical issues for municipalities that offer e-government services. Evidence exists that certain demographic-related access gaps are getting smaller and that trust levels are increasing as exposure to and comfort with technology grows. Managerial model e-government offerings may serve as high comfort points of entry for newer users. Participatory offerings, however, may retain more divide problems, as personal interactions with government staff and other citizens increase personal exposure to such things as criticism from those with opposing points of view. These e-participation divide issues of representativeness are also found in traditional participation: "Verba et al. (1995) found that citizens who participate politically at higher levels are unrepresentative of the larger public" (Weber et al. p. 29). As Watson & Mundy (2001) point out, "[r]ealistically, power will never be equally distributed within a democracy, but citizens are accustomed to a long trend of reforming acts that redress power imbalances. Indubitably, they will expect Internet technology to sustain, and maybe accelerate, this course" (p. 30).

Planner Views of the E-Participant

This study has been conducted in coordination with an annual audit of municipal planning department websites of U.S. cities with populations of more than 50,000 people. The audit examines the characteristics of adoption of e-government technologies. Planning departments have been selected because of their frequent engagement with citizens both in rezoning transactions and long-range neighborhood and city-wide planning efforts. Municipalities with populations of 50,000 or more are considered to be likely to, at a minimum, have a web presence for both their communities and their planning departments (when planning departments exist). There are 613 municipalities with populations of 50,000 or greater in the United States (US Census). Within that set, approximately 600 had planning-related websites as of 2007 (Evans-Cowley & Conroy, 2009).

In an effort to understand planners' perceptions of their e-participants, this study began with a review of the characteristics of the 600 planning-related websites to determine which had online participation tools. These tools include online surveys, wikis, interactive GIS, discussion forums, and streaming audio or video. There were 279 planning-related municipal websites that had such features noted. A link to an online survey created with SurveyMonkey.com© was sent to the email contact for each of these municipalities. Thirty-nine of the emails were returned as invalid addresses, typically due to a change in personnel, web address, or online forms; 27 of those returns were unable to be resolved.

The survey sent to the 252 valid email and form contacts addressed participation tools and e-participant perceptions using thirteen closed- and open-ended questions. Two weeks after the initial survey invitation was sent, a reminder was sent to the original or corrected email addresses. A total of 58 contacts responded, giving an approximately 23% response rate.

Online Participation Tools

The survey began with a verification question, asking if the department or municipality provided online participation tools. Although the municipalities had been screened based on responses to a related survey on e-government tools, 7% (4 total) of the respondents indicated that they did not offer any online participation tools. This may reflect that different individuals responded to the two surveys, a shift in a municipality's online offerings, or a misunderstanding of what may constitute online participation tools. The vast majority (75.4%) of respondents noted that their department or community offers some online participation tools, and 17.5% of the respondents said that they offer a wide array of online participation tools.

Given our interest in planners' perceptions of their e-participants, we then asked about registration and tracking of online participants. Registration and tracking would allow a municipality to know more about both their participants and the participants' tool usage habits. Registration might also give participants the ability to customize their experience of their municipality's website, to receive targeted emails based on interest areas, or to participate in an electronic bulletin board or wiki. Tracking gives a municipality insight into how and when participation tools are being used. However, these elements of a website can also strike unease in potential participants, raising questions of personal information security and government "big brother-ism." Over 84% of respondents noted that users do not have to register at all to make use of participation tools; the remaining respondents said that registration was necessary for only a select group of tools.

Usage rate tracking was used for some or all tools by 34.2% of the respondents. Almost 40% of respondents noted that their sites do not track usage rates of their online participation tools. The remaining respondents (26.3%) did not know if tracking was done for their tools. Citizens may object to tracking, often done through the use of

Table 1. Traditional versus online participation input comparison

Compared with traditional participation format...	AMOUNT of input received via online tools	USEFULNESS of input received via online tools
Online much more than traditional	2.8%	8.1%
Online some more than traditional	19.4%	13.5%
About the same	22.2%	**43.2%**
Traditional some more than online	**38.9%**	21.6%
Traditional much more than online	16.7%	13.5%

browser cookies, as an invasion of privacy or, possibly, out of concern for potentially malicious ware that may be downloaded to their computers. Decisions on tool tracking usage, however, are likely outside the decision scope of the planner, unless he or she also has web-development responsibilities.

E-Participant Perceptions

The next ten questions of the survey examined the respondents' views of their e-participants and the impacts that the provision of online participation tools have had. Specifically, respondents were asked to comparatively characterize the amount and usefulness of online input versus traditional input sources, as well as to characterize online participants versus those participating in a traditional venue (see Tables 1 and 2).

When asked about the amount of information their participants received, 55.6% of the respondents noted that they received either some or much more information from traditional input than from online participation tools, while 22.2% noted that they received about the same amount of information from each source. Online tools have been touted as a means for enhancing the citizen experience and their depth of participation in the planning process. The responses here are likely an indication that the tools provided to the citizens are ones in which the e-participant receives rather than provides information. Interactive GIS or streaming audio, for example, are ways in which a citizen can learn more about what is taking place—or is planned to take place—in his or her community. Based on the responses, such tools are not being used as a means of providing information about abandoned buildings or pedestrian danger spots. This may be a function of the provided tool, or it may be a comfort-level issue (e.g., with using web-based GIS tools) on the part of the e-participant.

When asked about the usefulness of the input received by their participants, over 43% of respondents felt that the usefulness of the input received from online sources was about the same as from traditional input sources. However, 35% of respondents felt that traditional input is somewhat or much more useful than that received from online sources. This points to a shortcoming in the client orientation (managerial model) evident in the e-government offerings of the surveyed communities. If the intent of the offerings is efficiencies, a priority is not placed on promoting a dynamic dialogue that would simulate face-to-face interaction. Therefore, the usefulness of a series

Table 2. Traditional versus online participant characterization

Compared with traditional participation format...	Participant knowledge / engagement
Online much more than traditional	8.1%
Online some more than traditional	16.2%
About the same	**54.1%**
Traditional some more than online	16.2%
Traditional much more than online	5.4%

of singular inputs may be lacking in comparison with a format that includes a group dynamic where ideas may be based on interaction.

One hope for ICT tools for citizen participation is that they may broaden the spectrum of participants, ideally engaging citizens who might not otherwise come to a public meeting or open house (Conroy & Gordon 2004). Komito (2007) has noted evidence that "new technologies intensify existing contacts within localities, especially by those already active in their localities" (p. 79). Unfortunately, the author also notes that evidence has not supported the role of e-government in increasing the number of people who choose to participate. When respondents in this study were asked if they felt that their e-participants were the same as the people who attended traditional meetings or were new participants, the results were mixed. While almost 38% felt that those using online participation tools were largely the same people or many of the same people who attended traditional formats, over 32% responded that the e-participants represented new participants. Almost 30% of respondents indicated that the e-participants represented about the same amount of both traditional meeting participants and new participants. While it is not possible to draw a conclusive insight from such results, it is evident that some new users are being served by the availability of online participation tools.

New users may not differ much from those with whom the respondents have typically engaged. According to the respondents, online participants are viewed similarly to their traditional venue counterparts (see Table 2). Fifty-four percent of respondents felt that online participants and traditional participants are about the same in terms of knowledge or engagement. Over 24 percent found e-participants to be much more or more knowledgeable and engaged than traditional participants, while over 21 percent felt that traditional participants had the knowledge or engagement edge. The respondents made no indication that citizen opinions are common in either traditional or online venues. However, the similarities in knowledge and engagement levels may signify that the digital divide is not immediately evident in this particular case. The divide may instead be between those who participate via any venue versus those who do not. It may also reflect what Jaeger & Thompson (2004, p. 101) refer to as "information poverty," or ignorance about the existence or value of an online offering.

Online Participation Impacts

Another important consideration with respect to online participation tools is to understand how the tools impact the participation process. Multiple studies have demonstrated "positive e-government impacts on data access and efficiency and productivity of government performance in both internal operations and external functions" (Holliday & Kwok, 2004, p. 551). Von Haldenwang (2004) goes further, with the anticipated positive contributions of ICT on citizen participation "based on the observation that the intensification of information and communication flows that characterizes e-government strengthens the capacity of public institutions as well as the

Table 3. Planner view of online participation tool impacts

	Sense of Government Transparency	Department Responsiveness	Perception of Responsiveness
Greater with online tools	**83.3%**	**48.6%**	24.3%
Greater with traditional format	8.3%	21.6%	16.2%
About the same	8.3%	29.7%	**59.5%**

transparency and openness of political processes" (Von Haldenwang, 2004, p. 427). The perception of planners regarding the influence of these tools on their citizenry provides new insight into the value each department may place on the tools, as well as their perceived limits.

Table 3 shows respondents' views of the impact of online tools versus traditional participation formats with respect to citizens' sense of government transparency and the public's perception of department responsiveness. We also asked respondents' views of their own department responsiveness. Respondents overwhelmingly noted (83.3%) that the online tools provide citizens with a greater sense of government transparency. This supports previous research and is linked to easier access to materials previously held in a physical office. Respondents themselves felt that their own responsiveness to citizens was greater with online tools (48.6%). Interestingly, nearly 60% of respondents felt that the public's perception of their responsiveness was independent of the online or traditional participation format. This may be the result of respondents' perceptions of unreasonable expectations on the part of the public, or that the public is satisfied in each venue. It may also be supportive of Wohlers' (2009) finding that " a majority or plurality of city officials also believes that the Internet has made no difference regarding increased incorporation of citizen feedback and citizen engagement in city politics" (p. 124). If respondents feel that the format of participation has no influence on feedback incorporation, perhaps they expect the citizenry to feel similarly.

The next question attempted to get more insight into the attitudes held by respondents regarding online participation tools. Table 4 shows the level of agreement that respondents had with five statements about the impact and influence of online participation tools on the respondent communities.

The overwhelming majority (77.4%) of respondents either strongly agreed or somewhat agreed with the assertion that the online tools allowed the community to reach individuals who would otherwise not participate. These individuals may be unwilling or unable to attend more traditional participation venues, or they may simply feel more comfortable with the online experience (Conroy & Gordon, 2004). Komito (2007) argues that "e-government systems can, if properly designed and implemented, involve citizens who have not previously been active in local community life" (p. 77). This study supports that finding, based on respondent views, although it does not address the more complex issues of participation equity and the digital divide. For example, one respondent offered the following cautionary note: "More sophisticated individuals are provided with an additional avenue to participate, however, given that our population is largely disenfranchised, I feel that traditional methods are best simply because of our community's demographic makeup."

Table 4. Planner agreement levels regarding online participation tool usage

Online participation tools...	Strongly Agree	Somewhat Agee	Neither Agree nor Disagree	Somewhat Disagree	Strongly Disagree
Allow us to reach individuals who would otherwise not participate	24.3%	**54.1%**	21.6%	0%	0%
Are all that is needed to do our community outreach	0%	5.4%	8.1%	8.1%	**78.4%**
Allow us to enhance the participation experience of our citizens	**48.6%**	32.4%	16.2%	2.7%	0%
Exclude a significant part of our citizenry	2.7%	**51.4%**	18.9%	21.6%	5.4%
Are most useful as a supplement to traditional participation approaches	**62.9%**	34.3%	2.9%	0%	0%

Another noted, "The online informational tools are helpful to the more computer savy [sic], but can be confusing for those who are not." Over half of the respondents noted that the tools exclude a significant part of their citizenry. The availability of these online tools, as noted in the literature discussion, may exacerbate the digital divide in a community. Providing publicly available centers (e.g., public libraries) may mitigate issues of access, but it does not address issues of participant self-efficacy or technical capacity.

While online participation tools provide a means for some citizens who might not otherwise participate to do so, most respondents (over 78%) strongly disagreed that they were sufficient as the only means by which they should do community outreach. On a related note, almost 63% of respondents strongly agreed that the tools are most useful as a supplement to traditional participation practices, and another 34% somewhat agreed with this role for the online participation tools. Therefore, while researchers highlight the potential of online participation to actively engage citizens and enhance democracy, practitioners' experiences indicate that they see a permanent role for conventional participation venues. This supports Von Haldenwang's (2004) assertion that "…ICT-based direct democracy should not be seen as an alternative to representative democracy but rather as a means to improve interest articulation and decision-making" (p. 427).

Most respondents (81%) somewhat or strongly agreed that online participation tools have enhanced the participation experience of their citizens. This is supported further by comments such as "Quicker response [sic] in providing written, comprehensive response"; "On-line info is really a convenience for certain segments of users and expedits [sic] getting info out into the community"; and "Improved response time to issues. Streamlines communuications [sic]." Respondent comments further highlight the view that the tools are a means to create efficiencies and participant/user convenience (Chadwick & May, 2003; Dutil et al. 2007). The values of the tools for potential increased community-building or alternate discussion venue have not been considered.

Finally, when asked about the impact that online participation tools have had on either each respondent's own or each respondent's department's relationship with citizens, more than half noted that the tools have positively but marginally changed how they interact with citizens. No one indicated that the tools had a marginally negative impact on interaction with citizens, although one respondent noted that there had been a radically negative change as a result of the tool. Over 26% of the respondents said that there had been a radical change, but in a positive way. One respondent noted that, "All work products are created with the thought that they will be on line." The electronic element of municipal planning has become, then, the new foundation of a municipality's community interface. Almost 16% of respondents noted that online participation tools have had little or no influence, positive or negative, on how the respondent or the respondent's department interacts with citizens. This might be expected for e-government offerings which are, for example, information- or managerial-based and lack a threshold level of citizen interaction.

FUTURE RESEARCH DIRECTIONS

Citizen participation tools in e-planning efforts are typically unidirectional, sequential, and information based (see, e.g., Evans-Cowley & Conroy, 2009). This characterization reflects the continued dominance of the managerial model of e-government. Insights from this study suggest both why this remains the case and point to the potential to move toward a participatory model.

Planners responding to our survey noted that the online participation tools improve responsiveness from both their own points of view and their perception of their users' points of view. Dimitrova and Chen (2006) conclude that "the higher the perceived usefulness of the new technology, the more likely it is to be adopted by the consumer. This proposition points to the fact that the decision to adopt a new technology service (e.g., electronic government) is based on a subjective perception on the part of the user" (p. 175). This appears to be a similar consideration for the provider. However, the online tools are considered "most useful" by respondents when they are supplementary to traditional approaches. This may be due in part to historical factors in the planning process, in which participation focuses on face-to-face interactions. Therefore, it is not surprising that the respondents, most of whom have been in their position for more than five years, would see traditional tools as the rightful dominant mode.

Another reason for the continued reliance on traditional participation tools rests with planners' perception of their citizenry. As one respondent in our survey noted, "[m]ore sophisticated individuals are provided with an additional avenue to participate, however, given that our population is largely disenfranchised, I feel that traditional methods are best simply because of our community's demographic makeup." It is unlikely that a move to a participatory model of e-government will be possible unless the government side of the interface (i.e., planners) see it as a way to engage the broader citizenry in a manner similar to face-to-face efforts. The generally positive view of how online participation tools have influenced respondents' relationships with their citizens, however, suggests that such a transition is possible. Future evaluations of this survey group's perceptions will continue to provide insight into whether the usefulness of the tools and their impact on citizen interaction changes as municipal planning online tools become more interactive.

U.S. communities continue to expand their offering of interactive planning tools, though technical and resource challenges limit the type of offering to those requiring less direct oversight (Evans-Cowley & Conroy, 2009). A municipality may have started its online presence with electronic versions of print materials, but it has typically moved over time to offering interactive online geographic information systems (GIS). One respondent in our survey noted, "[o]ur city is 97% internet active. On-line participation is used frequently." As interactive tools become easier to use and install and less expensive to adopt and maintain, we expect more municipalities to have a high percentage of online offerings.

Exclusion issues will likely broaden with the advent of the move to more consultative and participatory model of e-government. As noted by one of our respondents, "[t]he online informational tools are helpful to the more computer savy [sic], but can be confusing for those who are not." The e-participation tools may engage participation by a segment of the population that is already locally active; for example, the knowledge that such tools are available online may arise from announcements or fliers at meetings. As Komito (2007) points out, barriers to access "are diminishing over time ... and yet there still remain a large number of citizens who are disengaged from the policy process and, indeed, from their community in general" (p. 82). A concerted citizen training effort in both technology and citizenship may be a necessary component of the roll out of enhanced participation tools. The Internet is "an 'experience' technology, as it is difficult for people to understand how the Internet and Web work until they use them" (Dutton and Shepherd, 2006, pp. 447-448). Therefore, outreach efforts will need to go beyond informational mailings. We feel the future will be best served through a coordination between e-government functions and the efforts of existing community groups, as this will "enable citizens to explore electronic participation options when they use the e-government services" (Komito, 2007, p. 90).

CONCLUSION

Our study has examined the e-government offerings of municipal planning websites in light of planners' perceptions. A survey of planners has given insight into both the impact of online tools compared to traditional participation measures and the planners' views of the users of these tools. Two key findings have come out of our review.

First, the focus of website offerings remains dominated by the administratively oriented managerial model. In this model, enhancements to efficiency and transparency are seen as valuable outcomes of online tools. Participants are users or clients to be served rather than a population to be engaged. Participant satisfaction is valued, and has been shown to correlate with trust in government. This characterization of e-government users does not, however, attempt to transport the interactivity found in a face-to-face setting to the online realm. It retains traditional participation venues as the default for community engagement. This model highlights Wohlers' (2009) findings where "… only a small proportion of government Web sites [in the U.S. and Germany] provide residents with meaningful opportunities to gain tangible benefits or to directly interact with and get involved in government processes and decisions" (p. 123).

Second, while online tools may be introducing some new participants to the general planning process, there is recognition that existing tools are not reaching the broadest population. Expressed concerns stem from the experience, or lack thereof, that citizens have interfacing with technology. Although national surveys highlight increases in Internet use in general, this increase still may not be reaching members of underserved or disenfranchised populations. Traditional planning outreach focusing on personal connections takes pride in reaching such populations (Waldon, 2006). In conjunction with the managerial model dominance, this may indicate a latent bias of respondents for non-Internet forms of engagement.

The eventual success of e-government will be tied to citizens' willingness to use e-government tools and engage in the online realm of participation. It will require not simply advanced software but also a demonstration "that the use of such software has an impact on policy. Lack of efficacy and a consequent 'why bother' response is a significant issue" (Komito, 2007, p. 91). Planners also face challenges of participation in traditional venues. Incentives may be necessary to truly garner broad community-based participation, which may raise challenges fundamental to citizenship and democracy. There will, of course, be "some people who never want to use new technologies, so the alternative (face-to-face) offline government services should always be available" (Netchaeva, 2002, p. 476). There is no singular means of information dispersal or participation that will meet the full needs of a community.

The future of e-participation in planning may involve a radical shift from the efficiency mindset promoted by planners and subscribed to by users or it may involve only marginal enhancements to increase convenience and responsiveness. Citizens and the municipal staff with whom they interact will need to subscribe to the same vision of e-government and work together to achieve it. "The data suggests innovation in e-government is driven by legislative professionalism" rather than citizen demand" (McNeal et al., 2003, p. 65). While our findings focused on state level e-government offerings, they concur that the administrative focus has been dominant, and that "participatory goals are not presently a dominant factor" (McNeal et al., 2003, p. 66). It will only be possible to accomplish a radical shift if planners see the value of tools beyond an efficiency and responsiveness consideration.

REFERENCES

Al-Kodmany, K. (1999). Using visualization techniques for enhancing public participation in planning and design. *Landscape and Urban Planning, 45*, 37–45. doi:10.1016/S0169-2046(99)00024-9

Asgarkhani, M. (2007). The Reality of Social Inclusion through Digital Government. *Journal of Technology in Human Services, 25*(1/2), 127–146. doi:10.1300/J017v25n01_09

Becker, S. A. (2005). E-Government Usability for Older Adults. *Communications of the ACM, 48*(2), 102–104. doi:10.1145/1042091.1042127

Benyon-Davies, P. (2007). Models for e-government. *Transforming Government: People. Process and Policy, 1*(1), 7–28.

Brody, S. D., Godschalk, D. R., & Burby, R. J. (2004). Mandating citizen participation in Plan Making: Six Strategic Planning Choices. *Journal of the American Planning Association. American Planning Association, 69*(3), 245–264. doi:10.1080/01944360308978018

Chess, C., & Purcell, K. (1999). Public Participation and the Environment: Do We Know What Works? *Environmental Science & Technology, 33*(16), 2685–2692. doi:10.1021/es980500g

Cohen, J. E. (2006). Citizen satisfaction with contacting government on the internet. *Information Polity, 11*, 51–65.

Conroy, M. M., & Berke, P. R. (2004). What Makes a Good Sustainable Development Plan? An analysis of factors that influence principles of sustainable development. *Environment & Planning A, 36*, 1381–1396. doi:10.1068/a367

Conroy, M. M., & Evans-Cowley, J. (2006). E-participation in Planning: An Analysis of Cities Adopting On-line Citizen Participation Tools. *Environment and Planning C, 24*(3), 371–384. doi:10.1068/c1k

Conroy, M. M., & Gordon, S. I. (2004). Utility of Interactive Computer Based Materials for Enhancing Public Participation. *Environmental Planning and Management, 47*(1), 19–33. doi:10.1080/0964056042000189781

Day, D. (1997). Citizen participation in the planning process: An essentially contested concept? *Journal of Planning Literature, 11*(3), 421–434. doi:10.1177/088541229701100309

Dimitrova, D. V., & Chen, Y.-C. (2006). Profiling the Adopters of E-Government Information and Services: The Influences of Psychological Characteristics, Civic Mindedness, and Information Channels. *Social Science Computer Review, 24*(2), 172–188. doi:10.1177/0894439305281517

Dutil, P. A., Howard, C., Langford, J., & Roy, J. (2007). Rethinking Government-Public Relationships in a Digital World: Customers, Clients, or Citizens? *Journal of Information Technology & Politics, 4*(1), 77–90. doi:10.1300/J516v04n01_06

Dutton, W. H., & Shepherd, A. (2006). Trust in the Internet as an Experience Technology. *Information Communication and Society, 9*(4), 433–451. doi:10.1080/13691180600858606

Evans-Cowley, J., & Conroy, M. M. (2004). E-governance: on-line citizen participation tools for planners. [Chicago, IL: American Planning Association.]. *PAS Reporter, 525*.

Evans-Cowley, J., & Conroy, M. M. (2009). Local Government Experiences with ICT for Participation. In Reddick, C. (Ed.), *Handbook of Research on Strategies for Local E-Government Adoption and Implementation: Comparative Studies* (pp. 268–286). Hershey, PA: Information Science Publishing.

Fox, S., & Vitak, J. (2008). *Degrees of Access (May 2008 data)*. Presentation from the Pew Internet & American Life Project. Retrieved from http://www.pewinternet.org/Presentations/2008/Degrees-of-Access-(May-2008-data).aspx

Holliday, I., & Kwok, R. C. W. (2004). Governance in the information age: building e-government in Hong Kong. *New Media & Society, 6*(4), 549–570. doi:10.1177/146144804044334

Holman, N. (2008). Community participation: using social network analysis to improve developmental benefits. *Environment and Planning. C, Government & Policy, 26*, 525–543. doi:10.1068/c0719p

Horst, M., Kuttschreutter, M., & Gutteling, J. M. (2007). Perceived usefulness, personal experiences, risk perception and trust as determinants of adoption of e-government services in The Netherlands. *Computers in Human Behavior, 23*, 1838–1852. doi:10.1016/j.chb.2005.11.003

Islam, P. (2007). Citizen-centric E-Government: The Next Frontier. *The Kennedy School Review, 7*, 103–108.

Jaeger, P. T., & Thompson, K. M. (2004). Social information behavior and the democratic process: Information poverty, normative behavior, and electronic government in the United States. *Library & Information Science Research, 26*, 94–107. doi:10.1016/j.lisr.2003.11.006

Jones, S., & Fox, S. (2009). *Pew Internet Project Data Memo: Generations Online in 2009*. Memo from the Pew Internet & American Life Project. Retrieved from http://www.pewinternet.org/~/media//Files/Reports/2009/PIP_Generations_2009.pdf

Kalu, K. N. (2007). Capacity Building and IT Diffusion, A comparative assessment of e-government in Africa. *Social Science Computer Review, 25*(3), 358–371. doi:10.1177/0894439307296917

Kingston, R. (2007). Public Participation in Local Policy Decision-making: The Role of Web-based Mapping. *The Cartographic Journal, 44*(2), 138–144. doi:10.1179/000870407X213459

Komito, L. (2007). Community and inclusion: the impact of new communications technologies. *Irish Journal of Sociology, 16*(2), 77–96.

La Porte, T. M. (2005). Being good and doing well: Organizational Openness and Government Effectiveness on the World Wide Web. *Bulletin of the American Society for Information Science and Technology February/March*, 23-27.

Laurian, L. (2004). Public participation in environmental decision making: Findings from communities facing toxic waste cleanup. *Journal of the American Planning Association. American Planning Association, 70*(1), 53–66. doi:10.1080/01944360408976338

Lee, S. L., Tan, X., & Trimi, S. (2005). Current Practices of Leading E-Government Countries. *Communications of the ACM, 48*(10), 99–105. doi:10.1145/1089107.1089112

Lodge, J. (2003). Toward an e-commonwealth? A tool for peace and democracy? *The Round Table, 372*, 609–621. doi:10.1080/0035853032000150627

McNeal, R. S., Tolbert, C. J., Mossberger, K., & Dotterweich, L. J. (2003). Innovating in Digital Government in the American States. *Social Science Quarterly, 84*(1), 52–70. doi:10.1111/1540-6237.00140

Netchaeva, I. (2002). E-Government and E-Democracy: A Comparison of Opportunities in the North and South. *Gazette: The International Journal for Communication Studies, 64*(5), 467–477. doi:10.1177/17480485020640050601

Office of Management and Budget (OMB). (2007). *FY 2006 Report to Congress on Implementation of The E-Government Act of 2002*. Washington, DC: Office of Management and Budget.

Paul, S. (2007). A case study of E-governance initiatives in India. *The International Information & Library Review, 39*, 176–184. doi:10.1016/j.iilr.2007.06.003

Pew Internet & American Life Project (Pew). (2009). *Demographics of Internet Users*. Retrieved from http://pewinternet.org/Data-Tools/Download-Data/Trend-Data.aspx

Reddel, T., & Woolcock, G. (2004). From consultation to participatory governance? A critical review of citizen engagement strategies in Queensland. *Australian Journal of Public Administration, 63*(3), 75–87. doi:10.1111/j.1467-8500.2004.00392.x

Smith, A., & Rainie, L. (2008). *The Internet and the 2008 election*. Report from the Pew Internet & American Life Project. Retrieved from http://www.pewinternet.org/Reports/2008/The-Internet-and-the-2008-Election.aspx

Tambini, D. (1999). New media and democracy: The civic networking movement. *New Media & Society, 10*, 305–329. doi:10.1177/14614449922225609

Tan, S.-C., & Tan, A.-L. (2006). Conversational analysis as an analytical tool for face-to-face and online conversations. *Educational Media International, 43*(4), 347–361. doi:10.1080/09523980600926374

Tiamiyu, M. A., & Ogunsola, K. (2008). Preparing for E-Government: some findings and lessons from government agencies in Oyo State, Nigeria. *South African Journal of Library & Information Science, 74*(1), 58–72.

Tolbert, C. J., & Mossberger, K. (2006). The Effects of E-Government on Trust and Confidence in Government. *Public Administration Review*, (May/June): 354–369. doi:10.1111/j.1540-6210.2006.00594.x

Von Haldenwang, C. (2004). Electronic Government (E-Government) and Development. *European Journal of Development Research, 16*(2), 417–432. doi:10.1080/0957881042000220886

Waldon, R. S. (2006). *Planners and Politics: Helping Communities Make Decisions*. Chicago, IL: Planners Press.

Watson, R. T., & Mundy, B. (2001). A Strategic Perspective of Electronic Democracy. *Communications of the ACM, 44*(1), 27–30. doi:10.1145/357489.357499

Web Accessibility Initiative (WAI). (n.d.). *WAI-ARIA Overview*. Retrieved from http://www.w3.org/WAI/intro/aria

Weber, L. M., Loumakis, A., & Berman, J. (2003). Who participates and why? An analysis of citizens on the Internet and the mass public. *Social Science Computer Review, 21*(1), 26–42. doi:10.1177/0894439302238969

Wild, A., & Marshall, R. (1999). Participatory practice in the context of local agenda 21: a case study evaluation of experience in three English local authorities. *Sustainable Development, 7*, 151–162. doi:10.1002/(SICI)1099-1719(199908)7:3<151::AID-SD111>3.0.CO;2-0

Wohlers, T. E. (2009). The Digital World of Local Government: A Comparative Analysis of the United States and Germany. *Journal of Information Technology & Politics, 6*, 111–126. doi:10.1080/19331680902821593

Xie, B., & Jaeger, P. T. (2008). Older Adults and Political Participation on the Internet: A Cross-cultural Comparison of the USA and China. *Journal of Cross-Cultural Gerontology, 23*, 1–15. doi:10.1007/s10823-007-9050-6

ADDITIONAL READING

Arnstein, S. (1969). A Ladder of Citizen Participation. *American Institute of Planners Journal, 35*, 216–224.

Bekkers, V., & Homburg, V. (2007). The Myths of E-Government: Looking Beyond the Assumptions of a New and Better Government. *The Information Society, 23*, 372–382. doi:10.1080/01972240701572913

Bozinis, A. (2007). Internet Politics and Digital Divide Issues: The Rising of a New Electronic Aristocrats and Electronic Meticians. *Journal of the Social Sciences, 3*(1), 24–26. doi:10.3844/jssp.2007.24.26

Conroy, M. M., & Evans-Cowley, J. (2005). Informing and Interacting: The Use of E-government for Citizen Participation in Planning. *Journal of E-Government, 1*(3), 73–92. doi:10.1300/J399v01n03_05

Cuthill, M. (2004). Community Visioning: Facilitating Informed Citizen Participation in Local Area Planning on the Gold Coast. *Urban Policy and Research, 22*(4), 427–445. doi:10.1080/0811114042000296335

Eastin, M. S., & LaRose, R. (2000). Internet Self-Efficacy and the Psychology of the Digital Divide. *Journal of Computer-Mediated Communication, 6*(1). Retrieved from http://jcmc.indiana.edu/vol6/issue1/eastin.HTML.

Edmiston, K. D. (2003). State and local e-government: Prospects and challenges. *American Review of Public Administration, 33*(1), 20–45. doi:10.1177/0275074002250255

Evans-Cowley, J., & Conroy, M. M. (Eds.). (2005). *E-Government and Planning: Key Citizen Participation Issues and Applications.* John Glenn Institute for Public Service and Public Policy. Retrieved from http://hdl.handle.net/1811/519

Glicken, J. (2000). Getting stakeholder participation 'right': a discussion of participatory processes and possible pitfalls. *Environmental Science & Policy, 3,* 305–310. doi:10.1016/S1462-9011(00)00105-2

Jackson, L. S. (2001). Contemporary Public Involvement: toward a strategic approach. *Local Environment, 6*(2), 135–147. doi:10.1080/13549830120052782

Kaylor, C., Deshazo, R., & Van Eck, D. (2001). Gauging e-government: A report on implementing services among American cities. *Government Information Quarterly, 18,* 293–307. doi:10.1016/S0740-624X(01)00089-2

Koontz, T. M. (2005). We Finished the Plan, So Now What? Impacts of Collaborative Stakeholder Participation on Land Use Policy. *Policy Studies Journal: the Journal of the Policy Studies Organization, 33*(3), 459–481. doi:10.1111/j.1541-0072.2005.00125.x

Kumar, V., Mukerji, B., Butt, I., & Persaud, A. (2007). Factors for Successful e-Government Adoption: a Conceptual Framework. *The Electronic. Journal of E-Government, 5*(1), 63–76.

Kwan, M. P., & Weber, J. (2003). Individual accessibility revisited: Implications for geographical analysis in the twenty-first century. *Geographical Analysis, 35*(4), 341–353. doi:10.1353/geo.2003.0015

Lollar, X. L. (2006). Assessing China's E-Government: information, service, transparency and citizen outreach of government websites. *Journal of Contemporary China, 15*(46), 31–41. doi:10.1080/10670560500331682

Margerum, R. D. (2002). Collaborative Planning: Building Consensus and Building a Distinct Model for Practice. *Journal of Planning Education and Research, 21,* 237–253. doi:10.1177/0739456X0202100302

Padgett, D. A. (1993). Technological methods for improving citizen participation in locally unacceptable land use (LULU). *Decision-Making, Computers. Environment and Urban Systems, 17*(6), 513–520. doi:10.1016/0198-9715(93)90049-B

Potapchuk, W. R. (1996). Building sustainable community politics: Synergizing participatory, institutional, and representative democracy. *National Civic Review, 85*(3), 54–60. doi:10.1002/ncr.4100850311

Rose, M. (2004). Democratizing information and communication by implementing e-government in Indonesian regional government. *The International Information & Library Review, 36*, 219–226. doi:10.1016/j.iilr.2003.11.002

Shampa, P. (2007). A case study of E-governance initiatives in India. *The International Information & Library Review, 39*, 176–184. doi:10.1016/j.iilr.2007.06.003

Sirianni, C. (2007). Neighborhood Planning as Collaborative Democratic Design. *Journal of the American Planning Association. American Planning Association, 73*(4), 373–387. doi:10.1080/01944360708978519

Thomas, J. C., & Streib, G. (2003). The new face of government: Citizen-initiated contact in the era of e-government. *Journal of Public Administration: Research and Theory, 13*(1), 83–102. doi:10.1093/jpart/mug010

U.S. Department of Commerce. (2002). *A Nation On-line: How Americans are Expanding their Use of the Internet*. Retrieved from http://www.ntia.doc.gov/ntiahome/dn/nationon-line_020502.htm

Wellman, B., Salaff, J., Dimitrova, D., Garton, L., Gulia, M., & Haythornthwaite, C. (1996). Computer Networks As Social Networks: Collaborative Work, Telework, and Virtual Community. *Annual Review of Sociology, 22*, 213–238. doi:10.1146/annurev.soc.22.1.213

West, D. M. (2000). *Assessing e-government: The Internet democracy, and service delivery*. Retrieved from http://www.insidepolitics.org/egovtreport00.HTML

KEY TERMS AND DEFINITIONS

Comprehensive Plan: A document typically generated by a municipal planning department in conjunction with citizens and public officials that reviews community social, economic, and environmental trends and proposes a positive vision of the community based on land use, with related policies to achieve the vision.

Digital Divide: The gap between those who can access and benefit from digital technology and those who cannot. It may be affected by a multitude of factors, including income, age, race, physical and mental abilities, and geographical location.

E-Government: A coordinated effort by a government (local, state, federal) to provide technology-enhanced and often Internet-based tools for citizens, businesses, and employees to increase government efficiency and effectiveness.

E-Participant: A community member who utilizes online tools to participate in planning or other government activities.

Emoticon: A text or graphical portrayal of the tenor of a message as selected by its writer. A common emoticon is a smiley graphic or the use of colon + right parenthesis to indicate happiness or amusement.

Flaming: An often hostile exchange between Internet users in an online social dialogue venue. Exchanges may include insults and threats.

GIS (Geographic Information System): An electronically based mapping tool commonly used in planning that can capture, display, manage, and analyze spatial data.

Planning: A government activity focused on policies and projects related to community land. Typically includes housing, economic development, environment, land use, and transportation considerations.

Planning Department: An organizational unit within a municipal government that conducts planning activity. It is typically focused on determining appropriate use of land through adoption of general or comprehensive plan policies. A planning department is also responsible for the application of these policies, along with the requirements of codes and zoning regulations to permit approvals.

Wiki: An Internet-based webpage (or collection of webpages) to which users can contribute comments or edits to proposed content.

Chapter 12
Planners Support of E-Participation in the Field of Urban Planning

Mikael Granberg
Örebro University, Sweden

Joachim Åström
Örebro University, Sweden

ABSTRACT

The chapter questions what planners really mean when they display positive attitudes toward increased citizen participation via ICTs? Are they aiming for change or the reinforcement of existing values and practices? What are the assumptions that underlie and condition the explicit support for e-participation? In addressing these questions, this chapter draws upon a survey mapping the support for e-participation in the field of urban planning, targeting the heads of the planning departments in all Swedish local governments in 2006. The results show confusing or conflicting attitudes among planners towards participation, supporting as well as challenging the classic normative theories of participatory democracy and communicative planning.

INTRODUCTION

A common argument is that new information and communication technologies (ICTs) represent a vast array of technical improvements in our communication opportunities and that this would be a vital force in bringing about a closer relationship between citizens and their governments. The unprecedented degree of interactivity offered by new ICTs has the potential to expand the scope, breadth and depth of government interfacing with citizens and other key stakeholders during policy making. For e-participation to possess true meaning, however, these material capacities must be put to work in ways that reflect social intentions. Just as schools are not simply the buildings in which education takes place, but spaces organized around culturally specific notions of education, e-technologies cannot acquire a socio-political function without policy intentions (cf. Creasy et al. 2007; Coleman et al. 2008).

In the case of e-participation, there has been hardly any systematic work conducted aimed at identifying different normative positions. It is

DOI: 10.4018/978-1-61520-929-3.ch012

frequently assumed that all actors involved in e-participation share a common set of meanings about the nature of participation, technology and politics. This chapter examines to what extent planners support e-participation in planning as well as what they actually mean when they display positive attitudes. To meet this objective, we draw upon a survey questionnaire which maps attitudes towards e-participation in the field of urban planning and which was distributed by the authors in the spring of 2006 to the heads of planning departments in all 290 Swedish local governments.

While developments towards e-participation processes can be discerned in many countries, the prospects for implementing e-participation might be considered better in Sweden than in many other countries. The rapid expansion of the Internet, combined with an ongoing broadband expansion, implies that a technological platform now exists on which to develop applications that may strengthen citizen participation. According to 2005 statistics, 73% of the Swedish population between 18 and 79 years of age had access to the Internet at home, and, out of these, 66% had a broadband connection (Nordicom, 2006). Several state-backed studies on the national level in recent years have argued that the Internet should be used to promote government accountability and to increase public participation in politics and planning (SOU 2000: 1; SOU 2001: 48). This proposition is valid for various domains of political life, but public participation is regarded as particularly important for local planning processes: providing opportunities and encouragement for citizens to express their views on matters that affect their everyday lives.

The disposition of the chapter is as follows. First, the Swedish planning context is presented giving a background to the environment were planning is conducted. Second, a theoretical approach and two ideal typical models for the development of e-participation attitudes are developed. The e-participation models hold that opportunities for information and communication via digital technologies might affect attitudinal patterns, either by reinforcing existing values and practices or by contributing to change. In this section we also discussed some methodological issues. Third, the attitudinal survey, data, and measurement are described. In sections four and five, we describe the diffusion of e-participation support among Swedish urban planners in relation to individual and city characteristics. Furthermore we analyze the meaning of such support by examining whether the predominant beliefs and values found among "e-participation enthusiasts" are different from those of other planners, when enthusiasts are defined as those who state that Internet technologies "to a very large extent" should be used to increase citizen participation in policy making. Finally, the sixth section discusses the implications of the empirical evidence for the understanding and development of e-participation in planning. Accordingly, we will now continue with a presentation of the Swedish planning context aiming at giving the reader a basic understanding of the Swedish planning system and some of the changes it has gone through.

The Swedish Planning Context

Swedish local governments have a planning monopoly, the planning system is designed for local government, and each local government must have and maintain a comprehensive plan (Fog et al., 1989; Alfredsson & Wiman, 2001). The comprehensive plan should cover the whole municipal territory and guide planning but it is not binding. The next and more operational level of planning documents is the detailed development plan, which is a binding executive planning instrument. The general idea is that the intentions formulated in the comprehensive plan should be realized through detailed plans and finally manifests themselves through infrastructure and housing. The process between the initiation of planning and the formal planning decision through which the plan is adopted is punctuated by a number of formalized "control stations" (Nyström, 2003; Henecke & Khan, 2002). The law regulates how the process should be pursued and also which stages should be included. One important stage is the consultation stage, where citizens are given opportunities to study and react to planning proposals. The consultations address the demands for transparency and for opportunities for citizens to access, understand and reflect upon planning proposals (cf. Nilsson, 2003).

In the post-war period before the 1970s Swedish planning was primarily geared towards expansion and the production of housing and infrastructure. During this period planning was guided by instrumental rationality and based upon a trust in systematic, scientific, knowledge and with planners as experts (Khakee, 1989). During the 1970s new planning ideas focusing on citizen involvement and increased communication surfaced (Bohm, 1985). The planning and building act introduced in 1987 further strengthened local governments' role in planning in relation to state and other actors. The law also aimed at increasing citizen participation, influence and involvement in other guises than as voters, landowners or stakeholders (cf. Fog et al., 1989, Strömgren, 2007). Meeting this objective turned out to be problematic, however, and during the 1990s the law was revised to further strengthen the potential for citizens to take part in planning. Despite these efforts, lack of citizen participation continued to be a problem in Swedish urban planning. Some observers concluded that citizens' input was not truly regarded as important, but was seen simply as a step in the planning process that must be completed to comply with state regulations. The gap between rhetoric and practice seemed to be as great as ever in many planning processes (Khakee, 1999; Henecke & Khan, 2002).

Critics have continued to stress the lack of citizen participation. In general, planning was an activity characterized by expert knowledge and instrumental rationality; it was a process in which experts primarily communicated with other experts (Khakee, 1989, 2000; Granberg, 2004). One way information was dominant, communication and substantial interaction lacking. Observers of the debate on the national level have emphasized continuity rather than change and the predominance of representative democracy and the expert planner (Strömgren, 2007).

To understand this phenomenon more fully, and to set the scene for understanding the role of e-participation, we need to explore the issue of e-participation in planning through the theoretical lenses of planning, technology and democracy.

A Theoretical Approach and Methodological Issues

The roots of modern planning can be traced back to the late 18th century and enlightenment (cf. Camhis, 1979). This is important because the enlightenment tradition and its perception of knowledge and rationality have had an exceptional vigor in planning theory up to the present (Allmendinger, 2001). The focus has been on instrumental rationality and on systematic, scientific, knowledge and a focus on planners as experts.

Traditional planning has for a long period been subject to increasing critique. One aspect of the critique focused the perceived lack of awareness about the meaning and impact of politics (Lindblom, 1959; Etzioni, 1967; Wildavsky, 1987). Another aspect of the critique concerned the perception of relevant knowledge in the planning process which focused on expert-knowledge (Forester, 1989; Friedmann, 1987; Fischer, 1990). The critique stated that citizens' experienced-based knowledge was important for planning and needed to be included in the planning process through extensive interaction between planners and citizens (Fischer & Forester, 1993). This was, of course, paired with a critique of the role for citizen participation in traditional planning which was very limited and framed by the ideal of representative democracy.

During the eighties and nineties this critique developed into a planning paradigm of its own, the communicative paradigm in planning theory (cf. Forester, 1989; Fischer & Forester, 1993; Healey, 1997). Much of the inspiration came from the theory of communicative rationality as formulated by Jürgen Habermas, from American pragmatism but also from Foucault (Fainstein, 2000; Flyvbjerg, 1998). Communication is at the center of attention but also the perception of knowledge which is much inspired by the social constructivists (Healey, 1997). This paradigm entails a direct challenge to the supreme position of expert knowledge in traditional planning. Since the 1980s the tendency has been to perceive planning as a political and democratic activity (Forester, 1989, 1999). Accordingly, it must be understood in its political and democratic institutional setting (cf. March & Olsen, 1989). There are many theories as to what democracy means, two particular theories stands out as dominant (Held, 1987; Birch, 2001). These are "elite democracy" and "participatory democracy". Elite democracy has usually been linked to a "bureaucratic" or "instrumental" view of government and the participatory theory to a "pluralist" or "empowering" view of government. Citizen participation from the elitist perspective primarily focuses upon participating in public elections (Schumpeter, 1950). The assumption is that governments know best and that citizen participation would simply lead to inefficiencies in the decision-making process with detrimental effects on the quality of the decisions made. Participatory perspectives on democracy, on the other hand, criticize administrative decision making for its concentration on instrumental rationality and systematic scientific knowledge and its view of bureaucrats as experts. Participation is also considered to be good in itself, apart from the effects on political outcomes (Pateman, 1970; Barber, 1984). Accordingly, proponents of participatory policy making usually argue in favor of two important transitions. First, public officials must relinquish their elaborately constructed aura of expertise or, put another way, their reluctance to include lay citizens in policy deliberations and decisions. Second, one has to find the means to make citizen participation more manageable for regulators increasing collaboration between citizens and government (deLeon, 1992; Dryzek, 2000).

Today participation and active citizenship are widely celebrated by governments. While some commentators view this "participatory turn" in public policy as an important sign of revitalization and change, others are more skeptical, arguing that there are different views about how participation should and could be implemented. What divides participation into different objects, they argue, is primarily a matter of location in the structure of power. As the planning process to some extent is a struggle of power it provides different experiences to those who hold power and those who do not: those traditionally in control are more likely to adopt a defensive posture towards social change and are concerned with the protection of existing institutions from excessive and "uncontrolled" participatory input. In this line of reasoning, the attractiveness of participation for planners most likely does not reside in its potential to shift sovereignty from politicians and professionals to deliberating citizens. While the ethos of responsiveness put citizens at the center by inviting their input, there still is a desire to keep them away from the administrative work and actual decision-making centers (Blaug, 2002; Vigoda-Gadot, 2004; Fung, 2006; Åström & Granberg, 2007).

The relationship between contemporary governance and ICTs makes this question even more interesting. Potentially, the rise of the Internet can influence planners' predominant attitudes towards participation, strengthening the value or changing its purpose. Theorists of electronic democracy often argue that ICTs, for the first time, make possible more participatory and deliberative forms of political decision making and of transforming purposive-rational action by experts into communicative action by citizens (Shulman et al., 2003; Coglianese, 2004). Neil Postman (1993, p. 13), for instance, claimed that there is an ideological bias in every technology, a predisposition to construct the world in one way rather than another, to value one thing over another, to amplify one sense or skill or attitude more loudly than another. This bias of technology may lead to its use having influences of which the user may not always be conscious, and which may not have been part of its original purpose. At the same time the new technology can provide planners with the opportunity to ask new questions and shift their attention. The struggle over interpreting the democratic potential of ICTs is, however, far from over. Another view is that the new opportunities on the Internet will only serve to reinforce the grip of established political and administrative elites.

When it comes to factors influencing diffusion of e-participation support among planners, we will test these both at the individual and the city level. Both theories and empirical investigation in the field of the so-called "digital divide" have so far provided us with a better understanding of the Internet diffusion processes among and within societies (Norris, 2001). They have found that socio-economic and demographic factors, such as age, gender and education has a significant impact on attitudes towards, and use of, the Internet in modern societies. Less emphasized in this literature, however, are factors directly linked to ICTs, such as computer literacy and trust in new technologies, which may also explain the variance in attitudes towards e-participation (Åström, 2004; Hoff, 2004). From such a technology-oriented perspective, it might be expected that planners' own use of ICTs may produce greater incentives to move planning online. One argument is simply that technology users may believe more strongly in the opportunities offered by the new technology than non-users. Another argument is that the integration of ICTs in planning presupposes a change of behavior among those who are supposed to use it. It is therefore reasonable to believe that planners who already use the new technology to some extent, and have invested some "capital"

in it, are more likely to support its extended use than are those who have not been as involved with it. We will thus test the importance of individual characteristics by using variables of age, gender, education and e-literacy. The last variable, e-literacy, is measured by the respondents perception of their ICT competence on a four-point scale from very good (1) to not good at all (4).

Besides individual characteristics, e-participation attitudes may depend on city contexts. Based on previous literature on policy innovation and diffusion, some of our explanatory variables will focus on resources. Among resource variables, size often turns out to be important when it comes to innovation (Schmidt 1986; Damanpour 1996), not least when it comes to innovation in e-government (Åström 2004). Large local governments are often in a better position when it comes to specialist competence and "free resources". Their advantage can also be explained by a combination of costs and benefits. Bigger cities can profit from economies of scale both when it comes to lower costs for experimenting and risk-taking, and when it comes to gains made in cases of successful implementation. In this study size is measured by population 2005. A second potentially important resource-variable is the backend status of e-government. It is possible that the variation in e-participation support can be explained by the broader picture of the application of ICTs by government agencies. In this study the backend status of e-government is measured by an index consisting of eight items derived from the survey: agencies use (1) or lack of use (0) of geographic information systems (GIS), 3-D modeling, virtual reality simulations, permit tracking software, geographic positioning system (GPS), photo interpretation, remote sensing, and zoning and code enforcement management software.

Demand is another important factor in policy innovation (West, 2007; Tolbert et al., 2008). First, use of the Internet by planning agencies could be conceived of as a response to demand from an increasingly computer savvy populace (Norris et al., 2001; Åström 2004). We measure this by the availability of ICT- infrastructure in the area of the municipality 2002, which is closely related to Internet-usage. Second, we might expect that e-participation support is influenced by a participatory political culture and constituent demand for taking part in policy making (Yang, 2005). We test this by including voter turnout in the local governments 2002, as a measure of participation in politics.

Central to the argument in this study is that participation advances multiple purposes and values in contemporary planning. We are therefore forced to ask not only how much participation planners really want, but what purpose or value participation has. That is, what kind of citizen participation do local planners really promote? In order to address this question, two models of e-participation attitudes are outlined. The reinforcement model states that the emergence of new technologies primarily provides opportunities for handling large amounts of information and data, which paves the way for comprehensive planning and a possible strengthening of the professional/ expert planner in line with a neo-rational planning model (Sehestedt, 2001). Creating opportunities for direct e-participation in decision making, on the other hand, is considered to contradict the value of professionalism in planning because it forces public servants to satisfy citizens even when such actions run counter to the general public interest. Professionals become obliged to satisfy a vague public will, primarily channeled via political parties and elected representatives.

In our model for change, the opportunities for e-participation are changing the value and purpose of participation in the direction of participatory democracy and communicative planning (cf. Healey & Khakee, 1997; Sandercock, 1998; Forester, 1999; Albrechts & Denayer, 2001). In line with arguments raised by Sherry Arnstein (1969) this would include the right of individuals to be informed, to be consulted, and to express their views on governmental decisions (cf. Macintosh, 2003). This model also stresses the need for better representation of citizen interests directly in governmental decision making. In this line of thinking, the interactive capacity of the new technology represents an opportunity for citizens to engage more directly and individually in the planning process. Most importantly for the planners, this kind of interaction has the potential to build trust and therefore also to further legitimate the planning process and its output (Granberg, 2004).

Therefore a series of five statements in the questionnaire dealt more directly with the planners' value orientations. They begun with perceptions of expertise and the traditional institutions of representative democracy and then moved up to direct involvement of citizens in actual decision-making processes. Agreement with the statements was rated from 1 (strongly agree) to 4 (strongly disagree).

The theoretical discussion indicates that there is both a case for change and for continuity. In order to measure this, we will now look into the empirical evidence as we discuss the survey and map out Swedish planners' attitudes towards citizens' e-participation in planning. We will start with the methodological issues followed by a presentation of our results.

THE SURVEY

The analysis carried out in this chapter draws upon two types of data: (1) a survey questionnaire by the authors targeting the heads of the planning departments in all 290 Swedish local governments, and (2) a database covering a wide range of structural data on Swedish local governments developed by Leif Johansson at Lund University, Sweden. The survey, conducted during the spring of 2006 via the web followed by a traditional postal survey, resulted in a 67 percent response rate. Since it is important to know who is interested in e-participation, the characteristics of the 193 respondents and 97 non-respondents were compared. We found that planners in cities with strong ICT-infrastructure (72 percent) were more likely to respond than planners in cities with weak ICT-infrastructure (63 percent). Since the analysis in the next section will show that this variable is related to support for e-participation among respondents, it is important to keep this bias in mind when reading the tables.

The four-page survey contained mostly close-ended questions, giving room for comments on the last page. As discussed in the introduction, the primary dependent variable in our analysis is support for e-participation. This is measured by one single question, of which the exact wording is: "To what extent do you think that Internet technologies should be used to increase citizen participation in planning?" Respondents were asked to indicate their appropriate position on a four-point scale, ranging from a very little degree (1) to a very large degree (4). Based on this measure we compared the characteristics and orientations of those who we regard as e-participation enthusiasts, defined as those scoring highest on the scale (4), and e-participation non-enthusiasts (scoring 1-3).

The Results

One central finding of the survey is that 94 percent of the planners support e-participation in planning (34 percent to a very large degree, and 60 percent to a fairly large degree). First, this indicates that Sherry Arnstein's (1969) proposition that participation has achieved an unassailable status of motherhood, apple pie and spinach – "No one is against it in principle because it is good for you" – is still valid. Second, the strong support for the use of e-participation indicates that there is a strong faith in the positive democratic potential of ICTs among Swedish urban planners. This does not, however, tell us whether e-participation is neutral, in the sense that it affects all categories of planners to a similar degree. In other words, besides choosing another mode of participation, are the individual and city profiles of "e-participation enthusiasts" significantly different from those that remain more faithful to traditional channels of participation?

The analysis in figure 1 shows the percentage of planners that are most supportive of e-participation – "the e-participation enthusiasts" – within different social and technological groups. A slightly bigger proportion of men than women strongly support e-participation, but at the same time men are more often strong opponents of e-participation. Men thus seem to have stronger opinions, both positive and negative, than women. When it comes to age and education, the attitudes follow the expected pattern, but in this case the pattern is not particularly strong. Instead it is the technology-oriented variable that stands out. Planners with (perceived) more developed computer skills are more often strong supporters of e-participation than are planners with weaker computer skills.

The interrelationship between social and technological variables creates a problem here, since it is difficult to isolate the influence of each. By using multivariate cross-tabulations, however, we find that e-literacy has a positive effect for each of the other groups. The way technology is used by individuals at one stage or in one area thus seems to be of importance for the attitudes associated with the technology at a later stage or in another area. This may be interpreted as "a cumulative technological effect," or a process in which attitudes are gradually being adjusted to the available means. Interesting, though, is how the relationships between the social variables and e-literacy are played out. When it comes to gender, most notably, we find that the women planners are younger, that they have higher education and are more e-literate than men. Still, men are more enthusiastic about e-participation, which becomes even more evident in a multivariate analysis. For instance, we find that 67 percent of the e-literate men and only 36 percent of the e-literate women are e-participation enthusiasts. We may thus conclude that even if the most obvious divides concerning access and e-literacy are absent, other more subtle divides are emerging.

Figure 1. E-participation enthusiasts by individual characteristics, percentage and number

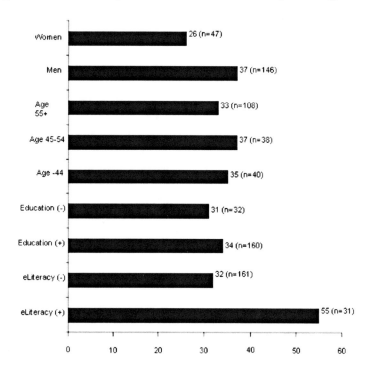

The results given in figure 2 shows that nearly all contextual factors analyzed here has the potential to explain the attitudinal variations. At least there are clear attitudinal differences by electoral turnout, e-government back-office developments and ICTs infrastructure. In some instances, relating the variables to each other through multivariate cross-tabulations reinforces these differences. For instance, 51 percent of the planners in a context of high electoral turnout and well-developed e-government back office are e-participation enthusiasts while this is the case for only 19 percent of the planners in a context of low turnout and less developed e-government back office. In other cases the variables are exchangeable, indicating that the impact does not come from each of the variables per se, but through their close association with one another. To examine this further a regression analysis was made, showing that the infrastructure of ICTs is the most important context variable explaining variations in attitudes (Beta .250 sig. .05).

The overall results in this section therefore suggest that the diffusion of e-participation support is primarily driven by ICT-oriented factors: on the individual level the respondent's computer competence and on the city level the availability of ICT-infrastructure in the municipality. It is likely that these developments directly influences how far planners can go to provide e-participation, and that they also produce greater incentives for the planners to do so, as the general public and they themselves becomes wired. By looking at e-participation support, without analyzing the meaning of such support, our data thus support the model for change, emphasizing organizations' adaptation to ICTs.

One conclusion from table 1 is that e-participation in planning is not related to representative democracy and party-based politics. A clear majority of the planners do not think that citizen influence should be channeled through elected representatives or political parties. Instead e-participation in planning is based on the idea of

Figure 2. E-participation enthusiasts by city characteristics, percentage and number

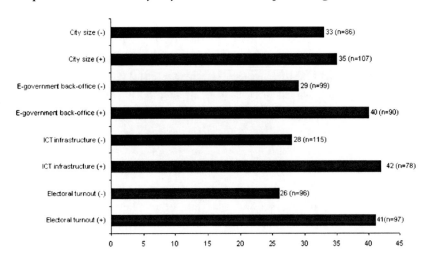

direct communication between planners and the individual citizen engaging in the planning process. On the one hand, this may be interpreted as the presence of communicative planning and participatory democracy ideals, where it is essential that planning should be done through interaction among those who have an interest in the outcomes of the planning process. Throughout history, many advocates of participatory or direct democracy have regarded parties with considerable suspicion, on the grounds that citizens should discuss issues and determine priorities within each community "uncontaminated" by partisan bias (Norris, 2005). On the other hand, this may be interpreted as a separation between planning and politics in line with the idea of a "rational planner," who serves as a technical expert outside the political sphere so that issues are "settled on their own merits."

Table 1. E-participation support and value orientations

	Model 1			Model 2	
	E-participation enthusiasts	E-participation non-enthusiasts	Mean diff.	Beta	Sig.
1. 'Citizen influence in planning should be channeled through political parties'	2.83	2.82	.01	.028	
2. 'Planning is a professional activity based on expert knowledge'	1.98	1.88	.10	-.001	
3. 'Citizen experienced-based knowledge is important for planning'	1.32	1.46	.14	.180	*
4. 'Planning receives its legitimacy through public dialogue'	1.40	1.59	.19	.158	*
5. 'Citizens should be able to participate directly in planning decisions'	3.09	3.14	.05	.063	
Note: Model 1: The figures represent the mean scores of e-participation enthusiasts and non-enthusiasts on scales from "strongly disagree" to "strongly agree" (1-4) without any controls. Model 2: The figures represent standardized beta coefficient in regression models measuring the impact of the e-participation support on statement agreement after controlling for gender, education and age. * = Sig. .05 and ** = Sig. .01					

E-participation in planning would then be about assessing which alternatives best meet politically established goals to make better recommendations to political decision makers about which course of action to follow. In this line of reasoning, the appeal of e-participation would not lie in its potential to shift sovereignty from elected representatives to a mass of deliberating citizens. Rather, it would be a means to better planning, as defined by expert planners.

The attitudinal pattern in relation to question 2-5 in table 1 leans more towards the second interpretation than the first, which means that it falls somewhere in between the ideal models of reinforcement and change. On the one hand, a clear majority of the planners display positive attitudes about receiving input from citizens and debating issues with them. On the other hand, it is obvious that expert knowledge is still very important and direct participation in decision making is still dismissed. This means that e-participation is not viewed as an alternative to expertise, as is often suggested in the normative literature in political science, public administration and planning literature, but can instead complement it by strengthening the planners' position. This, in turn, strengthens the notion that e-participation for planners is not linked to the normative, intrinsic reasons to favor participation, but is instead guided by an instrumental view of benefits. While planners want to preserve a substantial part of their autonomy, authority and power, problem solving rather than power sharing must be the purpose of involving the citizens. Potential benefits may include making the work of government more palatable to citizens, as well as making decision-making more effective. If citizens possess essential local knowledge that comes from close exposure to the context in which problems occur, for instance, their participation may supplement and strengthen planners' knowledge and make decisions more legitimate.

E-participation enthusiasts have somewhat different attitudes towards participation than other planners, especially when it comes to the value of citizens' experience-based knowledge and public dialogue. These differences mainly take the form of "the same, only more so," however, and do not challenge either power relations or the instrumental view of participation per se. Among e-participation enthusiasts there is no less emphasis on planning as a primarily professional activity based on expert knowledge, and no more positive attitudes towards direct citizen participation in decision making. In this context, the attitudinal differences between e-participation enthusiasts and other planners are therefore not likely to be the result of an ideological reorientation towards the classic values of participatory democracy and communicative planning. They are more likely to be the result of an instrumental calculation of costs and benefits in relation to different modes of participation. Those who perceive the administrative costs of e-participation to be low relative to instrumental benefits, such as the quality of input or the opportunities for increasing legitimacy, are more positive than others about involving citizens.

CONCLUSION

So what are the potential trends for the future? The evidence discussed above is somewhat ambivalent and a clear cut answer is of course not possible. The pressure towards amplified citizen participation is clearly increasing in the theoretical and political debate with calls for interactive decision making heavily engaging the citizens in policy processes from initiation to decision (cf. Blaug, 2002; Vigoda-Gadot, 2004; Fung, 2006). At the same time planning practices are institutionalized and planners' attitudes do not change rapidly. The tension between these two, the external pressure and internal practices, values and attitudes of planning, can pave the way for e-participation practices in planning that comes closer to the ideal of participatory democracy and communicative planning. The expert role often attributed to planners has considerable resilience and could perhaps withstand the increased pressure from normative arguments and technological potentials.

This study shows confusing or conflicting attitudes towards citizen e-participation among urban planners. Most respondents display positive attitudes towards the idea of receiving input from citizens and debating issues with them, but they hold negative attitudes towards a more direct involvement of citizens in decision making. Moreover, they are positive about including citizens' experience-based knowledge in planning, but they still view planning as a professional activity based on expert knowledge. While e-participation enthusiasts have somewhat different attitudes towards participation than other planners, these differences mainly take the form of "the same, only more so" and do not challenge the general pattern or current practice. Accordingly, the result does not indicate either the presence or the development of an attitude in line with the normative literature on participatory democracy and communicative planning. Instead, planners' attitudes towards the new opportunities of e-participation seem to be guided by an instrumental view of costs and benefits. Participation in planning seems to be more about problem solving and problem sharing than about real power sharing. This means that e-participation is at best an add-on to traditional planning processes and is not considered an alternative form of democratic deliberation and engagement.

This is not to claim that nothing changes. If the new technology is considered to lower the costs of public participation, the instrumental benefits of e-participation for democratic governance may be substantial. Citizens could become important resources for the solution of collective problems and the production of welfare. Along with an instrumental adoption of e-participation can also follow structural change. For instance, if electronic channels provide for more interactive communication between planners and citizens, online consultation and government polls has the potential to streamline the planning process, reducing reliance on intermediary bodies such as political assemblies and parties. The impact of new technologies on our representative structures would then be centered on the conciliating, mediating, or even adjudicating role of public servants; fostering the increasing reliance on channeling the democratic expression of opinion and preferences through bureaucratic processes (Snellen, 2001). If planners strengthen their role vis-à-vis local representatives this might increase the legitimating problems for political parties and representatives. Since planners do not have a legitimate role in handling political conflicts or competing interests and cannot be held responsible in elections, there is also a risk for strong particularistic interests to become represented as the actual "common good." According to some researchers (cf. Klijn & Koppenjan, 2000), the question in relation to these problems is not so much about whether we should have more collaborative forms of decision making, but about how to organize these forms and what role politicians should take in them. Challenging the distinction between politics and planning that still seems to be strong, they argue that politicians together with government officials must initiate and guide the societal discourses exploring the common interest.

We can also conclude that the classical arguments of participatory democracy and communicative planning somewhat fails to grasp what planners believe is most attractive about e-participation. The problem of understanding and developing e-participation is thus complex. On the one hand it involves challenging existing understanding on how democracy and planning work as well as challenging embedded structures of political power in communities. On the other hand it is about addressing the instrumental consequences and potential benefits of participation for modern governance. At its best, e-participation can work in synergy with representation and administration to yield more responsive democratic and effective practices. This, however, demands more forward-looking and theoretical imagination from practitioners as well as from researchers.

REFERENCES

Albrechts, L., & Denayer, W. (2001). Communicative Planning, Emancipatory Politics and Postmodernism. In Paddison, R. (Ed.), *Handbook of Urban Studies*. Thousand Oaks, CA: SAGE Publications.

Alfredsson, B., & Wiman, J. (2001). Planning in Sweden - Fundamentals Outlined. In Christoferson, I. (Ed.), *Swedish Planning - In Times of Diversity*. Gävle, Sverige: The Swedish Society for Town and Country Planning.

Allmendinger, P. (2001). Planning. In *Postmodern Times*. London: Routledge.

Arnstein, S. R. (1969). A Ladder of Citizen Participation. *Journal of the American Institute of Planners*, 35, 216–224.

Åström, J. (2004). *Mot en digital demokrati? Teknik, politik och institutionell förändring*. Örebro Studies in Political Science 9. Örebro, Sverige: Örebro University.

Åström, J., & Granberg, M. (2007). Urban Planners, Wired for Change? Understanding Elite Support for E-participation. *Journal of Information Technology & Politics*, 4(2), 63–77.

Barber, B. (1984). *Strong Democracy. Participatory Democracy for a New Age*. Berkeley, CA: University of California Press.

Birch, A. H. (2001). *The Concepts and Theories of Modern Democracy* (2nd ed.). London: Routledge.

Blaug, R. (2002). Engineering Democracy. *Political Studies*, 50, 102–116. doi:10.1111/1467-9248.00361

Bohm, K. (1985). *Med- och motborgare i stadsplanering. En historia om medborgardeltagandets förutsättningar*. Stockholm: Liber.

Camhis, M. (1979). *Planning Theory and Philosophy*. London: Tavistock Publications.

Coglianese, C. (2004). Information Technology and Regulatory Policy: New Directions for Digital Government Research. *Social Science Computer Review*, 22(1), 85–91. doi:10.1177/0894439303259890

Coleman, S., Åström, J., Freschi, A. C., Mambrey, P., & Moss, G. (2008). *Making eParticipation Policy – A European Analysis* (Demo-net booklet D14.4.).

Creasy, S., Gavelin, K., Fisher, H., Holmes, L., & Desai, M. (2007). *Engagement for Change: The Role of Public Engagement in Climate Change Policy*. London: Involve.

Damanpour, F. (1996). Bureaucracy and Innovation Revisited: Effects of Contingency Factors, Industrial Sectors, and Innovation Characteristics. *The Journal of High Technology Management Research*, 7(2).

deLeon, P. (1992). The Democratization of the Policy Sciences. *Public Administration Review*, 52(2), 125–129. doi:10.2307/976465

Dryzek, J. S. (2000). *Deliberative Democracy and Beyond*. Oxford, UK: Oxford University Press.

Etzioni, A. (1967). Mixed-scanning: A Third Approach to Decision-Making. *Public Administration Review*, 27(5), 385–392. doi:10.2307/973394

Fainstein, S. (2000). New Directions in Planning Theory. *Urban Affairs Review*, 35(4), 451–478.

Fischer, F. (1990). *Technocracy and the Politics of Expertise*. Newbury Park, CA: Sage.

Fischer, F., & Forester, J. (Eds.). (1993). *The Argumentative Turn in Policy Analysis and Planning*. Durham, NC: Duke University Press.

Flyvbjerg, B. (1998). *Rationality and Power: Democracy in Practice*. Chicago, IL: The University of Chicago Press.

Fog, H., Bröchner, J., Törnqvist, A., & Åström, K. (1989). *Det kontrollerade byggandet*. Stockholm: Carlssons.

Forester, J. (1989). *Planning in the Face of Power*. Los Angeles: University of California Press.

Forester, J. (1999). *The Deliberative Practitioner. Encouraging Participatory Planning Processes*. Cambridge, MA: The MIT Press.

Friedmann, J. (1987). *Planning in the Public Domain: From Knowledge to Action*. Princeton, NJ: Princeton University Press.

Fung, A. (2006). Varieties of Participation in Complex Governance. *Public Administration Review*, 66(1), 66–75. doi:10.1111/j.1540-6210.2006.00667.x

Granberg, M. (2004). *Från lokal välfärdsstat till stadspolitik. Politiska processer mellan demokrati och effektivitet*. Örebro Studies in Political Science 11. Örebro, Sverige: Örebro University.

Grönlund, Å. (2001). Democracy in an IT-Framed Society. *Communications of the ACM*, 44(1), 23–26. doi:10.1145/357489.357498

Gustafsson, G. (1981). KPP-projektet om kommunal demokrati och planering. (Rapport/Byggforskningsrådet 1981:5). Stockholm: Statens råd för byggnadsforskning.

Healey, P. (1997). *Collaborative Planning. Shaping Places in Fragmented Societies*. London: Macmillan.

Healey, P., & Khakee, A. (1997). *Making Strategic Spatial Plans: Innovation in Europe*. London: UCL Press.

Held, D. (1987). *Models of Democracy*. Cambridge, UK: Polity Press.

Henecke, B., & Khan, J. (2002). *Medborgardeltagande i den fysiska planeringen - en demokratiteoretisk analys av lagstiftning, retorik och praktik*. (Working Paper in Sociology. 2002:1/Report No. 36, November 2002). Lund, Sverige: Sociologiska institutionen/Avdelningen för miljö- och energisystem, Lunds universitet.

Hoff, J. (2004). Members of Parliaments' use of ICT in a Comparative European Perspective. *Information Polity*, 9, 5–16.

Khakee, A. (1989). Kommunal planering i omvandling 1947-1987. (Gerum nr 13, Gerum). Umeå, Geografiska institutionen, Umeå universitet.

Khakee, A. (1999). Demokratin i samhällsplaneringen. In E. Amnå (Ed.), Medborgarnas erfarenheter. (Demokratiutredningens forskarvolym V. SOU 1999: 113). Stockholm: Fakta info direkt.

Khakee, A. (2000) Samhällsplanering. Lund, Sverige: Studentlitteratur.

Klijn, E. H., & Koppenjan, J. F. M. (2000). Politicians and Interactive Decision Making: Institutional Spoilsports or Playmakers. *Public Administration*, 78(2), 365–387. doi:10.1111/1467-9299.00210

Kraemer, K. L., & Dutton, W. H. (1982). The Automation of Bias. In Danziger, J. N., Dutton, W. H., Kling, R., & Kraemer, K. L. (Eds.), *Computer and Politics: High Technology in American Local Governments*. New York: Columbia University Press.

Lindblom, C. E. (1959). The Science of Muddling 'Through'. *Public Administration Review*, 19, 288–304. doi:10.2307/973677

Macintosh, A. (2003). Using Information and Communication Technologies to Enhance Citizen Engagement in the Policy Process. In Promises and Problems of E-Democracy: Challenge of Online Citizen Engagement. Paris: OECD.

March, J. G., & Olsen, J. P. (1989). *Rediscovering Institutions: The Organizational Basis of Politics*. New York: Free Press.

Nilsson, K. (2003). Planning in a Sustainable Direction - The Art of Conscious Choices. Stockholm: Institutionen för infrastruktur och samhällsplanering, Kungliga Tekniska Högskolan.

Nordicom. (2006). *Nordicom-Sveriges Internetbarometer 2005*. (Nordicom-Sverige MedieNotiser Nr 2, 2006). Göteborg, Svergie: Göteborgs universitet.

Norris, D. F., Fletcher, P. D., & Holden, S. H. (2001). Is Your Local Government Plugged. In *Highlights of the 2000 Electronic Government Survey. Baltimore County*. Baltimore, MD: University of Maryland.

Norris, P. (2001). *Digital Divide? Civic Engagement, Information Poverty, and the Internet Worldwide*. Cambridge, UK: Cambridge University Press.

Norris, P. (2005). *Developments in Party Communications*. Washington, DC: National Democratic Institute for International Affairs.

Nyström, J. (2003). Planeringens grunder. En översikt (2nd Ed.). Lund, Sverige: Studentlitteratur.

Pateman, C. (1970). *Participation and Democratic Theory*. Cambridge, UK: Cambridge University Press.

Postman, N. (1993). *Technopoly: The Surrender of Culture to Technology*. New York: Vintage Books.

Pratchett, L. (1998). Technological Bias in the Information Age. ICT Policy Making in Local Government. In Snellen, I., & van de Donk, W. (Eds.), *Public Information in an Information Age*. Amsterdam: IOS Press.

Sandercock, L. (1998). *Towards Cosmopolis*. New York: John Wiley & Sons.

Schmidt, S. (1986). Pionjärer, efterföljare och avvaktare. Lund, Sverige: Kommunfakta Förlag.

Schumpeter, J. A. (1950). *Capitalism, Socialism and Democracy*. London: Allen and Unwin.

Sehested, K. (2001). *Investigating Urban Governance - From the Perspective of Policy Networks, Democracy and Planning*. (Research Paper no. 1/01). Roskilde, Sverige. The Department of Social Sciences. Roskilde University.

Shulman, S., Schlosberg, D., Zavestovski, S., & Courard-Hauri, D. (2003). Electronic Rulemaking: A Public Participation Research Agenda for the Social Sciences. *Social Science Computer Review, 21*(2), 162–178. doi:10.1177/0894439303021002003

Snellen, I. (2001). ICTs, Bureaucracies, and the Future of Democracy. *Communications of the ACM, 44*(1), 45–48. doi:10.1145/357489.357504

SOU. (2000). *En uthållig demokrati. Politik för folkstyrelse på 2000-talet. Demokratiutredningens betänkande (State Commission Report)*. Stockholm: Fritzes.

SOU. (2001). *Att vara med på riktigt - demokratiutveckling i kommuner och landsting. Bilagor till betänkande av kommundemokratikommittén (State Commission Report)*. Stockholm: Fritzes.

Strömgren, A. (2007). Samordning, hyfs och reda. Stabilitet och förändring i svensk planpolitik 1945-2005. Uppsala, Sverige: Acta Universitatis Upsaliensis.

Tolbert, C., Mossberger, K., & McNeal, R. (2008). Innovation and Learning: Measuring E-government Performance in the American States 2000-2004. *Public Administration Review, 68*(3), 549–563. doi:10.1111/j.1540-6210.2008.00890.x

Vigoda-Gadot, E. (2004). Collaborative Public Administration: Some Lessons from the Israeli Experience. *Managerial Auditing Journal, 19*(6), 700–711. doi:10.1108/02686900410543831

West, D. (2007). *Global E-Government, 2007*. Providence, RI: Center for Public Policy, Brown University.

Wildavsky, A. (1987). *Speaking Truth to Power: The Art and Craft of Policy Analysis*. New Brunswick, CT: Transaction Publishers.

Yang, K. (2005). Public Administrators' Trust in Citizens: A Missing Link in Citizen Involvement Efforts. *Public Administration Review, 65*(3), 273–285. doi:10.1111/j.1540-6210.2005.00453.x

ADDITIONAL READING

Bimber, B. (2003). *Information and American Democracy. Technology in the Evolution of Political Power*. Cambridge, UK: Cambridge University Press. doi:10.1017/CBO9780511615573

Birch, A. H. (2001). *The Concepts and Theories of Modern Democracy* (2nd ed.). London: Routledge.

Chadwick, A. (2006). *Internet Politics. States, Citizens, and New communication Technologies*. Oxford, UK: Oxford University Press.

Coleman, S. (2009). *The Internet and Democratic Citizenship. Theory, Practice and Policy*. Cambridge, UK: Cambridge University Press.

Conroy, M. M., & Evans-Cowley, J. (2006). E-participation in planning: an analysis of cities adopting on-line citizen participation tools. *Environment and Planning. C, Government & Policy, 24*, 371–384. doi:10.1068/c1k

Dryzek, J. S. (2000). *Deliberative Democracy and Beyond*. Oxford, UK: Oxford University Press.

Fountain, J. E. (2001). *Building the Virtual State. Information technology and Institutional Change*. Washington, DC: The Brookings Institution.

Norris, P. (2001). *Digital Divide. Civic Engagement, Information Poverty, and the Internet Worldwide*. New York: Cambridge University Press.

KEY TERMS AND DEFINITIONS

Citizen: An individual with rights to participate in political life (citizen rights also include civil and social rights). The term is integrated in citizenship which means to be a full member of society.

Communication: A two way flow of information, knowledge and experiences between entities (individuals, organizations, etc.) in society.

Influence: To have actual impact in decision-making. Accordingly, influence does not connote control over decision-making.

Participation: Taking active part in political life.

Chapter 13
Portals as a Tool for Public Participation in Urban Planning

Jens Klessmann
Fraunhofer Institute for Open Communication Systems (FOKUS), Germany

ABSTRACT

In this article it will be shown how different general types of portals can be utilized to foster public participation processes in urban and regional planning. First portals and the objectives of their use in the public sector are explained. This happens before the background of different concepts of administrative reform and a transition of government to an electronic manner. Then public participation will be described and different categories thereof are presented. This part forms the basis for the delineation of electronic participation in urban planning. Finally the already introduced general portal types will be applied to distinguish several kinds of participation portals.

INTRODUCTION

Public administrations utilize the advantages of portals in various dimensions. Electronic government supports a transformation of the way how governments and their administrations conduct their daily business. Public institutions try to achieve a more open style of communication with their citizens. Many, especially at the local level, even seek a stronger involvement of active residents.

Information and communication technology can support the involvement of citizens in many fields of government: from policy development through budget planning towards urban land use planning. The different activities can be concentrated and made accessible with the help of internet portals supporting participation. According to current research internet portals can be divided into four categories: Entry points, information pools, service centers and service clusters.

Entry points provide the user with A-Z listings to the websites of public administrations, local government, members of parliament, political organizations, relevant media and public institutions. Information pools make information from the public realm available in easy to understand language and different media formats. In the field of participation they might provide information about tools for publishing, participation and self-organizing for citizens like wikis, chats, blogs and news groups. Service centers can offer direct participation through different methods and tools. Citizens can get directly involved in budget planning or urban land use planning activities. Service centers could also provide e-voting processes of public bodies and political organizations. Service clusters give an overall access to many different participatory portals. They provide one-stop-access to all participation processes and support citizens by offering search and notification services (von Lucke, 2008).

In this chapter two examples of public participation processes relying in part on electronic portals in Germany will be depicted and categorized according to the above described model. Instances will be chosen from different administrative levels and thus differing geographical scopes.

The objective is to illustrate the potential of portals for supporting public participation. By depicting two current examples of internet portals in Germany the potential as a tool for participation is shown.

The research methods used include literature review and internet research. In order to identify, understand and classify selected examples of e-participation portals in Germany relevant literature will be reviewed. This work will cover the fields of participation and e-participation as well as internet portals. Internet research will be conducted to identify the examples, especially in the field of urban planning, for further analysis and to gain knowledge about best practices of e-participation portals. The objective of this design is to understand the potential of internet portals to provide an additional channel for participation in the field of urban planning.

BACKGROUND

Portals: Objectives and Types of Internet Portals

Internet portals play an important role in the modernization of the public sector. Their ability to provide a unified access to different services and IT-systems can enhance the service delivery of public sector organizations. Thus in this chapter the underlying conditions and goals for using portals in public authorities and the different types of such portals will be delineated.

In order to keep competitive as a city, region or nation and at the same time cut costs public authorities constantly have to re-define their core mission and procedures. Reducing red tape and thus becoming more attractive for citizens as well as businesses is a major goal. With the rise of the Internet in the mid 1990s the focus of electronic government initiatives became providing online information and digital public sector services. With more and more services being accessible through electronic media further raising the bar in electronic service delivery is increasingly difficult. The customers and clients of the public sector expect better services as they are used to online purchasing, banking, auctions and more complex digital process chains like in the automobile sector. Public authorities have realized the need to provide even more multifaceted electronic processes. For this to happen though, still separate back-end systems within the public administration have to be integrated (United Nations, 2008; Lenk, 2007).

One approach to allow for more flexibility is to separate the customer facing parts from the general staff of a public authority. Traditionally public administrations are organized vertically. Every officer fulfills both paperwork and customer care. Thus highly trained and specialized staff members often have to deal with more ordinary tasks like guiding customers through filling out forms. By separating the front office from the back office the officers can concentrate on their tasks without having to deal directly with customers. The front office staff handles most of the clients' requests. Some advantages of this setup are the following:

- The back offices can be relieved of repetitive and low level customer care tasks. Therefore specialized manpower is freed and can be utilized.
- The staff in the front office can be trained to serve the customer in the best way. A better customer service can be reached.
- One back office could provide services to front offices of different local authorities. Public administrations could form networks of production. Each knot in the network specializes on certain tasks and provides those for the other administrations front offices.
- One front office can cover different back offices or even all back offices of a single public authority or even several different administrations. Therefore customers do not need to know who is responsible for their concern within the public administration. The unified front office can provide support in all matters. Thus the customers can ideally solve all their governmental issues with one stop (Breitenstrom, Eckert, von Lucke, 2008).

The objective of such a One-Stop-Government is a combined delivery of services for the administrations customers. Citizens and businesses can access and use public services with few or only one visit to the public administration. The concept is derived from the private sector, where one-stop services are quite common. The customers should not have to deal with the structure of an organization in order to identify the correct contact responsible for their matter. Rather the organization appropriates a single point of contact. This point can either deal with the request itself or assist in accessing the correct resources. The point of contact can be an institution with staff supporting the customer's requests. It can also be organized as an electronic internet portal. Depending on the development stage portals can be set up from providing help in finding the correct resources up to allowing undertaking fully automated transactions themselves. Via internet portals different access channels to public administration can be used.

Citizens expect being able to access the public administration through different channels. The public sector is perceived as being like any other service provider. Thus flexibility and service mentality are expected. This service quality should be provided on all possible access channels, whether this is via phone, e-mail, the internet or in person (Bertelsmann Stiftung, 2005).

The electronic channel though is of special importance in a modern-day multi channel scenario. It can provide access to information, services and application systems. Intermediaries utilize the electronic channel in order to support users of other channels like phone or in person. The bundling of different access channels that can be provided by the electronic channel relies on certain access systems, called portals, which allow regulating who has the right to use which information.

Portals are herein defined as easy to use, customizable and safe access systems. They grant authorized users access to information, applications, processes and persons, which are available within the attached systems. According to the multi channel principle access can happen through different media and access channels (von Lucke, 2008).

Portals are not confined to providing access via one channel. Nonetheless the electronic channel is the most important channel. It can form the basis for access to information and services through all other channels. Portals within TCP/IP-based networks often can be accessed worldwide. Intermediaries from other channels like call center agents can utilize the electronic portals and thus enhance their own work.

Different functions can be associated with portals. They provide A-Z-like information about organizations, security and access checks, processing of applications, purchase orders and payment transactions. Further portals can provide access to and integration of back-office IT-systems. Portal technologies provide these functionalities. The main functions are management, access services, presentation, navigation and integration (von Lucke, 2008).

Portals can be categorized into different types. These different categories can be found at the same time, but they can also be used to describe the development of portals. From A-Z collections to integrated networks of portals with transactional services very different depths of service can be observed.

Portals functioning as entry points provide information pointing towards other resources where either information about ones request can be found, responsible agencies or even online transactions can be accessed. Consequently such entry points can be described as guideposts supporting citizens and businesses in finding the correct point of contact.

Portals designed as information pools yield not only guideposts to information at other websites, they furthermore provide information themselves. This content can be generated and structured by the portal carrier or an additional service entity. In the context of public sector portals such pools can be used to gather information about any upcoming administrative procedures.

The third kind of portals is called service centers. These portals give access to transactional administrative services. Customers can carry out electronic procedures for which they otherwise would have to visit a public agency in person. The digital services accessible have to be integrated into the portal. Service centers also include guideposts to other websites and sources of information as well as self-generated information.

The most advanced form of internet portals are service clusters. These are basically tightly integrated networks of service centers. In line with the One-Stop-Government paradigm users can access any information and transaction from any partaking service center. The customers thus do not have to deal with identifying the correct public entity and choose the corresponding portal. In a service cluster the point of access is not important. A comprehensive service cluster in the public sector would include each and every portal from all public authorities. Such an arrangement supports the reform of the current public administration. Processes could be connected much closer and with fewer barriers to data exchange. Of course the requirements of data privacy still have to be met (von Lucke, 2008).

Electronic Participation

Public participation is used in many different situations by power-holding bodies. The aims are diverse ranging from justifying decisions already made to giving power to the people. Different methods are used, like conducting surveys or organizing discussion forums. One of the main areas where participation is applied regularly is the domain of urban planning. Land uses and other urban issues often touch on the interests of many different parties. The special interests have to be resolved, thus many countries require public participation by law in urban planning processes. Urban planning authorities are mostly obliged to involve public stakeholders like non-governmental organizations. The involvement of individual persons received wide-spread recognition with the introduction of the Aarhus Convention by the members of the UN Economic Commission for Europe in 1998 (Kubicek, 2007; United Nations, 1998).

Public participation is often used in very general terms without further specification. In the scientific literature and subsequently in professional practice participation is differentiated into different degrees. Arnstein (1969) describes a ladder of participation on which she distinguishes eight rungs from nonparticipation to citizen control. In her view informing and consulting of citizens is more a form of tokenism than "real" participation. Only if public bodies engage in partnerships or even start to delegate power, the public can start to take part in decision making.

The different stages of participation according to Arnstein can be described as follows:

- **Manipulation:** People are made to believe, their input is valued, when in fact it is not. An example is the participation in committees without real influence.
- **Therapy:** Citizens are involved in public planning processes with the sole goal to "cure" them. Their different beliefs are to be adjusted during the procedure.
- **Informing:** Giving information to the people is considered as the basis for real participation. It is criticized, that informing is merely a one-way street.
- **Consultation:** Asking for the opinion of the public without providing a clear method of implementing their input into the decision making process is not considered to be full participation.
- **Placation:** Participants for the first time gain some degree of influence at this stage. An example is the participation in committees with real influence, but the citizen representatives can only form a minority opinion. The power holders do not share parts of their power.
- **Partnership:** The public and power holders at this stage agree to share power and figure out how the control can be distributed among the different stakeholders.
- **Delegated Power:** The sharing of power from the previous rung can lead to much power being delegated. Citizens may gain the majority opinion in certain planning processes.
- **Citizen Control:** Citizens are in control of certain public entities like schools or even a local authority like a district. This form of participation is rarely achieved, but citizens often strive for it.

One can argue that Arnstein does not demonstrate how the different rungs can be implemented. She criticizes the lack of "true" participation and describes what such involvement should look like in broad terms. Practitioners willing to thoroughly involve the public have to identify the correct methods themselves. Furthermore the model appears to leave no room for professionals and their expertise in urban planning or other fields where public action is called for. Thus it can lead to opposition by experts to involve the public. Win-Win Solutions have to be created, providing both, the layperson and the professional with an influential role (Goodspeed, 2008).

"Real" participatory processes should fulfill certain criteria in order to become a successful course of action of involvement. Macintosh & Whyte (2008) have developed six different dimensions for evaluating participation in public decision making. These are:

- **Representation:** The process should be supported by relevant key groups
- **Engagement:** The public should be encouraged to take part in the process and influence its outcome
- **Transparency:** From the start of each participation project it should be clear how the results will be made public and be included into the governmental processes.
- **Conflict and consensus:** Fora for deliberative discourse among the participants should be provided. Fostering a culture of friendly disagreement and collaborative consensus building is necessary for successful participation.
- **Political equality:** Different groups of society should be included equally in the decision making process.
- **Community control:** For successful participatory processes their results need

to really influence the governmental decisions.

With the rise of the internet a new channel for participation has opened up. Information technology has been used before, to enhance participation of the public. The ideas and comments of the people have been entered into electronic documents on personal computers for easier processing. Even plans on how to use television for participation processes were made. However, the internet is the medium which allows tools to be built for handling mass participation. Data highways make it possible to connect with the public. In addition to face-to-face meetings and questionnaires as well as paper or phone based surveys more possibilities for participation, for electronic participation are developed. In urban planning in Germany the start of electronic participation is attributed to the EU-sponsored project "Geographical Mediation". This project started in 1998 with the aim to allow participation within the implementation phase of a development plan in the city of Bonn. The groupware system consisted of a collaboration platform, discussion fora and a map viewing application. Since then many different participation projects have been implemented in Germany. Ten years later electronic participation is considered as an established channel for public involvement in urban planning (Märker & Wehner 2008).

Nonetheless the full scale of methods possible in electronic participation is often not utilized by public planning authorities. Many participation processes only climb up the lower rungs of Arnsteins "Ladder of Participation". Lots of information is given and possibilities for feedback are offered. In Germany, the input by citizens in formal participation processes has to be considered in the planning process, which allows for a form of consultation or placation. However, often the input at the officially considered stages is too late to have a decisive impact on spatial development. The main decisions about how a certain area should be developed have been made before, even if the draft of the development plan was not formulated.

Thus two different types of electronic participation are to be distinguished in urban planning. Electronic participation can be part of the formal planning processes or it is utilized for informal involvement. The German federal building code was adapted in 2004 to permit the use of information technology for public participation in formal processes (Märker & Wehner, 2008; BauGB, 2006). Participation in informal processes sets in at a much earlier stage. The involvement of the public can influence such decisions as whether a certain development should even take place. Participation in formal processes often has influence on the details of how a development should be implemented.

In planning processes which are supported by electronic participation different components for online involvement can be utilized. These also provide different degrees of participation. Some of the available tools help with information provision, fostering dialogue, moderation or assisting participants in forming and issuing opinions (Märker & Wehner, 2008).

The objectives connected with the application of electronic participation processes are diverse. On the one hand there is the mandatory use in formal processes and the willingness to offer some form of participation. On the other hand there are many other aspects driving the willingness of public authorities to involve different groups of society. Many public administrations are aware that citizens can provide highly specialized input. They are the experts on their life circumstances and thus know best about the issues within the city district they live in. These many different views and opinions often cannot be replicated by a few experts in urban planning. Public participation in general can help to access the knowledge of all these inhabitants. Electronic participation can especially support receiving input by many different stakeholders as it makes mass participation easier, faster and more efficient. Regarding the different rungs of participation and the described evaluation criteria thereof input by citizens should not only be gathered. In fact procedures have to be agreed upon how to incorporate the input into the planning process. This should be communicated to the stakeholders beforehand (Schulze-Wolf & Habekost, 2008; Märker & Wehner 2008).

Electronic participation can foster modernization of public administrations. Inviting the opinions and knowledge of many thousands of citizens through mainly electronic means causes a large flow of information. In order to process all this information the electronic participation system should be integrated into the internal working processes of the public sector. Consequently these processes should be adapted and digitized in order to allow for a flow of digital information without barriers. Communicating this aspect of public involvement within local authorities as a further advantage can be of special value. As a result public participation might be grasped as more than a democratic necessity. It can support the ongoing efforts in creating modern administrations of the 21st century.

Online Participation as Part of a Multi-Channel Strategy for Civic Involvement in Urban Planning

Citizens want to have access to their government through all possible channels. Depending on the preferences of the customers of public administrations it is necessary to provide the prevalent phone and face-to-face services as well as electronic access. This multi-channel strategy can also be found in public participation processes in urban planning.

The basis for such a strategy can be an efficient electronic channel. This channel can provide all sorts of information and transaction for the public administration and its customers. Portals can represent the interface for this channel, with which administrative staff in town hall or in public call centers or customers themselves can access the electronic services of the public sector. As an important part of public services urban planning issues in general can also be dealt with on the basis of a multi-channel strategy. The electronic channel serves all other channels and can be used directly through portals. Public participation processes form a special part of urban planning. With their objective of creating interaction between the public planning professionals and the affected people sophisticated types of portals are needed.

The division of portals into entry points, information pools, service centers and service clusters can be used to classify participation portals in urban planning. Some portals are being built specifically to fulfill the requests associated with one of the categories. Others probably have developed over time from entry points to service centers.

Urban planning participation portals which act as entry points would consist of different directories pointing towards other resources with information about or possibilities for participation. Such portals could comprise one of the described directories, all of them or other similar A-Z listings (Table 1). A participation portal as entry point could administrate a directory with different participation projects within a political subdivision or beyond that. Such a portal could also provide a register with links to different software systems specializing on the facilitation of electronic participation projects. In order to be able to identify authorities from the field of public participation be it practitioners or researchers a directory of eligible experts could support the urban planning community.

Participation portals in the form of information pools provide edited information about different aspects of public involvement (Table 2). This could be a guide about public participation in general. An information pool could deliver explanations on how formal and informal participation are handled within a political subdivision. The pool could give descriptions about the current formal and informal participation projects within a political subdivision. Further such a participation portal could provide information about how and where

Table 1. Possibilities for the implementation of entry points

Entry points
Directory of participation projects
Directory of software systems supporting electronic participation
Directory of experts in electronic participation processes

Table 2. Possibilities for the implementation of information pools

Information pools
How-to about public participation in general
Information about formal and informal participation in a political subdivision
Explanations about specific participation projects
Information how and where the public can get involved and influence the outcome of certain urban planning processes

the public can currently get involved and influence the outcome of certain urban planning processes.

Portals for public participation in urban planning in the category of service centers directly enable the involvement of the citizens. Different mechanisms for issuing ones opinion or sketching out ideas can be found within such a portal (Table 3). Furthermore a service center provides tools for fostering discussion among the participants. It should be possible to comment on previously made suggestions or to debate for example in chat rooms. Also of importance, such a portal should allow for collaborative involvement. Groups of citizens can find each other and agree to work together on a suggestion and develop it online in a collaboration environment (Dühr et. al., 2005).

Moreover a public participation portal as a service center should serve the means for tracking the whole process of civic involvement. It should be possible to follow different suggestions and the discussion thereof. More ambitious would it be to follow the outcomes of a participation project through the administrative and political system. On the one hand this could be achieved rather easily by the administration by creating status documents and posting them online. On the other hand it would be challenging to connect the electronic results of the participation process in the service center portal with the resulting internal documents used by the administration. This would probably ask for connecting the administrative document management system and the electronic meeting system of town halls with the participation portal. Such an arrangement could give citizens insight into what happens with their input. They could access the opinions and annotations of the public officers and the politicians. Thus the important transparency criterion for real participation according to Macintosh & Whyte (2008) would be fulfilled. Service centers could also be used in

Table 3. Possibilities for the implementation of service centers

Service centers
Public participation within a political subdivision can be undertaken directly within the internet platform
Suggestions made in a local participation process can be discussed by commenting and in chat rooms
The outcome of participation processes in a political subdivision can be tracked via the portal

electronic voting procedures. These are currently being developed in many different nations. So far only few political subdivisions have implemented e-voting.

One example of a portal with some of the characteristics of a service center is the citizen budget portal of the German City of Cologne. To allow citizens to appoint parts of a public budget for their local government is called citizen budgeting. It was first introduced in the cities of Porto Alegre in Brazil and Christchurch in New Zealand. Already in 1989 the Brazilian city established this form of citizen involvement in order to curb corruption and realize more transparent local politics.

In Germany the first citizen budgets were initiated at the end of the 1990s. Rather small towns were among the first to test the new form of citizen involvement. Accordingly smaller numbers of citizens were engaged in deciding whether to spend more money on bike paths, to improve playgrounds or to expand the neighborhood center. Since then the participation has broadened. The administrative bodies undertaking the steps to involve their citizens have learned from earlier experiences. Participation processes are implemented through different channels, with the electronic becoming more important. In person citizen meetings, paper-based surveys, proposals via phone hotlines or the consultation on the internet are all different methods for gathering public input (Bertelsmann Stiftung, 2004; Vorwerk, 2008).

Cologne initiated citizen budgeting in the year 2007 for the first time. With almost one million inhabitants it is the largest local authority to introduce such a measure to its residents. The according internet portal was heavily promoted and used to provide input. Although only three topics (Streets, public green and sports) were offered for discussion, about 10.000 suggestions were made. Submissions via phone or letters were entered by agents in the online-portal. The results of the internet-dialogue were analyzed, ranked, evaluated by the administration and decided upon by the municipal council. The final resolutions are displayed within the internet portal used for the whole participation process. The gathering of such a broad input would have been much more difficult without an efficient internet portal. (Vorwerk, 2008).

Another example of an urban and regional planning and public participation portal with characteristics of a service center can be found in the agglomeration of Frankfurt / Rhine-Main. This region created a regional planning association. This body is mandated by law to develop a regional land use plan in place of its member municipalities. The association decided to involve the public into the planning process by electronic means.

Public participation, mainly in the form of information provision, via the internet was first established in 1997 by the regional planning association. In 2004 in order to prepare the new land use plan for the first time the according mission statement was developed with public input. For this moderated online discussions were utilized. In 2007 a formal participation procedure according to the German federal building code was undertaken. Parallel to traditional participation means like workshops and statements given in person at the planning office it was possible to provide input online.

The planning association aimed to replicate the analog process of formal participation closely. Therefore it was possible to issue statements and commentary anonymously. Ease of use was of high priority in order to achieve a high percentage of online participation. Tips about how to get involved and make comments should help novices. The public could access the proposed maps of the future land use plan and the accompanying written explanations. Statements on the future regional land use plan could be given via electronic mail or online forms. The comments could be enriched with the help of graphic tools for use in the provided online maps. Comments thus could be located within the maps, tied to a certain geographical reference point. Contact

Table 4. Possibilities for the implementation of service clusters

Service cluster
A network of participation portals allows for the exchange of the results of participation processes.
A network of urban planning portals.

details of the responsible planning officers from the regional planning association were provided to answer questions in person. Additionally an online citizen office was implemented where citizens could ask questions. These were processed and answered by a professional editorial team (Richter & Blöhm, 2008).

The participation portal of the regional planning association of Frankfurt / Rhine-Main was also used to conduct informal participation processes, not mandated by the federal building code. Examples are the development of the mission statement, participation on landscape elements and information provision through weblogs.

Feedback on the realized electronic public participation was said to be positive. The digital channel can be used without time and spatial constraints, participants do not have to visit the planning office to issue their statements nor do they have to meet certain time windows. Of importance to the planning association was the possibility to carry out electronic participation with reasonable spending. Information could be provided easy and fast to the public through the online portal. Further the statements issued by e-mail or online forms could be processed more easily, as the back-end processes are mostly electronic ones. Nonetheless the electronic channel can only augment the existing participation channels. The success to the public participation process in the agglomeration Frankfurt / Rhine-Main was a balanced mixture of all channels (Richter & Blöhm, 2008).

The fourth general type of portals, the service cluster can also be applied to participation portals. Although there are, at least in Germany, no implemented examples it is probably a step into the future of urban planning portals. A network of portals built for involving the public could support exchanging the results of participation processes.

Portals can support public participation in the form of electronic participation in urban planning. Depending on the type of portal different degrees of participation can be fostered. Entry points can help by pointing towards participation possibilities. Information pools can provide valuable information about ongoing urban planning projects. According to Arnstein, this constitutes the foundation for true participation. The lack of a back-channel for the citizens though impedes true participation on the basis of this type of portals alone. The third and fourth category of portals can enable some forms of true participation in the sense of Arnstein understanding and Macintosh and Whyte's evaluation criteria. Service centers can provide tools for contributing ideas, comments and casting votes within and beyond active planning processes. They can further help to monitor the whole process and especially its outcome. Service clusters could support a better knowledge exchange about participation processes, a tighter integration of urban planning portals leading to higher visibility. The aggregation of results of participation processes could probably lead to a benchmarking of these procedures.

Nonetheless, it is important to conclude, portals are only tools for facilitating public participation by electronic means. They cannot substitute the will of public administrations and politicians to share power with the people. Without such a commitment of the stakeholders within the public sector participation processes cannot become true. This counts for all channels of public involvement including the electronic channel.

FUTURE RESEARCH DIRECTIONS

Portals as a tool for public participation in urban planning will become much more common in the future. Firstly the expected development path will probably follow the general trends of service oriented architectures in electronic government. In addition of such structures a network of participation portals could share data and allow for comparison of participatory processes.

Currently most portals for participation in urban planning processes are set up separately from other portals of the respective public body. Often already present information about building plans, land use plans or planning regulations is reused, but has to be included in the participation portal by hand. Similarly the results of the participation process often have to be transferred to other administrative IT-systems manually. In order to deal with fragmented data and systems in a more efficient way the approach of service oriented architectures (SOA) will be implemented more broadly. Such an approach supports the integration of different specialized applications without having to deal with requirements of various platforms (Breitenstrom, Eckert, von Lucke, 2008, p. 38).

The back office processes which are unified inside a participatory portal could thus be designed as services which can be easily used within other contexts. Such a redesign of the processes for electronic participation in urban planning should take place within a larger approach by a public authority towards a service oriented architecture. With a service for public participation the city can respond quickly and repeatedly to an ever changing urban dynamic (Zhu et. al., 2009).

As mentioned above the fourth general type of portals, service clusters could form the future of participatory portals. A network of portals built for involving the public could support exchanging the results of participation processes. A data exchange standard for participation processes would probably be necessary. By achieving a high degree of standardization the different results could be aggregated and compared. The cumulative results can support better public decision making by politicians and the administration. The comparison of participation processes can help to improve civic engagement activities. Additionally a network of urban planning portals would allow citizens to access all planning processes independent from their current location. Planning administrations can create new forms of service delivery, where different administrations specialize on certain aspects of urban planning and public participation processes. However urban planning is very location-sensitive in comparison to other public sector services. So a spatial division of labor can only be implemented within limits. Such a tighter integration and more efficient work balancing can strengthen the spatial planning sector within the service portfolio of the public sector. Furthermore, within such a network, topics like the explanation of how formal participation works could be tackled in cooperation by content sharing.

CONCLUSION

Applying the portal model by von Lucke to the realm of electronic participation is helpful in categorizing existing participation portals. The systematization provides insight into the state of electronic participation in general in Germany and the use of internet portals in order to support such action. Transferring ideas from the portal research, especially the activities on high performance portals in the public sector in general can probably help to identify development paths for urban planning in the information age.

Internet portals, as a special technical type of portals, can serve as the main access points to public participation processes. Depending on their development stage the portals can provide overview (entry points), information on participation in general or on specific participation processes (information pools). Furthermore they can give access to the electronic participation process itself (service center) or combine many different participation processes (service cluster). Examples, especially of the service center category can be found in Germany. An integrated network of portals providing different participation processes still has to be developed.

From the perspective of the people, electronic participation via portals can provide advantages. With more and better information and tools for online collaboration and commenting, public involvement can become more convenient. Thus more citizens might issue their statements and develop their own ideas how their built environment should evolve. For digital natives it might seem much more interesting to partake in political and administrative processes through online portals. As this group of society will grow over time the role of electronic participation in society will rise.

Nonetheless it is important to state: electronic participation and portals can nowadays and for the foreseeable future only support traditional means of civic engagement. First, there are groups of society who cannot use digital channels. Elderly and many handicapped persons on the one hand as well as illiterate and socially excluded persons have a difficult time using or even accessing the very prerequisite of the information age, the personal computer. Secondly, there are people who just prefer using more traditional channels like phone or face-to-face meetings for disposing their tasks. Third, and probably the main reason, the whole participation process has to aim for true participation. Depending on the general set up of a participation process, the engagement of the power holders and their willingness to relinquish some degree of control the outcome will be determined.

Public administrations might strive for online participation for other reasons besides sharing power with the people. Online participation can help to reform the public sector. Public involvement via internet portals can thus prove very advantageous for public administrations. Massive input from citizens, as was the case with the citizen budget in Cologne is easier to process if gathered electronically. With a tight integration of electronic participation systems into the back office systems of an administration the input can be directly inserted into different administrative ICT-systems. Geographic Information Systems can be enriched with citizen commentary. If a high degree of standardization of the input can be reached different planning alternatives could be weighed automatically. The information given by the citizens could also be included in document management systems in order to track its processing. The annotations made by the administrative body and the political committees to the original citizen statements could thus be attached to the digital representations of those statements. If parts of these back-end systems then are accessible via the electronic participation portal the administrative and political processes would become more transparent.

Overall, portals can play an important role in fostering public participation, especially electronic participation. Furthermore they can support the reform of public administration by enrichment of the decision making process with citizen input and allowing for easier processing of such massive input. Nonetheless, portals can only augment the whole participation process, as true participation is primarily not determined by technology but rather the willingness of the stakeholders in charge to set up a process leading to a certain degree of citizen control.

REFERENCES

Arnstein, S. R. (1969). *A Ladder of Citizen Participation*. Retrieved November 2, 2008, from http://lithgow-schmidt.dk/sherry-arnstein/ladder-of-citizen-participation.html

Baugesetzbuch (German Federal Building Code). (2006). Retrieved November 8, 2008, from http://www.gesetze-im-internet.de/bbaug/BJNR003410960.html

Bertelsmann Stiftung. (2004). *Kommunaler Bürgerhaushalt: Ein Leitfaden für die Praxis*. Retrieved November 22, 2008, from http://www.buergerhaushalt.org/wp-content/plugins/wp-publications-archive/openfile.php?action=open&file=5

Breitenstrom, C., Eckert, K., & von Lucke, J. (2008). *The EU-Services Directive – Point of Single Contact – Framework Architecture and Technical Solutions - White Paper Version 1.0*. Berlin: Fraunhofer-Institute for Open Communication Systems FOKUS. Retrieved September 15. 2008, from http://www.fokus.fraunhofer.de/en/elan/publikationen/Infomaterial/white_paper/DLR_1_0/index.html

Dühr, S., Bates-Brkljac, N., & Counsell, J. (2005). Public Digital Collaboration in Planning. In *Cooperative Design, Visualization, and Engineering* (pp. 186-193). Retrieved August 20, 2008, from http://dx.doi.org/10.1007/11555223_20.

Goodspeed, R. C. (2008). *Citizen Participation and the Internet in Urban Planning*. Unpublished master thesis, University of Maryland. Retrieved November 22, 2008, from http://goodspeedupdate.com/wp-content/uploads/2008/11/goodspeed-internetparticipation.pdf

Kubicek, H. (2007). *Electronic Democracy and Deliberative Consultation on Urban Projects - Putting E-Democracy into Context*. Report for the Congress of Local and Regional Authorities. Retrieved September 15, 2008, from http://www.ifib.de/publikationsdateien/Creative_final.pdf

Lenk, K. (2007). *Bürokratieabbau durch E-Government - Handlungsempfehlungen zur Verwaltungsmodernisierung für Nordrhein-Westfalen auf der Grundlage von Entwicklungen und Erfahrungen in den Niederlanden*. Report for the Information office d-NRW. Retrieved October 31, 2008, from http://egovernmentplattform.de/uploads/media/Lenk_Buerokratieabbau.pdf von Lucke, J. (2008). *Hochleistungsportale für die öffentliche Verwaltung*. Lohmar: Eul-Verlag.

Macintosh, A., & Whyte, A. (2008). Towards an evaluation framework for eParticipation. In *Transforming Government: People, Process & Policy*. New York: Emerald Group Publishing Limited. Retrieved September 16, 2008, from http://eprints.whiterose.ac.uk/3742/

Märker, O. (2008). E-Partizipation als Gesamtsystem. *Standort - Zeitschrift für angewandte Geographie, 32*(3), 80-83.

Märker, O., & Wehner, J. (2008). E-Partizipation - Bürgerbeteiligung in Stadt- und Regionalplanung. *Standort - Zeitschrift für angewandte Geographie, 32*(3), 84-89.

Richter, S., & Bloem, G. (2008). Ein Plan soll in die Region. *Standort - Zeitschrift für angewandte Geographie, 32*(3), 104-107.

Schulze-Wolf, T., & Habekost, T. (2008). E-Partizipation in der Raumordnung. *Standort - Zeitschrift für angewandte Geographie, 32*(3), 97-103.

United Nations. (2008). *United Nations e-government survey 2008: from e-government to connected governance*. New York: United Nations Publication.

United Nations, Economic Commission for Europe. (1998). *Convention on Access to Information, Public Participation in Decision Making and Access to Justice in Environmental Matters*. Retrieved November 1, 2008, from http://unece.org/env/pp/documents/cep43e.pdf

Vorwerk, V., Märker, O., & Wehner, J. (2008). Bürgerbeteiligung am Haushalt. *Standort - Zeitschrift für angewandte Geographie, 32*(3), 114-119.

Welzel, C., Falk, S., & Müller-Mordhorst, F. (2005). *Standortfaktor Verwaltung: E-Government und Kundenservice in Nordrhein-Westfalen*. Retrieved October 31, 2008, from http://de.sitestat.com/bertelsmann/stiftung-de/s?bst.Suche.nrw_standortfaktor.pdf&ns_type=pdf&ns_url=http://www.bertelsmann-stiftung.de/cps/rde/xbcr/SID-0A000F0A-E87D2A32/bst/nrw_standortfaktor.pdf

Zhu, D., Li, Y., Shi, J., Xu, Y., & Shen, W. (2009). (in press). A service-oriented city portal framework and collaborative development platform. *Information Sciences*. doi:10.1016/j.ins.2009.01.038

ADDITIONAL READING

Axelsson, K., & Melin, U. (2008). Citizen Participation and Involvement in eGovernment Projects: An Emergent Framework. In *Electronic Government* (pp. 207-218). Retrieved September 8, 2008, from http://dx.doi.org/10.1007/978-3-540-85204-9_18

Breitenstrom, C., Eckert, K., & von Lucke, J. (2008). *The EU-Services Directive – Point of Single Contact – Framework Architecture and Technical Solutions - White Paper Version 1.0*. Berlin: Fraunhofer-Institute for Open Communication Systems FOKUS. Retrieved September 15. 2008, from http://www.fokus.fraunhofer.de/en/elan/publikationen/Infomaterial/white_paper/DLR_1_0/index.html

Fung, A. (2006). Varieties of Participation in Complex Governance. *Public Administration Review*, (Special Issue), 66–75. doi:10.1111/j.1540-6210.2006.00667.x

Goodspeed, R. C. (2008). *Citizen Participation and the Internet in Urban Planning*. Unpublished master thesis, University of Maryland. Retrieved November 22, 2008, from http://goodspeedupdate.com/wp-content/uploads/2008/11/goodspeed-internetparticipation.pdf

Herzenberg, C., & Cuny, C. (2007). *Herausforderungen der technischen Demokratie: Bürgerhaushalt und die Mobilisierung von Bürgerwissen. - Eine Untersuchung von Beispielen in der Region Berlin-Brandenburg*. Retrieved September 9, 2008 from http://www.buergerhaushalt.org/wp-content/uploads/2007/10/herzberg-cuny2007.pdf

Kubicek, H. (2007). *Electronic Democracy and Deliberative Consultation on Urban Projects - Putting E-Democracy into Context*. Report for the Congress of Local and Regional Authorities. Retrieved September 15, 2008, from http://www.ifib.de/publikationsdateien/Creative_final.pdf

Lenk, K., Brüggemeier, M., & Dofivat, A. (2007). *Potentials of Electronic Government for Organizational Design in the Public Sector*. Retrieved June 29, 2009, from http://www.f3.htw-berlin.de/Professoren/Brueggemeier/pdf/Brueggemeier_Dovifat_Lenk_IRSPM_Paper_Potentials_eGov_100708.pdf

Lukensmeyer, C. J., & Torres, L. H. (2006). *Public Deliberation: A Manager's Guide to Citizen Engagement. Collaboration Series*. IBM Center for the Business of Government.

Mayo, E., & Steinberg, T. (2007). *The Power of Information*. Retrieved May 12, 2009, from http://www.cabinetoffice.gov.uk/media/cabinetoffice/corp/assets/publications/reports/power_information/power_information.pdf

Noveck, B. S. (2009). *Wiki government: how technology can make government better, democracy stronger, and citizens more powerful*. Washington, DC: Brookings Institution Press.

Organisation for Economic Co-operation (Ed.). (2009). *Focus on citizens public engagement for better policy and services*. Paris: Author. Retrieved July 6, 2009, from http://www.oecd.org/document/25/0,3343,en_2649_33735_42216857_1_1_1_1,00.html#how_to_obtain_this_book

Pratchett, L. (2009). *Empowering communities to influence local decision making - A systematic review of the evidence*. Communities and Local Government. London: Department for Communities and Local Government. Retrieved June 3, 2009, from http://www.communities.gov.uk/documents/localgovernment/pdf/1241955

Reinermann, H. (2000). *Der oeffentliche Sektor im Internet: Veraenderungen der Muster oeffentlicher Verwaltungen*. Speyer, Deutschland: Forschungsinstitut fuer Oeffentliche Verwaltung. Retrieved June 29, 2009, from http://www.foev-speyer.de/publikationen/download.asp?ID=206&REIHE=Spe&MB=N

Rinner, C., Keßler, C., & Andrulis, S. (2008). The use of Web 2.0 concepts to support deliberation in spatial decision-making. *Computers, Environment and Urban Systems*, *32*(5), 386–395. doi:10.1016/j.compenvurbsys.2008.08.004

Tambouris, E., Liotas, N., & Tarabanis, K. (2007). A Framework for Assessing eParticipation Projects and Tools. In *Proceedings of the 40th Annual Hawaii International Conference on System Sciences*. Washington, DC: IEEE Computer Society. Retrieved December 17, 2008, from http://portal.acm.org/citation.cfm?id=1255721

Tapscott, D., Williams, A. D., & Herman, D. (2007). *Government 2.0: Transforming Government and Governance for the Twenty-First Century*. Big Idea White Paper. New Paradigm Learning Corporation. Retrieved May 5, 2009, from http://business.twoday.net/static/foehrenbergkreis/files/20080126gov20.pdf

United Nations, Economic Commission for Europe. (1998). *Convention on Access to Information, Public Participation in Decision Making and Access to Justice in Environmental Matters*. Retrieved November 1, 2008, from http://unece.org/env/pp/documents/cep43e.pdf

Vickery, G., & Wunsch-Vincent, S. (2007). *Participative Web and User-Created Content - Web 2.0, Wikis and Social Networking*. Science & Information Technology. Paris: OECD. Retrieved May 3, 2009, from http://www.oecd.org/document/40/0,3343,en_2649_34223_39428648_1_1_1_1,00.html

KEY TERMS AND DEFINITIONS

Electronic Government: The handling of business processes in connection with actions by government and public administrations based on information and communication technologies.

Electronic Participation: Public participation processes supported by electronic means.

Entry Point: Guideposts with links to other resources where either information about ones request can be found, responsible agencies or even online transactions can be accessed.

Information Pool: Yields not only links to information at other websites, furthermore provides information itself. This content can be generated and structured by the portal carrier or an additional service entity.

One-Stop Government: The combined delivery of services for the public administrations customers. Citizens and businesses can access and use public services with few or only one visit to the public administration. The customers should not have to deal with the structure of an organization in order to identify the correct contact responsible for their matter. Rather the organization appropriates a single point of contact.

Portal: System providing access to information, applications, processes and persons available within the attached systems. Such a system should grant safe adaptable and easy access. It can be used through different media and access channels.

Public Participation: Participation of the people in the decision-making process of politics and public administration. Different forms of participation can be distinguished from non-participation through informing, consulting and partnerships to citizen control.

Service Center: These portals give access to transactional administrative services. Customers can carry out procedures online for which they otherwise would have to visit a public agency in person. The online services accessible have to be integrated into the portal. Service centers also include links to other websites and sources of information as well as self-generated information.

Service Cluster: Tightly integrated networks of service centers. In line with the One-Stop-Government paradigm users can access any information and transaction from any partaking service center. The customers thus do not have to deal with identifying the correct public entity and choose the corresponding internet portal.

Chapter 14
Can Urban Planning, Participation and ICT Co-Exist?
Developing a Curriculum and an Interactive Virtual Reality Tool for Agia Varvara, Athens, Greece

Vassilis Bourdakis
University of Thessaly, Greece

Alex Deffner
University of Thessaly, Greece

ABSTRACT

One of the recent main problems in urban planning is to find ways in order to employ practical, very broad and commonly used theoretical principles such as participation. An additional issue is the exploitation of the possibilities of new technologies. The process of developing a flexible three-part (common core, public and planners) curriculum in the case of Agia Varvara (Athens, Greece) in the framework of the Leonardo project PICT (2002-2005) showed that ICT (Information Communication Technologies) can help in participation, mainly because it constitutes a relatively simple method of recording the views of both the public and the planners in a variety of subjects (both 'open' and 'closed').

INTRODUCTION

One of the recent main problems in urban planning is to find ways in order to employ practical, very broad and commonly used theoretical principles such as *participation*. An additional issue is the exploitation of the possibilities of new technologies. In relation to the latter, one important aspect is the transformations of urban forms, urban processes and the perceptions of urban life though the technological advances (Fernández-Maldonado, 2004). However, this chapter focuses on the role of new technologies in public participation in planning.

The data are provided by the *PICT* (Planning Inclusion of Clients through E-Training) project which was a three-year (2002-2005) pilot project co-funded by the Leonardo da Vinci Programme of the European Commission. The project was inspired by the on-going debate about the relationship of people with their cities launched by the initial Charter of Athens (which took place in 1933, see Le Corbusier, 1943/1987) and re-defined by the

DOI: 10.4018/978-1-61520-929-3.ch014

New Charter of Athens adopted by European Council of Town Planners (ECTP) (1998/2003); and responded also to viewpoint that the involvement of communities in public decisions builds social capital and strengthens the civil society. The continuing debate on the participation of the public in official decisions is reflected on European policy, as expressed in the European Spatial Development Plan (European Commission, 1999), Local Agenda 21 and the Sustainable Development Framework of Gothenburg. Consequently, it has been widely accepted, at least in principle, in the European Union, that urban planning is part of the sustainable development process and as such requires consensus building through the engagement of citizens (PICT, 2006).

The chapter focuses on the curriculum developed for the Municipality of Agia Varvara in Athens, Greece. It has a population of approximately 30000 people with a multicultural identity and high unemployment rates. The developed curriculum consists of three parts: a common 'core' part that is shared by both planners and the public, and two distinct parts: one addressing the public and the other the planners. Each part consists of several modules, to cater for different learning levels, abilities and interests. The structure is flexible and the whole idea was to have a curriculum with a scientific, and not a 'journalistic' curriculum basis that could, at the same time, be simple but not simplistic.

The main objective of the chapter is to demonstrate if one of the main issues of urban governance, public participation in planning, can be helped through the use of ICT at the level of the community.

BACKGROUND

Participation in Urban Planning

There are different views of participation depending on the degree of involvement of the experts and the selection of criteria for the representation of the public. The most recent approach in relation to participation is collaborative planning which aims at fostering communication and collaborative action (Healey, 1997/2005), and especially at fostering partnerships (McCarthy & Lloyd, 2007). Although there is lack of experience of participation, and consequently of participatory culture in Greece, Agia Varvara has demonstrated participatory experiences in the past.

A useful '*schema of public participation*' is developed by Hampton, who claims that planning authorities should consider the means and techniques of public participation in planning in terms of three separate aims: dispersing information to the public, gathering information from the public, promoting interaction between policy-makers and the public. He identifies two major objectives behind the introduction of greater public participation in planning during the late 1960s: policy-making and decisions can benefit from better information about public preferences and residents' concerns, and public participation can draw people into a stronger and longer-term relationship with government and enhance their current and future ability to play a significant role in policy-making (Hampton, 1977 cited in Darke, 2000, pp. 391-392).

Hampton builds on a crucial idea from the government's Skeffington report (1969) that there is no single or simple category of 'the public', rather that there are many separate publics - a similar idea, albeit in relation to 'public opinion', is expressed by Pierre Bourdieu. Hampton also claims that planners should recognize that the *stakeholders* are distinguished in: major elites (e.g. local business groups, major employers, Chambers of Commerce, trade unions), minor elites (local interest groups, community associations, action groups, and public as a collection of individuals (Hampton, 1977 cited in Darke, 2000, p. 392).

The existence of *equal opportunities* constitutes one of the important conditions for success within Local Authorities, and, according to the Equal Opportunities Guide, the concerned groups of people are: disabled, racial minorities, women, gays and lesbians, part time and casual workers, elderly, gypsies and children. For the success of equal opportunities (at the level both of employment practice and service provision) there exist a series of prerequisites:

- long term goals
- practical action plans
- openness to learning from experience
- good relationship between members and officers
- positive attitude towards fellow workers and service users
- effective dealing with attacks
- encouragement and empowerment. (Brennan/ LGMB, 1991 cited in Darke, 2000, p. 409).

The *key principles for good practice in public participation* are the following:

- clear aims of participation at the outset
- insurances of the central role of local politicians at the programme
- link of motives, objectives and intentions of the participation programme with the appropriate techniques
- interpretation of the nature and implications of policies and plans for the users
- identification of the procedures for information collection from the public in order to evaluate and act (Alty and Darke 1987, cited in Darke, 2000: 410).

ICT and Participation in Planning

The use of ICT limits the temporal-spatial constraints that exist in the process of public participation (Kwan and Weber, 2003 cited in Conroy and Evans-Cowley, 2006, p. 371). When people are involved in planning it is more possible that they will support the elaboration and implementation of related policies and projects (Grant et al, 1996; Potaptchuk, 1996 cited in Conroy and Evans-Cowley, 2006, p. 372). 'The challenge for planners has been to improve the means by which people participate and to provide a more meaningful experience for citizens. ICT represents tools that can be used to improve citizen participation' (Conroy and Evans-Cowley, 2006, p. 372). Especially visual information can become a common language for all participants (Al Kodmany, 1999 cited in Conroy and Evans-Cowley, 2006, p. 373).

The challenge faced by municipal decision-makers, constrained as they are by limited budgets, relates to the 'daunting tasks of determining what to implement and when in the face of uncertainty regarding what new technologies may emerge once investment decisions and commitments have been made' (Kaylor et al., 2001, p. 294). Thus, the main problem is that poor software selection may lead to the need for costly fixes in the future (Conroy and Evans-Cowley, 2006, p. 382). In the particular case of citizens' visits to governmental websites, which 'have become a major new form of the traditional citizen-initiated contact', the usual situation is for one-way information obtainment rather than for two-way interaction (Thomas and Streib, 2003, p. 83). This observation is even more crucial if it is taken into account that in the citizens' use of governmental sites the emphasis is on everyday service delivery rather than large-scale policy (Thomas and Streib, 2003, p. 85). Regarding improved participation opportunities the issue that is raised is who is able to participate (Conroy and Evans-Cowley, 2006, p. 373). Thus, one crucial question posed is that if people who have more knowledge of ICT are less willing to participate, or, vice-versa, if people who are more willing to participate have less knowledge of ICT.

The answer to this question depends on the particular local context of a community (economic, social and demographic information) which can influence local planning processes (Conroy and Evans-Cowley, 2006, p. 379). However, the degree of possibility of e-government provision depends on the size of the community, i.e. the larger the community the greater the possibility (Kaylor et al., 2001).

ISSUES, CONTROVERSIES AND PROBLEMS

Community Planning

In the context of participation in planning an effective type of planning has been community (or even neighbourhood) planning. In this process, the importance of local economic development is reflected in a type of 'new localism': from outward- to inward-looking societies. This implies encouraging local ownership, increasing import substitution, encouraging local control of money, and localising work to meet local demand (Williams, 1999). The aim is to develop a sense for integrated local development (housing *and* public space *and* social-economic background).

Community planning should include a plan, but not necessarily a comprehensive one as Kelly and Becker (2000) claim. However, the three crucial questions that they raise are still valid: where are we, where can we go and where do we want to go – the latter refers to the citizens' involvement in making a plan.

Community planning should primarily focus on the needs of particular groups. Examples in the case study of Agia Varvara are the elderly and the gypsies (Roma): the first, along with housewives, were willing to participate in the PICT project but are IT (Information Technology) illiterate; on the other hand, young people are IT literate but did not seem willing to participate in PICT.

There exists a large variety of *general principles of community planning* irrespective of the approach one chooses: accept different agendas, limitations and varied commitment, avoid jargon, flexibility and focus on attitudes, follow up, go at the right place and go for it, human scale, involve all those affected and all sections of the community, local ownership of the process, maintain momentum and use mixture of methods, now is the right time, plan for your own process carefully and for the local context, prepare properly, process as important as product, professional enablers, quality not quantity, record and document, shared control, spend money, think on your feet, trust in other's honesty, use experts appropriately, facilitators, local talent and (carefully) outsiders, walk before you run, work on location (Wates, 2000, pp. 11-21). These principles should be adopted and adapted according to the particular case.

In the case of Agia Varvara it can be argued that the attempt was to follow the principles (not mentioned above): agree to the rules and boundaries, be honest, transparent and visionary yet realistic, build local capacity and communicate, encourage collaboration, have fun, learn from others, have personal motivation and take initiatives, respect the cultural context of others and local knowledge, be receptive to training, visualisation of result. The next step in community planning is to organize events, and this constitutes a cooperative and collaborative approach to planning that allows people from different walks of life to participate and work together creatively (Wates, 2008).

Urban Governance

The term government is confined to the formal structure of representatives and officials established to coordinate and oversee this function, while the term governance refers to the process of government and, more broadly, to the ways in which a society manages its collective interests. It includes functions that may be helped by government actions: strengthening institutions for collective decision-making, facilitating and forming partnerships designed to secure collective goals, ensuring the fair expression and adequate arbitration of a range of interests (Gilbert et al., 1996, p. 16). The revitalisation of local government towards the direction of governance is a cornerstone of city recovery but it needs to be done from bottom up:

from neighbourhood, where people know what is going on, to the city where politicians, businesses and civic bodies link up. Rotterdam and Spanish cities function as characteristic examples (Rogers and Power, 2000, pp. 264-265).

Greece continues to rely on formal mechanisms of administration. The actual role of the private sector and civic society has to be invented. As far as the third sector is concerned, the non-governmental organizations are underrepresented, and in most cases they constitute a one man/ woman show - the public sector is unable to press the state and vice versa (Economou et al., 2005).

The *role of local governments in the urban environment* consists of the following:

- They are the only bodies with the mandate, responsibility and potential to represent and act for the different and often conflicting interests
- Although they are the bodies with the greatest potential to take integrated approaches to the environmental and social challenges of urban areas they often have neither the legitimacy nor the capacity
- Even if this happens there will be effective action only if it involves leadership of elected officials and participatory and inclusive style of governing

For most issues of urban sustainability they should work with partners, other local governments and international networks (Gilbert et al., 1996).

The Good Functioning of Cities

According to the initial Charter of Athens, there are four main urban planning functions: housing, work, leisure, and transport (Le Corbusier, 1943/1987). 'Making cities work' is directly related to these functions and depends on best practice examples of:

- Arriving in the city (transport): most successful gateways and transport interchanges, first (and lasting) impressions really count, cities are not just places where people live but they are destinations that many people visit for brief period
- Getting around the city (transport): great challenge for most urban leaders: how to move people around in safety, comfort and speed, acute political trade-offs: pedestrian vs car, pollution vs clean air, communities vs roads, a matter not only of huge public investment but also of ideas and good operating practices
- Enjoying the city (leisure): ingenious approaches that are taken to parks, shopping malls and public spaces, large number of (usually) small-scale amenities that make a city fun to be in (Hazel and Parry, 2004, pp. 24-183).

Venice is a classic case study (even if few, if any, cities have canals), since its working principles can be applied to modern day cities (Hazel and Parry, 2004, pp. 17-23).

Two additional examples are:

- working in the city (work)
- living in the city (housing).

The *main lessons* learned from the study of the best practices in relation to 'making cities work' are that cities have to find a solution to the car (the simplest way is that road space has to be rationed since it is not a free public good), even the most spectacular developments have to be on a human scale, information is the key, it is people (often one individual) that make things happen. It is a cumulative effect of visionary ideas, sometimes small, that make cities work (Hazel and Parry, 2004, p. 187).

The relationship of planning and participation can be better illustrated through the example of ICT and the use of a particular case study.

SOLUTIONS AND RECOMMENDATIONS

The PICT Case Study of Agia Varvara, Athens, Greece

The three-year (2002-2005) pilot project co-funded by the Leonardo da Vinci Programme of the European Commission addresses the inclusion of clients in the planning process through e-training, the main objectives being to introduce key IT skills, fight technophobia and disbelief, improve communication skills and help everyone in acquiring an understanding of the built environment through spatial representations. Four case studies were carried out: Knowsley UK, Brabant Belgium, Abony Hungary and Agia Varvara Greece, each one addressing e-training from a different view point hence employing Computer Aided Design (CAD), time based design, Geographical Information System (GIS) and Virtual Reality (VR) technologies respectively. The PICT partners were:

- Knowsley Metropolitan Borough Council (Project Contractor), Liverpool John Moores University, School of the Built Environment, and the European Council of Town Planners (ECTP) from the UK
- PRISMA Centre for Development Studies (Project Coordinator), the Municipality of Agia Varvara in the Prefecture of Athens, and the University of Thessaly from Greece
- Hogeschool voor Wetenschap & Kunst Sint Lucas Architectuur from Belgium, and
- Budapest University of Technology and Economics, and WEBhu Kft. ICT Consultancy from Hungary.

The Greek case study took place in Agia Varvara, a southwestern municipality of Athens, covering 22Ha, with 62 urban blocks, 2163 buildings and a population of approximately 30000 inhabitants. Amongst the building stock are 31 1960's refugee apartment buildings accommodating 580 families. The main urban problems experienced are lack of free space and accessibility of communal space.

The population includes 15% Roma, refugees/immigrants from the Black Sea and economic immigrants and a large number of elderly households. The area is a low income one, with half of the households below EU poverty line, illiteracy level is reaching 40% (compared to an estimated 8% for Greece[1]), current school drop-out is 21% (compared to 3% for Greece), high unemployment rates (24%, half of it long term, compared to approximately 11% for Greece) and many single parent families (NSSG).

Part of the borough's redevelopment agenda was the renovation of two of its three main squares; the one located centrally next to a welfare centre for the elderly, whereas the other is an oblong area by a cemetery to the south-east border of the borough (Figure 1). Both squares were practically disused, with vegetation, street furniture and pavements in a sorry state, no grass, no lighting making it dangerous at nights, etc. The proposed redevelopment focused on the design of small parks, playgrounds, improvement of pedestrian movement and street lighting.

Following the procurement of the project and the final draft of the proposed design by the project architects, the research team developed a VR tool for the evaluation of the proposal and comparison to the existing conditions and layout through a custom made commenting and reviewing mechanism. Although not within the statutory terms and conditions of the typical public sector design and built contract, the project architects agreed that they would consider favourably residents' comments and try to integrate them to their final design.

The Developed Curriculum

The curriculum in Agia Varvara was developed by the authors, at the Department of Planning and

Figure 1. Plan view of Agia Varvara [through the ICT tool]

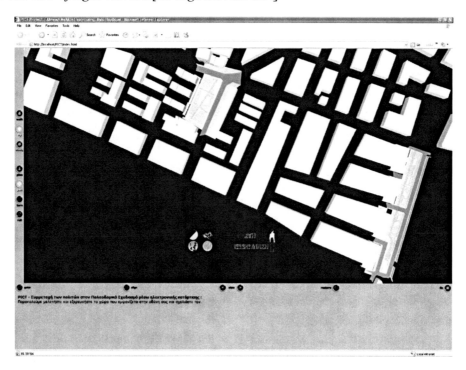

Regional Development, University of Thessaly, Volos, Greece. The curriculum consists of three parts: a 'core' part that is shared by both planners and the public, and two distinct parts: one addressing the public and the other the planners. Each part consists of several modules (further organised in units), to cater for different learning levels, abilities and interests guided by the needs survey. The structure is flexible and although the written text seems 'rigid' using an austere language, it functioned only as a basis for the oral presentations which were more 'free' and rather informal using everyday language. The rationale was to have a curriculum with a scientific, and not 'journalistic', basis that could at the same time be simple but not simplistic. Furthermore, the themes were developed in such a way as to allow members of the public to follow some sections of the curriculum addressed to the planners, and also planners to 'look back' at some sections of the curriculum addressing the public.

During curriculum development, the project team members (representatives of PRISMA Centre for Development Studies, the Municipality of Agia Varvara and the University of Thessaly) communicated extensively with the Local Consultative Committee to ensure relevance and acceptance of the learning approach. The main tools used are power point slides and hands-on experiencing in both portable and desktop PCs. As far as the learning material for the public is concerned, it relied heavily on visual information in an attempt to attract attention and stimulate discussion.

The main sections in the curricula modules refer to planning, participation, and ICT. The common *core part* has an introductory module which refers to general issues concerning the PICT programme. In total it has four teaching modules: Introduction to PICT, Planning, Participation, Methods and techniques of ICT. The thematic emphasis is put on sustainability. As far as ICT is concerned, the reason for including a common core part is that addressing older planners may not be at all different in terms of ICT skills needed to addressing similarly aged members of the public.

The *public's part* (which is the largest one) has five teaching modules in total: Introductory Themes to Urban Planning, Participation, 'Key skills' in ICT, GIS, Virtual Reality. The various 'key skills' for the less educated members are necessary before embarking in any substantive learning on the subject of public involvement in planning. This is in accordance with the project aim of empowering local communities. The thematic emphasis is put on 'making cities work', a section which includes many pictures of real international examples. As far as ICT is concerned, the emphasis is in understanding information presented, photomontages, drawings, renderings, video and most important the ability to interact with the computer based VR models (Figure 2).

The *planners' part* (which is the smallest one) has five teaching modules in total: Advanced Themes in Urban Planning, Participation, GIS, CAD and Virtual Reality. There is no particular thematic emphasis. As far as ICT is concerned, the focus is on operating the various ICT tools, in terms of building new environments, 3D data formatting, converting, translating information from different platforms, etc.

The Developed ICT Tool

The ICT tool developed employs the Virtual Reality Mark-up Language (VRML97), the undisputed standard for visualizing and communicating three dimensional information on the web. The project was completed before the successor to VRML97, X3D was finalised and modelling/visualisation tools released. Considering that most 3D modelling animation software can read and write VRML97 files, the decision taken enhanced overall compatibility, flexibility and accessibility of the produced data. The project can be experienced through a web browser equipped with the appropriate plug-in (in this case the Cortona3D Viewer from Parallelgraphics™) enabling anyone from any computer (home, office, internet café, etc) to have immediate access to the VR model. The environment features a single-user set-up; each user enters the virtual space on his/her own, can organise his/her way of experiencing the space, spend as much time as he/she feels necessary to comprehend the layout and give feed-back to the designers.

Creating an urban model for VR use is a complex task and a series of technical issues need addressing. The topic has been covered extensively (Dokonal et al, 2002; Bourdakis, 2004; Bourdakis, 2008) hence the building steps will be briefly presented. First of all availability of digital maps at an accepted quality level and scale (preferably 1:2000) should be examined, lack of them means a very expensive surveying process is necessary or of similarly costly licensing fees must be paid. Stereo aerial photogrammetric techniques should

Figure 2. Teaching sessions for the public

be avoided due to both cost and effort involved. Street level digital photographic survey data, if collected properly, should be a sufficient source of information together with digital video of the area (for documentation purposes). Appropriate PC based tools (AutoDesk Imagemodeler™, EOS Systems Photomodeler™) should be used in order to build the textured 3D model of the area. Three Levels of detail (LOD) should be employed for the three distinct levels of modelling–urban block and general terrain, buildings and landmarks, street detailing. Modelling should focus on street level interaction including enough information for acceptable bird's eye view navigation. Finally, a careful balance of 3D detail versus texture data (digital imaging) should be achieved in order to create a streamlined, relatively low bandwidth model experience–a 10MB urban model is not accessible to everyone as it requires fast ADSL lines to download in an acceptable timeframe, a fairly powerful computer with a fast processor, above average graphics performance and large amount of memory to function properly.

As far as the PICT case study is concerned, the relatively small area affected and the scale and type of redevelopments designed, generated a series of problems in terms of the low degree of abstraction in geometric modelling acceptable and the high level of realism and extensive use of textures necessary to improve navigability and sense of immersion. The non-geometric representations for street furniture and mainly trees affect the overall impression a visitor gets on the volume, foliage and subsequently shadow expected on parts of the redeveloped area. This opposes the simplicity and non-photorealistic nature of VR applications instrumental in creating the intended synthetic/digital reconstruction/artificial feel on the visual experience (Figure 3).

Concluding, the developed ICT tool did not support shadow casting, direct and volumetric lighting and many more features available in vi-

Figure 3. Bird's eye view of the VR model

sualisation software instrumental for introducing the unwanted high level of realism that typically fascinate inexperienced users/visitors. It is a design tool and as such a higher than usual level of abstraction is necessary, varying according to the task at hand and the expertise/training of the participants.

Data security issues complicated the design of the VR tool in regards to the commenting mechanism and in particular to the recording of comments on a remote database stored at a University of Thessaly server. This demanded complex coding addressing securely save and retrieval of data. At the time of developing this tool VRML97 was the only viable solution. Tests with open source as well as commercial game engines failed in data logging and secure data exchange. If the whole project was carried out now, it seems that the open source game engine of Blender would be capable of completing the task and would provide a much better rendering engine supporting better texturing, shadow casting and an overall inclusive lighting model.

Following the initial completion of the ICT tool and the development of the curriculum, pilot studies were conducted in two schools of the area with secondary school students aged 12-16. The authors were aware that the sample and its orientation were not representative of the area population, but due to time constraints and lack of early support from the municipality it was not possible to conduct a pilot study in a more representative sample. As such, the focus was on the usability and robustness of the ICT tool and less on the actual curriculum. As already stated, the curriculum was overly analytical in its base and the tutors/researchers were adapting to the audience, focus and interests in a subtractive manner, by simply "omitting" material. Minor fine-tuning of the interaction mechanisms was necessary as the results were very promising.

Interaction Modes for the 2 User Groups (Public and Planners)

Exploring a Virtual Environment is an unfamiliar experience for any newcomer, often even for experienced IT professionals. Navigation (metaphors and actual process of "moving"), way finding and interaction (as in selecting, picking, exploring objects) in digital spaces is an active area of research for almost two decades (Ellis, 1993; Charitos, 2008). A widely adopted solution often employed in the gaming industry is the placement of a semi-translucent 2D map of the area on a heads-up display (HUD) following the user as he/she moves about the VR model. Initial experiments with the residents showed that in this particular setup the amount of detail available at street level is sufficient to orient and familiarize oneself and considering their low level of education and familiarity with map reading renders the abovementioned technique inappropriate. Subsequently, a compass was employed in the HUD together with the basic controls mainly for the advantage of the planners and other relevant groups of professionals not familiar or local to the area.

The feedback mechanism employed in the design is worth analysing further. The user has access to a commenting tool that he/she can use to place comments in the exact spatial location that he/she considers appropriate. Such an experiment was first carried out during the Ijburg development project in the Netherlands in 1997 (see Bourdakis, 1997). The Agia Varvara commenting mechanism employed colour-coded objects/markers in space and featured two types of comments (Figure 4):

- Negative visualised as red exclamation marks and
- Positive in the form of a green conical shapes

The comments, either in audio format (visitors in a high bandwidth access point featuring high

Figure 4. Street level perspective with the commenting interface

performance computers – not implemented in the Agia Varvara setting) or textual (visitors in low bandwidth access points) linked to the appropriately coloured node. It is important to stress that all comment nodes are visible and accessible to new visitors, so that the overall "feel" and trends developing among the visiting public is clearly visible and passed on to the next visitor as he/she can comment not only on the merits of the proposal but also on the previously placed comments.

The ensuing "dialogue" or rather asynchronous discussion is of vital importance as far as the public participation process is concerned. The themed threaded discussion attained is very interesting to follow, however care must be taken not to overload the model with comments and develop a methodology for "storing" and "streamlining" comments so that new users are neither overwhelmed nor oblivious of the previous topics and issues discussed. This approach creates a visual feedback to the designers as well as future visitors since a bird's eye view of the area presents the differently coloured nodes segregated on the problematic or satisfactory areas of the model respectively hence one can easily follow a "discussion" in a visual manner (Figure 5).

The Planners' Specific Tools

The tool proposed for the planners' interaction with the available data was identical to the one for the public described above with the addition of being able to visualize planning and other spatial data related of the area under examination. Such data can be pollution, traffic, building regulation, permissible building volumes, etc. For example, maximum allowed building heights and volume can be visualised via a semi-translucent grey block model, generated automatically based on the regulations permissions for each urban block and super-imposed on the existing urban model. Such a tool highlights structures not in line with the current legislation (protruding from the semi-translucent grey blocks) as well as areas where

Figure 5. Proposed versus existing layout with comments

building activity is potentially expected–considering current legislation and existing building stock without counting age, condition and other property status.

It should be stressed that, setting up the 3D datasets and underlying databases, collecting and editing texture data managing other collected data and finally building the urban model and the VR application were issues not dealt by the planners as the authors strongly believe that the research team responsible for the creation of the model should have an overview and be in charge of such process.

Keeping digital models of all categories and technologies – be it a website, a 3D model interactive or not, a QuicktimeVR™ panorama, or other – up to date, is the achilles heel of digital datasets. Based on authors' 15year experience on urban digital modelling (Bourdakis, 2008), it is suggested that once the prototyping stages are completed in a appropriately structured and organised manner, the updating process could be automated and a local team could be responsible for the subsequent changes (typically commercial property changes in name, logos, shop front details, colours, etc., as well as new buildings both residential and commercial). However, this all depends on the structure and capabilities of local groups as far as setup of a team responsible for keeping digital models up to date is concerned. This calls for systematic surveying of the area, photographic documentation of alterations, necessary 3D modelling editing, digital photographic image processing and overall VR model and underlying database updating.

The Results Drawn

The VR application as well as being available on the internet, was also installed at the welfare centre for the elderly by the main square under redevelopment. Following a period of almost two months, data were collected from two sources; the mediated[2] digital commenting mechanism that recorded 87 entries and a dozen semi-structured interviews following one-to-one sessions on the VR tool. Keeping in mind that one subject would sometimes comment more than once, the estimated overall number of people that interacted with the system (mediated or interviewed) was around 65. The discussion that follows is based on the comments recorded and not on the estimated number of users since anonymity was protected and the comments recorded decipher this information. It should also be noted that very few visited the VR tool site via the internet although this could be due to lack of appropriate publication, marketing and communicating inefficiencies from the local authority.

Starting with, 28% of the comments were made by men and 72% by women — not a representative distribution and a possible indication that women tended to comment more than once. As far as age of the participants is concerned, 16% were up to 25yrs old, 21% at the 26-35 age group, 28% 36-45, 15% at the 46-65 and 20% were over 65, closely reflecting the age distribution in the area. The majority of the users had completed secondary education (46%) a quarter higher and 20% was at primary education level only. There were also a large percentage of illiterate people (8%) closely matching the census data of the area reported earlier (NSSG 2001). In terms of occupation, the vast majority were working or studying 80%, 13% retired and 7% unemployed.

As far as the typology of comments is concerned, 9% were positive, 31% negative and 60% were classified by the authors as suggestive – considering the context of the comments irrespective of the actual type of comment selected. This can be seen as a failure of the system employed since only positive and negative comments could be added; future applications should either feature a single category/shape for all comments, or employ at least 3 categories. The major suggestions regarding the redevelopment of the two squares were to add some form of water (fountain, small lake, springs, etc), create a playground, change the street furniture (concrete benches were considered too "cold", dustbins too large, too many and made of the wrong material), improve security (with suggestions ranging from fences for the grass covered areas to improved lighting), more/different plants, special needs facilities (access ramps were not shown in the ICT tool as they were not provided in the drawings but due to EU regulations would have been included anyway during the construction), general comments on traffic, pedestrian paths, etc. It is apparent that citizens did examine in great detail the proposed solutions, invested time in understanding the projects and commented responsibly on the design. Hence it is important for such exercises to be addressed with great care by the decision makers in not only incorporating the proposed ideas in the final design, but also in feeding back to the citizens.

FUTURE RESEARCH DIRECTIONS

Two emerging trends, as far as the theoretical background is concerned, have been observed.

Firstly, that although governance can be applied at the municipal level (Hannemann, 2008), it can also constitute an important factor of competitiveness (Boddy and Parkinson, 2003), something that has an effect for the city as a whole at the national or even the international level. Secondly, planning re-emerges at the neighbourhood level (Barton et al. 2003), something that has its roots in the 1920s (Perry, 1929/ 2000), with an ecological emphasis (Friedman, 2007).

Integrating public consultation in the urban planning implies that the public sector procurement methods employed do cater for such process. In the Greek legislation, this is not the case hence the procurement methods must be radically altered and all parties involved must be consulted eventually reaching to a mutual agreement. This can be a long term process to accomplish which will have knock on effects on a series of other procurement methods, rules and legislation employed by the public sector. It is nevertheless a prerequisite for the development of meaningful processes leading to successful outcomes and confidence-building from the public.

Considering the complexities, effort needed and expenditure involved in creating a digital urban model – in other words the backbone of an interactive VR urban planning tool – one should consider lower level of interaction and lower cost alternative technologies such as photographic panoramas. Depending on the level/type of projected urban intervention, low bandwidth and limited interactivity but higher visual quality technologies such as QuickTimeVR™ may be beneficial, however if new constructions and elaborate investigation/exploration is expected, the modelling limitations will be highlighted and the suitability is questionable. Additionally the issue of interacting with the VR application should be thoroughly evaluated and mediation should be addressed.

CONCLUSION

The principle of participation in urban planning can be made more practical through the implementation of governance. Thus, urban governance interrelates with community planning leading to 'making cities work'.

In relation to the (generally positive) role of ICT in public participation in planning, the following conclusions can be drawn based on the research done so far: the use of ICT limits the temporal-spatial constraints in participation, participation increases the possibility of support in policies and projects, ICT represents tools (especially visual) that can improve citizen participation, the quality of software relates to lower costs in the long term, the larger the community the greater the possibility of e-government provision, concerning the public's use of ICT the emphasis is on everyday service delivery rather than large-scale policy, and, finally, a crucial question that is raised is who is able to participate - the answer to this question depends on the particular local context of a community. The process of developing a flexible three-part (common core, public planners) curriculum in the case of Agia Varvara (a relatively small community in Athens, Greece) in the framework of the Leonardo project PICT showed that ICT can help in participation, mainly because they constitute a relatively simple mean of recording the views of both the public and the planners in a variety of subjects (both 'open' and 'closed').

As far as the ICT tool and its implementation is concerned, it is important to carefully evaluate the potential versus the cost of employing different ICT methodologies and applications in the evaluation stages of a design project. Local authorities in general, and in Greece in particular, are having great difficulties in grasping the effort involved in developing such tools and consequently are unable to invest the necessary funds for this to keep up to date, or even to materialize.

The VR tool produced for Agia Varvara is sufficiently accurate, relatively efficient in time and effort spent to develop it and provides a high density of visual information to the viewer/visitor. The ability of each user to interact with the model, switch between alternatives (existing and proposed) and most importantly add comments (direct or mediated) to particular points within the model is enhancing communication by creating a pseudo-multi-user environment without the extra complexity, resources and problems involved.

Feeding back the proposals and comments to the experts and the local authority is vital as is the verification of the process that should follow with the local authorities and the planners in an open public consultation/validation meeting. This last stage is instrumental in building up the necessary entrustment links between the two since the public is willing to (and will) help only if and to the extent that it "feels" and "sees" that this tool is helping in improving living conditions. Consequently, it could be argued that employing a VR planning tool has paid off in terms of acceptance, use and quality/usability of the attained results.

REFERENCES

Al-Kodmany, K. (1999). Using visualization techniques for enhancing public participation in planning and design. *Landscape and Urban Planning*, *45*, 37–45. doi:10.1016/S0169-2046(99)00024-9

Alty, R., & Darke, R. (1987). A city centre for people: involving the community in planning for Sheffield's central area. *Planning Practice and Research*, *2*(3), 7–12. doi:10.1080/02697458708722679

Barton, H., Grant, M., & Guise, R. (Eds.). (2003). *Shaping Neighbourhoods: A Guide for Health, Sustainability and Vitality*. London: Spon Press.

Bourdakis, V. (1997). Virtual Reality: A Communication Tool for Urban Planning. In Asanowicz, A., & Jakimowitz, A. (Eds.), *CAAD-Towards New Design Conventions* (pp. 45–59). Bialystok, Poland: Technical University of Bialystok.

Bourdakis, V. (1998). Navigation in Large VR Urban Models. In Heudin, J. C. (Ed.), *Virtual Worlds Lecture Notes in Artificial Intelligence* (pp. 345–356). Berlin, Heidelberg: Springer-Verlag.

Bourdakis, V. (2004). Developing VR Tools for an Urban Planning Public Participation ICT Curriculum; The PICT Approach. In B. Rudiger, B. Tournay & H. Orbaek (Eds.), Architecture in the Network Society: eCAADe2004 Proceedings (pp. 601-607). Copenhagen: eCAADe.

Bourdakis, V. (2008). Low Tech Approach to 3D Urban Modelling. In M. Muylle (Ed.), Architecture. 'in computro', eCAADe 26 Proceedings (pp. 959-964). Antwerpen: eCAADe.

Brennan, R./ LGMB [Local Government Management Board] (1991). The Equal Opportunities Guide. Luton: LGMB.

Charitos, D. (2008). Precedents for the design of Locative Media as hybrid spatial communication interfaces for social interaction within the urban context. In Saariluoma, P., Isomaki, H. M., & Isomaki, H. (Eds.), *Future Interaction Design II*. New York: Springer Verlag.

Conroy, M. M., & Evans-Cowley, J. (2006). E-participation in planning: an analysis of cities adopting on-line citizen participation tools. *Environment and Planning. C, Government & Policy*, *24*(3), 371–384. doi:10.1068/c1k

Darke, R. (2000). Public participation, equal opportunities, planning policies and decisions. In P. Allmendinger, A. Prior, & J. Raemaekers (Eds), (2000). Introduction to Planning Practice (pp. 385-412). Chichester: John Wiley and Sons.

Dokonal, W., & Martens, B. (2002). Round Table Session on 3D-City-Modeling. In *20th eCAADe Conference Proceedings* (pp. 610-613). Warsaw, Poland: eCAADe.

Economou, D., Coccossis, H., & Deffner, A. (2005). Athens, a Capital City Under the Shadow of the State: 'Too Many Cooks Spoil the Broth. In Hendriks, F., & van Stipdonk, V. (Eds.), *Urban-regional Governance in the European Union: Practices and Prospects* (pp. 83–99). The Hague: Elsevier.

ECTP [European Council of Town Planners] (1998/2003). *The New Charter of Athens 2003: The European Council of Town Planners' Vision for Cities in the 21st century*. Lisbon.

Ellis, S. R. (1993). *Pictorial Communication: Pictures and the Synthetic Universe, Pictorial Communication in Virtual and Real Environments*. London: Taylor & Francis.

European Commission. (1999). *European Spatial Development Perspective: Towards Balanced and Sustainable Development of the Territory of the European Union*. Luxembourg: Office for Official Publications of the European Communities.

Fernández-Maldonado, A. M. (2004). *ICT-related Transformations in Latin American Metropolises*. Delft: Delft University Press.

Friedman, A. (2007). *Sustainable Residential Development: Planning and Design for Green Neighborhoods*. New York: McGraw-Hill Professional.

Gilbert, R., Stevenson, D., Girardet, H., & Stren, R. (1996). *Making Cities Work: The Role of Local Authorities in the Urban Environment*. London: Earthscan.

Grant, J., Manuel, P., & Joudrey, D. (1996). A framework for planning sustainable residential landscapes. *Journal of the American Planning Association. American Planning Association*, *62*, 331–345. doi:10.1080/01944369608975698

Hampton, W. A. (1977). Research into public participation on structure planning. In Coppock, J. T., & Sewell, W. R. D. (Eds.), *Public Participation in Planning* (pp. 27–42). New York: Wiley.

Hannemann, A. (2008). *Strategic Urban Planning and Municipal Governance*. Saarbrücken, Germany: Verlag Dr. Muller Aktiengesellschaft & Co.

Hazel, G., & Parry, R. (2004). *Making Cities Work*. Chichester, UK: Wiley-Academy.

Healey, P. (1997). *Collaborative Planning: Shaping Places in Fragmented Societies*. Basingstoke, UK: Palgrave Macmillan.

Kaylor, C., Deshazo, R., & Van Eck, D. (2001). Gauging e-government: a report on implementing services among American cities. *Government Information Quarterly*, *18*, 293–307. doi:10.1016/S0740-624X(01)00089-2

Kelly, E. D., & Becker, B. (2000). *Community Planning: An Introduction to the Comprehensive Plan*. Washington, DC: Island Press.

Kwan, M. P., & Weber, J. (2003). Individual accessibility revisited: implications for geographical analysis in the twenty-first century. *Geographical Analysis*, *35*, 341–353. doi:10.1353/geo.2003.0015

Le Corbusier. (1943/1987). *The Athens Charter*. Athens: Ypsilon.

McCarthy, J., & Lloyd, G. (2007). *From Property to People? Partnership, Collaborative Planning and Urban Regeneration*. Aldershot, UK: Ashgate.

NSSG (National Statistical Service of Greece). (n.d.). *2001Census*. Retrieved May 2009 from http://www.statistics.gr/gr_tables/S1101_SAP_1_TB_DC_01_03_Y.pdf PDF

Perry, C. (2000). Neighbourhood Unit: From the Regional Survey of New York and its Environs: *Vol. VII. Neighbourhood and Community Planning*. London: Routledge. (Original work published 1929)

PICT (Planning Inclusion of Clients Through E-training). (2006) *A Guide to Good Practice: People, Planners and Participation. Can ICT help?* Retrieved July 2009 from http://vr.arch.uth.gr/pict-dvd/PDF/wp8_guide/GUIDE_ENGLISH.pdf

Potapchuk, W. R. (1996). Building sustainable community politics: synergizing participatory, institutional, and representative democracy. *National Civic Review*, *85*(3), 54–60. doi:10.1002/ncr.4100850311

Rogers, R., & Power, A. (2000). *Cities for a Small Country*. London: Faber and Faber.

Thomas, J. C., & Streib, G. (2003). The new face of government: citizen-initiated contact in the era of e-government. *Journal of Public Administration: Research and Theory*, *13*(1), 83–102. doi:10.1093/jpart/mug010

Wates, N. (Ed.). (2000). *The Community Planning Handbook: How People Can Shape Their Cities, Towns and Villages in Any Part of the World*. London: Earthscan.

Wates, N. (2008). *The Community Planning Event Manual: How to Use Collaborative Planning and Urban Design Events to Improve Your Environment*. London: Earthscan.

Williams, C. (1999). Local economic development. In Allmendinger, P., & Chapman, M. (Eds.), *Planning Beyond 2000* (pp. 176–187). Chichester, UK: John Wiley and Sons.

ADDITIONAL READING

Agia Varvara, I. C. T. *Tool*. (n.d.). Retrieved July 2009 from http://fos.prd.uth.gr/vrml/uth/pict/

Allmendinger, P., & Chapman, M. (Eds.). (1999). *Planning Beyond 2000*. Chichester, UK: John Wiley and Sons.

Allmendinger, P., Prior, A., & Raemaekers, J. (Eds.). (2000). *Introduction to Planning Practice*. Chichester, UK: John Wiley and Sons.

Allmendinger, P., & Tewdr-Jones, M. (2001). *Planning Futures: New Directions for Planning Theory*. London: Routledge.

AutoDesk Imagemodeler. (n.d.). Retrieved July 2009, from http://usa.autodesk.com/adsk/servlet/index?siteID=123112&id=11390028

Barton, H. (Ed.). (2000). *Sustainable Communities: The Potential for Eco-neighbourhoods*. London: Earthscan.

Blender Manual. (n.d.). Retrieved June 2009 from http://www.blender.org/

Chiu, M.-L. (Ed.). (2005). *CAAD TALKS 4: Insights of Digital Cities*. Taipei: Archidata.

Conroy, M. M., & Berke, P. R. (2004). What Makes a Good Sustainable Development Plan? An analysis of factors that influence principles of sustainable development. *Environment & Planning A*, *36*, 1381–1396. doi:10.1068/a367

Conroy, M. M., & Evans-Cowley, J. (2005). Informing and Interacting: The Use of E-government for Citizen Participation in Planning. *Journal of E-Government*, *1*(3), 73–92. doi:10.1300/J399v01n03_05

Conroy, M. M., & Gordon, S. I. (2004). Utility of Interactive Computer Based Materials for Enhancing Public Participation. *Journal of Environmental Planning and Management*, *47*(1), 19–33. doi:10.1080/0964056042000189781

Cortona3D Viewer. (n.d.). Retrieved July 2009, from http://www.cortona3d.com/cortona

Final Deliverables, P. I. C. T. (2006). Retrieved July 2009 from http://vr.arch.uth.gr/pict-dvd/

Girardet, H. (1999). *Creating Sustainable Cities*. Totnes, UK: Green Books.

Girardet, H. (2004/2008). *Cities People Planet: Liveable Cities for a Sustainable World*. Chichester, UK: Wiley-Academy.

Healey, P. (2006). *Urban Complexity and Spatial Strategies: Towards a Relational Planning for Our Times*. London: Routledge.

Kitchin, R. (1998). *Cyberspace; The World in the Wires*. Chichester, UK: John Wiley and Sons.

Lang, S. B. (2007). Novel Approaches to City Modeling: Generation and Visualization of Dynamic Complex Urban Systems. In J. Kieferle & K. Ehlers (Eds.), Predicting the Future, eCAADe25 proceedings (pp. 343-350). Frankfurt, Germany: eCAADe.

Lynch, K. (1972). *What Time is This Place?* Cambridge, MA: MIT Press.

Moggridge, B. (Ed.). (2007). *Designing Interactions*. Cambridge, MA: MIT Press.

Quicktime, V. R. *Takes Media for a Spin.* (n.d.). Retrieved July 2009, from http://www.apple.com/quicktime/technologies/qtvr/

Reeves, D., & Littlejohn, A. (1998). *Virtual Reality as a Tool in Community Participation*. Report, University of Strathclyde.

Rogers, R. (1997). *Cities for a Small Planet* (Gumuchdjian, P., Ed.). London: Faber and Faber.

Systems Photomodeler, E. O. S. (n.d.). Retrieved July 2009, from http://www.photomodeler.com/

Von Uexkull, J., & Girardet, H. (2005). *Shaping Our Future: Creating the World Future Council*. Totnes, UK: Green Books.

Wajcman, J. (2002). Addressing Technological Change: The Challenge to Social Theory. *Current Sociology, 50*(3), 347–363. doi:10.1177/0011392102050003004

Web3D Consortium. (n.d.). *Open Standards for Real Time 3D Communication*. Retrieved July 2009, from http://www.web3d.org/

Whyte, J. (2002). *Virtual Reality and the Built Environment*. Oxford, UK: Architectural Press.

KEY TERMS AND DEFINITIONS

Community Planning: Planning focusing on community and aiming to develop a sense for integrated local development.

ICT (Information Communication Technologies): Computer based technologies (mainly software) that facilitate the production, processing and communication of information

Public Participation in Urban Planning: The process that can enhance the ability of the public to play a significant role in policy-making and can benefit the planners from better information about public preferences and residents' concerns.

Urban Governance: Governance refers to the process of local government in a city and to the ways in which a local society manages its collective interests.

VR (Virtual Reality): Technology enabling the user to interact with a non-real synthetic/simulated environment via appropriate computer-user interfaces. The original term was coined in 1989 by Jaron Lanier. Different levels of "immersiveness" are accepted, ranging from desktop VR to fully immersive body suit solutions.

ENDNOTES

[1] Last decade's substantial increase of economic immigrants that have not yet been assimilated in the Greek educational system. Pre 1990 value was approximately 4%

[2] An IT professional (a lady in her 40s) was employed in order to help and in many cases cheer up and motivate people in the welfare centre and other locals into using the VR tool.

Chapter 15
The Role of Local Agencies in Developing Community Participation in E-Government and E-Public Services

Bridgette Wessels
University of Sheffield, UK

ABSTRACT

This chapter discusses the way in which understanding of participation in e-services has evolved through a social learning process within planning and implementation processes. The chapter traces the development of methodologies, partnerships and design constituencies in pilot projects that inform the development of inclusive e-services. It draws on case studies of e-services between 1995 and 2009 to show how planning processes become embedded in cycles of learning and development. E-services involve change in services as well socio-technological change and relate to change in forms of participation. This has led to the development of partnerships to plan and implement e-services and to the development of research and design methodologies that foster participation in the design and use of e-services.

INTRODUCTION

The aim of this chapter is to discuss how local agencies found that they had to create partnership approaches and experiment with research and design methodologies in the planning and design of e-services. The design and planning of e-services involves both technological and organizational change, which has resulted in local service providers forming partnerships to plan and implement e-services. The development of e-services raises issues beyond technological and organizational change to the consideration of how people participate in the design and use of new services. Those involved in the planning of e-services have had to adapt to address these issues. This has led to the development of partnerships to plan and implement e-government and e-services and adaptations of participatory approaches in designing e-services. The structure of the chapter is: first the background to e-services is discussed; then the issues of design and partnerships are considered. This is followed by a section exploring the rethinking of participation before an outline of the methods employed in the case stud-

DOI: 10.4018/978-1-61520-929-3.ch015

Figure 1. The themes of eGovernment (Cornford et al. 2004)

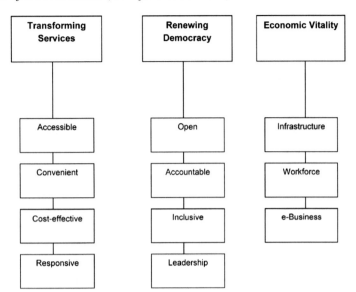

BACKGROUND: E-SERVICES

The development and use of information and communication technologies (ICT) by the public sector is seen in the domains of e-government and public sector e-services. The aim of developing e-services goes beyond e-enabling existing services in that governments and administrations seek to modernize services, for example, the eGovernment programme in England (2002) sought to:

1. Transform services by making them more accessible, more convenient, more responsive and more cost-effective.
2. Renew local democracy by making councils more open, more accountable, more inclusive and better able to lead their communities.
3. Promote local economic vitality through a modern communications infrastructure. (The National Strategy for Local eGovernment, 2002, Office of Deputy Prime Minister and UK Online and the Local Government Association).

The above outline encompasses many of the themes in the development of e-services more generally. The figure below gives a schematic overview of the themes of e-services.

These themes illustrate the ways in which e-services revolve around potential benefits that can be gained through accessible and improved participation in services. For instance, if services are more convenient to use with feedback loops then service provision can be more responsive and cost effective. By transforming services in this way, the accessibility, convenience and resulting responsiveness can be harnessed to efforts to renew local democratic processes creating a more open and accountable approach to service provision through improved communication and access to local politicians (Dunleavy et al. 2002). By developing e-services Local Authorities and their partnerships contribute to regional strategies that seek to build ICT infrastructure and skills base within regional economic plans (Wessels, 2008b).

A priority in developing e-services in the public sector is that these services have to be available to everyone, and they need to meet the needs of citizens. The development of e-services requires that developers gain an understanding people's access to, and ability to use, ICT as well as their service needs (Alford, 2002). This concern relates to ideas of a digital divide, which can be defined as the:

... differentiation between Internet-haves and have-nots [that] adds a fundamental cleavage to existing sources of inequality and social exclusion in a complex interaction that appears to increase the gap between the promise of the information age and its bleak reality for many people around the world (Castells, 2001, p. 247.).

When policy makers started to think about using ICT in public services in the early to mid 1990s they were aware of the risk of excluding some people from services (Wessels, 2007). The concern was that some people do not have access to ICT in their everyday lives, and even if access could be obtained, people might not have the skills to use ICT (Wessels, 2007). More recently, policy makers have widened the scope of the idea of a digital divide and have expanded the scope of ICT in inclusion strategies. This is evidenced in the European Union Riga Ministerial Declaration (2006) on e-inclusion. The Declaration defines e-inclusion as both 'inclusive Information and Communication Technologies (ICT) and the use of ICT to achieve wider inclusion objectives and policies aiming at both reducing gaps in ICT usage and promoting the use of ICT to overcome exclusion' (p. 2). The focus on e-inclusion and the priority of the universal provision of public services reinforces the need for e-services to be user-centric and based on the needs of service users. Furthermore, support needs to be available to ensure that people can use ICT and are digitally literate in using e-services. The background of ICT usage in Europe shows that 57% of individuals living in the EU do not regularly use the Internet (Riga Dashboard Study, 2007). Divisions in use are clearly seen in relation to age, educational levels, and employment status. For example:

- 10% of persons over 65 use Internet, against 68% of those aged 16-24.
- 24% of persons with low education use the Internet, against 73% of those with high education.
- 32% of unemployed persons use the Internet against 54% of employed persons.
- In relation for accessibility of ICT for those with disability (15% of the EU population) only 3% of public web sites complied with the minimum web accessibility standards (Riga Dashboard, 2007, p. 3).

These statistics show how levels of ICT usage can link with situations of exclusion, such as unemployment and lack of opportunities due to low levels of education, disability and ageing. This context of providing services means that planners and developers have to consider ways of gaining an understanding of a wide range of user requirements for e-services – from those with little education to people with disability as well as the more affluent users with good ICT skills who expect quality services (Alford, 2002). These issues are demanding for planners, however, another dimension of planning e-services is that they are new and emergent; hence planners, implementers, and providers are still learning how to develop them, evaluate them and measure their cost and benefits.

Issues of Design and Partnership in E-Services

E-services are based on the interactivity over a network of information sources and service provision (Bellamy and Taylor, 1998). In the planning of services it is important to distinguish between ICT and e-services. ICT is seen as a technology, which is digital, convergent, and involves computers and communication technologies. It includes a combination of video, sound and text and can involve video conferencing, scanning, interactive tracking and smart card technologies, touch screens and so on. ICT requires content to be interpolated on it, in that it has no intrinsic content or context of its own (Wessels, 2007, p. 3). A definition of e-services is that they involve the use of ICT to deliver various services. The factors that contribute to the contextual definitions are the ability of organisations to provide access for the public to information and services via ICT, which promises an enhanced service for the public (ibid.). The distinction between ICT as technology and 'e-services' helps to locate the planning area under consideration in this chapter, which is the development of service provision using ICT and not the technological components of ICT as such.

A defining feature of ICT as a technology is that it is more malleable than many other technologies. This means that design – in the form of interpretation and adaptation – is closely related to situated practices and use (Dittrich et al., 2002). The way in which users appropriate the technology, adapt it, or reject it feeds back into cycles of design, planning and use (Pierson et al., 2008). The ways in which ICT applications are shaped is therefore highly situational, which means that the successful deployment of ICT applications – and therefore their development – is dependent on the contexts of use. The development and deployment of ICT therefore takes this situational character into account (Dittrich et al., 2009; Stewart, 2007).

The constituents of e-services, namely work practices and organisation, ICT, and the practices of service users are highly situational. The interaction of these dimensions requires multi practice design constituencies, methods for the interlacing of design and use, and an ethos of reflective development of work practices that cover service organisation as well as technology. The development of e-services involves understanding how different situational factors affect development and use of the technology. Research and development of e-services often interacts with situated design and use processes, and learning from it to further develop e-services. Dittrich et al. (2009) calls this 'situated innovation' in which ICT in e-services are appropriated, adapted, and/or rejected in situ. Furthermore, it is through the development and use of pilot projects that knowledge is created in the ongoing development of e-services.

To develop networked e-services planning groups have found that they need to have a wide range of expertise in order to address service provider and end-user perspectives and needs within an e-service (Wessels et al., 2008). To this end the planning process involves forming partnerships that underpin and provide e-services. These partnerships have developed from some of the early 4th Framework European Commission funded Telematics projects (between 1995 and 1998), which consisted of five or six country partners, each contributing user needs research and technological development skills in the implementation of pilot projects (Wessels, 2007). Other types of partnerships to emerge include those based on regional strategies that bring together local service providers, technologists and researchers to build single access networked e-services across a region (Wessels, 2008). Some partnerships focus more the particular needs of a specific user group, such as those with disability for example (Wessels, et al. 2008). Other types of partnership are those aimed at empowering users, for example feminist led partnerships or those wanting to encourage grassroots political participation in local democratic processes (Harrison and Wessels, 2005).

In the development of pilot projects within different types of partnerships, planners, developers and researchers gain understanding of e-services. The way in which e-services are gaining shape and evolving is through the 'learning by doing' (Arrow, 1962) and 'social learning' processes of project-based work (Williams et al., 2005). Work in the early Telematics projects between 1995 and 1998 shows that there is a lack of knowledge of how to research user needs (Wessels, 2007). In further pilot projects and through the implementation of services (with evaluation and learning, and further development loops) project partners also found that they had to be able to link design and use together so that projects in e-services could constantly improve to meet the needs of users. This led to partnerships experimenting with various forms of participant design that is fed into ongoing planning processes (Dittrich, et al, 2009). From these pilot projects, planning partnerships could start to gain some understanding of user needs and design-in-use in pilot developments of e-services. This involved a significant shift in perspective because project methodology moved from trying to capture user needs to realising that users needed to participate not only as service users but also in the development and design of services (c.f. Suchman 2000, 2002; Schuler and Namioka, 1993; Stewart, 2007; Pierson et al., 2008).

Rethinking Participation

Developers gain understanding in planning of e-services through the lessons they learn in the take-up and use of services that involve ICT infrastructure and interfaces, back office re-engineering and end-user practices. Given the malleability and interactivity of ICT, the way e-services are shaped is influenced by different users groups (Castells, 2001). Thus a key part of planning and designing e-services involves understanding what service users want - and will use. This dynamic informs one of the key characteristics of the changes to planning and implementation, which is the move to more user-centric methodologies in design processes (Dittrich et al., 2002). In particular this has lead to an overarching trend in research and development cycles that aims to foster participation in the design and use of e-services (Wessels, et al., 2008; Pierson et al., 2008; Jirotka, and Goguen, 1994). In differing ways participation is now understood as an important dimension in the successful planning of e-services. In the first instance, planners have to focus on the development of services that could foster more participative interaction between service providers and service users in order to meet need and improve their responsiveness. This involves developers 'designing in' participation into service design and extending their work into more participative methodologies (Norman and Draper, 1986; Dobson et al., 1994; Stewart, 2007).

Within design communities of ICT there is considerable debate about how to research user needs and include users in design-in-use methodologies. Suchman (2002) points out however that there is little evidence of how to do this. One approach that seeks to include users is that of 'co-construction' where technologists and user groups co-construct ICT system design. This moves research and development away from the traditional notion of a technical expert 'capturing user needs' (Jirotka and Goguen, 1994; Crabtree et al., 1999). For example, Dobson et al. (1994) argue that co-construction approaches facilitate a symbiotic integration of user requirements and design. This allows ICT systems to be constantly refined during activities with the aim of developing something relevant to users, organizations, and institutions. There is an array of specific methodologies to foster participation in design such as user-centered design, participatory-design (PD) and meta-design. User-centered design (Norman and Draper, 1986) focuses on user activities and processes but it does not enable users to develop the system beyond that fixed in the design process itself. PD approaches (Schuler and Namioka, 1993) expand user-centered design by involving users in design more deeply by empowering them to propose and generate design alternatives themselves, as co-designers. PD, however, does not support ongoing change in applied ICT. Meta-design tries to address this by creating open systems that are modified by users. It is a co-adaptive process in which users are co-developers within an integrated design space of technologies and supported learning facilities (Fischer and Giaccardi, 2004). These methodologies seek to include users in the design and development of ICT based systems. However, they do not consider how this research and design can be integrated into planning and implementation processes.

To bring in user centred methodologies in addressing participation and inclusion, planners and developers develop partnership approaches. Molina (1995) recognises that the design and implementation of services based on ICT involve building design constituencies. Molina (1995) stresses the inseparability of technical and social constituents in technical development. He argues that expertise is needed in technological, organisational and user needs research to develop socio-technical systems. To bring these sets of expertise together he points out developers create 'sociotechnical constituencies' in which knowledge can be exchanged as systems are developed. Many of the practical partnerships discussed above take the form of a constituency. However, Molina's constituencies tend to come together around particular projects or programmes and do not necessarily integrate end-user communities into their core organisation.

To address the inclusion of users in design and in the planning and implementation processes, Wessels, et al. (2008) extend Molina's concept of sociotechnical constituencies into a broader and more dynamic 'social formation methodology'. Wessels et al.'s methodology is based on the premise that technology is socially shaped and it includes a design constituency, an ethnographic sensibility and a dynamic positioning of research in design and use processes. They define a design constituency as:

the [formal] development of a constituency of social actors that has a remit to facilitate inclusive and democratic socio-technical change through multi-disciplinary, multi-perspective and multi-positional approaches to design. The core constituency is made of a variety of user groups, social researchers, organizational change researchers, designers of hardware, software and user-interfaces, policy makers and any related interest groups (Wessels et al, 2008, p. 36).

The constituency can be formed either in response to a specific socio-technical design project or for longer term socio-technical change seen, for example, in the ongoing organizational and ICT changes in e-services. The members of the constituency develop a research design through internal discussion and in dialogue with other relevant parties. The design of the research has an ethnographic dimension in that it seeks to understand the different user perspectives in networked and interactive e-services. Furthermore, the research also seeks to address e-services from within the various perspectives of users, technologists, and service providers; research is therefore conducted from within each of these areas. The findings of the user centred research and design is integrated within the design constituency to inform planning and implementation processes. The trajectory of the methodological lifecycle can vary with the nature of the design 'problem', but it does have the potential to develop a more iterative approach to design, so developing a design-in-use ethos, or at least a reflective and user-designer feedback process during design and development. The aim is to balance the agency of all groups in socio-technical change, such as user groups, social and organizational researchers, ICT system designers, service providers and planning and implementation groups (Wessels et al., 2008).

Methodology and Case Studies

The argument made in this chapter is based on a series of case studies that provide a longitudinal analysis of change in planning and developing e-services in the public sector over a 14 year period (1995 to 2009). In each study researchers spent between 1 and 2 years in the field. They combined participant observation with interviews, document analysis and surveys. The case studies cover the development of:

- Telematics services in London (UK), focusing specifically on the exemplar European Commission (EC) funded project Advanced TransEuropean Telematics Applications for Community Help (ATTACH) running from 1995 to1998
- User generated web-services in Ronneby (Sweden), Lewisham (UK) and Bologna (Italy) in the EC funded Dialogue Project running from 1999 to 2001
- E-services for multi-agency health and social care focusing on Newcastle (UK) in the Economic and Physical Research Council (UK) funded Advanced Multi Agency Service Environments (AMASE) project running between 2001 and 2004
- Local E-government in England and Wales funded by the UK Office of Deputy Prime Minister running between 2002 and 2003
- Regional e-inclusion programme focusing on South Yorkshire (UK) funded through various sources of EC, national, and regional funds from 2002 and 2009

These case studies are predominantly based in urban areas, although the South Yorkshire and Ronneby cases also include rural areas

Case Studies

1995- 1998: Early Approaches to Planning and Developing E-Services: The Case of Telematics Services in London

This section covers the early ways in which developers use pilot projects to gain an understanding of e-services. The Telematics projects, which were all funded by the EC, represent some of the early approaches in developing e-services. The projects in London were:

- 'Delivery and Access to Local Information and Services' (DALI) project, which sought to develop user-driven multimedia e-services for comprehensive services
- 'Telematics Applications and Strategies Combating Social and Economic Exclusion' (PERIHERA) project, which addresses training and employment for groups at risk of exclusion
- 'Citizens Access Networks and Services' (CANS), which focuses on the needs of older, disabled citizens and migrant groups through skills and language training
- 'Advanced TransEuropean Telematics Applications for Community Help' (ATTACH), which sought to developing interactive multimedia service kiosks with local administrations, police and local organizations for community services.

These projects have common aspects, such as a concern with accessibility of services addressing the needs of people in society who may be excluded for various reasons, including elderly, disabled or ethnic groups and those marginalized by lack of employment.

The developers of these pilot projects and other service providers formed a partnership called 'Telematics for London'. The members of the partnership thought that each of these projects promised, in quite different ways, to become effective and efficient carriers of vital ingredients of democracy and improve services. The services could, for instance:

- receive information about citizens' concerns,
- support decentralization of bureaucracy,
- help to develop one-stop approaches to access to information and services,
- contribute to the development of market testing approaches,
- develop more effective machinery in support of community groups, and
- provide a useful resource for the emerging citizens' juries and similar approaches (Wessels, 2007, p. 104).

To avoid exclusion from these new types of pilot services, the London group felt that there is need for initial simplicity and user-friendliness as well as making sure services would address inequality (Wessels, 2007). The group started to foster partnership approaches in the development of e-services and sought to build networks that might enable multi-agency services to be delivered to help counter exclusion.

The project teams had to learn how to build e-services based on multi-agency networks that were user-driven. In the ATTACH project (which acts as an exemplar) the team utilized established project methodologies to capture user needs, build a prototype, test and develop the technologies, pilot and implement new services (Wessels, 2000; Wessels, 2007). However, from the outset of the project the team was aware of the difficulties of capturing user needs for totally new services. The team first undertook a conventional piece of survey research but service users found it difficult to articulate their needs because they could not envisage what e-services would look like. To get richer picture of user needs, the team set up a network of local service providers and undertook user needs workshops. The team also conducted interviews and mock-ups with potential end users. This gave the team basic user needs such as the technology needs to be easy to use and reliable and the information needs to be up to date. Despite having very little real insight of user needs, the team decided to build and test a prototype in public settings.

The use and evaluation of prototype kiosks made the team realize that the project was not a "technology one but a service one" (Project manager). The evaluation showed that people could use the technology but they wanted access to quality information and services that met their needs. This feedback meant that the team had to focus on the services that the technology would provide access to. To achieve this, the team had to get agencies to share information, re-engineer access to services and develop joined up services.

This meant that the team with local planners had to develop an iterative research and design process involving service providers, researchers and end user groups in the development of new services. To get a deeper understanding of end-user needs, the Local Authority funded 10 in-depth focus groups to explore service user needs (sampling across age, socio-economic status and ethnicity) that built on their existing community research. The key finding of the research was that people wanted information online to be 'joined-up' and targeted to specific situations as well as access to off-line personal advice and services. To gain insights into how services could evolve as e-services, the ATTACH team also organized service provider information days for service providers from the public, private and voluntary sectors. The aim of these days was to raise awareness of e-services with the objectives of fostering the sharing of information and improving access to services to build one-stop e-services (Wessels, 2007).

The knowledge gained by these methodologies led the ATTACH team and the Local Authority planners to build a community online information service that is adapted to a range of user access points. This resulted in eight Local Service Centres sited across the Borough, which offer online information and offline personal advice and support. Other access points are public computer access points in libraries, community centres and schools. Residents can also access the information via home computers. From the ATTACH project (and the other London projects) developers realized that the planning of e-services involved networks of actors and iterative design-in-use methodologies. Without these two components it is difficult to design and deliver e-services that are inclusive and meet specific needs. The Local Service Centres supported by the online information system are now established as a core part of the Borough's service strategy. User feedback surveys show that they are meeting the needs of local people, with the level of service satisfaction rated at excellent. They are proving to be a cost effective way of providing services as they are reducing the cost of face-to-face support and phone inquiries by 40% whilst still offering face-to-face support when needed.

1999 – 2001: Developing Inclusive Approaches in E-Services for Community Participation: The Case of the Dialogue Project

From the early pilot projects described above, planners started to consider ways of fostering participation by providing interactive services to generate inclusion and enhance local democracy. One EC funded pilot project called the Dialogue project sought to develop community participation through the Web. The EC project team from Ronneby (Sweden), Bologna (Italy) and Lewisham (London, UK) wanted to see how interactive ICT could foster grass-root participation. The team needed to explore various methodologies for developing participation in ICT based communication. The project team in each country developed a partnership with local people and the local authority to see how they could develop participative fora. The team in Sweden focused on 'women writing on the Web'. The team in Italy and the UK focused on children taking part in video-conferencing to discuss issues and cultural events across these nations. The UK group developed local interactive policy fora in which local residents (in Lewisham) discussed their concerns and political perspectives. The development of these services was achieved through adapting of participatory design to enable these user groups to shape the technology and service for their practices, needs and desires (Dittrich et al., 2002).

The aim of the project was to train citizens to use the Web, with 'use' understood in a broad way through the metaphor of 'communication as a network, as a cyberspace, and as participative'. This ethos of the project was: "we teach you the techniques; you teach us the content to which they can be applied" (project worker, 1999). Feminist researchers working in the project in Ronneby (Sweden) took a strong participatory design and inclusive development approach by working with the women to empower them to develop their own services. The researchers and the women based their development work on Virginia Woolf's argument that women need their own space to write. From this basic idea the researchers and the women in the community developed four virtual writing rooms. Their design-in-use process created different rooms based on the ways in which the women wanted to write and communicate: room one is for participants' personal presentations; the second room is the collaborative writing of the 'Virtual Cookbook'; the third room is for poems and short stories; and the fourth room is a chat room (Harrison and Wessels, 2005). The evaluation of the project shows that the women, who had no experience of ICT, felt they had become competent users and developers of Website design and use. They felt that they had learnt how to have a voice in the public sphere and that they were better connected with other women, and they had learnt to self-organize. All the women felt empowered by the experience (Dialogue Project Report, 2002).

Using a participatory design approach the researchers in Lewisham (UK) fostered participation in local democratic processes through 'On-line Live Chat'. The researchers worked with existing community fora and local intermediaries to create local democracy websites with interactive chatrooms. Members of each forum posts concerns on the website and then debate issues online, once a discussion is closed the issues raised are sent to the local council. Thus opinions of local people expressed in the chat rooms are fed into formal council debates and procedures and when local politicians have considered the issues, they feedback to residents through the Local Authority web site. At the start of the project, many of the residents were not sure that local politicians would listen to them, however, as the project progressed, they found that their concerns were heard and addressed to create change at the local level. This aspect of the Dialogue project improved local participation in democratic processes and improved trust between local people and their local politicians and councillors (Dialogue Project Report, 2002).

Another aspect of the Dialogue project is the development of videoconferencing for children. Videoconferencing was developed for participation by schoolchildren in Lewisham (UK) and Bologna (Italy). The children from these places took part in videoconferences with each other and they discussed issues that are important to them. These range from the fabric of their everyday lives, their hobbies and school-life to issues such as bullying, racism and migration. The schoolchildren say that they got used to using the technology and found it easy to use. In the evaluation of this aspect of the project, the children report that they feel that 'connecting through the screen' with other children helps them understand what life is like in Bologna and Lewisham, which widens their understanding of other cultures (Dialogue Project Report, 2002).

The work in the Dialogue project shows how researchers using participatory design involving users developed different ways to involve people and to foster participation. By involving local people and local intermediaries in the design process they could inform the planning of e-services by suggesting the types of services people want (Stewart, 2007). This project shows that by getting people involved in the design process helps to promote participation in services.

2001 – 2004: Beyond Participatory Design to Social Formation Methodology: The Case of the Multi-Agency Services for Children With Disability

Participatory design helps in the development of e-services, particularly in the area of user interfaces and in the identification of service needs. It does not, however, necessarily facilitate the ongoing development of e-services based on networks. The rise of networked e-services coincided with the quest for multi-agency services, which required planners to gain understanding of holistic multi-agency approaches to care (c.f. Leat et al., 2002). In this context policy frameworks were also pushing for service-users to participate in the development of services (e.g. the UK's Children's National Service Framework, 2002). These factors prompted the development of a 'social formation methodology' as described above, which incorporates the various perspectives of actors involved in the planning, design, implementation and use of multi-agency e-services in community based settings. The area in which this methodology was developed and used is in community health and social care services for children with disabilities (Wessels et al., 2008).

A central part of a social formation methodology is the 'design constituency'. In the case of services for children with disability the constituency was made up of a practitioner group made up of 6 community key-workers with backgrounds in nursing, social work and education; 3 health and social care policy makers with a children's services planning remit; 6 university researchers with backgrounds in systems design, organizational change management and sociology; and 2 private sector ICT designers. In providing multi-agency services for children in the community, research and design has to be able to involve a variety of perspectives of highly skilled actors and understand the myriad of sensitive and complex social activities. To this end, the research for design and planning included interviews with service providers and practitioners across all the agencies in the multi-agency team, namely nurses and doctors, social workers and educational health and social care workers. This was followed by a one-year participant observation of the work done by key-workers in the multi-agency centre. The understanding gained of the work in these two phases was analysed within the constituency, and the resulting analysis then formed the foundation for a series of focus groups and workshops to explore the issues of user access and needs, working practices and ICT, and policy making that are integral in the planning of services (Wessels et al., 2008).

The qualitative methods provide an ethnographic sensibility (Wessels and Craglia, 2007) that enables members of a design constituency to become more reflective of the research process. It raises awareness of different actor perspectives within e-services and their resources and learning processes, and through this awareness the design constituency and its stakeholders can plan and produce change reflexively (Wessels, et al., 2008). The methodology includes dissemination workshops that aim to consolidate the relationships built up in the constituency and the knowledge embedded within the community, and ensures that this is filtered into further planning and development. The planning of new services involves the re-configuration and adaptation of new and existing tools, a re-thinking of practices and skills, and re-imagining service provision within a social formation of developer and user groups (Wessels et al., 2008). The development of the social formation methodology within this project informed the UK National Framework for Multi-agency Environment (FAME) program. This program has produced a generic framework and a software demonstrator to support multi-agency development and operation including multi-agency children's services in the UK.

2002 – 2003: Cycles of Learning in Planning and Implementation: The Case of E-Government in England and Wales

Cornford et al's (2004) study addresses the planning and implementation of e-government in England and Wales between 2002 and 2003 and it shows how the change process is shaped by specific local contexts as well as generic development processes. Local e-government partnership develop paths through a complex change programme involving not only changes in technologies and business processes but also changes in working practices, culture, and attitudes that influence rates of participation by citizens and businesses in e-services. Following on from the lessons learnt through such projects as described above, e-government planners seek to develop a 'joined up' approach to e-services to provide a seamless delivery of services from various providers. Cornford et al.'s (2004) study found that the main path to joining up service is through partnership working. Drawing on a Local eGovernment Survey as well as their own research, Cornford et al. point out that 94% of local authorities set up e-government partnerships to help them implement e-government (Cornford et al., 2004). The aim of these partnerships was to change the delivery of services away from single services individually accessed to joined-up services with single access.

To plan and manage this change, partnerships undertook an iterative change management process. Initially they took an established approach to change management based on a model of change involving a process of:

1. the development of a vision and establishing aims and objectives
2. a planning phase to identify and secure the resources to achieve the aims and objectives
3. the implementation of the plan
4. monitoring to establish whether the objectives have been met and identify other outcomes

In relation to e-government, these general change processes also need to incorporate changes in a socio-technical system. Socio-technical change requires the simultaneous configuration and co-ordination of 'technical', 'organisational' and 'social' aspects of a new system. For example, e-government requires the integration of the four dimensions of change, which are: the configuration of the technologies; the re-engineering of business processes; changing working practices; and the alignment of active participation by individuals and businesses as customers, interlocutors, clients and citizens (Cornford et al., 2004). In socio-technical change programs including e-government this means that the traditional stages of change management cycles described above, which although significant, need to be matched in importance with the progression between each stage (Orlikowski and Hofman, 1997). It is in the progression between stages that the learning and preparation for the planning of ongoing development and implementation occurs. This means that the transitions between the phases are significant in planning processes and these involve:

1. strategic planning that involves a process of moving from a vision to a plan
2. transition phase of mobilizing of resources that links planning with implementation
3. evaluation of the transition between implementation and monitoring
4. these processes feed into a learning process that takes monitoring to back to envisaging to inform further planning and development cycles (Cornford et al. 2004)

One of the key dimensions in this cycle of developing e-government is consultation with stakeholders and the public. 95% of local authority e-government partnerships undertook consultation with stakeholder groups and the public (Cornford et al., 2004). In the development process the groups consulted include: other partners in the Public Sector; partners in the Private, Community and Voluntary Sector; Local Strategic Partnerships; stakeholders in the local community including public consultation; and local business. Most of the authorities have a separate corporate e-government strategy, which are publicly available (82% of Local Authorities) (Cornford et al., 2004). Although the way in which local authorities undertook consultation varied, they nonetheless found that consultation improved the development and take-up of e-services (Cornford et al., 2004p., 44). The need to understand the needs of service users is important in ensuring that services meet needs and are accessible and useable. Furthermore, the improvement of take-up of services also enhance the ability of service providers to be responsive to service users, thus improving performance indicators in responsiveness of services (ibid. p.45)

The development of e-government in the England shows how the development of e-services requires the space for ongoing learning from the implementation process that feeds into the subsequent planning of the next stages of the evolvement of e-services. Local authorities and their partners are learning to foster participation in the planning, research, and implementation processes of e-services. They consider issues of accessibility and participation in the design of e-services, and find that consultation with stakeholders and the public is important in ensuring take up of services and in improving the responsiveness of service providers to their user groups (Cornford et al., 2004).

2002- 2009: Regional Strategy for E-Inclusion: The Case of South Yorkshire

So far the cases studies show how different planning strategies and research and development methodologies have been adapted and developed as e-services have evolved. However, planning groups are pushing participation further as they seek to construct strategies for e-inclusion (Wessels, 2008). For example, South Yorkshire (UK) is planning to develop regional e-services to foster e-inclusion by facilitating participation of those living in the region. The planning process has to address issues of a digital divide in a region that underwent de-industrialization because it lost its main industries of coal and steel that resulted in high levels of poverty and unemployment. The region is seeking to regenerate through information society policy and activity. To do this it needs to build capacity within the region. It has built a high quality infrastructure and a strong emerging New Media sector, and it is now seeking to find ways to raise the levels of ICT knowledge of local business and people so that they can participate in the emerging economy and shape the social and political landscape of the region (Wessels, 2008).

The population of South Yorkshire maps trends in low take-up of ICT, for example:

- people over the UK retirement age is 18.9%
- Pakistanis (36%); mixed race (19.5%); Black Caribbean (10.0%)
- 24% are not economically active
- 9.3% claim Incapacity Benefit and Severe Disability Allowance

These regional figures provide an indication of the issues of participation in economic terms, which when looked at in relation to the broader population statistics show a region with a diverse population in relation to age, employment and ethnicity. The regional strategy combines economic and social policy to plan for an inclusive digital region.

In relation to economic policy, the development of the Cultural, Creative and Digital Industries as a key sector in the area (http://www.sypartnership.org.uk/coredocs.php) is important. The sector employs 60,000 people across the region and there is a strong business base in many of the key sub sectors, with over 2,700 employees involved in Visual Arts, over 1,300 employees in Audio Visual and 600 employees in Books and Press. The sector is characterised by new starts and rapid growth, and a comprehensive programme of creative and digital flagship projects with the ability to catalyse the growth more widely are being developed in the region including:

- E-Campus, Sheffield
- Digital Media Centre, Barnsley
- Digital Knowledge Exchange element of Doncaster Education City
- A range of digital SME initiatives around CENT@Magna, Rotherham
- Business Innovation Centres in Barnsley, Doncaster, North East Derbyshire, Bassetlaw, and Chesterfield (South Yorkshire Partnership, 2006).

The economic policy in relation to these industries includes the regional investment in fast, high-width broadband and other digital technology. This means that the region has a good ICT infrastructure and is using ICT to transform itself from an industrial region to an information society and knowledge economy. This is the economic base for developments in policy for e-inclusion.

The economic planning for the region moves to a focus on e-inclusion to address social cohesion and participation in a regional information society. The regional policy makers have developed a partnership for ICT strategy, development and implementation called the South Yorkshire Public Sector E-Forum. This network of partners is made up of the 4 Local Authorities (Sheffield, Barnsley, Rotherham and Doncaster) and other local organisations from the public and voluntary sectors. The forum links closely to e@syconnects, which is a public sector partnership consisting of the South Yorkshire local authorities, health authorities, emergency services (Ambulance, Fire and Police), voluntary sector organisations, Yorkshire Forward (Regional Development Agency), South Yorkshire Passenger Transport Executive, Job Centre Plus and a myriad of other organisations who work together to offer joined-up services.

The regional partnership, with 10 years of planning, research and development experience has a deep and grounded understanding of developing e-enabled projects from a community base and citizen perspective. It uses the e@syconnects partnership to undertake innovative projects that have a citizen centric approach. Using knowledge gained from focus groups, interviews and observation of service provider settings and domestic settings, the partnership implement a range of pilot projects. They undertake evaluation of the projects that feed into the development of new services. The partnership builds services on a range of channels including interactive digital television (DiTV) and mobile phones as well as home computers and public access computers and kiosks.

The region's easy-to-use system enables people to access the information and services from a range of service providers through one point of access. For example, e@syconnects links with health services to allow people to book appointments with their doctor 24/7 through the Internet, mobile telephones and DiTV. This approach reduces the demands upon service providers enabling tangible benefits to be realised by both the citizens who use the services and the service partners who collaborate to offer these new services. The online appointments service has reduced missed doctors' appointment by 30%, improving efficiency of doctors' surgeries. Service users report that they find it easier to contact their GP than via the phone and one unexpected benefit emerged when an older woman said that she "always checks the doctors appointments before going to bed to be sure that an appointment is available" should she need one, which she found very reassuring (Wessels, 2008). Evaluation of service costs in 2007 shows that the transaction costs are reduced by 20%.

Building on this knowledge, the e-forum and e@syconnects seek to address e-inclusion and participation by integrating ICT in the everyday life of local people, local business and local services. It seeks to facilitate "a genuinely connected community in which every sector of society - citizens, service providers, businesses - are personally empowered to improve their lives and strengthen their communities" (e-forum member).

The basis for this vision is to use personalised ICT where people want it and need it most – in the home, streets, shops and personal interaction points in the community - to ensure existing ICT initiatives are meaningfully connected, communicated and accessible to all. The key challenge is to ensure that everyone has the skills to make ICT an integral part of their lives and to participate in inclusive change in the region. This means that the strategy must involve people across ages from children to the elderly, across educational and (non) employment capacities, as well as ethnic minority groups. The main argument is that "the technology exists. The services are online. The community vehicles are in place. The missing element is people. Rather than introduce new 'Gizmo's', we believe the *real* Digital Challenge is to empower people to shape their own lives" (e-Forum, ICT Director).

The plan has two core components, which are Digital Outreach Teams (DOTs) and a Digital Directory where personal ICT outreach and education links with e-services that are accessible to all. The Digital Outreach Teams (DOTs) will teach people about the potential of ICT for their own lives. Local people will be coached in their homes, bingo halls, community centres, libraries, post offices and new media start-up businesses and so on. The coaching will range from 'novice to advanced' in ICT knowledge and literacy. The DOTs will be made up of volunteers from community programmes such as: Age Concern; National Youth Volunteers; other National Initiatives, and volunteers from public, private and voluntary sectors. They will receive ICT and community training as well as being security checked and registered with the e-forum. The teams will have a clear identity and branding in the community and they will support inclusion by:

- Sustaining existing progress
- Providing equal opportunities
- Innovate service design and delivery
- Feedback to redesign workflow
- Anticipate and plan for future challenges and trends.

The second part of the strategy is a social delivery model that connects with the work of the DOTs to provide a structural underpinning for future developments by the communities themselves. This is based on a Digital Directory, which seeks to gather feedback from people to inform the improvement of services and to enable local priorities to be linked to mainstream services. The Directory as an information source will help connect citizens with citizens and community organisations. It will provide companies with listings of neighbourhood resources, and is a resource to advertise existing initiatives and to promote e-transactions. In total the strategy involves DOTs who will foster and deepen coordination and integration between community and voluntary organisations by 'cross selling' their expertise and educating the community about their efforts. The Directory will underpin this initiative by acting as an online knowledge base for information sharing (e-forum ICT Director). The e-forum envisages the DOTs and the Digital Directory working together to:

- Give everyone skills to make technology integral to life
- Give everyone opportunity to benefit from ICT initiatives
- Eradicate top-down, fragmented service provision
- Ensure existing ICT initiatives are connected
- Create citizen centric governance ideas
- Drive service transformation through feedback
- Enhance develop of public service culture
- Connect neighbourhoods across the region. (e-Forum, ICT Director)

This case shows how the development of partnerships for the planning and implementation of e-services has evolved over time. A central dimension in planning e-services is the creation of partnership approaches and with their development planning and implementation has adapted various research and design methodologies into the development cycles of e-services. This project is still at the strategic stage and has yet to be funded but its points to the ways policy makers are drawing on past experience and are envisaging and planning the further development of e-services.

Solutions and Recommendations: Partnerships and Design Constituencies in Planning Processes

The interactive character of ICT and the development of complex networked e-services are found in public settings, work settings, home settings and in personal settings. The development of e-services raises interesting questions regarding who should be included in the design of the technology and services. ICT is to some degree structuring the ways in which people can participate in society and the policy trend to include users in the shaping e-services raises questions of design methodology and the planning process. The complexity generated through networks of users and networked technology demands a methodology that goes beyond traditional requirements engineering and beyond traditional forms of action research. A more sophisticated methodological approach needs to be able to address multi perspectives and positions of a variety of users and needs to build a flexible and secure network with accessible user-interfaces. To achieve this requires an interdisciplinary and multi perspective/positional approach, which has led to the development of design constituencies within social formation methodologies. Furthermore, the social learning involved in the implementation process is feeding into cycles of development in which planners gain knowledge and understanding of the evolvement of e-services.

This approach differs from traditional user needs research that often involves engineers going into a particular environment to study users and their patterns of behaviour. Researchers observe social settings to produce finely grained descriptions of social activity and the meanings of that activity (Randall et al. 1994). This approach however produces a dichotomy of 'experts' who design and implement the technology and the users who have to adapt and learn to use the technology. This can be difficult for some people who experience problems using or accessing new services due to the design of the technological interface (Suchman, 1987). It can also lead to people having to 'work around' (Cornford and Pollock, 2003) the technology because the design of the system restricts usage. Both of these consequences can limit participation in e-services.

Another research approach to change and planning change is action research, which seeks to change a set of circumstances by research informing strategies for change. In a benign sense, action research can support the efforts of those who are excluded by research that identifies their needs and concerns that can inform the planning and design of services. It can also work for governments and corporate business to further their goals and interests perhaps at the expense of wider populations especially if policy does not cover inclusion in its framework. Neither of these approaches really touches on inclusive design by users and designers in relation to planning processes. Part of the aim of inclusive design is that it fosters more participation in services and it helps to ensure that services meet the needs and aspirations of users. Participant design addresses the concerns of inclusion by fostering more open and democratic design processes. There are however difficulties with this approach especially when trying to integrate it within the planning of broader socio-technical change. The use and development of design constituencies within e-service partnerships builds in a more flexible, context sensitive approach to socio-technical change and in the planning of e-services. The constituency forms a learning agency in that greater understanding between perspectives and knowledge can be achieved, which in turn allows for more informed decision-making by the actors involved in planning socio-technical change.

FUTURE RESEARCH DIRECTIONS

Early developments shows that traditional user needs research could not identify user needs. They also found that without building in user feedback cycles into design and development processes they could not maximize participation in e-services. With basic e-services in place partnerships started to work with the ideas of participant design, which helped to foster participation in the design and use of e-services. The quest for user shaped multi-agency services meant that developers expanded participant design into social formation methodologies with design constituencies. These constituencies include a range of users and developers and using a variety of methods they designed and implemented e-services. The knowledge gained by these constituencies is fed into planning groups as they develop e-services further. This iterative approach is evident in the implementation of e-government. A distinctive factor that emerged in this context is the importance of learning within cycles of planning and implementation.

Another important factor in planning e-services is that public sector e-service partnerships are recognising the importance of e-inclusion in the development of information society policy. To this end strategic planning by regional partnerships is now focusing on participation and inclusion across populations within regions. In this context planning is part of the participation and inclusion agenda because users and the outreach teams feed information into the directory that forms the user-driven structure for e-services. This information also feeds directly into policy-making group to inform further developments of e-services. In the social learning process within the planning process of e-services and e-government knowledge is accrued and continues to be developed through feedback mechanism instituted through the various planning partnerships, which is now being expanded into including regional populations.

CONCLUSION

The case studies show how local government, public services and voluntary agencies have had to go through a social learning process regarding the planning of e-services. A key part of the planning process is the development of partnerships to plan, design and implement e-services. The expertise required in developing e-services is varied and diverse which is triggering the formation of partnerships. A key-contributing factor of these partnerships to planning is how they gain knowledge of users and their needs and how services can foster participation. This ability to access and understand users and their needs has evolved as e-services have become more established. In conclusion e-services partnerships and forms of participative research and development are an important aspect in cycles of planning e-services.

REFERENCES

Alford, J. (2002). Defining the Client in the Public Sector: a social exchange perspective. *Public Administration Review*, 62(3), 337–346. doi:10.1111/1540-6210.00183

Bellamy, C., & Taylor, J. (1998). *Governing in the Information Age*. Buckingham, UK: Open University Press.

Cornford, J., & Pollock, N. (2003). *Putting the University Online: Information, Technology and Organizational Change*. Buckingham, UK: Open University Press.

Cornford, J., Wessels, B., Richardson, R., Gillespie, A., McLoughlin, I., & Kohannejad, J. (2004). *Local e-Government: Process Evaluation of Electronic Local Government in England*. London: ODPM.

Dialogue Project Report. (2002). Retrieved from http://www.beepknowledgesystem.org/Search/ShowCase.asp?CaseTitleID=182&CaseID=574

Dittrich, Y., Eriksen, S., & Wessels, B. (2009). *From Knowledge Transfer to Situated Innovation: cultivating spaces for co-operation in innovation and design between academics, user-groups and ICT providers*. Karlskrona, Sweden: Blekinge Institute of Technology.

Dittrich, Y., & Floyd, C. C., & Klischewski, R. (Eds.). (2002). Social Thinking, Software Practice. London: MIT.

Dobson, J. E. Blyth., A. J. C., Chudge, J., & Strens, R. (1994). The ORDIT approach to organisational requirements. In M. Jirotka & J. Goguen (Eds.), Requirements Engineering: Social and Technical Issues (pp. 87-106). London: Academic Press.

Dunleavy, P., Margetts, H., Bastow, S., Callaghan, R., & Yared, H. (2002). *Progress in implementing e-Government in Britain: Supporting Evidence for the National Audit Office Report Government on the Web II*. London: LSE Public Policy Group and School of Public Policy UCL.

European Commission. (2006). *The Riga Ministerial Declaration on eInclusion*. Retrieved from http://ec.europa.eu/.../events/ict_riga_2006/doc/declaration_riga.pdf

European Commission. (2007). *Measuring progress in eInclusion, Riga Dashboard*. Retrieved from http://ec.europa.eu/.../einclusion/docs/i2010_initiative/rigadashboard.pdf

Harrison, J., & Wessels, B. (2005). A new public service communication environment? Public service broadcasting values in the reconfiguring media. *New Media & Society*, 7(6), 861–880. doi:10.1177/1461444805058172

Jirotka, M., & Goguen, J. A. (Eds.). (1994). *Requirements Engineering: Social and Technical Issues*. London: Academic Press.

Leat, P., Seltzer, K., & Stoker, G. (2002). *Towards Holistic Governance: The New Reform Agenda.* Basingstoke, UK: Palgrave.

Mansell, R., & Steinmueller, W. (2000). *Mobilizing the Information Society: strategies for growth and opportunity.* Oxford, UK: Oxford University Press.

Members of the High Level Group on the Information Society. (1994, May26). *Europe and the Global Information Society: Recommendations to the European Council* (The Bangemann Report).

Molina, A. H. (1995). Sociotechnical Constituencies as Processes of Alignment: the rise of a large-scale European Information Initiative. *Technology in Society*, *17*(4), 385–412. doi:10.1016/0160-791X(95)00016-K

Norman, D. A., & Draper, S. W. (Eds.). (1986). *User-centered System Design: New Perspectives on Human-Computer Interaction.* Hillsdale, NJ: Lawrence Erlbaum Associates, Inc.

Orlikowski, W., & Hofman, D. (1997). An Improvisational Model of Change Management: The Case of Groupware Technologies. *Sloan Management Review*, 11–21.

Pierson, J., Mante-Meijer, E., Loos, E., & Sapio, B. (Eds.). (2008). Innovating for and by users. COST Office: COST Action 298.

Randall, D., Hughes, J., & Shapiro, D. (1994). Steps towards a partnership: Ethnography and system design. In Jirotka, M., & Goguen, J. (Eds.), *Requirements Engineering: Social and Technical Issues* (pp. 241–258). London: Academic Press.

Schuler, D., & Namioka, A. (Eds.). (1993). *Participatory Design: Principles and Practices.* Hillsdale, NJ: Lawrence Erlbaum Associates, Inc.

South Yorkshire Partnership. (2006). *Progress in South Yorkshire, Sheffield.* Sheffield, UK: Sheffield Regional Development.

Stewart, J. (2007). Local Experts in the Domestication of Information and Communication Technologies. Communication. *The Information Society*, *10*(4), 547–569. doi:10.1080/13691180701560093

Suchman, L. (1987). *Plans and Situated Actions. The Problem of Human-machine Communication.* New York: Cambridge University Press.

Suchman, L. (2000). Located Accountabilities in Technological Production. Retrieved October 31, 2008, from http://www.comp.lancs.ac.uk/sociology/soc0391s.html

Suchman, L. (2002). Practice-Based Design of Information Systems: Notes from the Hyperdeveloped World. *The Information Society*, *18*, 139–144. doi:10.1080/01972240290075066

Wessels, B. (2000). Telematics in the East End of London: New Media as a Cultural Form. *New Media & Society*, *2*(4), 427–444. doi:10.1177/14614440022225896

Wessels, B. (2007). *Inside the Digital Revolution: policing and changing communication with the public.* Aldershot, UK: Ashgate.

Wessels, B. (2008). Creating a regional agency to foster eInclusion: the case of South Yorkshire, UK. *European Journal of ePractice,* (3), 3-13.

Wessels, B., & Craglia, M. (2007). Situated innovation of e-social science: Integrating infrastructure, collaboration, and knowledge in developing e-social science. *Journal of Computer-Mediated Communication*, *12*(2). Retrieved from http://jcmc.indiana.edu/vol12/issue2/wessles.html. doi:10.1111/j.1083-6101.2007.00345.x

Wessels, B., Walsh, S., & Adam, E. (2008). Mediating Voices: Community Participation in the Design of E-Enabled Community Care Services. *The Information Society*, *24*(1), 3–39. doi:10.1080/01972240701774683

Williams, R., Stewart, J., & Slack, R. (2005). *Social Learning in Technological Innovation: Experimenting with Information and Communication Technologies*. Cheltenham, UK: Edward Elgar.

ADDITIONAL READING

Barney, D. (2004). *The Network Society*. Cambridge, UK: Polity Press.

Bell, D. (2001). *An Introduction to Cybercultures*. London: Routledge.

Bijker, W. E., Hughes, T. P., & Pinch, T. (Eds.). (1987). *The Social Construction of Technology System*. Cambridge, MA: MIT Press.

Button, G. (Ed.). (1996). *Technology in Working Order. Studies of Work, Interaction, and Technology*. New York: Routledge.

Castells, M. (2001). *The Internet Galaxy: Reflections on the Internet, Business and Society*. Oxford, UK: Oxford University Press.

Coleman, S. J., Taylor, J., & Van der Dunk, W. (Eds.). (1999). *Parliament in the Age of the Internet*. Oxford, UK: Oxford University Press.

Foot, K., & Schneider, S. M. (2006). *Web Campaigning (Acting with Technology)*. Cambridge, MA: MIT Press.

Fox, S. (2005). *Digital Divisions: There are Clear Differences Among Those with Broadband Connections, Dial-up Connections, and No Connections at all to the Internet*. Pew Internet & American Life Project.

Graham, S., & Marvin, S. (2001). *Splintering Urbanism, Networked Infrastructures, Technological Mobilities and the Urban Condition*. London: Routledge. doi:10.4324/9780203452202

Hacker, K., & van Dijk, J. (Eds.). (2000). *Digital Democracy: Issues of Theory and Practice*. London: Sage.

Haddon, L. (2004). *Information and Communication Technologies and Everyday Life*. Oxford, UK: Berg.

Hartswood, M., Procter, R., Slack, R., Voss, A., Butcher, M., Rouncefield, M., & Rouchy, R. (2002). Co-realisation: Towards a Principled Synthesis of Ethnomethodology and Design. *Scandinavian Journal of Information Systems*, *14*(2), 9–30.

Jones, S. (1995). *CyberSociety: Computer-mediated Communication and Community*. Thousand Oaks, CA: Sage.

Kluver, R., Jankowski, N., Foot, K., & Schneider, S. M. (Eds.). (2007). *The Internet and National Elections: A Comparative Study of Web Campaigning*. London: Routledge.

Lievrouw, L., & Livingstone, S. (2006). *The Handbook of New Media*. London: Sage.

Loader, B. (Ed.). (1997). *The Governance of Cyberspace, Politics, Technology and Global Restructuring*. London: Routledge. doi:10.4324/9780203360408

Lyon, D. (2001). *The Information Society: Issues and Illusions*. Cambridge, UK: Polity Press.

MacKenzie, D., & Wajcman, J. (Eds.). (1985). *The Social Shaping of Technology. Milton Keynes*. UK: Open University Press.

Mansell, R., & Silverstone, R. (1996). *Communication by Design. The Politics of Information and Communication Technologies*. Oxford, UK: Oxford University Press.

McLoughlin, I. (1999). *Creative Technological Change: The Shaping of Technologies and Organisations*. London: Routledge. doi:10.4324/9780203019870

Webster, F. (Ed.). (2003). *The Information Society Reader*. London: Routledge.

Wellman, B., & Haythornthwaite, C. (Eds.). (2002). *The Internet in Everyday Life*. Malden, UK: Blackwell Publishing. doi:10.1002/9780470774298

Wyatt, S., Henwood, F., Miller, N., & Senker, P. (Eds.). (2000). *Technology and In/equality: Questioning the Information Society*. London: Routledge.

KEY TERMS AND DEFINITIONS

Digital Divide: The way how access to the Internet can add to existing inequalities, namely due to lack of access to and lack of skills in internet usage.

E-Government: The delivery of government services using electronic means.

E-Inclusion: A term devised by the European Union, which refers inclusive Information and Communication Technologies and the use of ICT to achieve wider inclusion objectives.

E-Services: The integration of ICT in the delivery of services electronically. The factors that contribute to the contextual definitions of e-services are the ability of various organizations to provide access to information and services for the public via ICT, the promise of which is an enhanced service for the public.

Local Agencies: Organisations that deliver services locally. These may be local offices of national public sector organisations and regional organisations as well as local voluntary agencies.

Participant Design: Seeks to involve users in design more deeply by empowering them to propose and generate design alternatives themselves, as co-designers.

Partnerships: The configuration of public sector service providers, voluntary agencies and private sector companies to develop and deliver e-services.

Chapter 16
ICTs and Participation in Developing Cities

Alexandre Repetti
Ecole polytechnique fédérale de Lausanne EPFL, Switzerland

Jean-Claude Bolay
Ecole polytechnique fédérale de Lausanne EPFL, Switzerland

ABSTRACT

Developing cities are experiencing substantial gaps in urban planning. They are due to approaches and instruments that do not correspond to the realities of the developing city including the prevalence of informal sector and slums, urban governance problem, and few resources. Information and communication technologies (ICTs) now offer enormous possibilities to use information flows, communication, and land-use models better. ICTs offer solutions that take greater account of informal activities, enable discussions with civil society and Internet forums to take place, etc. ICTs can enhance the planning of developing cities, if conditions are right. The chapter provides a review of the situation in developing cities. It analyses the challenges and potential of using ICTs to improve urban planning. Lastly, it puts forward key conditions for the successful and relevant implementation of ICTs in order to create the best conditions for real technological added value.

INTRODUCTION

The objective of this chapter is to discuss the potential use of information and communication technologies (ICTs) in urban planning of cities outside Europe and North America that have the highest growth rates and the most chaotic processes of urbanisation.

New urban planning instruments are becoming available with the worldwide spread of ICTs. They make it possible to adopt innovative e-planning approaches, strengthen communication between urban stakeholders, and make communication available at various stages of the planning process.

At the moment, urban planning in developing cities has serious flaws and the most notable consequence of this failure is slums. In order to cope with urban planning in developing cities, it is crucial that we understand the mechanisms that contribute to these failures as well as the situation as far as informal activities and land tenure are concerned. One of the reasons for this is the lack of communi-

DOI: 10.4018/978-1-61520-929-3.ch016

cation and information exchange between urban stakeholders.

In the future, it will be essential that we are able to put forward innovative solutions for urban planning. ICTs can play a crucial role, improving communication and information flow. However, ICTs will not be equal to this challenge on their own. They must be accompanied by more extensive approaches that meet the key conditions of tackling the challenges of urban planning in developing cities.

The chapter begins with a definition of the concepts of developing cities and slums, which are the most obvious manifestation of planning failures in developing cities. It also presents the case of the city of Thiès in Senegal, which will be used as an example to illustrate the theories we develop. The second part provides more detailed background information. It presents the planning instruments that are generally used in developing cities, the failure of these instruments and the reasons behind this failure. The second part also reviews some innovative approaches that deal better with slums and other informal conditions that are a general feature of developing cities. Part III focuses on future trends. It discusses the potential use of ICTs in developing cities and the key conditions for successful ICT implementation based on experiences conducted in West Africa and Latin America. The chapter ends with our conclusions about the potential of ICTs for urban planning in developing countries.

DEFINITIONS AND CASE STUDY

Developing Cities

World population was estimated at 6.7 billion on January 1st 2009. Three out of four people live in countries with an intermediate or low Human Development Index (below 0.8) and every second person lives in a city.

Cities outside Europe and North America are experiencing a population explosion. They are known as developing cities and double in size every 25 years on average and every 15 years in some regions (UN-Habitat, 2008). Megacities, such as Mumbai, Djakarta, Sao Paulo, Cairo or Lagos are the most impressive examples of rapid urban growth, but smaller cities face even higher growth rates and are home to more people overall.

Managing this high rate of change in urban sprawl and demography is not only limited by a lack of financial and human resources. Developing cities also have to cope with seasonal residents, informal market activities, a complex land-tenure situation, uncertain real estate conditions, corruption, complex governance questions, and diverse socio-political practices. These realities result in chaotic urbanization, social disparities, limited or dilapidated infrastructure, insufficient access to basic services, and numerous governance conflicts.

Slums

Slums are the clearer evidence of the failures of urban planning in developing cities. Slums are the result of various conditions including their precarious habitat, and their lack of ownership security and infrastructure. They are the materialization in habitat of a broader phenomenon - the informal sector. It is characterised by conditions that do not obey administrative rules and the law, but which are acknowledged and generally tolerated. The co-existence of informal and formal conditions and habitat is the core feature of developing cities.

Slums are mainly a result of uncontrolled urban sprawl. With high demographic growth, residents and new migrants look for attractive accommodation. They make arrangements to settle on the outskirts of town with farmers, neighbours, villages or local dignitaries although these do not formally own the land. They build temporary shelters or real houses as their own quick solutions for urban integration, disregarding any building laws or land allocation plan. Urban authorities often tolerate these new slums as they are tangled up in urban governance conflicts and cannot provide better alternatives.

Slums are the site of various manifestation of urban poverty and marginalization, as well as a population segregation by income, land ownership, infrastructure, living standards, access to goods and services, criminal activity, corruption, environmental degradation, and financial and city management deficits. Slums are also often social melting pots and hotspots of cultural creativity.

The issue of land ownership is fundamental. Most house owners do not own the land on which they have built their house (Durand-Lasserve & Royston, 2002). In certain cases, customary forms of land occupancy exist and the plot is allocated to a family by the local community. On rare occasions, this solution is legally recognized. Generally though, land occupancy is deliberately ignored by the authorities in favour of existing administrative, financial and regulatory procedures, often based on European legislation imposed just after the colonial era (Bolay, 2006).

Poor citizens recognize the importance of infrastructure and urban services for their wellbeing, but they do not have minimum requirement to move into areas where this infrastructure is available. Rather, they develop an informal infrastructure network. They rely on their own solutions to have access to basic services at a local and often household scale. In most cases, the result is a collection of individual solutions that are either illegal or exist by informal agreement. They use the water from wells and canals, although this is not always suitable for drinking. They install homemade electricity networks to avoid paying for the power they use and thus disregard all safety precautions. Roads are of poor quality and most alleyways are only wide enough for pedestrians and motorcycle.

Thiès, Senegal

Thiès is a regional administrative centre and the third largest city in Senegal (350,000 inhabitants). The economy of the city is mainly based on industry and trade. This medium-sized city faces several classical problems of underdevelopment such as steep demographic growth (it had 200,000 inhabitants in 1990), a weak economy, lack of infrastructure, informal and unsanitary settlements, poverty and environmental degradation. However, the city is a leading regional trade centre and boasts an important network of public and private infrastructure and services.

Since the introduction of decentralization policies in Senegal (1996), urban planning has come under the responsibility of local government. The state, the regional government, local associations and NGOs play a particularly important role in various sectors of local development, some of them with the support of multinational backers. Thus land-use planning involves a wide variety of stakeholders. The reality of urban planning in Thiès provides evidence of planners' difficulties of managing a fast-growing city (Repetti & Prélaz-Droux, 2003). Some of the reasons that might explain this situation are a lack of information and technical skills to prepare decisions as well as the means to implement them, a lack of knowledge and coordination between the actors, and administrative services (technical) that are caught between the central state and the decentralized authorities and mostly limit themselves to managing their current business.

In reality, the official planning instruments are hardly ever used. A set of classic planning instruments exists, some of them inherited from the period prior to decentralization. Most do not meet the needs of the various actors and have not been fully negotiated. In 1999, a participatory forum arose following a demand from the municipal authorities for an experimental project that made use of new urban planning methods. In order to facilitate the exchange and storage of information, an ICT application has been specifically developed and made available to the main stakeholders in urban planning (elected representatives, technical services, representatives and services of the administration, one association and one NGO) since summer 2000 (Repetti & Prélaz-Droux, 2003; Repetti et al., 2006; Soutter & Repetti, 2009). The participatory forum brings together elected representatives of the City of Thiès and of the suburban communities, state representatives, administrative services responsible for local development and a couple of associations and NGOs. The forum's general objective is to strengthen the development of the urban area through information-sharing, consultation and joint decision-making.

Setting out with this objective, the SMURF instrument was integrated into the process to support appropriation of land-use and participatory planning. More specifically, the SMURF instrument aimed at improving the supply of information about land use, monitoring urban development and supporting the exchange of information between the various stakeholders involved. SMURF was integrated into the activities of the development forum, which were coordinated by the city council. The director of technical services collected the data updates, checked them for errors and presented the adjusted to the forum once or twice a year. The instrument was installed on about 30 computers in the urban area at the city hall, the regional hall, at local and regional administrative services (technical), associations and NGOs, and at a cyber café that provided the public with access to the information.

BACKGROUND

In order to understand the potential of ICTs in developing cities' planning processes, it is necessary to analyse the specificities of developing cities and the limits of planning approaches when confronted with these specificities. As planning in developing cities generally resorts to relatively classic approaches, we will review these approaches in order to understand their failures, especially the failures in communication. We will then provide an overview of innovative approaches and discuss the importance of information and communication for the implementation of new planning models that are better adapted to the situation in developing cities.

Classic Urban Planning

Adapting the results of research by Borja & Castells (1997), Sachs (1997) and ADEME (2006), we can define six main challenges to planning an urban community:

- Providing space for economic activities and habitat,
- Building the necessary urban infrastructure,
- Ensuring social integration and an equitable distribution of goods, services and other benefits,
- Ensuring environmental sustainability,
- Improving the quality of life, and
- Guaranteeing good governance.

Land-use planning must face up to these challenges through visionary strategies and plans that support an appropriate organization of infrastructure and activities on the land.

Classic urban planning foresees a long-term plan for the city in the shape of guidelines or master plans. In practice, the master plan is the result of concrete analysis of land use, activities, economy, social conditions and environmental issues. It is composed of guidelines, land-use regulations and a plan that specifies the limits of the various regulation zones. The master plan is generally completed with more detailed plans defining the guidelines, regulations and limits for smaller areas (Randolph, 2004; Berke et al., 2006).

Master plans consist of simple rules that are as mandatory and constant as possible on matters of land allocation, precise boundaries, land-use restrictions, land-use density, etc. These rules are set by law. They simultaneously define the strategic objectives and the manner of achieving them through the organization land. Master plans limit the uncertainty and control the future of land use, in order to create an overall urban project.

Classic urban planning is complemented by land-ownership procedures. The planning is done through the land-ownership register, which constitutes the legal inventory of parcels of land, land-ownership regimes, land owners, or usufructuary and land-use constraints. Land ownership rules are specific to each particular context as a result of local traditions, historical specificities and postcolonial policies.

Urban equipment and infrastructure are generally planned as the logical consequence of the master planning. The biggest infrastructural projects, however, necessitate a revision of the master plan, as they have a wider impact on the general urban project (Fritsch et al., 2008). Urban equipment and infrastructure respond to the needs, norms, and regulations of planning. They take into account land allocation and land-ownership. The funding of equipment and infrastructures is shared between public bodies, private developers, land owners, and consumer fees, as the result of the application of the law, case-by-case discussions, and funding opportunities.

Classic urban planning approaches have been tested many times. They have been validated in many European and North American cities, where informal habitat and land tenure remains marginal, even if the trend towards discriminatory and inequitable urbanization is ubiquitous in both rich and poor countries.

Developing Cities, Slums, and Informal Conditions

The urban reality is a complex pattern of human-environment interactions. In developing cities, this pattern is even more complex than in European and North American cities, due to the importance of informal realities, for which the slum is the clearer manifestation. The slum begins with poor people and informal settlements setting up. It results in precarious shelters, lack of infrastructure and disengagement of urban managers, with consequences on water supply, energy supply, waste disposal, access to schools and health centers, environmental risks, etc. Finally, the outcomes are increased health problems, illiteracy and informal urban planning that reinforce poverty (Bolay, 2006).

Certainly, the informal conditions are wider ranging than the simple question of shelter. It also includes economic activities that are not taxed or regulated according to institutional and legal norms. Informal economic activity is mainly made up of small economic channels, generated by low-level traders or small-scale activities. But it also includes larger, regular enterprises with permanent employers (ILO, 2002). This informal sector bypasses tax as well as social security and environmental regulation. In developing cities, the informal sector often involves more than half of the city's labour force; for example 51% of non-agricultural employment takes place in the informal economy in Latin America, 65% in Asia, and 72% in Sub-Saharan Africa (ILO, 2002).

Nevertheless, there are no obvious solutions since legal activities are not neatly divided from informal or illegal. In the example of Thiès, our investigations (Repetti & Prélaz-Droux, 2003; Tepe, 2005) show that farmers allocated land without any consideration for legal procedures. However, due to an obsolete master plan, it was no longer possible to allocate land following legal procedure. Therefore, informal land allocation was regarded positively by most state bodies. Employees of the service responsible for land-ownership even participated in the land allocation, albeit not in an official capacity. We encountered many similar cases in other developing cities.

Informal land tenure and poverty have consequences for all aspects of urban management.

Figure 1. Diagram of the complex human-environment interactions in a slum

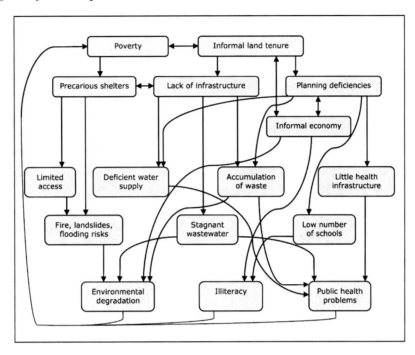

Our analysis in Thiès and other cities (Repetti & Desthieux, 2006) shows that they are at the very root of informal urban conditions, as summarized in Figure 1.

Even confronted with severe problems of applicability, we observe that in practice most cities rely on classic planning approaches. For technical, economic and ideological reasons, there is an apparent inability to look beyond the usual approaches. In the context of developing cities, most investment takes place in the formal districts of the city. Most interventions in slums are targeted at urgent problems and implementing remedial solutions linked to the existing social and technical conditions. Winarso & Mattingly (1999) demonstrate that most developing cities have a master plan, but that few are adequate and realistic. In the case of Thiès, the administration was always referring to the master plan, even if it was not updated and became obsolete. The land allocation procedures got lost in the bureaucratic system that could not deliver any formal authorization without a new master plan. Thus all urban planning referred more to what should have been done than to the actual situation.

This situation may lead to two opposite tendencies. On the one hand, the denial of the informality by urban managers results in a suppressive policy aimed at ignoring and destroying whatever infrastructure or housing is created outside official regulations and standards. This approach remains the most widespread practice, as in Thiès for instance. On the other hand, denial resulted in the adoption of alternative policies aimed at reorganizing and rehabilitating slums on the basis of what resident communities have undertaken themselves (Gaye, 1996; Mukhija, 2001).

Local, national and international policies have steadily evolved from suppressive approaches aimed at eradicating slums to an integration of the poor sectors of the population. Authorities offer support as facilitators rather than as infrastructure-builders. At best, this strategy has resulted in improved legislation, collective infrastructure and services. At worst, it has exacerbated corruption and forced the poor to become micro-entrepreneurs responsible for their own livelihoods. However desirable some aspects of this transition may be, it means that the majority of the urban poor are still living in highly vulnerable conditions (Bolay, 2006).

The Limits of Classic Planning in Developing Cities

Classic urban planning encounters limitations in developing cities, especially where there are numerous slums and informal activities.

First, analysis of land use, activities, the economy, social conditions and environmental issues is problematic, especially where there are informal activities and land occupation. It requires time, skills and a deep understanding of the socio-economic realities of the inhabitants, including slum residents that are often more or less intentionally ignored by the authorities. Therefore, master plans do not generally address conditions in slums. Often it suggests a complete redevelopment of the slums, or considers the slum as an uncertain part of the plan. Consequently, there is a gap between the planned urban project and the realities of city-dwellers. Most cities have an unrealistic or irrelevant master plan, an out-of-date land-ownership register, and do not take account of informal realities. In reality, many developing cities do not rely on master plans and land-owner registers, or only partially. Many administrative bodies prefer having out-of-date or obsolete plans rather than attempting to manage the informal setting.

Attempts have been made to propose a complete diagnosis for slums, and then developing infrastructure and land-tenure registers. Such approaches have failed almost without exception (Bolay, 2006), mostly because they do not question the urban model that generates the slums and do not address the slum residents' needs, expectations and available resources.

Second, land-ownership registers are not often up-to-date in developing cities, and many conflicts concerning owners and property rights are not solved. Classical instruments rely on the register of land ownership and land-owner requirements. Their implementation generally breaks down if there are not clear land-owner responsibilities.

In many countries, there is even a co-existence of two land tenure regimes: the traditional and the colonial regimes. They co-exist in different ways in each situation, depending on local tradition, colonial history and post-colonial policy. In Thiès (as in most parts of Africa) for instance, land is generally an inalienable collective heritage; land management is controlled by the first cultivator, land is shared between the community members according to the needs, but it belongs to everybody. In most African situations, a colonial land-ownership based on private property replaced the traditional regime, but the population didn't respond to the call to register, resulting in the superposition of two land management regimes and a great deal of confusion.

Land-ownership confusion and informal settlements are central problems in planning. Without recognized land tenure, it is impossible to impose the rules defined by the master plan. They make tax collection tricky as slum dwellers are generally not registered, avoid paying the tax and regularly do not pay for water or electricity. They make access to mortgages impossible; in informal conditions, a bank will rarely give credit without collateral, a private provider will rarely develop infrastructure without a guarantee of durability, and a public provider will rarely develop infrastructure in a zone that is not formally recognized.

Finally, observers recognize the barriers planners face in managing urban development in developing countries. Latouche (2001) such as Sudhira & Ramachandra (2009) highlight the realities of urban development, which brings with it social and environmental problems, especially exclusion, poverty and pollution. Wright (1996) stresses that urban planning is conditioned by economical rules. He demonstrates that infrastructure planning follows criteria such as financial feasibility, ratios of cost and profit, and return on investment. This urban realpolitik complicates investment in infrastructure for slums and in the informal sector, where there is no tax collection and it is difficult to measure the economic added value. De Graaf & Dewulf (2002) note an important difference between the plan and the inhabitants' real situation, as well as gaps in the planners' diagnosis. One main reason that is invoked to explain the limits of classical planning approaches in developing cities or their restriction to some central districts is the poor communication between decision-makers and urban residents, especially poor residents. It clearly appears that the quality of interactions between urban decision-makers and representatives of the various residents' factions influences the quality of any urban planning process.

Alternative Approaches Leading to Success

To address the limits of classic planning approaches, there have been some modifications and reactions in how situations are analysed over recent years. These alternative approaches are generally more dynamic and often derived from economic management principles with reinforced information and communication.

The urban reality is regarded as a vast and complex system of human-environment interactions. In a system like this, planning must address heterogeneity and complexity (Vilmin, 1999). It is thus necessary to understand the variety of socio-economic realities, needs, and expectations in order to organize planning actions (Le Galès, 1998; Bolay & Cissé, 2001). Furthermore, developing cities follow complex planning logics, with overlapping administrative, social, economic and informal forces. In these conditions, how planning is organized is the result of interactions between social, political, and economic stakeholders (Repetti & Prélaz-Droux, 2003).

At the same time, the participation of stakeholders in the planning process is being reinforced. The participation techniques include participatory forums, participatory appraisal techniques (Chambers, 1994; ICLEI, 1996), rapid diagnosis (Santandreu, 2001), public GIS (Abbot et al., 1998), research-action (Ndione et al., 1993), self-development (Gaye, 1996), societal diagnosis (Noisette, 1996), and other participatory diagnostics. This participation often brings results in more adapted and appropriated projects. It raises, however, the whole question of urban governance, of the involvement of the informal sector, and the role of the different city stakeholders.

There is also a shift in the urban planning approach from the master plan towards instruments that are more adapted to complex and unobvious realities through the integration of a strategic global scheme and local planning dynamics. It addresses urban strategic project, which corresponds to flexible tools for management and communication established at the conurbation level (Carmona & Burgess, 2001). The strategic urban project is accompanied by management tools for analysis and monitoring, as well as control of the strategic objectives (Allen, 2001; Repetti et al., 2006). In practice, this approach is increasingly used to complement the classic approach since the legal base of the master plan remains mandatory.

In addition, urban planning instruments increasingly involve powerful databases and information systems. To deal with the considerable volumes of information, there is a need for tools that generate an overview of the goals and an appropriate level of synthesis (Klosterman, 2001; Geertman & Stillwell, 2003; Geertman & Stillwell, 2009). Decision-makers and managers must be able to access all the relevant data without getting lost in details or puzzled by information that does not provide a clear picture (Repetti & Desthieux, 2006; Allen, 2001; Joerin et al., 2001). Planning support systems are used in urban management to allow more objective and rational planning.

Finally, policies have steadily evolved from repressive approaches aiming at eradicating slums and controlling migrants to policies that assimilate slum residents. From this perspective, the state, in its role as facilitator, offers services and acts as a coordinator of policies and actions in the urban area. This approach has often resulted in improved legislation, and collective infrastructure and services. However, it has also exacerbated corruption and led to the poor becoming micro-entrepreneurs who are responsible for their own livelihoods.

Urban managers do have to develop appropriate methods to cope with slums and the informal sector. They must develop truly integrated urban projects to construct roads and drainage systems, install collective services, regulate land-ownership, relocate families affected by construction projects, promote crafts and other employment, and support grassroots community organizations. They have to accept the conditions of slum residents who have organized themselves to cope with the most pressing problems prior to the launch of rehabilitation projects, and who have built schools, paths, drains, water wells, or/and electric networks. It also requires financial backers that are prepared to grant credit on a different basis, accommodating poor people's capacity to pay loans back and the possible credit guarantees of the informal sector. This is the kind of logic that is essential for sustainable urban improvement in developing cities.

FUTURE RESEARCH DIRECTIONS

Planning in developing cities is clearly multi-faceted, multi-dimensional, and touches on fundamental governance questions. The right response involves a variety of approaches that include a range of aspects, with a reinforced communication that associates the representatives of all urban residents, and at different scales: intra-urban, peri-urban, regional. It must include coordinated contribution by specific public, private and community-based stakeholders, taking account of the expertise of professional groups such as planners, engineers, architects, economists, social scientists, and administrators.

One important line of improvement of urban planning involves ICTs. As stated in the UN Global Report on Human Settlements (Un-Habitat, 2007), improvement strategies related to rapid urban development can be achieved through improved access to information, with the involvement - and possibly the empowerment - of slum dwellers. Several authors (Bell, 1986; Boon, 1992; Camble, 1994; Castells, 1996; Sturges & Neill, 1998; Chapman & Slaymaker, 2002) have referred to situations where information was a key condition for success or where the lack of information had a negative impact on urban planning. Our experience in Thiès and other cities demonstrated that communication is a central aspect in planning. When the dynamics of planning are lead by informal stakeholders rather than by official planners, the communication among these stakeholders is a condition for a coherent urban vision to emerge (Repetti & Prelaz-Droux, 2003).

ICTs in Developing Cities

ICTs are instruments and technologies that have been developed through the combination of telecommunication, computer sciences and audiovisual applications. They are unequally distributed around the world and this creates an imbalance in access to information, to technology, and to infrastructure.

The cell phone is certainly the most successful application of ICTs in developing cities. Due to the lack of landline telephone infrastructure, mobile phones represent a good alternative. According to the International Telecommunication Union (ITU), the number of mobile phone subscriptions in Africa rose from 15 to 274 million between 2000 and 2007, passing from 1.9 to 28.4 mobile phone subscriptions per 100 inhabitants. In 2007, Asia had 37.6 mobile phone subscriptions per 100 inhabitants and Latin America 66.7. According to the GSM Association, networks in Africa covered

10% of the population in 2000 and 60% in 2007. In addition, users in developing countries have invented a variety of ways to share cell phones among several users. For instance, 97% of all Tanzanians say they can access a mobile phone, whereas there are only 20.6 subscriptions for 100 inhabitants.

Internet has had greater difficulty establishing itself in developing cities. In 2007, 5.5% of the African population was Internet users, 14.4% in Asia and 26.1% in Latin America, while this reached 70% to 80% in Western Europe and the US. In 2007, Senegal reported one computer, 0.3 Internet connections and 6.6 Internet users per 100 inhabitants. The imbalance is significant between rich and poor, between literate and illiterate, and between cities and rural villages.

Facts and figures highlight the spread of ICTs, but also the gap between rich and poor countries and neighbourhoods.

With the success of cell phones, Internet and other ICTs, we could assume that developing cities have the opportunity to adapt the technology to their specific needs, to propose new technologic components and new Internet or GSM applications. However, the reality appears somewhat different. Developing cities almost only use standard technology and make products under patents registered in industrial countries.

Developing cities are currently facing enormous challenges including an explosion of their population, limited resources, poverty, and governance problems. In such conditions, ICTs are extremely promising, but must take account of the specific difficulties if they are to be implemented with success and add value.

The worldwide information society is a reality that cannot be ignored. It is accompanied by a dynamic and growing telecommunication economy. Furthermore, the appropriation and impact of ICTs are largely determined by the specific social and political context. As information control is a form of power, it inevitably interferes with the expected results of ICT development projects. Thus, the question is not to decide if ICTs are good, but rather how to make optimum use of ICTs in a given context.

ICTs are powerful tools that are efficient when they are used well. In this way, appropriate ICT applications have been developed in the fields of health or education to reduce poverty. ICTs can have a positive impact on spreading democracy and on economic development. But more than anything, the objectives of ICTs resemble the objectives of the media - to allow access to information, to generate knowledge, and to promote mass communication. Used well, ICTs are an excellent tool to support knowledge acquisition, analysis, decision-making and social discussion. Like any communication tool, ICTs are double-edged and can have severe social repercussions when access is withheld (exclusion and isolation), in other words, they are sometimes unexpected and not always positive.

What Role for ICTs in Developing Cities Planning?

We have demonstrated that the sharing of information among decision-makers and stakeholders is a key factor in increasing the efficiency of developing cities management. ICTs today are powerful tools that can potentially improve data storage, structuring, consultation and publication. Geographic information systems (GIS) and planning support systems (PSS) have confirmed their potential to support and improve urban planning processes. New instruments are available for planning such as agent-based modelling (An et al., 2005), Google maps (Gibin et al., 2009), new GIS modules, GSM applications, and many others. Yet, there are still too few ICT success stories in the planning of developing cities.

One challenge is to adapt technology to a particular context. ICT applications must be adapted to the specific nature of dealing with the informal sector such as a lower level of IT infrastructure and skills, the difficulty experts, decision-makers and social representatives have in handling the instruments, and the need for them to be part of an integrated approach to urban planning. According to Klosterman (2001) and Geertman & Stillwell (2003), support systems for urban planning are on the whole not fulfilling the expectations of planners and managers and are still underused, especially in developing cities.

One other challenge is to reinforce the link between technology and planning needs. Chambers (1997) raised the concern that contents are often secondary in information-based development projects since priority is given to the technology and information infrastructure. Moreover, information is often introduced only into institutional environments, which favours politically or individually driven decisions instead of decisions based on diagnosis and facts (Heek, 2003).

If these challenges can be overcome, experience shows that the provision of adequate information and information-handling capacities contribute to more informed decision-making and can enhance local communities' involvement in planning. Castells (1996) proposes that good management of metropolitan areas in the information age will depend on the capacity for overall management and coordination. Borja & Castells (1997), as well as Berghäll & Koncitz (1997) demonstrate the effectiveness of new combinations of greater information, new instruments, and participation. Such instruments can help to overcome problems, improve the allocation of resources, and mitigate some of the negative impacts of poorly managed developing cities.

The first role for ICTs in planning consists of supporting data storage as well as spatial or statistical treatment (Klosterman, 2001; Geertman & Stillwell, 2003). The design of the data to be stored is crucial. Too much data leads to information overload and reduces the readability of the system. But a lack of vital data limits ICTs' potential contribution to decision-making and diagnosis. Depending of the objectives, the planning priorities, strengths and weaknesses of a particular city and other factors, the set of data must be well-balanced and adapted to the needs.

The second role for ICTs consists of monitoring urban development and controlling the planning scenarios. Urban planning in developing cities is an open and dynamic process, influenced by numerous internal and external constraints and by uncertainty. ICTs can analyze and monitor urban development through urban indicators, strategic objectives, and comparison with other cities. It can evaluate planning scenarios or planning measures to check that the developments fulfil the objectives and expectations. For example, indicators for the effectiveness of planning prove the value of the urban planning assessment in Bangalore (Sudhira & Ramachandra, 2009).

The third role for ICTs is linked to models. Models are used to support planning in many sectors of urban development such as mobility, water management, habitat, pollution, etc. ICTs have enormous potential to produce the required data and run the models. For example, integrated land-use and transport modelling for Caracas (Venezuela) highlighted the impact of various planning scenarios on traffic flows (De la Barra, 2001).

The fourth role for ICTs lies in exchanges between urban planning stakeholders. This role is special in developing cities because there are many stakeholders - both formal and informal - carrying out urban development actions. ICT-based information exchange must reach the many stakeholders involved in planning affairs and support their participation. This type of information forum is able to ensure that comprehensive information is available. Furthermore, it can support the exchange of ideas, constraints and needs, thereby developing a common ideal that binds stakeholders together.

The fifth role consists of providing the public with information.

Information about the project and opportunities to influence the project design are the starting points for genuine stakeholder involvement. However, their support for the project will ultimately be a question of costs and benefits, the degree of involvement in the project planning process, whether they are considered as social partners, and overall urban governance. Regarding the involvement of urban citizen, we can refer to Turner (1976) and state once more that the participatory process is important because citizens can control the decisions, contribute to the design and administration of their city and will be more likely to accept conditions that come about as a result of their decisions. The importance of the project is not what it delivers in material terms, but its overall impact on citizens' lives.

ICTs in Practice

Our experience in Thiès illustrates the positive role that ICTs can play in the planning processes of a developing city. The SMURF instrument (Repetti et al., 2006) has been developed for the needs of managers of medium-sized African cities and follows a complete methodological approach (diagnosis, instrument development, instrument evaluation, and impact assessment). The instrument does not automatically try to identify the best planning scenario, but aims to reinforce the stakeholders and their interrelationships through various supplementary sub-objectives, e.g. improving knowledge of land and development projects, serving as a data-exchange platform, instrumentalizing participatory and collaborative structures through cartographic support, monitoring changes in land-use using indicators, and informing and consulting the public. The instrument does not try to revolutionize the stakeholder system by dogmatically imposing participation. It aims to get public authorities, land technicians, business representatives, and citizens' rights defenders to coordinate their actions as much as possible through a formalized space for exchange.

For example, the comparison of the city of Thiès with other West African cities (Repetti & Desthieux, 2006) revealed that Thiès had high health infrastructure coverage (health index 70%) and poor education coverage (education index 20%) when comparing with other cities with similar UN-Habitat indexes and indicators. Further, GIS analysis that collated school statistics with demography showed that some schools had to accommodate more than 80 pupils per classroom whereas other schools in other districts had less than 40. By providing such information, ICTs can play a crucial role in identifying strategic projects and evaluating their potential benefits.

The use of the instrument in Thiès led to improved knowledge and better coordination between decision-makers and stakeholders. Participation is used as a way of gathering data, which is made possible as long as users perceive the added value of using and maintaining the proposed technology. The field studies show that even if all projects and scenarios are not included in the technologic instrument for evaluation, each decision-maker and stakeholder will take the common diagnosis and strategies into account during his own particular planning process.

In contrast to existing projects in this field, the proposed approach is focusing on building consensual information and strategic objectives rather than on evaluating scenarios. In our experiments, participation is used to constitute a knowledge base that supports decision-making and consensus building. The process results in a better level of knowledge for all participants and in a strong consensus around the diagnosis of the actual situation and the strategic objectives of local development. The role of the forum is central as some information cannot be shared through a database.

We overcame the technology challenges by means of an ICT application that had been specifically developed for a developing city. The instruments had a positive impact on planning capacities such as the elaboration of a strategic plan and the exchange of information between stakeholders. The Thiès experience clearly demonstrates the potential of ICTs to contribute to the sharing of information about land-use.

Thiès is one experiment among many others that tries to develop up-to-date technologies for urban planning. It demonstrated that ICTs have enormous potential to support planning in developing cities, but also that the challenge of planning in developing cities goes beyond simply developing powerful technologies.

A Global Negotiation That Includes the Informal Sector

As has been described above, developing cities face many challenges including a high number of seasonal inhabitants, slums, informal activities, complex land-tenure systems, real-estate speculation, corruption, complex governance regimes, and various socio-political legacies. Too many projects have failed because they did not take sufficient account of these complex formal and informal processes. Thus, it is essential to accompany the technological side of projects with an integrated approach that will help representatives of all interest groups to take part in the discussion.

The integrated approach aims at including marginalized social groups in city planning through representation. It also aims to gain a better understanding of social realities in the city and thereby contribute to the project's objectives and feasibility. One symptomatic example is when scientific experts were asked to implement a presidential decision to supply subsidized housing for the poorest urban residents in Bolivia (Bolay, 1998). The conclusion was that it was necessary to find financial backers who would be ready to supply credit on a different basis by adapting loans to residents' ability to repay them and to how quickly they would be able to complete the construction work. This required a complete change in perspective.

This approach also includes organizing the participation and representation of the various social groups, and formalizing this process in the form of a contract. It then involves negotiating the project conditions in detail including pre-payment, tax collection or the charge for a given service.

There are still very few ICT applications in developing cities that help to involve the local population or other social representatives in a project process; this is due to limited ICT access, illiteracy, cost and other issues. Participatory GIS or planning support systems are often far too difficult for people to access and are thus reserved for experts. Many cities have elaborate websites that provide no space for constructive debate about urban affairs. Still, ICTs are potentially an appropriate medium to organize the participatory presentation of a project or a debate with local people or their representatives.

Designing Specific ICT Applications for Developing Cities

Enabling ICTs in developing cities must overcome limited infrastructure, incomplete data about the informal sector, limited resources, and considerable disparities in education. In such conditions, instruments that were initially designed for European or US cities are not always transferable.

In developing cities, ICT applications must primarily overcome a lack of infrastructure and data. The largest cities generally provide high speed Internet access for the administration and political authorities, but smaller cities may not have any connection or only a slow analogue one. A demand for Internet access has led to the spread of private connections and cyber cafés, even in slum areas. Today, the quality of services, computer skills, level of literacy and income is correlated to social segregation; most cities still have a digital divide between their richest and poorest areas.

ICT use must also cope with uneven data quality. Good quality data may be available for some districts or sectors, whereas for others no information is available. The information is often out-of-date or only partially accurate. In slums and in the informal sector, civil society,

NGOs and associations often take over from the administration, but their data is not centralized. Additionally, some data reflects informal situations and cannot be combined with other official data. For example, in Thiès (Senegal), land plots were allocated "by night" with the silent agreement of all parties including the administration, but this was not entered in the land-ownership register, nor did it give rise to a formal title deed. ICTs must cope with this type of conditions and therefore mesa-information is essential. Technology will have to cope with stakeholders who do not have transparent information. ICT applications must also adapt to their users. Authorities and social representatives generally don't have any computing or specific planning skills. This means that specific visualization and monitoring tools must be developed.

Access and Training

Finally, the success of ICTs depends on accessibility. Infrastructure development is a priority for the International Telecommunication Union and private companies. It includes basic infrastructure, connection quality, data flows, and accessibility.

With constant improvements in infrastructure, education and training become increasingly important factor to ensure that people in developing cities can make the best of being digitally connected.

There are many public and private initiatives that promote better access and training. There is still a demand for applications and initiatives that will incite poor urban dwellers to access the Internet.

Beyond ICTs

From a critical perspective, ICTs will not solve all the problems of planning in developing cities on their own. It is therefore essential to make progress on other fronts at the same time.

Firstly, there is an urgent need to develop concerted policies aimed at combating slums and unequal urban development in general. Political authorities must tackle the problem and put forward long-term strategies. This includes mobilizing all stakeholders, especially the slum dwellers and their representatives.

Secondly, there is a need for a change in administrative practice. Instead of expecting inhabitants to adapt to inappropriate bureaucratic practices that encourage social exclusion and corruption, alternatives must propose new approaches that are appropriate to the situation.

Thirdly, the growing informal sector should be redesigned so that it is a driving force for development that is both socially and environmentally beneficial. This requires structural economic change so that these activities can be controlled and regulated.

Fourthly, the credit and mortgage system is an essential factor in the production of slums. There is therefore a need for alternative forms of credit that are socially and financially appropriate.

CONCLUSION

There is great potential to make ICTs available that can be used in urban planning and they are now spreading quickly. In the particular context of developing cities, which are experiencing a major crisis, powerful ICTs offer a real opportunity to develop new planning approaches that are adapted to the complexity and constraints of such situations. The potential of ICTs depends primarily on knowledge support, by proposing relevant information on land use and activities. It also involves communication interfaces that allow a shared system of information, the development of a common spatial language, and for this information to be shared between stakeholders. It then moves on to support systems for planning and decision-making that includes specific information-processing modules to make analysis easier. Finally, it requires new exchange spaces, real and virtual, to encourage the exchange of information between urban stakeholders, decision-makers and the users of a city's facilities, no matter if the stakeholders are implied through political, technical, economic or associative functions.

In the context of globalization and creation of a worldwide information society, ICTs are an attractive for everyone. The number of Internet users in developing countries increased tenfold over the last five years and almost every city in the world has cell phone coverage. This boom in technology will not solve every development problem. The effectiveness of technological instruments, however powerful they may be, remains dependent on how well they are used. Despite their potential benefits for urban planning and some negative or unexpected experiences, the question is however not if ICTs can be relevant for developing cities but how we should make best use of these technologies to satisfy the needs of managers and the expectations of citizens.

The implementation constraints of ICTs are primarily technical - cost, accessibility, a focus on needs - and secondly methodological, - adapting them to the local situation, encouraging effective participation, genuine transparency in terms of urban governance principles. Finally, the impact of ICTs remains dependent on urban development policies. No instrument is better than the local policies that use it.

REFERENCES

Abbot, J., Chambers, R., Dunn, C., Harris, T., de Merode, E., & Porter, G. (1998). Participatory GIS: opportunity or oxymoron? *IIED PLA Notes, 33*, 27–33.

ADEME. (2006). *Réussir un projet d'urbanisme durable*. Paris: Le Moniteur.

Allen, E. (2001). INDEX: software for community indicators. In Brail, K. R., & Klosterman, R. E. (Eds.), *Planning support systems* (pp. 229–261). Redlands, CA: ESRI Press.

An, L., Lindsernman, M., Qi, J., Shortridge, A., & Lui, J. (2005). Exploring complexity in a human-environment system: an agent based spatial model for multidisciplinary and multiscale integration. *Annals of the Association of American Geographers. Association of American Geographers, 95*(1), 54–79. doi:10.1111/j.1467-8306.2005.00450.x

Bell, S. (1986). Information systems planning and operation in less developed countries, part 1: planning and operational concerns. *Journal of Information Science, 12*(5), 231–245. doi:10.1177/016555158601200503

Berghäll, E., & Koncitz, J. (1997). Urbanisation and sustainability. In OECD (Ed.), Sustainable development OECD policy approaches for the 21st Century (pp. 117-127). Paris: OECD Publications.

Berke, P. R., Godschalk, D. R., & Kaiser, E. J. (2006). *Urban land use planning*. Urbana, IL: University of Illinois Press.

Bolay, J.-C. (1998). Habitat des pauvres en Amérique Latine. In Rossel, P., Bassand, M., & Roy, M.-A. (Eds.), *Au-delà du laboratoire, les nouvelles technologies à l'épreuve de l'usage*. Lausanne, France: PPUR.

Bolay, J.-C. (2006). Slums and urban development: Questions on society and globalization. *European Journal of Development Research, 18*(2), 284–298. doi:10.1080/09578810600709492

Bolay, J.-C., & Cissé, G. (2001). Urban environmental management: new tools for urban players. In KFPE (Ed.), Enhancing research capacity in developing and transition countries (pp. 175-182). Bern, Switzerland: Geographica Bernensia.

Boon, J. A. (1992). Information and development: some reasons for failure. *The Information Society, 8*(3), 227–241.

Borja, J., & Castells, M. (1997). *Local and global: management of cities in the information age*. London: Earthscan.

Camble, E. (1994). The information environment of rural development workers in Borno State, Nigeria. *African Journal of Library Archives and Information Sciences, 4*(2), 99–106.

Carmona, M., & Burgess, R. (2001). *Strategic planning and urban projets*. Delft, The Netherlands: Delft University Press.

Castells, M. (1996). *The rise of the network society*. Oxford, UK: Blackwell Publishers.

Chambers, R. (1994). Participatory rural appraisal: challenges, potentials and paradigm. *World Development, 22*(10), 1437–1454. doi:10.1016/0305-750X(94)90030-2

Chambers, R. (1997). *Whose reality counts: putting the first last*. London: ITDG Publishing.

Chapman, R., & Slaymaker, T. (2002). *ICTs and rural development: review of literature, current interventions and opportunities for action*. ODI Working Paper 192. London: Overseas Development Institute.

De Graaf, R., & Dewulf, G. (2002). Interactive urban planning, hype or reality. In *3rd International Conference on Decision Making in Urban and Civil Engineering*, London.

De la Barra, T. (2001). Integrated land us and transport modeling: the Tranus experience. In Brail, R. K., & Klosterman, R. E. (Eds.), *Planning support systems* (pp. 129–156). Redlands, CA: ESRI Press.

Durand-Lasserve, A., & Royston, L. (2002). *Holding their ground. Secure land tenure for the urban poor in developing countries*. London: Earthscan.

Fritsch, M., Repetti, A., Vuillerat, C.-A., & Schmid, G. (2008). Integrated and participatory land management. In *11th Interpraevent Conference*, Dornbirn.

Gaye, M. (1996). *Entrepreneurial cities*. Dakar, Senegal: ENDA.

Geertman, S., & Stillwell, J. (2003). Planning support systems: an introduction. In Geertman, S., & Stillwell, J. (Eds.), *Planning support systems in practice* (pp. 3–22). London: Springer.

Geertman, S., & Stillwell, J. (Eds.). (2009). *Planning support systems best practices and new methods*. London: Springer.

Gibin, M., Mateos, P., Petersen, J., & Atkinson, P. (2009). Google Maps Mashups for local public health service planning. In Geertman, S., & Stillwell, J. (Eds.), *Planning support systems best practices and new methods* (pp. 227–241). London: Springer. doi:10.1007/978-1-4020-8952-7_12

Heeks, R. (2003). E-government in Africa: promise and practice. *Information Policy, 7*(2-3), 97–114.

ICLEI. (1996). *The Local agenda 21 planning guide*. Toronto, Canada: ICLEI.

ILO. (2002). *Women and men in the informal economy: a statistical picture*. Geneva, Switzerland: International Labor Organization.

Joerin, F., Rey, M. C., Desthieux, G., & Nembrini, A. (2001). Information et participation pour l'aménagement du territoire. *Revue Internationale de Géomatique, 11*(3-4), 309–332.

Klosterman, R. E. (2001). Planning support systems: a new perspective on computer-aided planning. In Brail, R. K., & Klosterman, R. E. (Eds.), *Planning support systems* (pp. 1–23). Redlands, CA: ESRI Press.

Latouche, S. (2001 May). En finir une fois pour toute avec le développement. *Le Monde Diplomatique*, 6-7.

Le Galès, P. (1988). Regulation, territory and governance. *International Journal of Urban and Regional Research, 22*(3), 482–506. doi:10.1111/1468-2427.00153

Mukhija, V. (2001). Enabling slum redevelopment in Mumbai: Policy paradox in practice. *Housing Studies, 18*(4), 213–222.

Ndione, E., de Leener, P., Jacolin, P., Perier, J.-P., & Ndiaye, M. (1993). *La ressource humaine, avenir des terroirs*. Paris: Karthala.

Noisette, P. (1996). Le marketing urbain: outils du MT. In Decoutère, S., Ruegg, J., & Joye, D. (Eds.), *Le partenariat public-privé* (pp. 261–281). Lausanne, France: PPUR.

Randolph, J. (2004). *Environmental land-use planning and management*. Washington, DC: Island Press.

Repetti, A., & Desthieux, G. (2006). A relational indicatorset model for urban land-use planning and management: methodological approach and application in two case studies. *Landscape and Urban Planning, 77*, 196–215. doi:10.1016/j.landurbplan.2005.02.006

Repetti, A., & Prélaz-Droux, R. (2003). An urban monitor as support for a participative management of developing cities. *Habitat International, 27*, 653–667. doi:10.1016/S0197-3975(03)00010-9

Repetti, A., Soutter, M., & Musy, A. (2006). Introducing SMURF: a software system for monitoring urban functionalities. *Computers, Environment and Urban Systems, 30*, 686–707. doi:10.1016/j.compenvurbsys.2005.06.001

Sachs, I. (1997). *L'écodéveloppement: stratégie pour le XXie siècle*. Paris: Syros.

Santandreu, A. (2001). Rapid visual diagnosis, a rapid, low cost, participatory methodology applied in Montevideo. *Urban Agriculture Magazine, 5*, 13–14.

Soutter, M., & Repetti, A. (2009). Land Management with the SMURF Planning Support System. In Geertman, S., & Stillwell, J. (Eds.), *Planning support systems best practices and new methods* (pp. 369–388). London: Springer. doi:10.1007/978-1-4020-8952-7_18

Sturges, P., & Neill, R. (1998). *The quiet struggle: information and libraries for people of Africa*. London: Mansell.

Sudhira, H. S., & Ramachandra, T. V. (2009). A spatial planning support system for managing Bangalore's urban sprawl. In Geertman, S., & Stillwell, J. (Eds.), *Planning support systems best practices and new methods* (pp. 175–190). London: Springer. doi:10.1007/978-1-4020-8952-7_9

Tepe, I. (2005). *Le lotissement à la périphérie de Thiès*. Dakar, Senegal: Ecocité.

Turner, J. F. C. (1976). *Housing by people*. New York: Pantheon Books.

UN-Habitat. (2007). *Global report on human settlements 2007: enhancing urban safety and security*. London: Earthscan.

UN-Habitat. (2008). *State of the world cities 2008-2009*. Nairobi, Kenya: UN-Habitat.

Vilmin, T. (1999). *L'aménagement urbain en France: une approche systémique*. Paris: CERTU.

Winarso, H., & Mattingly, M. (1999). *Local participation in Indonesia's urban infrastructure investment programming: sustainability through local government involvement?* Bandung, Indonesia: Bandung Institute of Technology.

Wright, D. W. (1996). Infrastructure planning and sustainable development. *Journal of Urban Planning and Development, 122*(4), 111–117. doi:10.1061/(ASCE)0733-9488(1996)122:4(111)

ADDITIONAL READING

Borja, J., & Castells, M. (1997). *Local and global: management of cities in the information age*. London: Earthscan.

Brail, R. K., & Klosterman, R. E. (2001). *Planning support systems*. Redlands, CA: ESRI Press.

Castells, M. (1996). *The rise of the network society*. Oxford, UK: Blackwell Publishers.

Geertman, S., & Stillwell, J. (2003). *Planning support systems in practice*. London: Springer.

Geertman, S., & Stillwell, J. (2009). *Planning support systems best practices and new methods*. London: Springer.

KEY TERMS AND DEFINITIONS

Classic Urban Planning: Urban planning based on a master plan composed of guidelines, land-use regulations and a plan that specifies the limits of the various regulation zones.

Developing Cities: Cities outside of Europe and North America, where the growth rate is the highest and the urbanization the most chaotic.

Informality: Organization and administration mode that do not follow the administrative rules and the law, but that is generally recognized and tolerated.

Slums: Poor neighborhoods in developing cities characterized by precarious habitat, lack of ownership security and lack of infrastructure.

Urban Governance: Systemic interrelations resulting from the balance in power and liability between the city authorities, the central government, the local stakeholders and other governmental or non-governmental stakeholders, and their capacity to effectively manage the urban affairs with recourse to participatory mechanisms.

Chapter 17
Public Participation in E-Government:
Some Questions about Social Inclusion in the Singapore Model

Scott Baum
Griffith University, Australia

Arun Mahizhnan
National University of Singapore, Singapore

ABSTRACT

Singapore's E-government model is considered to be among the best in the world. Over the past decade the Singapore government has constantly developed and re-developed its on-line presence. International comparisons have consistently rated Singapore as one of the most advanced E-government nations. However, despite significant progress towards full E-government maturity, some issues of full public participation remain. It is these issues which this chapter discusses. In particular, it will consider the ways in which a digital divide within the Singapore model has emerged, despite specific policies to address such a problem.

INTRODUCTION

A trend towards reforming the public sector has emerged in many countries in recent years spurred, primarily by the aspirations of citizens around the world, who are placing new demands on governments. The success of government leaders is increasingly being measured by the benefits they are creating for their constituents, namely, the private sector, citizens and communities. These 'clients' of government demand top performance and efficiency, proper accountability and public trust, and a renewed focus on delivering better service and results. (United Nations 2008, p. xii).

Technology is energizing grassroots politics of all stripes: call it powering up. (Alex Perry 2001).

The powering up of grassroots organizations and local citizenry has been an enduring though not always an intended aspect of what is called the ICT revolution. Where once politics at all levels was about merely electing officials and leaving it to them to do what is best for the country until the next election, the contemporary political landscape in many countries is now a two-way street whereby communication and consultation between the electorate and the elected is an ongoing process. The ability of local citizens to provide feedback to those who are elected to lead and to deal with government bureaucracy in more efficient ways is now greater than before. The revolution that is leading to this transformation of government systems is electronic government or E-government.

E-government refers to the use of ICTs such as the internet and mobile phone as a platform for exchanging information, providing services and transacting with citizens, businesses, and other arms of government. The more common type of E-government model focuses on providing easy access to citizen centered services and generating efficiencies in government administration. However, it is widely acknowledged that a mature and robust E-government is not simply technologizing the business of government. Rather 'it is about government harnessing IT to redefine its social technologies in order to remain relevant in a more participative, more interactive and more informational era' (Allen et al. 2001, 94). Moreover, as noted by advocates of E-government, developing a successful E-government sector is associated with a range of beneficial outcomes including the potential to foster strong and robust political debate, enhanced civil society and strengthened relations between citizens and those who govern (Martin and Byrne 2003).

Following the increasing roll-out of E-government programs there has been a commensurate increase in academic debate and research focusing on a range of aspects of E-government. While those on the supply side have focused on issues of the supposed cost savings, increased efficiency and improved public face of government the focus of those dealing with the emergence of E-government from the consumer's point of view have tended to focus on the impacts across a range of social areas (Abbott 2001, Silcock 2001, Bains, 2002, Van Der Meer and Van Winden 2003, Jho 2005).

From a social inclusion paradigm, it is questions of accessibility that are of most importance. Accessibility issues relate not only to the degree to which the required hardware is available across society, but also the extent to which potential users have the capability to access and understand online content and services. This relates to the widely discussed digital divide and has serious implications if only certain groups in society are able to access online services and information (Silcock 2001). Interest in the digital divide grew in prominence during the mid-1990s and today it remains an important component of public policy debate and encompasses a range of social, economic and political factors (Helbig et al. 2009, Warschauer 2003). Often discussion about the digital divide concentrates on the interaction between individuals, technology and society and tends to present on a technological determinist argument. From this point of view the argument is that once on-line there is no gap and that everyone can utilize the internet and benefit from the information society. In terms of broad social inclusion goals it is a broader multi-dimensional view of the digital divide that is needed. This broader focus looks at not only the technological questions but also questions the extent to which everyone can utilize E-government content once on-line (DiMaggio and Hargittai 2001)

For a start, the provision of a basic level of telecommunications infrastructure is an important step in the E-government process as inadequate provision may well hinder the widespread adoption of E-government services (Graham 1998, Healey and Baker Consultants 2001). The success of E-government will also depend significantly on the extent to which online content is usable, relevant and up-to-date. Potential users will be turned away if online content does not meet their needs and if information is frequently out of date. Without wide acceptance and usage by the public, the potential for a growing digital divide within the citizenry is high. Furthermore, in addition to access to hardware and infrastructure, if a broader concept of the digital divide is considered, then a complex collection of factors need to be accounted for (Warschauer 2003). Social inclusion is one such factor.

It is within the context of understanding the social inclusion issues surrounding E-government that this chapter is set. It aims to make a contribution to the growing literature dealing with the development of E-government and its broad social implications by considering the development of E-government in the Republic of Singapore. In what follows the chapter first considers the broader development of an E-government presence in the Singapore context and the extent to which local citizens are accessing E-government content. The chapter then examines the issue of social inclusion and the Singapore model before considering future trends and conclusions.

BACKGROUND

E-Government in Singapore: The Genesis of a Revolution

On a global scale, Singapore is among the leaders in the adoption of information and communications technologies and ICTs have now become an integral part of most facets of daily life in the country. The E-government project undertaken by Singapore could be seen as the leading edge of an ICT revolution that has been sweeping across the country. Unlike many other governments which tend to react to changes in the operating environment, the Singapore government typically initiates changes from within, as a preemptive measure to prepare for the future. The E-government initiatives are clearly in this category of forward planning measures.

Long before the term 'E-government' came into common usage, the Singapore government had set up a special body – the National Computer Board (NCB) – in 1981 to spearhead Singapore's entry into the Information Age. NCB's stated mission was "to drive Singapore to excel in the information age by exploiting IT extensively to enhance our economic competitiveness and quality of life." Thus began a series of transformational initiatives that collectively could be called an ICT revolution in Singapore. Among these visionary blueprints for ICT development, special attention was directed at the government's own transformation into E-government. This chapter examines those specific initiatives that pertain to the development of the E-government in Singapore.

The computerization of the government was at first introduced in stages and in limited arenas. This was in a way part of the "start small, ramp up fast" philosophy of the government in initiating new and unfamiliar technologies. However, once a certain level of confidence and competence was reached, the government accelerated its innovations on a wide scale. The first such "whole of government" approach was taken with the introduction of the E-government Action Plan 1 (eGAP1) in 2000.

eGAP1 identified three critical and fundamental relationships as its target: Government to Citizens; Government to Business; and Government to Employee. Based on these relationships, the government set out five strategic thrusts:

- Reinventing government through continuous rethinking of all aspects of government to explore the nature and quality of government interactions with its citizens, businesses and employees.
- Delivering integrated electronic services centered on customers' needs.
- Being proactive and responsive by adopting a 'sense and respond' approach.
- Using ICT to build new capabilities and capacities for achieving quantum leaps in service delivery.
- Innovating with ICT by embracing enterprise and experimentation.

It is interesting to note that these statements by a government includes words and phrases that are uncharacteristic of bureaucracies: the notion of a customer instead of a member of the public; sense and respond instead of waiting for persistent and categorical requests; quantum leaps instead of step by step; and experimentation instead of certainty of outcome. The lexicon of Singapore's E-government already suggested a sea change in the bureaucracy's way of thinking. The effect of this mindset change was felt in the implementation of the plan.

Two examples illustrate how far the government has succeeded in bringing citizens into its e-fold. Singaporeans can now interact online with the Government 24 hours a day, seven days a week. The eCitizen Portal, www.ecitizen.gov.sg, provides a single access point to government information and services. An indication of the reception to this new service is that hit rates for this portal increased from 240,000 per month in 2001 to 24 million hits per month in 2004. All Singapore residents above 15 are eligible for the Singapore Personal Access (SingPass). It is a nation-wide personal authentication system for accessing government for e-services.

With the successful completion of eGAPI, the government launched its second plan in 2003 (eGAPII) with the objective of achieving three higher level goals: Delighted Customers, Connected Citizens and Networked Government. Again, the emphasis on getting citizens to use e-services more widely and more easily and to engage citizens in consultation on government matters indicates a strong desire towards social inclusion.

In the wake of eGAPII, the Singapore government released its third and most ambitious plan yet – the iGov2010. This 5-year plan, stretching between 2006 and 2010, is the most crucial in moving towards a fully integrated model of E-government. As with previous strategic plans, the iGOV2010 plan aims to fulfill several goals including: increasing the reach and richness of e-services (improving the quality of e-services); increase the citizen's mindshare in e-engagement (improve online information and encourage greater participation); enhancing capacity and synergy in government (improve the internal running of the bureaucracy); and enhancing national competitive advantage (transform industry sectors and foster a pro-business environment) (iGOV2010 Project Steering Committee 2006). Among other things iGOV2010 aims to move towards full E-government maturity whereby agencies act as seamless entities, information is shared and citizens are increasingly active in the business of government.

E-Government for Citizens

While the Singapore E-government project involves portals for business (Government-to-Business or G2B) and internal portals for the bureaucracy (Government-to-Employees or G2E), it is the citizen portal which is most important for the issue of social inclusion. An underlying focus of Singapore's E-government framework is the acknowledgment of interactions that exist between the government and citizens and the potential for these to be improved and streamlined under a well developed E-government. In short, the government recognizes the need to make the business of government more citizen-centric. It is ironic that in many cases politicians who are elected and bureaucrats who are paid to "serve" the people often end up as civil masters instead of being civil servants. They place their own administrative convenience above citizens' needs. But the concept of citizens as paying customers, much as in the business world, tries to balance this relationship in favor of the citizen.

In its attempt to cater to the wide-ranging needs of the citizen, the Singapore Government has taken the approach of delineating key domains of the ordinary citizens' life and building information and service clusters for them. The eCitizen portal provides one-stop online information and

Figure 1. Singapore e-citizen web portal (Source: www.ecitizen.gov.sg)

services that are intuitively grouped along those domains. The eCitizen Centre is currently home to a significant array of services and the further development of mobile technology has meant that E-government services can now be accessed using handheld devices, thereby significantly increasing accessibility. The services available are categorized into 7 towns which cater for various aspects of social and economic life:

- Culture, recreation and sports;
- Defence and security;
- Education, learning and employment;
- Family and community development;
- Health and environment;
- Housing; and
- Transport ad travel (see figure 1).

Citizens can pay taxes and fines, apply for licenses, obtain information relating to community facilities and download government publications through the online portal. Using mobile technology, citizens can check traffic conditions, pay traffic related payments such as fines, receive notification of statistical releases and receive information on health and well-being. In addition, for those unfamiliar with the new technologies, the e-citizen portal provides assistance to use online services through links to citizen-connect centers and e-citizen helper outlets. Users are also encouraged to personalize their e-citizen experience by configuring their personal homepage with alerts regarding payments and renewals.

In its effort to engage its citizens and enable them to interact actively, the Singapore government has opened up the Internet as one of the channels available for public feedback. The E-government Action Plan II itself is the result of a consultative process involving both members of the public as well as the business community.

The Feedback Unit, the traditional channel for the general public, now offers an online portal – http://www.reach.gov.sg – for the citizens to give their views on national policies. REACH stands for "reaching everyone for active citizenry@ home", a rather labored acronym but in its short form does covey the intent of the government to reach out to its citizens instead of just waiting for feedback from it s citizens. It was set up in 2006 as part of the restructuring plans to move the Feedback Unit beyond gathering public feedback, and to become the lead agency for engaging and connecting with citizens. Within the government consultation portal, citizens can comment on discussion papers posted by government departments, take part in e-polls, join in discussion forums and engage with feed back groups. The initiation of m-gov (mobile government) as part of iGOV2010 has also meant that citizens can make use of mobile technology and send SMS (short message service) containing feedback to dedicated feedback numbers. Two other initiatives with a specific focus that are increasingly used by the general public are the portals for "Cut Red Tape" and "Cut Waste Panel". The Singapore government introduced these two initiatives as part of its effort to gather public feedback on cost-cutting measures and improving competitiveness. The Cut Red Tape program aims to streamline bureaucratic regulations in three areas: Life as a citizen, Working as a public officer and Doing business. Similarly, the Cut Waste Panel has been formed to improve on the government's delivery of public services in its drive to ease fiscal pressures.

Getting On-Board the E-Society

As mentioned earlier, the availability of hardware is one of the first issues in terms of the social inclusion, E-government debate. Very early on the Singapore government recognized the importance of ensuring wide spread availability and take up of computer and internet technologies. Computer ownership has risen significantly with almost 80 percent of households owning a computer in 2007, up from 60 percent in 2000. Of this 80 percent,

Figure 2. Access to a computer at home, 2000 to 2007 (Source: Infocomm Development Authority, infocomm usage surveys, various years)

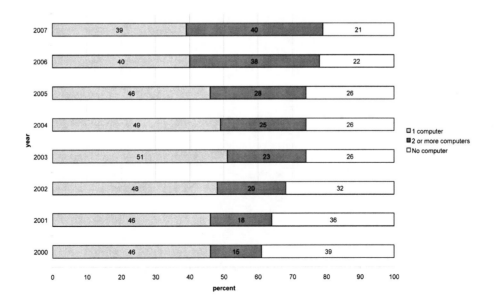

almost half had two or more computers, a significant change from 2000 when only approximately 15 percent had two or more computers (figure 2). These are even more significant when earlier data is also considered. Figures for 1988, (the earliest year data is available) suggest that only 11 percent of households had a personal computer, with this percentage doubling by 1992 and doubling again (40 percent) in 1997 (Singapore Department of Statistics 1997).

While hardware is important, the supply of and access to telecommunications infrastructure is also a critical factor. Within its goals of creating an intelligent nation, the Singapore government's infocomm strategy has continually invested in crucial infrastructure through a range of related programs. Key foci have been the widespread computerization of civil service, programs that have established electronic data interchanges and the establishment of Singapore ONE, the world's first nationwide broadband infrastructure (iN2015 Infocomm Infrastructure, Services and Technology Development Sub-Committee 2006).

This constant infrastructure investment has paid off with Singapore being ranked highly in the World Economic Forum's Global Information Technology Report (Dutta et al. 2006), which considers the extent of a country's readiness and usage of infocomm in business, government and society. The percentage of households with Internet access (broadband and dial-up) has increased from 50 percent in 2000 to almost 75 percent in 2007 (IDA 2007), while the rate of broad band penetration (the number of broadband subscriptions divided by the number of households) has increased at a faster rate. Figure 3 shows that penetration has raised from 33 percent in 2003 to 77 percent just four years later in 2007.

The mere availability of hardware and infrastructure, however, do not necessarily lead to active engagement by the citizen. Though Singaporeans have proven themselves to be very active on the

Figure 3. Household Broadband Penetration, 2003 to 2007 (Source: Infocomm Development Authority, Statistics on telecom services, various years)

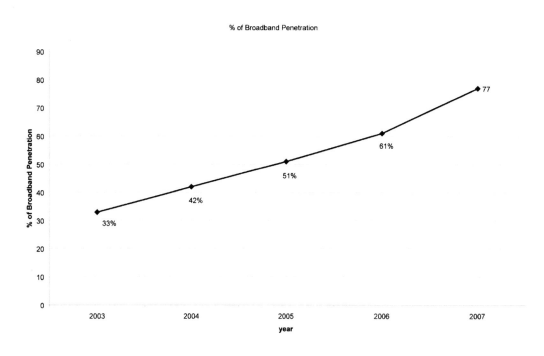

net, E-government related activities represent relatively small proportions of internet activity. According to the government's own survey in 2007 and 2008, only 12 percent of internet users accessed information from government organizations (Table 1); 10 percent made online payments to government (IDA 2008) and 7 percent completed and lodged forms (IDA 2008). However, despite this small take-up of online government resources, those who do access government online content are largely satisfied. E-government perception surveys found that of a sample of on-line users, a large proportion (86 percent) were satisfied or extremely satisfied with the quality of government on-line services (IDA 2007).

Being Off-Line: The Digital Divide and Singapore's E-Government Revolution

While being on-line and accessing services appear, in the Singapore context, to be having varying degrees of success, the most important social question around successful E-government relates to the issue of the digital divide. A move to a mature fully integrated E-government requires more than simply an online presence. Importantly, it requires a high level of citizen integration into E-government processes. In what has become known as the Socially Inclusive Governance Framework (United Nations 2005), there is recognition that a fully mature E-government presence requires a multi-pronged approach to promoting meaningful ICT-led access, with a particular focus on promoting access to marginal groups. Silcock (2001, 101) identifies some of these issues when she asserts that the extent to which E-government will make a difference and add value ... will depend on three factors: strong leadership, to ensure that the public sector workforce is ready to meet the challenges *ahead; management of the digital divide, to ensure that already excluded groups do not become further disadvantaged;* and well managed innovation (emphasis added).

The digital divide that does exist between those able and unable to access online services has the potential to limit the effectiveness of E-government plans. Put simply, the information poor are unlikely to be able to successfully access government services online even if they wanted to. Furthermore, it is ironic that those citizens most likely to benefit from access to government information and services may not be able to. Across the world, novel programs have been developed to try to overcome this growing digital divide. However, its existence will remain a barrier to developing the full potential of the E-government sector. Within Singapore, programs have included providing public access to information technology through community centers and libraries, an almost free PC scheme for poor families and equipping the populace with basic IT literacy.

Despite the programs to aid all Singaporeans to get online, there will remain some potential users who will be unable or unwilling to access online resources, thereby potentially limiting the reach of policy. A review of responses to the annual household infocomm usage survey conducted by the Infocomm Development Authority of Singapore provides an indication of those who are most likely to be out of reach. Table 2 provides the most important reasons for not having access to a computer at home, while table 3

Table 1. Main internet activity, 2007

	Internet activity	2007
1	Sending receiving emails	69%
2	General web browsing	35%
3	Instant messaging	22%
4	Getting information about goods and services	21%
5	Checking account information	20%
6	Bill paying	15%
7	Education or training activities (e-learning)	15%
8	Downloading music	15%
9	Transfer funds between accounts	13%
10	Getting information from govt websites	12%

presents reasons for lack of internet access. In 2007 approximately 26 percent of Singaporean households did not have access to a computer at home. The majority of these stated that the main reason they did not have a computer was that they had no need. Significant markers of the digital divide — cost, lack of skills and age — were also important reasons. In the 2007 survey 29 percent of respondents who did not have a computer claimed that they lacked the necessary skills to use a computer, while 11 percent stated that the purchase price was too high. Eight percent said that old age was a barrier. For internet access, the main reason for the lack of connectivity was that the respondent had no need (55 percent in 2007). A lack of skills (11 percent) and cost (8 percent) were also important.

These findings reflect the results of an earlier Singapore internet use study which illustrates that non-users resemble, across several socio-economic characteristics, the profiles of the information poor in other societies and nations. Like elsewhere, young people are likely to be more net savvy than their parents or older generations, leading to a divide between parents and the next generation (Kuo et al. 2002). While this profile of the digital divide is a characteristic of ICT use generally, it also has a specific dimension in terms of E-government; users of E-government services tend to be those most often thought to be already net-savvy. The 2006 and 2007 Singapore E-government perception surveys suggested that the typical profile of users was young (20 to 39 years) and typically white-collar workers. Those not using e-services tend to be 50 or older, likely to be blue-collar workers, housewives and retirees (IDA 2006, 2007). Among the reasons given for not using e-services, the following were the key: No need to use Government e-services; Not familiar (with e-services); Do not know how to access internet/ not familiar with internet; Preferred personal contact; Did not have computer access.

The surveys do not provide much detail and it is difficult to pinpoint why exactly many feel they do not need to use e-services. It could be that the offline alternatives are readily available, easier or even that they don't use any government service at all.

Though statistics are not available, it is reasonable to raise two related issues as to why some segments of the Singapore population may be excluded from E-government.

First, the economic divide. Though the price of computers has been falling down over time, it is still not within the reach of every family in Singapore. Compounding this problem, there is a tendency the world over – and in Singapore more so than many other countries – whereby newer generation of computers and software are replacing older ones far too frequently for the general

Table 2. Main reason for not having access to a computer at home, 2007

Main reason	2007
No necessity to use	46%
Lack of skills	29%
Too costly to purchase a computer	11%
Other	6%
Old age is a barrier to learning computer skills	7%
Children are too young	2%
Total	100%

Table 3. Main reason for not having internet access at home, 2007

Main reason	2007
Lack of interest/ no need	55%
Lack of knowledge/ skills/ confidence	11%
Have access elsewhere	9%
Other -lack of equipment	14% 7%
Costly equipment	8%
Subscription to the internet is too costly	3%
Concern about exposure to inappropriate or harmful content	1%
Total	100%

population to cope with. It is not just a matter of increasing costs but also greater difficulties in re-learning the use of the computer with each new software or hardware. Even for those who are familiar with computers, this has become a serious problem and all the more so for the less- or un-initiated. Thus, the Singapore government may inadvertently widen the digital divide simply by upgrading the software and hardware without corresponding uplift within the general population. Even in the business sector, there have been numerous examples of how firms have not been able to cope with different browsers in use and have had to invest in costly upgrades and diversifications. Interestingly, there has not been much public debate about this issue nor has there been any publicly available research data to support or discount this view. However, few can deny the inevitable digital divide between those who can constantly upgrade and those who cannot, unless there is a fundamental remedy. This is both a cost and a computer literacy issue.

Still within the domain of the digital divide, there is also the issue of the language divide, whereby local language and cultural constraints may adversely impact on online usage (Vassilakis, 2004, Metaxiotis and Psarras, 2004, Maenpaa 2004, Kovacic 2005). Globally, English is the most common language of the Web (OECD 1997). Within its E-government framework, the Singaporean government has made a decision to make English the sole official E-government language. This is in sharp contrast to the policy which in the analog world makes allowances for four official languages—the Mother-Tongue languages, as spoken by the three major ethnic groups, namely the Chinese language, the Malay language and the Tamil language, together with the English language. There was no public debate as to why this should be so nor was there any public research to justify this major policy change. It simply seems to be based on an assumption that eventually everyone in Singapore will be English literate as all school going children in Singapore study English as their first language. However, this will be a long time in the future. It also flies in the face of a number of successful non-English language E-government applications and other internet sites world-wide (see Yong 2003).

While many Singaporeans do speak and read English, there is a minority for whom English is not their first language, not even a familiar language and therefore for whom access to an English-only citizen portal is not appropriate. Figure 4 presents census data relating to the percentage of literate resident population who speak only a non-English language (Chinese, Malay, Tamil, other non-official language). The divide in English language use is stark with older residents being much less likely to speak English than younger groups. Decomposing these figures we see that of the one million people who speak one language only, 57.2 percent speak Chinese only, 6.2 percent speak Malay only and 1.5 per cent speaks Tamil only. For those aged over 65 years, 70.4 percent speak Chinese only, 9.1 percent speak Malay only and 4 percent speak Tamil only (Singapore Department of Statistics 2001).

There is, however, a question of practicality that could be raised: Should all E-government services be available in all four official languages in order to be inclusive? Even in the past, not every service was made available in all four languages but some like the Income Tax return forms, which is still one of the most widely used service, were. The least the government has to do is to go through a thorough analysis of what e-services would be language sensitive and what additional measures need to be taken to redress the problem. It is not evident that the Singapore government has done this in the past or plans to do this in the future.

FUTURE RESEARCH DIRECTIONS

Clearly then there still remains some potential digital divide. A significant question relates to what this means in terms of reaching a fully

Figure 4. Percent speaking only other non-English language

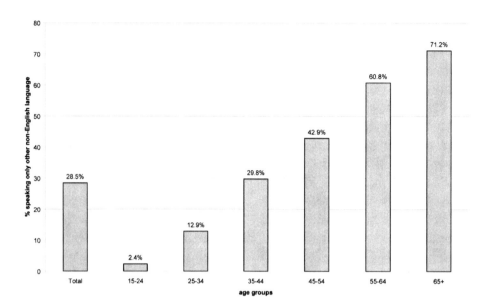

mature and integrated E-government? Regardless of what programs are put in place there will always remain some residual accessibility issues. Some of this will be in terms of resistance by some citizens to transact online even when other issues of infrastructure and content have been successfully dealt with as is the case in Singapore. This potentially is only a serious social issue in cases where online transacting is virtually the only option available. The more serious problem for Singapore's E-government strategy relates to the issue of the language divide. Although some relate the issue of a language divide to developing countries (Barnard et al. 2004), it is likely that Singapore will also face potential constraints introduced by the language barrier. Given that a large proportion of citizens aged 65 years and above speaks (and presumably reads) a language other than English means that despite programs to encourage all Singaporeans to get online, the true potential of online services will be lost in terms of these people.

CONCLUSION

This chapter has discussed the advent of information and communication technologies and in particular the emergence of E-government as a new form of government-citizen relationship. Specifically, the paper has focused on the development of an online presence by the Singaporean government and the strategies and outcomes associated with this increased use of online facilities. As an example of world best practice the Singapore model is held by many to be exemplary. Many of the issues associated with successfully moving to full E-government maturity have been dealt with and the three strategic plans that have been the driving force behind E-government development have proven to successfully reach many of their goals. The development of a functional e-citizen portal, e-business portal and a significant online component within the civil service has gone a significant way towards building a system that is world class and one which delights customers and connects citizens.

Despite this, the Singapore model is still evolving. The third E-government strategy iGOV2010 recognizes that full E-government maturity is still a future goal to be achieved. Although the performance to-date has been impressive and many issues associated with access, content and trust have been dealt with, there still remain some impediments including the existence of a digital divide characterized by age and language. Within this context, it is clear that there will always remain some residual non-utilization. However, the significant constraint faced by Singapore is the digital divide associated with a citizenry which has a significant proportion of mono-lingual non-English speakers. While such a language divide remains, the potential for a fully mature E-government may be limited and questions of a fully inclusive E-government process will remain.

REFERENCES

Abbott, J. (2001). Democracy @ Internet.asia? The challenges to the emancipatory potential of the net: Lessons from China and Malaysia. *Third World Quarterly*, *22*(1), 99–114. doi:10.1080/01436590020022600

Allen, B., Juillet, L., Paquet, G., & Roy, J. (2001). E-governance and government online in Canada: Partnerships, people and prospects. *Government Information Quarterly*, *18*, 93–104. doi:10.1016/S0740-624X(01)00063-6

Bains, S. (2002 April). Wired cities. *Communications International*, 21-25.

Barber, B. (1998). *A place for US: How to make Society Civil and Democracy Strong*. New York: Hill and Wang.

Barnard, E., Cloete, L., & Patel, H. (2004). Language and technology literacy barriers to accessing government services. In Traunmüller, R. (Ed.), *Electronic Government* (pp. 37–42). Berlin: Springer.

DiMaggio, P., & Hargittai, E. (2001). *From the Digital Divide to Digital Inequality: Studying Internet Use as Penetration Increases*. Woodrow Wilson School, Centre for the Arts and Cultural Policy Studies, Princeton University.

Dutta, S., Lopez-Carlos, A., & Mia, I. (2006). *Global Information Technology Report, 2005-2006*. New York: Palgrave-Macmillian.

Graham, S. (1998). The end of geography or the explosion of space? Conceptualizing space, place and information technology. *Progress in Human Geography*, *2*, 165–185. doi:10.1191/030913298671334137

Healey and Baker Consultants. (2001). *European E-locations Monitor*. London: Healey and Baker Consultants.

Helbig, N., Gil-Garcia, J., & Ferro, E. (2009). Understanding the complexity of electronic Implications from the digital divide literature. *Government Information Quarterly*, *26*, 89–97. doi:10.1016/j.giq.2008.05.004

iN2015 Infocomm Infrastructure, Services and Technology Development Sub-Committee. (2006). *Totally Connected, Wired and Wireless*. Report by the Infocomm Infrastructure, Services and Technology Development Sub-Committee, Singapore: IDA.

IDA. (2006). *E-government Customer Perception Survey Conducted in 2006*. Retrieved from http://www.ida.gov.sg, date accessed 20 July 2006

IDA. (2007). *Annual E-government Customer Perception Survey Conducted in 2007*. Retrieved from http://www.ida.gov.sg/Publications/20071001171301.aspx

IDA. (2008). Annual survey on Infocomm Usage in households and Individuals for 2007. Retrieved from http://www.ida.gov.sg/doc/Publications/Publications_Level2/20061205092557/ASInfocommUsageHseholds07.pdf iGOV2010 Project Steering Committee. (2006). *From Integrated Service to Integrated Government*. Singapore: Ministry of Finance, Infocomm Development Authority of Singapore.

Jho, W. (2005). Challenges for e-governance: protests from civil society on the protection of privacy in E-government in Korea. *International Review of Administrative Sciences, 71*(1), 151–161. doi:10.1177/0020852305051690

Kovaic, M. (2005). The impact of national culture on worldwide egovernment readiness. *Informing Science Journal, 8*, 143–158.

Kuo, E., & Choi, A. Mahizhnan, A. Peng, L-W, Soh, C. (2002). Internet in Singapore: A Study on Usage and Impact. Singapore: Times Academic Press.

Maenpaa, O. (2004). *E-government: Effects on Civil Society, Transparency and Democracy*. Presented at International Institute of Administrative Sciences, 26th Congress of Administrative Sciences, Seoul.

Martin, B., & Byrne, J. (2003). Implementing E-government: Widening the lens. *Electronic Journal of E-Government, 1*(1).

Metaxiotis, K., & Psarras, J. (2004). E-government: new concept, big challenge, success stories, *Electronic Government, an International Journal, 1*(2), 141-151.

OECD. (1997). *Webcasting and Convergence: Policy Implications*. Paris: OECD. Retrieved from www.oecd.org/dsti/sti/it/cm/prod/e_97221.htm

Perry, A. (2001, June 4). Getting Out the Message. *Time Magazine, 157*(22). Retrieved August 29, 2002, from http://www.time.com/time/interactive/politics/changing_np.html

Silcock, R. (2001). What is E-government? *Parliamentary Affairs, 54*, 88–101. doi:10.1093/pa/54.1.88

Singapore Department of Statistics. (2001). *Singapore Census of Population 2000: Statistical Release 2—Education, Language and Religion*. Singapore: Department of Statistics.

United Nations. (2005). *UN Global E-government Readiness Report 2005: From E-government to E-inclusion. Department of Economic and Social Affairs, Division of Public Administration and Development Management*. New York: United Nations.

United Nations. (2008). *UN Global E-government Readiness Report 2005: From E-government to Connected Governance. Department of Economic and Social Affairs, Division of Public Administration and Development Management*. New York: United Nations.

Van Der Meer, A., & Van Winden, W. (2003). E-governance in cities: A comparison of urban information and communication technology policies. *Regional Studies, 37*(4), 407–419. doi:10.1080/0034340032000074433

Vassilakis, C. (2004). Barriers to Electronic Service Delivery. *e-Service Journal, 4*(1), 41-63.

Warschauer, M. (2003). *Technology and Social Inclusion: Rethinking the Digital Divide*. Cambridge, MA: MIT Press.

Yong, S. L. (Ed.). (2003). *E-government in Asia: Enabling Public Service Innovation in the 21st Century*. Singapore: Times Media P/L.

ADDITIONAL READING

Aldrich, D., Bertot, J. C., & McClure, C. R. (2002). E-government: Initiatives, Developments, and Issues. *Government Information Quarterly, 19*(4), 349–355. doi:10.1016/S0740-624X(02)00130-2

Andersen, K. V., & Henriksen, H. Z. (2005). The First Leg of E-government Research: Domains and Application Areas 1998-2003. *International Journal of Electronic Government Research, 1*(4).

Basu, S. (2004). E-government and developing countries: an overview. *International Review of Law Computers & Technology, 18*(1), 109–132. doi:10.1080/13600860410001674779

Borras, J. (2004). International Technical Standards for e-Government. Electronic. *Journal of E-Government, 2*(2), 75–80.

Buckley, J. (2003). E-service Quality and the Public Sector. *Managing Service Quality, 13*(6), 453–462. doi:10.1108/09604520310506513

Dugdale, A., Daly, A., Papandrea, F., & Maley, M. (2005). Accessing E-government: Challenges for Citizens and Organizations. *International Review of Administrative Sciences, 71*(1), 109–118. doi:10.1177/0020852305051687

Fountain, J. (2001). *Building the Virtual State: Information Technology and Institutional Change*. Washington, DC: Brookings Institution.

Gilbert, D. B., & Littleboy, D. (2004). Barriers and Benefits in the Adoption of E-government. *International Journal of Public Sector Management, 17*(4), 286–301. doi:10.1108/09513550410539794

Hazlett, S. A. H. (2003). E-government: The Realities of Using IT to Transform the Public Sector. *Managing Service Quality, 13*(6), 445–452. doi:10.1108/09604520310506504

Holden, S. H., Norris, D. F., & Fletcher, P. D. (2003). Electronic Government at the Local Level - Progress to Date and Future Issues. *Public Performance and Management Review, 26*(3), 1–20.

Kearns, I., Bend, I., & Stern, B. (2002). *E-participation in Local Government*. London: Institute of Public Policy Research.

Lenk, K., & Traunmuller, R. (2002). The Aix declaration on e-government: Public governance in the 21st Century. In K. Lenk and R. Traunmuller (Eds.), Electronic government: first international conference, EGOV 2002, Aix-en-Provence, France, September 2-6, 2002. New York: Springer.

Metaxiotis, K., & Psarras, J. (2004). E-government: new concept, big challenge, success stories. *Electronic Government, an International Journal, 1*(2), 141-151.

Tambini, D. (2000). *Universal Internet Access: A Realistic View*. London: IPPR/Citizens Online Publication.

Teicher, J., Hughes, O., & Dow, N. (2002). E-government: A New Route to Public Sector Quality. *Managing Service Quality, 12*(6), 384. doi:10.1108/09604520210451867

Thomas, J., & Streib, G. (2003). The New Face of Government: Citizen-Initiated Contacts in the Era of E-Government. *Journal of Public Administration: Research and Theory, 13*, 83–102. doi:10.1093/jpart/mug010

Tolbert, C., & Mossberger, K. (2006). The Effects of E-Government on Trust and Confidence in Government. *Public Administration Review, 66*(3), 354–369. doi:10.1111/j.1540-6210.2006.00594.x

Van Den Berg, L., & Van Der Meer, A. Van winden, W., & Woets, P. (2006) E-Governance in European and South African Cities. Hampshire, UK: Ashgate.

Wang, L., Bretschneider, S., & Gant, J. (2005). Evaluating Web-Based E-Government Services with a Citizen-Centric Approach. *International Conference on System Sciences, 2005. HICSS '05*.

Warkentin, M., Gefen, D., Pavlou, P., & Rose, G. (2002). Encouraging Citizen Adoption of e-Government by Building Trust. *Electronic Markets, 12*(3), 157–162. doi:10.1080/101967802320245929

KEY TERMS AND DEFINITIONS

E-Governance: The use of ICTs such as the internet and mobile phone as a platform for exchanging information, providing services and transacting with citizens, businesses, and other arms of government.

Social Inclusion: The use of policies and programs to reduce inequality, exclusion and disadvantage.

Public Participation: The belief that those who are effected by a decision have a right to be involved in the decision making process.

Digital Divide: The gap between people with effective access to digital and informational technology and those with very limited or no access at all.

Civil Society: Comprises the totality of voluntary civic and social organizations and institutions that form the basis of a functioning society and is contrasted with the formal structures of the state and commercial/market institutions

Section 3
Innovations and Challenges in Urban Management

Chapter 18
Integrating ICT into Sustainable Local Policies

Antonio Caperna
Università degli Studi Roma Tre, Italy

ABSTRACT

This chapter analyses the Information and Communication Technologies (hereafter referred to as ICT) phenomenon, the opportunities it offers, the potential problems, and the relationship with local policies. It moves on the actions needed to develop, within the Agenda 21 process, a framework able to define some fundamental features for a new spatial theory in the information age, which will eventually consider Information and Communication Technology not just a simple tool, but a crucial aspect of a sustainable policy, capable, if well addressed, to mitigate various current or emerging territorial challenges such as literacy and education, public participation in the planning process, social and geographical divide, institutional transparency, etc.. This chapter will illustrate a framework able to assist politicians and planners in planning a sustainable development through ICT.

INTRODUCTION

As claimed by several authors, we live in the information age (Masuda, 1981; Castells, 1996, 2002). An era where knowledge and information have become key factors in the growth of contemporary society triggering socio-political and economical as well as cultural and spatial changes (e.g. the emergence of the space of flows, Castells, 1996).

On one side, new political and environmental challenges inspired by the acceptance of the sustainable development principles have induced governments and public authorities to open up access to environmental information as a means to improve public participation in environmental decision making and awareness. On the other side the growth of ICT is a tool that not only constitutes an industry in its own right but which also pervades all sectors of economy, where it acts as integrating and enabling technologies. ICT have a profound impact on society, and their production and use have important effects on the development of economic, social and environmental areas, promoting new

DOI: 10.4018/978-1-61520-929-3.ch018

questions discussed by theorists and planners. But the extent of ICT in everyday life and its strong relationship with socio-cultural and economic aspects produce a complex equation which is difficult to understand and solve.

This chapter aims to explain some fundamental aspects about ICT, and to offer a framework that will allow: planning a sustainable policy; addressing ICT in a sustainable way; developing an analytical process of understanding environmental information use, and supporting public access, improving awareness and participation processes. This assumption is based on the current trend within public authorities to use ICT as a major delivery medium.

BACKGROUND

Case Study

The context is represented by a district near Rome: Parco Naturale Regionale dei Monti Lucretili. It is a mountain landscape including 13 Municipalities with a total extension of 18 hectares and about 35,000 inhabitants. Politically it is governed by the Park Agency. The administrative structure of this agency is a consortium formed by the representatives of the 13 Municipalities and by 9th and 10th Mountain Communities.

The territory has an agricultural and naturalistic vocation (Mantero & Giacopini, 1997). The naturalistic richness of the Park lies in the particular configuration of the pre-Apennine landscape, where the proximity to the sea has contributed to the formation and coexistence of biotypes determined by different microclimates, the latter given to different exposures and influenced by the variation of the circulation of the air masses within the mountain group. The current vegetation features on Lucretili Mountains derive from a wavering series of events produced by human action leading to the change of the original aspects. Vineyards, cherry cultivation, and olive groves characterize these piedmont areas, with skilfully built terracing dating back to Roman times (Mantero & Giacopini, 1997). The vegetation is composed of large woods (beech, chestnut and hornbeam) and mountain pastures.

The labour market showed, in the last decade, a reduction in workers. A detailed analysis, shows that there has been a strong reduction for small and medium commercial concerns that is not compensated by the increase of other sectors as firms, utility companies, and tourism (Caperna, 2007). The local economy is characterized by agriculture: about 75% of the entire territory is agricultural with excellent products as olive oil, wine, orchards and citrus orchards; 10% sowable land, and 34% pasturage (Caperna, 2007). Finally activities linked to agriculture as bed & breakfast or trekking are proliferating (Mantero, 2000).

In synthesis the data describe an ageing of the population and economical difficulties. On the other side there are some aspects that should be meant as an opportunity: I mean natural treasures represented by natural landscape, food culture, few excellent agricultural products and the possibility of developing sustainable tourism.

With regard to development of tourism, we can talk about sustainable tourism if there is balance between environmental, social, and economical aspects of tourism, and the need to implement sustainability principles in all segments of tourism (WTO, 2004).

According to these principles, sustainable tourism should (WTO, 2004): make optimal use of environmental resources; respect the socio-cultural authenticity of host communities, conserve their built and living cultural heritage and traditional values, and contribute to inter-cultural understanding and tolerance, and ensure viable, long-term economic operations, providing socio-economic benefits to all stakeholders that are fairly distributed, including stable employment and income-earning opportunities and social services to host communities, and contributing to poverty alleviation.

This means that sustainable tourism requires the informed participation of all relevant stakeholders, as well as strong political leadership to ensure wide participation and consensus building. Achieving sustainable tourism is a continuous process and it requires constant monitoring of impacts, introducing the necessary preventive and/or corrective measures whenever necessary.

General Overview

The challenges we face today, and those we will have to cope with in the future, require new ways of thinking about and understanding the complex, interconnected, and rapidly changing world in which we live and work. Thus, the proposed model will be structured according with three basic elements that are mutually interconnected: non-linear dynamic system (NDS) theory, international agreement, and sustainable development.

Nonlinear Dynamics System (NDS)

First of all, I'd like to introduce the concept of paradigm. The term was formulated by Thomas Kuhn in The Structure of Scientific Revolutions. He argued that science does not progress through a linear accumulation of new knowledge, but undergoes periodic revolutions, also called paradigm shifts, in which the nature of scientific inquiry within a particular field is abruptly transformed. According to Kuhn, we can talk in terms of old scientific paradigm called Cartesian, since its main characteristics were formulated by Descartes, Newton and Bacon. This paradigm is characterized by the fact that considers the world and its elements as mechanical structure and we can explain the phenomena in terms of linear causality. This means that any problem can be divided into components, and that the dynamics of the whole could be understood from the properties of the parts. In addition, this method is characterized by the fact that an event is independent from human observer.

But at the early of 1900 it was necessary to furnish answer about some anomalies of the Physics and nonlinear behavior showed by simple systems. In these context, the classical laws based on mechanical and linear principles, weren't able to predict the evolution of a structure becoming increasingly ineffective to address modern problems (Kofman, 1993; Kelly, 1998; Hjort & Bagheri, 2006).

In this regard, NDS provides a new theory-driven framework to think about, understand, and influence the dynamics of complex structures, issues, and emerging situations. The NDS theory has generated a new paradigm that tries to model and to study phenomena that undergo spatial and temporal evolution. These phenomena range from simple system to complex structures as population dynamics or biological organisms, and it is applied to a wide spectrum of disciplines including physics, chemistry, biochemistry, biology, economy and even sociology.

This paradigm is based on relationship between three terms: system, dynamic, and nonlinear.

Among the several existing definitions of system, there are some that appear more commonly than others, although in slightly different shapes. The term system was adopted for the first time by the biochemist Henderson (Lilienfeld, 1978) concerning living organisms as well as social structures, since system is an assemblage of interrelated parts that work together by way of some driving process. These are the features of a complex system: a large number of interacting components, whose aggregated activity is nonlinear and typically exhibits self-organization under selective pressures; components are different each other; the relationships' structure change continuously; inputs derived from each component and its internal variables form sub-systems.

Nonlinearity means that a change is not based on a simple proportional relationship between cause and effect. Therefore, such changes are often abrupt, unexpected, and difficult to predict.

Dynamics refers to forces or processes that produce change inside a group or system.

Subsequently the development in several fields provided a theoretical support able to assert this new paradigm called holistic or systemic: a new way of thinking about our world as the art and science of linking structure to performance, and performance to structure, often for purposes of changing structure (relationships) so as to improve performance (Richmond, 1993, Hjort & Bagheri, 2006).

Moreover it includes the following five criteria (Capra, Steindl-Rast, Matus, 1992), that provide a paradigmatic passage from: the part to the whole; structure to process; objective science to epistemic science, or epistemology (the understanding of the process of knowledge is to be included explicitly in the description of natural phenomena); building to network as metaphor for knowledge; truth to approximate descriptions.

The system approach is not only a tool for studying innovation processes, but also a conceptual framework for innovation policies and strategies (Edquist, 1997). Schienstock and Hämäläinen (2001) suggest that a system approach-based is an appropriate way to enhance the regional innovation policy. This innovation policy pays particular attention to the communication, cooperation and networking processes among firms and supports organizations aiming to tackle all areas of systematically weak performance in the regional innovation system (Schienstock and Hämäläinen, 2001).

International Agreement

The political background is supported by several international documents, in particular by Agenda 21 (A21) adopted by more than 170 governments at the United Nations Conference on Environment and Development (UNCED) held in Rio de Janeiro (Brazil) 1992. It is a comprehensive action plan to be taken globally, nationally and locally by organizations of the United Nations system, governments, and major groups in every area in which humans impact on the environment. It aims to address the pressing problem of the contemporary world, and to do this, it's been structured in four sections such as social and economic dimensions, conservation and management of resources, strengthening the role of major groups, and means of implementation.

In particular, chapter 28 describes the local authorities' initiatives in support of A21, because many of the problems and solutions being addressed by A21 have their roots in local activities. In this regard, each local authority should start a dialogue with its citizens, local associations, productive systems (business, industrial, farms, etc.) in order to formulate the best strategies. The process of consultation would increase household awareness of sustainable development issues. So the participation and cooperation of local authorities will be a determining factor in fulfilling its objectives.

The political lines proposed by this document aims at indicating a more sustainable environment and bottom-up processes. In this regard, the A21 dedicates a whole chapter to the planning and management of land resources. The text recognizes that the increasing pressures on land and land resources are creating conflicts and an unsustainable use of land and natural resources. Thus, in order to promote a sustainable development, it's necessary to examine all uses of land in an integrated manner through the coordination of the sectorial planning and management activities concerned with the various aspects of land use and land resources.

Subsequently from A21 other documents as Aarhus Convention, Ferrara and Åalborg Charters have arisen. With regard to these international agreements, we can underline some fundamental elements. Aarhus Convention (UN/ECE, 1998) demands to improve access to information and public participation in decision-making, suggesting that environmental information will progressively become available in electronic databases, which are easily accessible to the public through public telecommunication networks (UN/ECE, 1998).

Åalborg Charter (1994) is an European initiative launched at the end of the first European Conference on Sustainable Cities and Towns which took place in Åalborg, (Denmark), that provides a framework for the delivery of local sustainable development and calls on local authorities to engage in local A21 processes.

Finally, the charter of Ferrara is an Italian initiative that aims to promote the adoption of local A21, and to spread knowledge, best practices, promotion of information sharing to local authorities, and to develop research on environmental problems.

Sustainable Development

There are over 200 definitions of sustainable development (SD), and various graphic schemes have been developed to portray the multi-dimensional nature of sustainability (Adam 1993; Elliott 2004).

Over the past three decades there have been many events that have contributed to the development of the concept of sustainable development. Many people consider 1962 as the seminal year in which people began to understand how closely linked environment and development are. Rachel Carson's Silent Spring, a book that collects researches on toxicology, ecology and epidemiology, suggested that agricultural pesticides were built at catastrophic levels and there was a link between the damages to animal species and those caused to human health. It shattered the assumption that environment had an infinite capacity to absorb pollutants.

Even though from then on there was a growth of researches about the connection between human population, resource exploitation, economical pattern and environment, only in 1980 the World Conservation Strategy defined the word "development" as "the modification of the biosphere and the application of human, financial, living and non-living resources to satisfy human needs and improve the quality of human life", and in 1987 the United Nations World Commission on Environment and Development published "Our Common Future" (also known as Brundtland Report). It ties problems together and, for the first time, gives some direction for comprehensive global solutions, and it also popularizes the term "sustainable development" giving the following description: "development that meets the needs of the present, without compromising the ability of future generations to meet their own needs" (WCED 1987).

Even though there is no accepted definition about sustainable development, there is a substantial agreement about the three pillars of sustainable development: economical, environmental, and social aspect.

So a sustainable system must achieve distributional equity, adequate provision of social services including health and education, gender equality, and political accountability and participation.

It has been underlined that sustainable development must also take into account the institutional policy (Brandt, 1980; Cernea, 1987), and cultural environments (Cernea, 1987; Korten, 1990) of the local governments in which such efforts are initiated.

Clearly, these three elements of sustainability introduce many potential complications to the original simple definition. The goals expressed or implied are multidimensional, raising the issue of how to balance objectives and how to judge success or failure.

Each of these three areas is commonly referred to as a system: economic system, environmental system, and social system. According to Balaton Group's report on sustainability indicators: "The total system of which human society is a part, and on which it depends for support, is made up of a large number of component systems. The whole cannot function properly and is not viable and sustainable if individual component systems cannot function properly... sustainable development is possible only if component systems as well as the total system are viable. Despite the uncertainty of the direction of sustainable development, it is necessary to identify the essential

Table 1. Evolving views of planning and information technology

1960s	System Optimization	"Planning as applied sciences" Information Technology are considered as providing the information needed for a value and politically neutral process of "rational" planning
1970s	Politics	"Planning as politics" Information Technology are considered as inherently political, reinforcing existing structures of influence, hiding political choices, and transforming the policy-making process
1980s	Discourse	"Planning as communication" Information Technology and the context of planners' technical analyses are often considered less important than the ways in which planners transmit this information to others
1990s	Intelligence	"Planning as reasoning together" Information Technology considered as providing the information infrastructure that facilitates social interaction, interpersonal communication, and debate that attempts to achieve collective goals and deal with common concerns
Source: R.K. Brail and R.E Klosterman (eds), 2001, Planning Support Systems: Integrating Geographic Information System, Models and Visualization Tools. U.S.A., ESRI Press		

component systems and to define indicators that can provide essential and reliable information about the viability of each and of the total system" (Bossel, 1999, p.6).

ICT and Planning

The transition from an industrial society to a knowledge society poses several challenges for the planners and politicians. If the concept of the information society has been successfully developed over the last 30 years by a number of distinguished proponents, such as Bell (1974), Masuda (1981) and Castells (1996, 2002), there is again a great difficulty in the development of new planning theories, methods and models. So, if on one side the introduction of computers into planning in 1960s was part of a fundamental transition allowing the collecting, storing and analysis of myriad of data, on the other side the diffusion of information technology and the ever increasing social complexity have showed an inappropriate methodological approach about the planning matter. In Table 1, we can see how computers were assumed to play an important role in collecting and storing the required data trying to provide, from time to time, system models that could describe the present and project the future through the identification of the best plan within the range of possible alternatives (Harris & Batty, 1993; Brail, 2001).

But, as stated above, many problems have been faced through an old methodological approach by the planners as they found it difficult to define new methods and models concerning this matter. In an article published in 1991, Manuel Castells addressed urban planners asking "The world has changed: can planning change?". Indeed, most of the political and economical processes that are affecting our present world were already in motion during the early 1990s. Again Castells (1997) explained the situation: *"Confronted by this whirlwind of social and spatial transformation, the intellectual categories that constituted the foundation of planning in general, and of city planning in particular, have been made obsolete"* and *"the danger for the profession is to face this transformation defensively"*. Moreover, he underlined that as in all major processes of social change there are extraordinary opportunities to be seized, but also serious costs for those institutions and individuals unable or unwilling to adapt. Thus, there is an obvious danger of digging the trenches of cultural resistance and resisting change by refining old concepts, or by embarking in a process of self-reflection in which planning itself becomes the goal, rather than the means. While in the professional world, the harsh reality of bureaucracies, politics and markets will leave little room for intellectual escapism, in the academic planning field the building of fantasy worlds made of abstract categories, or the attempt to justify planning by inventing a new academic discipline around an ad hoc theoretical foundation, could substitute for the harsh task of reinventing what to do out there, in an increasingly complex world. It must not be.

In the same way, in their seminal work "Telecommunications and the city", Graham and Marvin (1996) denounced the lack of consideration for telecommunication issues in urban studies and planning, stressing on the invisibility and intangibility of ICT infrastructures, the conceptual challenge regarding the increasing space-time complexity that digital technologies bring, and the conservatism of urban and regional planning disciplines, still focused on the industrial city concepts and models. Aurigi (2005) analysed the dichotomy between the continuing rapid development of the information society and, on the other side, the city planners and managers' need to reconsider their level of awareness– and their degree of control – of the impact Information Technology has on contemporary cities.

Moreover, Talvitie (2003) argued that ICT, as the main driving force of the development of the information/knowledge/network society should be more specifically taken into account in urban and regional planning, whereas he also pointed out the details about the need for further research on the spatial impact of ICT. Obviously, in addition, the programmes of planning education and further training should be updated, as should the legal provisions for planning. With regard to the Italian situation, in my previous research ICT and urban planning (2005) I analysed the use of ICT in a little sample of medium size cities. Several planners and decision-makers were interviewed and the result was a situation of substantial uncertainty about the policies and the planning processes.

We must take into account the fact that, in most cases, the actors involved, especially local governments, are just beginning to wrestle with the wider economical and societal implications of the information revolution (Evans, 2002; Cohen, van Geenhuizen, Nijkamp, 2001), and with the wide range of possible interventions which can lead to different ICT policies for cities. Van Winden in his research (2003) clearly showed the fact that cities are searching for appropriate strategies to deal with new realities. In his comparative example he showed how the various goals referred to aspects of urban environment: in particular, it emerged clearly that a local access policy can give a substantial contribution to sustainable urban development and the promotion of social equity when these goals are complemented with educational and social components. In other cities, it has been noted that where the private sector was favoured, there has been less social equity and a more conflictual relationship between private and public sectors.

So, this tumultuous situation needed a political and theoretical effort to address and balance the latent socio-economical conflict, in particular between public and private interests, and to answer several crucial questions such as: the changing nature of space and time and their impact on urban environment; a "fluid" socio-economical and cultural scenario that produces a substantial uncertainty in the planning processes, and the extreme difficulty to predict possible scenarios; the difficulty, most of all for local decision makers, to define a comprehensive policy capable of escaping the traditional way of thinking and where ICT will be considered not just a technical problem, but a central question in new policy models; the substantial methodological lag of planners to understand how to conceptualize and incorporate the impacts of ICT; sustainability issues, or how to promote a sustainable low-impact socio-political and economical growth and incorporate ICT into urban and regional planning.

The concept of network society (Castells, 1996) can summarize the complexity of all these questions and the power of the forces which influence the spatial structure. A system where the power of changing, as demonstrated by a remarkable number of social and economical studies (Caso, 1999; Castells, 1996; Graham & Marvin, 1996; Mitchell, 1995; Roberts et al., 1999; Sassen, 2000), has created a strong relationship between several environments (social and cultural, as well as economical) and the substantial request of answers from local governments, taking into consideration the fact that the complexity of these processes generates a high level of uncertainty, especially concerning local policies (van Geenuizen, 2001; Caperna, 2005).

In the last few years we have registered contradictory situations: on one side, a strong role of ICT as fundamental element in the political programs; but on the other side, we have met many uncertainties related to: a fragmented knowledge and absence of coordinate policies with no effective final goal; value of strategic choices, especially referred to the local government; insufficient and inadequate knowledge and/or resistance from planners, decision makers and stakeholders about the spatial impacts of the new technology and the high levels of uncertainty.

The impact of ICT needs to be analysed at different spatial levels, considering that it can't be conceptualized as a new type of infrastructure, catering for the transport of data or information, as well as a power able to bring on socio-cultural and economical transformations.

As I stated above in the text, complexity can be considered as a new paradigm able to provide a methodological support for politicians and planner. So the main challenge for a sustainable spatial planning is trying to furnish knowledge and an organization concerning this complex and interactive environment. Organized complexity means thinking in terms of multiple options and promoting interdisciplinary models. On the basis of what is set out above, I will delineate a framework which incorporates ICT into the planning process as a tool capable of supporting virtuous actions, in democracy, knowledge, transparency and participatory design, etc.

Prerequisites for a Sustainable Approach

According to the general background showed above, in this section we will outline a framework able to permit a sustainable planning approach through the use of ICT. The Park Agency is working according to this framework. Actually it is in the early stages.

Several times in this chapter I have underlined the necessity to develop an effective policy which takes into account the intrinsic complexity of the contemporary world, at the same time able to incorporate the ICT impact in a sustainable way. According to this, I have identified the following prerequisites: vision, governmental culture, resources and skills.

The first prerequisite is to imagine a vision of the community's future. The local Government should cooperate with all citizens and stakeholders to establish a clear, challenging vision about the future of local community. It seems to be an indispensable prerequisite because it can support the local Government in choosing the common direction where to concentrate all the efforts needed to promote a socio-cultural and economical growth without a dispersion in terms of human and economical capital. As all the effective policies, it needs a long-term perspective supported by a dynamical agenda with concrete, easily comprehensible goals that can be adapted according to the evolution of events. Finally, the citizen has to feel that the ICT contains the cultural values which make a community alive and that it's possible to promote and support a sustainable development through the use of ICT due to its technological and economical aspects.

The second prerequisite is a governmental culture. This means the necessity to promote, through an on-going relationship with research centres and universities, a new ruling class able to adopt a new approach based on holistic structure and where a social contract commits the politicians to the respect of ethical values in conformity with the vision. Only in this way it will be possible to develop a new administrative architecture, where ICT becomes an essential tool for promoting political actions, a bottom-up democracy, and for generating social inclusion and a sustainable economy that pays attention to the local requirements. In this context, another interesting question concerns the responsibility. To put a plan into practice and promote sustainability, it's necessary that everyone is aware of his own responsibility.

The third prerequisite concerns the capability of the local government to obtain and attract resources, both public and private. In both cases a governmental action is indispensable to combine several factors as the aptitude to promote local plans using EU funds for development of local economy or infrastructure, or to support local companies and encourage them in new investments. The local government should encourage all relevant actors to report how much they invest and how they work to combine ICT development with sustainable development.

The fourth, and final prerequisite, concerns the necessity to increase the knowledge and skills about new methodological and complex approaches to planning. Previously we have discussed about this necessity, originated by a substantial inadequacy of the educational structures. Of course this aspect relates to the national system. Furthermore, according to Castells, we can give indication about the education of city and regional planners in the information age. It must contain a wide range of issues such as: the overarching issue of sustainability; the planning of urban and metropolitan infrastructures; the reconstruction of cultural meaning in spatial forms and processes; the shift towards local and regional governments as decisive instances of governance.

Briefly, a strong political action, which of course takes into account and balances several variables, is fundamental. All the efforts must be directed towards the vision through a holistic approach able to manage and combine the complexity of urban and territorial structures with a shared future scenario.

A Framework to Apply ICT to Local Policies

Goals

According to Cullingworth and Nadin (2006), spatial planning is concerned with "the problem of coordination or integration of the spatial dimension of sectoral policies through a territorially-based strategy". The European Commission (1997) defines spatial planning as methods largely used by the public sector to influence the future distribution of activities in space. In this context the key role of spatial planning is clear in: promoting a more rational arrangement of activities; reconciling competing policy goals; creating a more rational territorial organization of land uses and the linkages between them; balancing the demands for development with the need to protect the environment and to achieve social and economic development goals; promoting sustainable development and improving the quality of life.

Thus, one of the main problems in recent spatial planning is how to make very broad and commonly used theoretical interrelated principles, such as sustainability, governance and ICT, more practical and operative.

The goal is to establish sustainable rural communities by balancing economic development, and environmental protection according to the carrying capacity of the land. As illustrated before, the area has a great agricultural and natural potentiality, and in the past there has been a strong link between human presence and natural environment. Thus, this plan treats conservation of the natural environment and development of the human community as equally important. We must take into account that rural, in this context, means open or sparsely populated areas as well as villages and small towns, and environmental means all the context, social and cultural, economical as well as physical. (Sargent F.O., Lusk P., Rivera J., Varela M., 1993).

The challenges of this plan are twofold: on one side to develop the ability of the local community to manage a sustainable environment, a viable economy, and other aspects that make up the natural and rural ecosystem and, on the other side, to accept the challenge and the opportunity offered by ICTs.

In particular, we are developing the following actions:

Internal to the territory. It consists in developing a plan, structured on ICT, and directed to increase the: management and monitoring capability of the natural environment; local services, assuring more efficiency and transparency of local governments and agencies; efficiency in the communicational exchange between all the actors supporting the traditional communication media; diffusion of computer literacy skills among the population; participatory policies and participatory planning processes.

External to the territory. It consists in facilitating the integration of the territory in the regional environmental network, and more generally in the space of flows. In addition to what stated above, the digital technology will provide more economic and cultural opportunities to this territory.

Hereinafter a framework will be illustrated.

Structure of the Proposed Framework

The elaborated procedure aims at furnishing a general framework able to support a sustainable policy plan through ICT tools, in which it will be possible to both manage and storage structured data, as well as obtain information and knowledge in several formats for all the citizens as well as for several stakeholders or other agencies.

According to this target, the framework consists of: Political Action Plan (PAP) to delineate all the actions needed for a new local policy; and a Technical Plan (TP). This plan concerns the capability to develop a conceptual model (anthropospheric system) able to translate the real environment into a set of interactive sub-systems. Thus, an interactive information system denominated Environmental Data-flow Analysis and Visualizing Information (EDAVI) will be illustrated. This specific system will be able to give support in management and analysis willing to increase the efficiency of services, land resource management, monitoring, etc. It will permit to address the policies.

Political Action Plan

According to international documents and a vast literature (Florida, 1995; Storper, 1997; Scott, 2000; Cooke et al., 1997, 1998; Camagni, 2002; Tura & Harmaakorpi, 2005), we can see a region as a reasonable entity in assessing economic growth and socio-institutional adjustment. Accordingly, sustainable competitiveness is the main source for the success of an economic actor, and the competitiveness of the economic actors is strongly related to their adaptability to the emerging techno-economic environment (Tura and Harmaakorpi, 2005; Schienstock and Hämäläinen, 2001 citing Abramovitz, 1995 and Lipsey, 1997).

Thus, the necessity to combine a political and socio-economic growth with sustainable development principles, means the planning of an institutional environment able to support theoretical and technological changes. According to this view, the main objective of winning local policy is based on promoting productivity and innovativeness by reconfiguring the socio-institutional settings (Tura and Harmaakorpi, 2005).

With regard to this, Storper (1995) says that the sources of competitiveness are determined in a non-market environment rather than in a market environment, including the untraded interdepen-

Figure 1. Actors involved in the implementation of local policy. (Copyright © 2009, IGI Global, the distribution of IGI Global in print or electronic formats without written permission is prohibited)

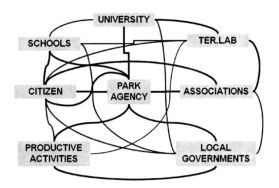

Figure 2. Political action plan. It consists in planning actions that are in relationship to one another, and where ICT can reinforce this action. (Copyright © 2009, IGI Global, the distribution of IGI Global in print or electronic formats without written permission is prohibited)

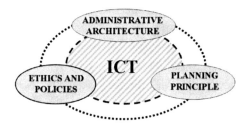

dencies as the institutional, social, cognitive and cultural conventions being formed during history in a region. Actually, according to several scholars (Kostiainen, 2002; Tura and Harmaakorpi, 2005) the Park Agency is working hardly to develop a system of innovative networks and institutions (Figure 1) located within this geographic area, with regular and strong internal interaction that will promote the innovativeness of the territory and its management and monitoring.

The Park Agency is the political core and the qualified structure in charge of the management and monitoring of the territory. TerLab is a Territorial Laboratory created to elaborate and develop educational and research projects.

In order to facilitate these actions, it's necessary to adopt a policy that will be a governance, or to encourage relationships between numerous actors and give a clear vision able to address soft factors of competition as identity or institution as much as hard factors such as tax level (Tura and Harmaakorpi, 2005; Boschma, 2003; Storper, 1997).

The Park Agency is developing a political action plan stressing on three main topics (Figure 2): ethic and policies, planning principles, and adjustment of local administrative architecture.

Thus, the administrative adjustment concern, at back office level, knowledge management. Today it is not realistic not to have this capability for both technological and human resources. Only a balanced support by these two elements can assure an optimal functionality and avoid wastes of money. A bad use of new technologies does not facilitate the public administrations; on the contrary, it can cause a rise in costs or an unbalanced relationship between benefits and costs; efficiency and agile bureaucracy. One of the most important goals should be improving the efficiency. Many studies suggest that ICT can bring a substantial efficiency, but it demands a strong leadership and a clear policy supported by an appropriate administrative structure. Oversized structures increase the costs, make the functionality heavy and produce an inefficacious informational flow and a slow decisional process. The processes bound for a more sustainable development could be optimized through prescriptive rules, or defining clear and simple norms able to give inputs for the local socio-economical growing processes. This should be a priority area, because through this action supported by ICT, it's possible to assure a correct and balanced development of the several components, social, economical as well as natural environment.

At the same time, it is necessary to re-organize and increase the efficiency relative to the front office needs. This implies an efficient use of ICT as tools able to furnish bigger and better: online E-service based on citizens' demand, cultural and business life events; support to the traditional services and channels, human, organizational and physical; E-democracy, or greater accountability, openness, transparency, accessibility, and participation into planning processes.

According to these policy actions, it is necessary to improve internet local access. This must be planned through a careful discussion in order to balance and properly combine financial and socio-political factors. Indeed, one of the main problems in this context is the relationship between public and private sectors and the best way to combine it. The private company which offers internet access does not consider a profitable business deal to invest in little communities that are geographically far away from main towns. Thus, in our case, it's possible to use the UMTS structure and/or develop agreement between communities and companies interested in increasing the number of customers. Another opportunity should be offered by the development of a few digital piazzas, or the use of a public library as library and digital piazzas with computer equipment and also the opportunity to attend computer and internet courses.

This is particularly important, because through a careful analysis and intelligent projects it should be possible to reinforce and encourage various synergic political (e.g. transparency, e-government, etc.), economical (e.g. e-commerce), and socio-cultural (e.g. preservation of local culture, health service, etc.) effects. Every single area must have a clear goal according to the vision, and the goals must be followed up and evaluated during the time. Without this kind of focus, the concept of ICT and sustainable development could easily be reduced to empty catchwords to which everyone pays lip-service, while no concrete results are achieved.

Technical Action Plan

This plan concerns the capability to develop a conceptual model (anthropospheric system) which could translate the real environment into a set of interactive sub-systems. It is based on other similar experiences (Haklay, 2002; Bossell, 1999) and structured according to complex systems theory, and it describes the local environment and the dependencies of human and natural activities through six sub-systems structured according to qualitative and quantitative indicators.

The model takes into account the financial aspect and the fact that many other experiences failed because it has been impossible, for local communities, to manage and update the databases as it often turned out to be a really expensive operation. Thus, I have decided to examine only those data strictly necessary to develop an efficient informational system able to provide the basic and necessary data-set. The framework tries to take into account several factors or trends which have influenced the political and social environment in the last few decades such as: the concept of sustainable development as a framework to integrate environmental considerations with economical development; the attempt to "modernize democracy" and urge a more participatory and inclusive mode of decision making and to combine it with more efficiency in the local administrations; helping the local economy, through ICT, and promoting the interaction between local and global entities; more opportunities for young people according to the notion of "information age", acknowledging the crucial importance of information and knowledge in the economy.

In this regard, the anthropospheric system is structured on the following six sub-systems: right of the individual, or the promotion of equity, social inclusion, health, job opportunities, etc.; social system, or the capability to increase social cohesion through sanitary system, welfare, ability of economical system to distribute local wealth, etc.; governance, or the capability of the local authorities to furnish transparency and effectiveness in the management of local government; infrastructure and services, or the incidence of the infrastructural web on the park territory, the availability of services, and the capability of the policy system to assure a sustainable development concerning all the components; economical system, or production and consumption, commerce, local market, job opportunities, income, consumption models, etc.; natural resources and environment, or safeguard of ecosystem, waste of natural resources, pollution, etc.

Then a data set of significant indicators has been identified. This data set must furnish the description of the system. It consists of three sub-systems, or anthropospheric sub-system, economical sub-system and environmental sub-system. In general, the data sets must be flexible, not expensive, including only the required data.

The indicator selection process is designed to satisfy three objectives: to identify the goals of the community; to understand and describe the environment (natural, socio-economical and urban); to identify a set of potential sustainability indicators.

In order to accomplish these objectives, several focus groups and a research team will be formed.

The community advisory committee, supported by researchers, will evaluate these potential indicators and combine them with a set of criteria culled from the literature. It is desirable that the indicators have the following characteristics: validity: the capability to measure an aspect of community sustainability; measurability: it should be possible to express indicators in terms of nominal, ordinal, interval, or ratio data; reliability. We must assess if the data can be reliably obtained: i.e., are the scope and quality of the data consistent from one period to another? responsiveness (sensitivity): it's important that the indicators respond quickly and reliably to those social, economic, or ecological changes which may significantly influence a community; data availability: relevant data must be obtained within a period of time sufficient to determine trends; comprehensibility: the indicators must be easy to understand for the potential end users.

Data availability may not be a long-term constraint since ways of generating information may be developed over time, or sources of data previously unknown to the community may be discovered. As for what concerns comprehensibility, in a limited number of cases, there is a role for indicators that may be technical in nature (e.g., a biodiversity index) and therefore not immediately understandable for laymen, a difficulty that can usually be overcome by explaining the indicator in a clear and simple language.

Now we are able to develop an Environmental Data-flow Analysis and Visualizing Information (EDAVI).

Some of the most important policy advances linking the information society and sustainable development have occurred in the areas of access to information and participation since 1992, when the 10th Principle of the Rio Declaration called upon signatory governments to encourage actions addressed at facilitating the public's right of access to information, stating that participation and justice in decision-making are instrumental in protecting the environment and in integrating environmental values into development choices. In that occasion, global policy-makers acknowledged the importance of information to decision-making declaring that "in sustainable development, everyone is a user and provider of information considered in the broad sense. That includes data, information, appropriately packaged experience and knowledge. The need for information arises at all levels, from that of senior decision-makers at the national and international levels to the grass-roots and individual levels" (Chapter 40 of Agenda 21).

Worldwide demand for access to information has grown in recent years, driven by the increase of ICT becoming a primary factor capable of promoting better decisions for decision-makers, empowering citizens to hold a more active role in decision-making and participatory processes.

Most of the early researches (Kersten, Gregory, Mikolajuk, Gar-On Yeh, 1999; Halder, Willard, 2003) on linkages between sustainable development and information society focused on the potential for information systems to assist decision-makers in monitoring and evaluating the state of the environment. The above introduced system, EDAVI, seeks to increase the quality, efficiency and accountability of decision-making processes through applications that systematically use anthropospheric system information. In this regard EDAVI development seeks to: enhance the use of harmonized environmental data sets through improving data availability; facilitate access to data both for decision makers as well as citizens. This will encourage and support participatory democracy and planning; ensure that data are internally consistent and that different data sets match.

We must take into account that information processing is, according to Herbert Simon (1973), the essence of decision making. So it becomes important trying to organize this complexity because any effective attack on problems of organized complexity must usually be mounted on several fronts and a conceptually sound, practical, and well-executed information strategy will, as Mason and Mitroff (1981) pointed out, be at the heart of every successful attack. Appropriate, high quality, easily accessible information is, therefore, vital to the success of any planning and management process. However, it's very important to remember that an excess of information becomes expensive and difficult to manage. It's therefore highly desirable to have a clear perspective on which information is essential, which is useful but not vital, and which is extraneous, even though this selection is difficult to achieve when the planning process is not clear about the goals.

EDAVI is a "light" environmental system based on GIS open source core. It aims at supporting the local authorities in the management and monitoring actions, and at assisting the display and query of information through a user-friendly interface especially for users without or with limited computer skills. As said above, it's extremely important to balance qualitative and quantitative data and to calibrate them to real necessity and financial capacity of the territory.

EDAVI display capabilities are also utilized for the purpose of public participation which is legally required in preparing development plans. It is important to gain feedback from the public for the review of plans. As said before, the core of this system is based on Geographical Information System (GIS). The introduction of the use of GIS provide an exciting potential for geographic information to be used more systematically and

Figure 3. Conceptual framework. From the citizen vision to an anthropospheric system able to describe the environment in its main components. Subsequently we have a data set of indicators analysed by GIS system. These analyses can be communicated to end users via internet (Copyright © 2009, IGI Global, the distribution of IGI Global in print or electronic formats without written permission is prohibited)

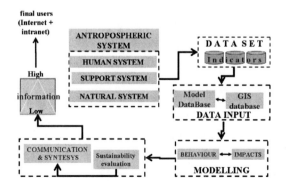

by a greater diversity of disciplines than even before. GIS has proved to be an invaluable tool for evaluating alternative solutions to urban planning problems. Planning database can be extensively interrogated to generate several alternative solutions to urban strategic planning problems. Various scenarios which take into account the socio-economic characteristic, the constraints of physical development, availability of land and land suitability for different kind of development can be generated within GIS or through the incorporation of modelling packages and other planning support systems.

EDAVI aims at being not just a planning support system for decision makers in planning and monitoring the area, but a communication tool as well. It permits, through the data set, to describe the countries' environment through an environmental information system, where information system includes natural environment, basic infrastructure, social and economical data for a holistic decision-making. The system so developed provides access to a comprehensive database which includes spatial data and analysis results, policies and guidelines for the purpose of planning and reviewing where development plan preparation is concerned. Of course a web-based GIS application is expected. Web-based GIS is seen as an evolving approach that provides large opportunities to GIS users in enhancing their involvement in planning and management for a more efficient and well organised project. It is seen as a good means to encourage public participation as well as a gateway to data integration and sharing between the planning agencies involved, especially through distributed data access.

Public participation in local planning must be a strategic choice for the local policy, and a tool to improve transparency and democracy. The use of internet will contribute to this process as it will facilitate the access to information giving, amongst other things, the possibility to consult the maps representative of the territory. For the purpose of coordinating the development in Monti Lucretili territory, the use of GIS for planning and monitoring is managed by the park agency and the GIS database is made up of integrated information provided by the local authorities directly involved in planning and monitoring of development in the territory. As such, the Park Agency, consisting of local members, will have full control of informational system. The web developed application, apart from encouraging data sharing between the various local authorities, will be, in the next future, an effective channel for the integration of the available data sets. The use of web-based GIS as a new medium to exchange data and gather information will brings numerous benefits: it will provide a simple means for the local authorities, and exchange of spatial information on-line. Data integration through the distributed GIS database is expected to minimize redundancy in data preparation as well as overlapping of information, apart from reducing time consumption and cost for database development.

However, it should be noted that the implementation of these technologies involves far more than hardware and software decisions. Effective implementation rests on a thorough and systematic evaluation encompassing planning, operational, organizational, institutional, personnel, financial and technical aspects. More research and attention need to be directed towards organizational and institutional issues, as well as developing the technology for planning and management purposes.

FUTURE RESEARCH DIRECTIONS

Over the last 10 years, digital technology has changed the way the world works, plays, communicates and shops. These changes have been so pervasive that it has become very hard to formulate a whole scenario, especially referred to the city and its complexity. This chapter has highlighted the fact that we are far from defining a clear picture of the impact of ICT on urban environment but, at the same time, it aims at furnishing a theoretical contribution in the direction of a balanced relationship between ICT impact and a sustainable spatial planning process.

In the next 5-10 years much more people than today will be connected, and in this "wired new world", digital connectivity becomes essential to achieve social and economic sustainability in cities of both the developed and developing world. We will use Internet for all aspects of our lives, and this will involve more efficiency and knowledge from decision-makers and planners. No longer will government web sites be sufficient. Computing is moving from a tool for the rich and educated to a standard part of life. Governments must use these tools because people will expect more from their governments in terms of transparency, efficiency, and services.

I have taken into consideration several researches that analyzed the impact of ICT on several fields of urban environment, and we have also seen which situations can produce negative scenarios.

However, a lot of research work still needs to be done in the field of ICT and spatial planning. Hereafter, I list a number of research topics and questions as: ICT role in global and local socio-economical changes and its effects on urban environment; on the rural environment, due to its peculiar socio-economical features, in which way we can combine a potential economical growth with a sustainable development that preserves the local identity; which role the policy (especially local policy) must take and how to balance the possible contrast between public and private sectors; planners' education, or in which way it's possible to update the planners' role. Now planners need to think hard both how to grasp and interpret the complex changes that cities are going through in the information age, and how to turn this understanding into action, successfully addressing the problem of setting strategies for the regeneration of the increasingly "digital" city.

The complexity of these issues and their transversality require a multidisciplinary and co-operative approach. The new challenge for urban planners and politicians is certainly to promote policies and planning procedures able to guarantee access to digital connectivity to all people, and a good level of telecommunication services for the local urban economy, and to combine these aspects with the relationship between private companies and public policies. This will be the main challenge.

The factor most likely to be crucial in the emergent "digital city" will be the ability of municipalities and technological entrepreneurs in general to mobilise a wide spectrum of knowledge and expertise to address tensions such as those highlighted here. The digital city has to deal with complexity just as the 'traditional' city did.

CONCLUSION

In this chapter I have explained that the issue of ICT and its impact on urban environment is a complex and multifaceted one. The flexible nature of ICT permits to adapt it to several environments producing strong relationships with economical and social aspects. For what concerns its possible development, we have attended to several positions: from visionary positions that have hypothesized the end of physical city or the death of distance, to a position that considered ICT just as an economical tool capable of increasing the economical growth, to come with scholars that have underlined the fact that ICT does not have a neutral nature. Anyway, it seems to be indispensable to incorporate it in urban policies and in urban planning processes. This is an inevitable and necessary topic that we must tackle.

A careful analysis of the planning literature shows a complex situation with many uncertainties. But it's clear that ICT development concerns, in particular, the metropolitan areas where the interaction between information technologies and the metropolis creates a mutual reinforcement. Within this scale – global scale – we can observe new relationships between these "global cities" and their surroundings (Castells & Hall, 1994; Sassen, 2001), new social and economical dynamics. On the other side, we have a local scale which concerns the medium-little size cities. In this set there are the country towns and very small villages that are cut out from economical opportunities. In this last context ICT can provide an appropriate support to increase a socio-cultural and economical growth as long as there is a citizens' vision about what the city is and what it will be, a clear policy able to address the strategic choices, and a planning system to put a plan into practice.

As far as policies matter, I have noted that urban ICT policies, in particular referring to local policies, are still at an early stage and in many situations I have observed that the political action is much too limited or lacking a clear strategy. Sometimes this unclear situation implies a waste of public money, and encourages economical or social imbalance. So, while the need of a national policy about ICT is clear, the role of local policy makers becomes crucial. With regard to this topic, many scholars (Malecki, 2002; Evans, 2002; Graham, 2002; Cohen, van Geenhuizen, Nijkamp, 2001) hold the role of local governments as important and critical to mitigate the above mentioned negative aspects and exploit the positive aspects of urban areas.

ICT is not a neutral technology. It can play both positive and negative roles according to the existence of a clear policy. I have identified some prerequisites and then according to the framework it is essential that the political actions should be structured following these matters:

adoption of a holistic approach: economical system addressed by a strong governance able to manage the conflict between public and private sectors; democracy: political choices should be public and transparent; information and communication: all the main information should be of public domain, using traditional communication channels as well as ICT channels; public participation: decision makers must promote public participation particularly in relation to the main decisions and promote participatory design; responsibility: public and private bodies should interact. So it's necessary to define distinctly the responsibilities every subject holds.

In this way ICT can have the capability to furnish new energy and provide answers to few fundamental needs and improve the quality of life. Finally, I would like to underline the fact that the framework compiled by no means constitutes a complete list but rather a starting point for any local institution aiming at incorporating ICT into its policies and planning processes.

REFERENCES

Abramowitz, M. (1995). The origins of the postwar catch-up and governance boom. In Fagerberg, J., Versbangen, B., & von Tunzelmann, N. (Eds.), *The dynamics of technology, trade andgrowth*. Brookfield: Edward Elgar.

Adams, B. (1993). Sustainable development and the greening of development theory. In Schuurman, F. J. (Ed.), *Beyond the Impasse – New direction in development theory* (pp. 207–222). London: Zed Books.

Adams, W. M. (2001). *Green Development: environment and sustainability in the Third World*. London: Routledge.

Agenda 21: Earth Summit - The United Nations Programme of Action from Rio. *(1993)*.

Alexander, C. (1964). *Notes on the Synthesis of Form*. Cambridge, MA: Harvard University Press.

Alexander, C. (1988). A City is Not a Tree. In Thackara, J. (Ed.), *Design After Modernism*. London: Thames and Hudson.

Alexander, C., Ishikawa, S., Silverstein, M., Jacobson, M., Fiksdahl-King, I., & Angel, S. (1977). *A Pattern Language*. New York: Oxford University Press.

Aurigi, A. (2005). *Making the Digital City: The Early Shaping of Urban Internet Space*. Aldershot, UK: Ashgate Publishing.

Baker, S. (2006). *Sustainable development*. New York: Routledge.

Bell, D. (1973). *The Coming of Post-industrial Society*. New York: Basic Books.

Bertuglia, C. S., Giuliano, B., & Mela, A. (1998). Introduction. In Bertuglia, C. S., Giuliano, B., & Mela, A. (Eds.), *The city and its Sciences* (pp. 1–92). Heidelberg, Germany: Physica-Verlag.

Betty, M., & Longley, P. (1994). *Fractal Cities*. London: Academic Press.

Boschma, R. (2003). *The competitiveness of regions from an evolutionary perspective*. Paper presented at the Conference of Regional Studies Association, Pisa, Italy, 12–15 April 2002.

Bossell, H. (1999). *Indicators for Sustainable Development: Theory, Method, Applications. A Report to the Balaton Group*. Winnipeg, Manitoba, Canada: International Institute for Sustainable Development.

Bossell, H. (2007). *Systems and Models Complexity, Dynamics, Evolution, Sustainability*. Books on Demand GmbH.

Brail, R. K. (2001). Planning Support Systems: A new perspective on computer-aided planning. In Brail, R. K., & Klosterman, R. E. (Eds.), *Planning Support Systems: Integrating Geographic Information System, Models, and Visualization Tools*. New York: ESRI Press.

Brandt Commission. (1980). *North-south: A programme for survival*. London: Pan Books.

Brown, L. D. (1991). Bridging organizations and sustainable development. *Human Relations, 44*(8), 807–831. doi:10.1177/001872679104400804

Bunch, M. J. (2000). *An Adaptive Ecosystem Approach to Rehabilitation and Management of the Cooum River Environmental System in Chennai, India*. PhD Thesis, University of Waterloo, Waterloo, Canada.

Camagni, R. (2002). *On the concept of territorial competitiveness: sound or misleading?* Paper presented at the 42nd Congress of the European Regional Science Association (ERSA). Dortmund Germany, 27–31 August 2002.

Caperna, A. (2005). *ICT per un progetto urbano sostenibile*. Milan: Tesionline.

Caperna, A. (2007). *Linee per un Piano d'Azione per lo Sviluppo Sostenibile del territorio del Parco Regionale dei Monti Lucretili*. Paper presented at the meeting of a.d.a. on the Environmental State Report, Rome, Italy.

Caperna, A., & Salingaros, N. A. (2006). *La complessità come nuovo approccio metodologico alla progettazione urbana ed architettonica*. Paper presented at the meeting of Istituto Nazionale di Urbanistica (INU) on Urban Environment and Complexity, Genoa 22-23 June 2006 (Italy).

Capra, F., Steindl-Rast, D., & Matus, T. (1992). *Belonging to the Universe: New Thinking About God and Nature*. New York: Penguin Press Science.

Caso, O. (1999). The city, the elderly, and telematic. Design aspects of telematic applications in a residential neighbourhood. In *Transformation, 2*. Delft University Press.

Castells, M. (1992). The world has changed: can planning change? *Landscape and Urban Planning, 22*, 73–78. doi:10.1016/0169-2046(92)90009-O

Castells, M. (1997). The education of city planners in the information age. Berkeley Planning Journal, No. 12.

Castells, M. (2000). *The rise of the network society*. Oxford, UK: Blackwell Publishers.

Castells, M. (2002). *The rise of the network society. The information age: economy, society and culture* (*Vol. 1*). Oxford, UK: Blackwell Publishers.

Castells, M., & Hall, P. (1994). *Technopoles of the World: The Making of Twenty-first-century Industrial Complexes*. London: Routledge.

Cernea, M. (1987). Farmer organizations and institution building. *Regional Development Dialogue*, *8*(2), 1–24.

Checkland, P. (1999). *Soft Systems Methodology: a 30-year Retrospective*. Chichester, UK: John Wiley & Sons.

Checkland, P., & Holwell, S. (1998). *Information, Systems and Information System -Making Sense of the Field*. Chichester, UK: John Wiley & Sons.

Cohen, G. (2003). *Urban future and ICT policy: perceptions of urban decision makers*. PhD Dissertation, Free University, Amsterdam.

Cohen, G., van Geenhuizen, M., & Nijkamp, P. (2001). *Urban Planning and Information Communication Technology*. Amsterdam: Timbergen Institute.

Cooke, P., Uranga, M., & Etxebarria, G. (1997). Regional Innovation Systems: Institutional and Organisational Dimensions. *Research Policy*, *26*, 475–491. doi:10.1016/S0048-7333(97)00025-5

Cooke, P., Uranga, M., & Etxebarria, G. (1998). Regional Innovation Systems: An Evolutionary Perspective. *Environment & Planning A*, *30*, 1563–1584. doi:10.1068/a301563

Coward, L. A., & Salingaros, N. A. (2004). The Information Architecture of Cities. *Journal of Information Science*, *30*(1), 101–112.

Cullingworth, B., & Nadin, V. (2006). *Town and Country Planning in the UK* (14th ed.). London: Routledge.

De Angelis, G. (1996). *I Monti Lucretili*. Roma: Provincia di Roma.

Drewe, P. (1998). *In search of new spatial concepts, inspired by Information Technology*. Report for the Ministerie van Volkshuisvesting, Ruimtelijke Ordening en Milieubeheer.

Drewe, P. (2000). *ICT and Urban Form. Urban Planning and Design - Off the beaten track*. Report for the Ministerie van Volkshuisvesting, Ruimtelijke Ordening en Milieubeheer.

Drewe, P. (2002). The Network City – from Utopia to new paradigm. *Atlantis*, *13*(5).

Drewe, P. (2003). *ICT and urban form. Old dogma, new tricks*. Delft: Delft University of Technology.

Edquist, C. (1997). Systems of Innovation Approaches (Their Emergence and Characteristics). In Edquist, C. (Ed.), *Systems of Innovation: Technologies, Institutions and Organizations*. Washington, DC: Pinter Publishers.

Elliott, L. (2004). *The global politics of the environment*. New York: Palgrave Macmillan.

Erickson, B., Lloyd-Jones, M. T., Nice, S., & Roberts, M. (1999). Place and Space in the Networked City: Conceptualising the Integrated Metropolis. *Journal of Urban Design*, *4*(1).

ESCAP. (1992). Social development strategy for the escap region towards the year 2000 and beyond. Bangkok. United Nations, ST/ESCAP/1124.

European Commission. (1997). *Compendium of European planning systems. Regional Development Studies. Report 28*. Luxembourg: Office for Official Publications of the European Communities.

Evans, R. (2002). E-commerce, competitiveness and local and regional governance in Greater Manchester and Merseyide: a preliminary assessment. *Urban Studies (Edinburgh, Scotland)*, *39*(5-6), 947–975. doi:10.1080/00420980220128390

Felleman, J. (1997). *Deep Information: The Role of Information Policy in Environmental Sustainability*. Greenwich, CT: Ablex Publishing Corporation.

Florida, R. (1995). Toward the Learning Region. *Futures*, *27*(5), 527–536. doi:10.1016/0016-3287(95)00021-N

Fraga, E. (2002). Trends in e-Government: How to Plan, Design, and Measure e-Government. In *Government Management Information Sciences (GMIS) Conference*, June 17, Santa Fe, New Mexico, U.S.A.

Frankhauser, P. (1994). *La Fractalité des Structures Urbaines*. Paris: Anthropos.

Gallent, N., Juntti, M., Kidd, S., & Shaw, D. (2008). *Introduction to Rural Planning*. New York: Routledge.

Graham, H. (2007). *Seeking sustainability in an age of complexity*. Cambridge, UK: Cambridge University Press.

Graham, S. (2002). Bridging urban digital divide Urban Polarisation and Information and Communications Technologies (ICT). *Urban Studies (Edinburgh, Scotland)*, *39*(1), 33–56. doi:10.1080/00420980220099050

Graham, S., & Marvin, S. (1996). *Telecommunications and the City*. London: Routledge. doi:10.4324/9780203430453

Graham, S., & Marvin, S. (2001). *Splintering urbanism. Networked infrastructures, technological mobilities and the urban condition*. London: Routledge. doi:10.4324/9780203452202

Griffin, T., Harris, R., & Williams, P. (2002). *Sustainable Tourism*. London: Butterworth-Heinemann.

Haklay, M. (2001). *Public Environmental Information Systems: Challenges and Perspectives*. Ph.D. thesis, Department of Geography, University of London, London.

Haklay, M. (2002). Public Environmental Information - Understanding Requirements and Patterns of Likely Public Use. *Area*, *34*(1), 17–28. doi:10.1111/1475-4762.00053

Halder, M., & Willard, T. (2003). *Information Society and Sustainable Development: Exploring the Linkages*. Paper presented at the World Summit on the Information Society.

Harris, B., & Batty, M. (1993). Locational Models, Geographical Information and Planning Support System. *Journal of Planning Education and Research*, *12*, 184–198. doi:10.1177/0739456X9301200302

Hester, U., Mesicek, R., & Schnepf, D. (2004). *Prerequisites for a Sustainable and Democratic Application of ICT*. Presentation at the World Summit on Information Society.

Hillier, B. (1996). *Space is the Machine*. Cambridge, UK: Cambridge University Press.

Hjorth, P., & Bagheri, A. (2006). *Navigating towards sustainable development: a system dynamic approach*. Amsterdam: Elsevier Ltd.

Honadle, G., & Van Sant, J. (1985). *Implementation for sustainability: Lessons from integrated rural development*. West Hartford, CT: Kumarian Press.

Jacobs, J. (1961). *The Death and Life of Great American Cities*. New York: Vintage Books.

Katok, A. B., & Hasselblatt, B. (1999). *Introduction to the Modern Theory of Dynamical Systems*. Cambridge, UK: Cambridge University Press.

Kelly, K. L. (1998). A systems approach to identifying decisive information for sustainable development. *European Journal of Operational Research*, *109*, 452–464. doi:10.1016/S0377-2217(98)00070-8

Kersten, G. E., Mikolajuk, Z., & Gar-On Yeh, A. (1999). *Decision Support Systems for Sustainable Development: A Resource Book of Methods and Applications*. Boston: Kluwer Academic Publishers.

Kofman, F., & Senge, P. M. (1993). Communities of commitment: the heart of the learning organizations. *Organizational Dynamics, 22*, 5–19. doi:10.1016/0090-2616(93)90050-B

Korten, D. (1990). *Getting to the 21st century: Voluntary action and the global agenda*. West Hartford, CT: Kumarian Press.

Kuhn, T. S. (1962). *The Structure of Scientific Revolutions*. Chicago, IL: University of Chicago Press.

Leitner, C. (2003). *E-Government in Europe: The State of Affairs*. Maastricht, The Netherlands: European Institute of Public Administration.

Lilienfeld, R. (1978). *The rise of system theory*. New York: John Wiley.

Lipsey, R. G. (1997). Globalisation and national government policies: An economist's view. In Dunning, J. H. (Ed.), *Governments, globalisation, and international business*. London: Oxford University Press.

Malecki, E. J. (2002). Hard and soft networks for urban competitiveness. *Urban Studies (Edinburgh, Scotland), 39*(5-6), 929–945. doi:10.1080/00420980220128381

Mantero, D. (2000). *Escursioni nel Parco Naturale Regionale Monti Lucretili. Provincia di Roma*. Roma: Italia.

Mantero, D., & Giacopini, L. (1997). *Guida al Parco Regionale Naturale dei Monti lucretili*. Roma: Regione Lazio, Italia.

Mason, R. O., & Mitroff, I. I. (1981). *Challenging Strategic Planning Assumptions*. New York: John Wiley.

Masuda, Y. (1981). *The information society as Post-industrial society*. New York: World Future Society.

Meadows, D. H., Meadows, D., Randers, J., & Behrens, W. W. III. (1972). *The limits of Growth*. New York: Universe Books.

Mitchell, W. J. (1998). *City of Bits: Space, place, and the infobahn*. Cambridge, MA: The MIT Press.

OECD. (2008). Information Technology Outlook 2008. Paris: Organisation for Economic Co-operation and Development.

Richmond, B. (1993). Systems thinking: critical thinking skills for the 1990s and beyond. *System Dynamics Review, 9*(2), 113–133. doi:10.1002/sdr.4260090203

Richmond, B. (1994). System dynamics/systems thinking: let's just get on with it. In *International Systems Dynamics Conference*, Sterling, Scotland, 1994.

Salingaros, N. A. (1998). Theory of the urban web. *Journal of Urban Design, 3*, 53–71. doi:10.1080/13574809808724416

Sargent, F. O., Lusk, P., Rivera, J. A., & Varela, M. (1993). *Rural Environmental Planning for Sustainable Communities*. Washington, DC: Island Press.

Sassen, S. (2000). *Cities in a World Economy*. Thousand Oaks, CA: Pine Forge/Sage Press.

Sassen, S. (2001). *The Global city*. Princeton, NJ: Princeton University Press.

Schienstock, G., & Hämäläinen, T. (2001). *Transformation of the Finnish innovation system. A network approach. Sitra Reports series 7*. Helsinki: Hakapaino.

Scott, A. J. (2000). *Regions and World Economy. The Coming Shape of Global Production, Competition and Political Order*. New York: Oxford University Press.

Sharma, S. K. (2004). Assessing E-government Implementations. *Electronic Government Journal, 1*(2), 198–212. doi:10.1504/EG.2004.005178

Sharma, S. K. (2006). An E-Government Services Framework. In Khosrow-Pour, M. (Ed.), *Encyclopedia of Commerce, E-Government and Mobile Commerce* (pp. 373–378). Hershey, PA: Information Resources Management Association.

Sharma, S. K., & Gupta, J. N. D. (2003). Building Blocks of an E-government – A Framework. *Journal of Electronic Commerce in Organizations*, *1*(4), 34–48.

Simon, H. (1973). Applying information technology to organizational design. *Public Administration*, *33*, 269–270.

Storper, M. (1997). *The Regional World: Territorial Development in a Global Economy*. New York: The Guilford Press.

Talvitie, J. (2001). *Incorporating the Impact of ICT into Urban and Regional Planning* (p. 10). Stockholm: European Journal of Spatial Development.

Talvitie, J. (2003). *The Impact of Information and Communication Technology on Urban and Regional Planning*. Report for Helsinki University of Technology, Department of Surveying.

Tura, T., & Harmaakorpi, V. (2003). Social Capital in Building Regional Innovative Capability: A Theoretical and Conceptual Assessment. Conference Report. In *43rd Congress of the European Regional Science Association (ERSA)*, Jyväskylä, Finland, 27–30 August 2003.

Tura, T., & Harmaakorpi, V. (2005). Measuring Regional Innovative Capability. In *Conference Report, 45th Congress of the European Regional Science Association (ERSA)*, Amsterdam, The Netherlands, 27–30 August 2005.

UN Global E-government Readiness Report. (2005). *From E-government to E-inclusion, UN-PAN/2005/14*. New York: United Nations.

UN/ECE. (1998). *Convention on Access to Information, Public Participation in Decision-Making and Access to Justice in Environmental Matters*. Aarhus, Finland: ECE Committee on Environmental Policy.

UNEP/UNCHS. (2008). *Building an Environmental Management Information System (EMIS). Handbook with Toolkit*. SCP Source Book Series.

Van der Meer, A., & Van Winden, W. (2003). E-governance in cities: A comparison of urban ICT policies. *Regional Studies*, *37*(4), 407–419. doi:10.1080/0034340032000074433

Van Geenhuizen, M. (2001). ICT and Regional Policy: Experiences in The Netherlands. In Heitor, M. (Ed.), *Innovation and Regional Development*. London: Edward Elgar Publishing.

Van Winden, W. (2003). *Essays on Urban ICT Policies*. Doctoral dissertation, Erasmus Universiteit, Rotterdam.

WCED (World Commission on Environment and Development). (1987). *Our common future*. Oxford, UK: Oxford University Press.

World Conservation Union IUCN. (1980). *World conservation strategy: Living resource conservation for sustainable development*. Gland, Switzerland: IUCN.

World Tourism Organization (WTO). (2004). *Committee on Sustainable Development of Tourism, 2004*.

ADDITIONAL READING

Adams, N., Alden, J., & Harris, N. (2006). *Regional Development and Spatial Planning in an Enlarged European Union*. Aldershot, UK: Ashgate.

Allenby, B. (2005). *Reconstructing Earth: Technology and Environment in the Age of Humans*. Washington, DC: Island Press.

Arnfalk, P., Erdmann, L., Goodman, J., & Hilty, L. (2004). *The future impact of ICT on environmental sustainability*. Paper presented at EU-US seminar: new technology foresight, forecasting & assessment methods, Seville 13-14 May 2004

Baker, S. (2005). *Sustainable Development*. London: Routledge.

Brown, L. R. (2003). *Eco-Economy - Building an Economy for the Earth*. New York: W. W. Norton & Co. Publisher.

Bruckmeier, K., & Tovey, H. (2008). *Rural Sustainable Development in the Knowledge Society*. Aldershot, UK: Ashgate.

Camagni, R. (2002). *On the concept of territorial competitiveness: sound or misleading?* Paper presented at the 42nd Congress of the European Regional Science Association (ERSA), Dortmund, Germany, 27–31 August 2002.

Campagna, M. (2005). *GIS for Sustainable Development*. New York: CRC Press. doi:10.1201/9781420037845

Caperna, A. (2005). *La percezione del ruolo della ICT in rapporto alle politiche locali nel contesto della rete città strategiche*. TIPUS Lab. Retrieved from http://www.pism.uniroma3.it/wp-content/uploads/2009/02/ict-rete.pdf

Caperna, A. (2005). *Introduzione alla Information Communication Technology (ICT)*. TIPUS Lab. Retrieved from http://www.pism.uniroma3.it/9-introduzione-alla-information-communication-technology-ict/

Carver, S., Cornelius, S., & Heywood, I. (2006). *An introduction to Geographical Information Systems*. Upper Saddle River, NJ: Prentice Hall.

Counsell, D. (2003). *Regions, Spatial Strategies and Sustainable Development (Regional Development & Public Policy)*. New York: Routledge.

Coward, L. A., & Salingaros, N. A. (2004). The Information Architecture of Cities. *Journal of Information Science*, *30*(2), 107–118. doi:10.1177/0165551504041682

Dresner, S. (2008). *The Principles of Sustainability*. London: Earthscan Ltd.

Galsson, J. (2007). *Regional Planning: Concepts Theory and Practice in the UK*. London: Routledge.

Gorman, M. E. (2005). Earth systems engineering management: human behavior, technology and sustainability. *Resources, Conservation and Recycling*, *44*, 201–213. doi:10.1016/j.resconrec.2005.01.002

Haklay, M. (2001). *Conceptual models of urbans environmental information systems – toward improved information provision*. Working Paper, Centre for Advanced Spatial Analysis, University College London.

Halder, M., & Willard, T. (2003). *The Information Society and Sustainable Development*. Manitoba, Canada: International Institute for Sustainable Development.

Harmaakorpi, V. (2004). *Building a Competitive Regional Innovation Environment – the Regional Development Platform Method as a Tool for Innovation Policy*. Doctoral Dissertation. Espoo, Finland: Helsinki University of Technology.

Harmaakorpi, V., & Pekkarinen, S. (2003). *Defining a Core Process in a Regional Innovation System – Case: Lahti Age Business Core Process*. Paper presented at the Conference of Regional Studies Association, Pisa Italy, 12–15 April 2003.

Healey, P. (2006). *Urban Complexity and Spatial Strategies: Towards a Relational Planning for Our Times*. New York: Routledge.

Information Society and Agriculture & Rural Development. (2006). *Linking European Policies. European Commission*. Luxembourg: Office for Official Publications of the European Communities.

Kingston, R. (2002). The role of e-government and public participation in the planning process. In *XVI Aesop Congress Volos*. Retrieved from http://www.ccg.leeds.ac.uk/democracy/presentations/AESOP_kingston.pdf

LaGro, J. A. (2007). *Site Analysis: A Contextual Approach to Sustainable Land Planning and Site Design*. Hoboken, NJ: John Wiley & Sons.

Laurini, R. (2001). *Information Systems for Urban Planning*. London: Taylor and Francis.

Mortola, E. (2003). *Architettura, comunità e partecipazione: quale linguaggio? Problemi e prospettive nell'era della rete*. Roma: Aracne.

Nelson, R. R., & Nelson, K. (2002). Technology, institutions, and innovation systems. *Research Policy*, 31, 265–272. doi:10.1016/S0048-7333(01)00140-8

Otjens, A. J., & van der Wal, T. Wien J.J.F. (2003). *ICT tools for participatory planning*. Paper presented at European Federation for Information Technology in Agriculture, Food and the Environment (EFITA) Conference, 5-9. July 2003, Debrecen, Hungary

Peraro, F., & Vecchiato, G. (2007). *Responsabilità sociale del territorio. Manuale operativo di sviluppo sostenibile e best practices* (Angeli, F., Ed.).

Rambaldi, G., Corbett, J., Olsen, R., McCall, M., Muchemi, J., Kwaku Kyem, P., et al. (Eds.). (2006 April). Mapping for change: practice, technologies and communication. Participatory Learning and Action, 54.

Rotondo, F., Selicato, F., & Torre, C. (2001). *A Collaborative Approach To An Environmental Planning Process: The Lama Belvedere Urban Park in Monopoli*.

Sassen, S. (2007). *A Sociology of Globalization*. New York: W. W. Norton & Co.

Shiblaq, F. (2008). *ICT in Rural New Zealand*. Berlin: VDM Verlag.

Tongia, R., Subrahmanian, E., & Arunachalam, V. S. (2005). *Information Communication Technology for sustainable development*. Bangalore, India: Allied Publishers Ltd.

Wilson, A. G. (2006). Ecological and urban systems' models: some explorations o similarities in the context of complexity theory. *Environment & Planning A*, (38): 633–646. doi:10.1068/a37102

Wilson, A. G. (2008). *Urban and regional dynamics – 2: an hierarchical model for interacting regions*. Working Paper 129. London: Centre for Advanced Spatial Analysis, University College London.

Wong, C. (2006). *Indicators for Urban and Regional Planning*. New York: Routledge.

KEY TERMS AND DEFINITIONS

Framework: An instrument providing a general spatial framework for a town or a territory Information and Communications Technology (ICT). We can refer the relationship, through digital technology, between information and communication and how we can storage, analyze, and to spread information. It includes any communication device or application, encompassing: radio, television, cellular phones, computer and network hardware and software, satellite systems and so on, as well as the various services and applications associated with them, such as videoconferencing and distance learning. About the ICT sectors, in 1998, OECD member countries agreed to define the ICT sector as a combination of manufacturing and services industries that capture, transmit and display data and information electronically.

Infrastructure: Basic services necessary for development to take place (roads, electricity, water, education, health facilities, communications, etc.).

Land Use: The way land is used or developed.

Local Authority / Local Government: The lowest tier of elected government.

Local Plan: A plan that sets out detailed policies and specific proposals for the development and use of land in a district and guides most day-to-day planning decisions.

Local Policy: A comprehensive view about a territory able to address the strategic choices of a community.

Spatial Development: Changes in the distribution of activities in space and the linkages between them through the conversion of land and property.

Stakeholder: An institution, organization, or group that has some interest in a particular sector.

Strategic Planning: Preparation of a strategy identifying the broad patterns of growth; it is generally long-term and comprehensive, bringing together social, economic and spatial considerations.

Sustainable Development (SD): According to the WCED, SD is "development that meets the needs of the present without compromising the ability of future generations to meet their own needs." The essence of this form of development is a stable relationship between human activities and the natural world, which does not diminish the prospects for future generations to enjoy a quality of life at least as good as our own.

Chapter 19

Architectures of Motility:
ICT Systems, Transport and Planning for Complex Urban Spaces

Darren J. Reed
University of York, UK

Andrew Webster
University of York, UK

ABSTRACT

This chapter engages with contemporary approaches to urban planning by introducing an analytic strategy rooted in the sociological approach of Science and Technology Studies. By demarcating a 'social frame' and comparing this to the established 'engineering frame' through different 'architectures', the chapter reveals hitherto unrecognised features of the implementation of an intelligent transportation system called BLISS (the Bus Location and Information SubSystem). Through the 'mobilities' conceptual approach, the relationships between various aspects, including the urban space, the experience of passengers, drivers and managers, and component technologies, are revealed as forming an 'assemblage' of conflicting features, that at the same time move toward a form of 'stabilization'. The underlying point, is that we need to engage not only with the technical difficulties of technology implementation in the city, but also with the contingent and experiential processes of those who use, and are affected by such implementations.

INTRODUCTION

The notion of e-planning in urban contexts is, in important ways, linked to facilitating and disciplining the movement of people, their mobility within urban space. The Strategic Research Agenda of The European Research Forum for Urban Mobility (EURFORUM) identifies four main components of a proposed 'urban mobility system': 1. users' needs and behaviours; 2. the urban structure; 3. integrated mobility services (usually based on Information Communication Technologies ICTs); and integrated transport systems. The first two relate to transport demand, the second two to transport supply. The agenda therefore prioritizes 'users and user-related organisations' in the development of effective and sustainable transport policy and practice. E-planning is increasingly geared towards understanding the

DOI: 10.4018/978-1-61520-929-3.ch019

impact of policy measures and system innovations on mobility behaviour.

In this chapter we suggest how e-planning can be strengthened through undertaking a sociological analysis of mobility based upon the study of the design and implementation of an Intelligent Transport System (ITS) in a northern city in the UK. Key to our analysis is an understanding of the way different configurations of urban arrangements, technology interventions, and user experiences result in varying 'degrees of mobility' (Urry, 2007) for different people. We discuss this understanding in relation to the social theory of 'mobilities' that provides an analytic frame for developing a nuanced understanding of the complex relationships between physical spaces, behaviour, experience and technologies (Kaufman, 2002). Our substantive focus is on the bus, seen as a crucial public transport in urban centres; and much of our story is about matters of detail that might seem mundane. In fact, however, any future e-planning will need to understand how innovation will only become normalised if it articulates with everyday existing arrangements and practices.

A defining feature of transport management systems is that while they are integrated into multiple aspects of the urban environment and a key component of e-planning, they are largely invisible as a form of "ubiquitous computing" (Weiser, 1993). Much of this chapter seeks to reveal the details the features of the ITS central to the process of reconstructing (or 'reassembling' (Latour, 2005)) the system as an understandable social object. These descriptions lead off from moments of 'breakdown', wherein the system is seen to fail. Importantly, these failures are seen as such by the users and stakeholders of the system, rather than the consequence of an objective assessment on the part of the researcher, as some sort of evaluator or arbiter of worth. They are also characterized by multiple features, at once technical, perceptual, and social. They are far from mere 'technical failures'. To bring out these cross cutting features, we conceive of these breakdowns as occurring at points of intersection or interaction between differing 'architectures'. Our motivation here is to engage with engineering understandings, yet disrupt taken for granted notions of the term. The descriptive elements are then combined with theoretical and conceptual resources to form a 'deep' reading of the system and its social consequences. Our conclusions suggest the need for a perceptual or paradigmatic shift in this aspect of e-planning (within design, as much as research). The resulting sensitivity to the multifaceted features of the ITS, and other such large-scale, ubiquitous, urban technologies that feature in e-planning, should carry forward into research and design activities by orienting practitioners to the socio-technical, temporal and essentially interpretive aspects of the technical system itself.

The description and discussion are based upon a three-year study of the design and implementation of the system. The research involved a combination of interviews, focus groups, and fieldwork observations. Regular semi-structured interviews were carried out with the primary project collaborator at the Network Management team as well as with related staff from the transport section, complemented by informal telephone interviews throughout the fieldwork stage of the project. Interviews and day-long discussions were also carried out with members of the technology development team, and members of the bus company's operational management team. These formal interviews (n=10) were underpinned by a more informal strategy, whereby results of the analysis were presented in successive feedback and discussion meetings. Interviews were also conducted with bus drivers during the latter part of the fieldwork to determine their views on the system and the changes it would bring to their work practices. Focus groups (made up from 25 passengers) were carried out throughout the implementation of the system (n=4). Initially these were characterized by facilitated discussion centering on the passengers' practices and attitudes towards bus travel in the York area and scenario and personal based material to explore perceptions of utility in such systems (Cooper, 2004). We also carried out a design exercise based upon the Participatory Design method and asked our participants to design travel information systems. At regular intervals in the implementation a researcher undertook field studies within the city. These were in part to record the status of the installation, but also used to collect photographic and video data of the interface. When appropriate, passengers standing at bus stops were engaged in informal discussion about the passenger information panels (PIP).

BACKGROUND

Perry (2003) argues that urban planning is concerned with the intersection between two types of 'plans': the planned city or "la ville crée" and the 'lived' or 'everyday' practices of people in the city or 'la ville spontanée'. Accordingly 'planning mediates between the *freedom* of the city as ville spontanée and the orderly production of the city as ville crée" (p.143). Into this formulation we must add the mediating role of ICTs. Such technologies fit with a 'new urbanism' approach to urban planning which 'promotes a revitalized vision of high-density, transit-and-pedestrian-friendly neighborhoods as an antidote to faceless suburban sprawl' (Campbell and Fainstein, 2003 p.10), yet their introduction is often undertheorised, and its role in the relationship between the city as physical constraint and its meaningful social construction goes unquestioned. At the same time, technology implementation gathers apace.

The confluence of urban planning and ICTs brings the potential for two distinct orientations. The first we will describe in terms of the technological architecture or engineering frame; the second is the social frame, that incorporates the technological architecture into the social architectures of passengers and driver interpretations, organizational structure and relationships between different stakeholder groups. In relation to planning theory the dynamic nature of ICTs 'intervention' in the city extends and questions traditional 'comprehensive planning' notions because the city space becomes 'fluid' and dynamic shaped by the technology itself.

From a technological point of view, the integration of ICT into urban planning is seen as a means to better manage the movement of people and transport through the city environment. Planning then becomes a contingent matter of movement and flows and moves beyond a simple materialist notion of space management. The 'interpretive turn' in urban planning theory (Healey, 2002/1997) (also known as 'communicative planning', see Woltjer, 2002) rests on the social construction of the city and the need to recognize the 'collaborative' realization of meaning through contingent practices in and through the cityspace. Hence "planning work is both embedded in its context of social relations through its day to day practices, and has a capacity to challenge and change these relations through the approach to these practices; context and practice are not therefore separated but socially constituted together" (p. 499). The urban is for Graham (1999) a matter of local agency, and with the introduction of ICTs these local and contingent influences are unpredictable. As McGuigan (1999) argues, 'There is no inevitability in the making and deployment of technologies" (p. 1). Graham asks why there is little recognition of local agencies within urban studies, what he calls 'maneuvering space' (ibid, p. 11). He decides that this is because of a dominant 'technological determinism' in play, which assumes a 'grand narrative' of objective, and 'revolutionary' moves towards an 'information society'. Hence '[i]n this rush to describe this re-creation of the world, actual telecommunications-based developments in real contemporary cities are rarely analyzed in detail" (p.12).

In 2005, the UK Department for Transport (DfT, 2005) published its 'Ten Year Plan', a central feature of which was the allocation of over £20m to support the pilot introduction of real time information (RTI) initiatives on bus and public transport services, funding for which Local Authorities were invited to compete. However, public transport RTI is a long way behind that for roads. There is a very low level of geographical coverage for real time information, particularly for buses, and generally the quality is inconsistent and complaints are relatively high (DfT, 2005).

RTI in a public transport context, the DfT believed, would have a number of otherwise elusive benefits. RTI might for instance improve the moment-by-moment coordination of public transport, preventing for instance the perennial problem of 'bunching' whereby buses on a particular route will over a period of hours converge. Crucially, in addition to service management improvements it is believed that better access to information – on a real time basis – will encourage a modal shift away from private transportation. In turn, modal shift is seen, not only in the UK but globally, as an important means of local authorities meeting their air quality and environmental targets

One of the private sector initiatives that responded to this new funding was the BLISS (Bus Location and Information Sub System) system, which had been under development through the combination of discrete elements brought together through the merger of competing companies during the late 1990s. Wherever such systems are introduced they have to contend with differing local urban settings, and the physical and technical legacies they carry, such as existing urban traffic management programmes that focus especially on the car. As elsewhere, one key initiative of this particular City's transport planning department has been the development of the park-and-ride services to lower the number of cars coming into the centre by providing frequent bus services from car parks on the outskirts of the town.

The system as a whole provides passenger information at bus stops in terms of predicted arrival times of buses, automated traffic management through the prioritising of late buses through traffic light junctions (so providing more consistent bus running times), and traffic report data, enabling responsive management and the development of more effective timetables. Its most public facing element is the electronic PIP positioned at some (though not all) bus stops. The system is made up of various components, shown in Figure 1. A comprehensive list of the components can be found in the appendix.

In brief, the design of BLISS is distributed across various sites and components: equipment at bus stops - the PIPs; that found on buses (GPS receivers and onboard controllers that receive information from the ticket machine); and traffic lights (additional 'controllers). Computers act as 'consoles', showing simple map based tracking of vehicles, at various places (including the bus drivers' cafeteria, and the city network management offices); an "instation" is located at the Council buildings and main servers receiving and relaying information about bus movements to the supplier and bus operator. The server is now located remotely in London, though initially was sited at the Council. Traffic lights are fitted with additional 'controllers'. With respect to information that is of interest to bus operators, BLISS offers the operational and management software 'Busnetlive', which provides real time and historical performance information on the movement of buses. The software can be used for audit and reporting purposes. Such information is of interest because bus operators can be penalised financially by the Traffic Commissioner if their reliability levels fall below a certain threshold.

This sketch is a technology-based description of BLISS, one that can be found in the 'System Specification' document provided by the designers and implementers of the system. This document is a fixed reference point, an 'immutable mobile' (Latour, 1987), that provides for a common understanding of BLISS. It also provides a contractual basis for judging what the system is, and what the suppliers are expected to provide. Yet it does

Architectures of Motility

Figure 1. Components of BLISS

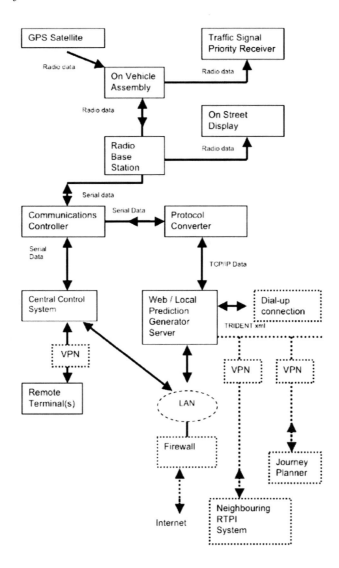

not include how the system is actually used, nor the details of the urban context into which it is added and the various other computer systems already in place. Crucially it does not detail the implementation process nor take account of the interpretations attributed to BLISS over the period of implementation.

The system is, as noted above, typically hidden but becomes visible when there is "breakdown" in its 'readiness-to-hand' (Urry, 2007; Wright & Monk, 1991; Heidegger, 1962). These moments often occur at points where the different architectures intersect. Such breakdowns prompt a form of questioning of the system. In the sociological literature there are a number of instances when breakdown is used as means to carry out analysis. Garfinkel (1967) for example proposed to deliberately undermine taken for granted expectations in social life, through what he called 'causing trouble', as a means to reveal how order is achieved revealing the methods by which people make sense of their world. Dewsbury (2000) identifies 'asignifying rupture' (p.477) as a useful analytical moment by saying, "only when there is a disturbance of some sort do we appear to ourselves as agents, with beliefs and desires directed toward goals that require some particular action" (Dewsbury, 2000: 477). In this case it is the person who is under question in terms of the 'performativity' (Jacucci et al, 2005; Michael, 2000) of a particular role. Taken together, moments of breakdown can then instigate questions about performativity and the meaning and nature of a technology.

Moments of Breakdown

We first then introduce three moments of breakdown that were identified in the research, we will then go on to detail our conception of the different architectures involved and then discuss the relation between our examples of breakdown and architecture. Breakdown was seen:

1. In the cab: the BLISS system fails to recognise individual buses and so the prediction of bus arrival and the prioritisation of bus movement and provision of information are undermined.
2. At the bus stop: the information panel displays confusing and inconsistent information.
3. In the control room: the calculation of bus running 'efficiency' undermines the bus manager's position.

These range from tangible instances when BLISS simply did not function (in the cab), through an instance of perceptual breakdown (at the bus stop), to an interpretative breakdown, in that BLISS was seen to contradict appropriate 'working' parameters (in the control room) for particular stakeholders. We order them in this way so as to broaden the applicability of the conception of breakdown to include technological, experiential, perspectival, and finally configurational questions. We first need to sketch out our architectures, so as to detail how these moments of breakdown occur at points of intersection between architectures.

Intersecting Architectures

The term architecture is typically applied to the 'total built environment' and refers to form and function within material or spatial constraints. The architecture of software systems on the other hand refers to the structure of components and their interrelationships over time. These traditional uses of the term usually relate purely to the technical object (building, features such as traffic lights, computer systems etc.). The metaphor allows for a material appreciation of BLISS that, while essentially an invisible (pervasive) computer system, is situated in the physical contingencies of an historic town.

The word architecture also speaks to 'the realization and reification of structures of relevance' (Chambers Dictionary, 2008). This allows us to extend architecture as a metaphor to include people and places as something like components and the dynamic relationships of people to the system over time in terms of their attributions of meanings and values.

We can, then, think in terms of four architectures: the architecture of BLISS (understood as a distinct thing); the contextual architecture into which BLISS was placed (other systems, the physical environment etc.); the temporal architectures in play across the system-human divide; and the architectures of meanings attributed to it by passengers, bus drivers, traffic managers etc. In each case we pursue relationships within and between architectures as moments of intersection through selected examples based on two key technical objectives, 'bus priority' and 'predicted arrival time'.

Architecture of BLISS

As we have seen, BLISS is a highly complex system that incorporates a range of components. Some are installed on buses, while others are situated at particular places. This may be in a control room, at a bus stop, or at a traffic light.

We saw the various technical aspects of BLISS in Figure 1. The different components are themselves self-contained technologies and manufactured by different companies. In addition to configuring these new components, BLISS has to articulate with existing elements. For example the PIPs are installed at pre-existing bus shelters and these physical constructions comprise backdrop for the information screen. An interesting design choice utilises the existing physical bus shelter to affect ownership by the bus passengers. The PIPs are installed to be seen by passengers standing facing the oncoming traffic, and hence invisible to the drivers of the buses. This was a deliberate move, according to the suppliers, to alleviate driver's concerns that they were being monitored *by the customers*. Another appropriated technology is the ticket machine on the bus, which acts as an interface between the driver and the onboard controller. As we will see, this was initially unsuccessful, and caused system failure.

Contextual Architecture

BLISS was introduced into a situation containing a vast array of existing material and non-material features. These included already installed and working computer systems, organisational arrangements and practices, the traffic infrastructure, including the physical traffic lights; and the matrices of documentary artefacts, including bus schedules and human and equipment resource scheduling. All of which have a history of development and a level of permanence.

The city under study has a population of 137,500 people and is served by a number of national bus companies. The transport market is dominated by one company that has 62000 employees in the UK and US. The company operates 20 routes in the city, is competitive and looks to develop its service in a number of ways. This includes the introduction of novel 'streetcar' buses, which emulate traditional city trams, built by Wrightbus based upon a Volvo B7LA engine and chassis.

As an historic centre, the city is partially surrounded by a medieval wall with 'gates' through which buses must traverse. There are strict building controls that limit the possible approaches to traffic management such as the creation of additional bus lanes. BLISS was seen as a way to circumvent these restrictions through electronic means, a viable alternative to physical adaptations. The city is characterised by straight roads extending from the centre cutting the suburbs into triangular segments. Movement from one segment to another is impeded by rivers, which are only bridged nearer the centre. The net effect is that there is a great deal of traffic on these radial roads. At the same time, the city is characterised by tourism. At different times of the year there is an influx of visitors and hence there is a need for adequate public transportation services. Around the outskirts of the city are positioned a number of 'Park and Ride' stations at which visitors (and everyday commuters) can park their cars and then travel into the city centre by bus. It was on one of these 'Park and Ride' routes that the information and control system was first tested.

To simplify matters we can think of some aspects of this contextual architecture as subordinate to BLISS (e.g. the existing equipment on buses) incorporated as a component. Other features can be seen as acting in parallel in a 'sibling' relation (e.g. other traffic management systems). Still further features can be seen to be super-ordinate (e.g. the urban environment). It is important to maintain recognition of the irreducible interconnectedness of the features. So, for example, the ticket machine, which is in a subordinate relationship to BLISS, and that communicates with the 'onboard controller', is configured in line with the physical bus stops on a route that is itself situated in relation to the existing road system and hence based upon the super-ordinate architecture of the city.

Temporal Architectures

There are two identifiable temporal aspects of the BLISS system: that concerned with the everyday moment-by-moment management of buses; and that concerned with providing information to potential passengers at the bus stop. As we will see however, these aspects are not unrelated, and both rely upon the tracking of buses. These new temporal architectures are built upon existing architectures, including the historical production of the timetable by the bus operator and the formulation and distribution of chapter timetables to passengers.

Existing Temporal Structures: (I) Bus Management

Any bus service has a public facing 'timetable' and an administration oriented 'running board'. This second set of timings is an administrative tool and includes not only the points in time that a bus should be at a particular bus stop (as in the public facing timetable), but also emphasizes various 'timing points' when a bus should arrive and rest at a particular point on a route. If a bus arrives earlier than the stipulated timing point, it has to wait. In addition the running board is organised to show a series of individual 'bus journeys' in sequence, and may include those from different routes, such that a bus driver following a particular running board first drives one route, and then journeys to another route and drives the bus on this route. The running board therefore includes such timings as 'walking time' between different routes, rest periods (including lunch break and short "cab breaks") and acts as the main standard by which the organisation, in terms of ongoing management of the bus service, is judged. However it is only the 'timing points' that are used to calculate whether a bus is 'running to time' and are the only measures of the service's efficiency. The timing points for the total bus service are registered with the UK government, and as mentioned earlier, there is a financial penalty levied when buses do not maintain a particular level of efficiency (at the time of the study this was that 95% of the buses were running within a five minute variance of the scheduled times). While it is the bus operator who creates these standardised timings (the timing points are typically the same for each weekday and for each weekend day), the guarantee to meet them on every occasion can cause problems due to varying changes across different days in the week, and different points in the year. In addition, ad hoc influences such as road works or weather conditions may dramatically influence the buses' progress. The skill in writing a schedule and the accompanying running board is in balancing all these changes while writing a standardized timing document. In addition, as has already been intimated, the different running boards must complement each other and fit together to form a complete service, that has multiple routes, and tens of drivers: they must fit together to provide an adequate service, while at the same time meeting the contracted working hours of different sets of drivers (part time, full time etc.).

Historically, the bus operator has moved from a 'pencil and paper' system to one based upon various monitoring tools. For example, on-street supervisors sample the arrivals at a particular timing points and report these back to the management. It cannot be emphasized enough however, how random these activities are. In addition, supervisors are typically ex-drivers and are seen to be 'on the driver's side'. Historically, aside from customer feedback and complaints, there have been few other indicators of an individual bus's progress and hence the efficiency of the bus service.

Existing Temporal Structures: (Ii) Bus Information

The schedules mentioned above are publicized through paper leaflets and on bus stop noticeboards. One set of resources that sits alongside the published timetabled schedules is the local knowledge of regular passengers. This can be based on repeated journeys taken at the same time each day, or may be in terms of mental representations of service regularities, such that a particular route has a bus every ten minutes. The published timetables take advantage of this kind of local knowledge by firstly incorporating regularities such as this and also by publishing a service as having a regular character. Indeed often the printed timetables will give specific times at the beginning and end of the table and then denote regular intervals in the intervening period. These local understandings are a form of local meaning and result in repeated practices.

GPS equipment calculates the location of the bus every thirty seconds. This information is transmitted to the BLISS instation and compared to a bus schedule and a 'degree of lateness' is calculated (from 1 to 'more than 5' minutes). Figure 2 describes this process. The degree of lateness is transmitted back to the bus, and as it reaches a particular location the ascribed degree of lateness is relayed to equipped traffic lights. The traffic light, based on communication with the Traffic Congestion Management System (TCMS), decides whether or not to give 'priority' to the approaching bus and then decides to do this either by changing the lights more quickly ("fast-to-green") or holding the lights for longer ("stay-on-green"). At the same time the GPS information is transmitted across the city to all PIPs. Each PIP listens for relevant information based on bus route and journey number and with this calculates the 'predicted arrival time' for a particular bus. This is displayed to waiting passengers.

We now turn to consider such aspects as a form of architecture of meaning and practice.

Architectures of Meaning and Practice

The BLISS computer system relies upon, and affects a large number of people who comprise sets of 'stakeholders', which includes passengers, drivers, the designers, the Network Management team and bus managers. In addition there are other road and city 'users': pedestrians, car drivers, and passengers of other bus services. We consider here four main stakeholders: the network managers, the bus operator, the bus drivers, and the bus passengers.

Network Managers

The story of the implementation of BLISS plays out against a backdrop of ascribed meanings and expectations. For example it was assumed by the

Figure 2. Process and temporal dynamic of the BLISS

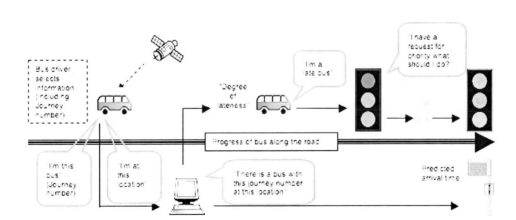

local council's transport Network Management team that the incremental implementation of the system elements would be enough to persuade the bus management of the "benefits" of BLISS in that it would show that the system would indicate the potential for consistency of journey time. They hoped that the private bus operator would recognize these benefits and agree to help finance the maintenance of the system over time. The benefits imagined by the Network manager were not realised because the limited initial implementation did not have a noticeable effect on the efficient running of the total bus service, in part because of the limited number of equipped traffic junctions and the range of technical problems, but more importantly because the bus operations team had a different notion of quite what the benefits might be.

Bus Operators

The bus operator understood the potential benefits of the BLISS system in terms of customer satisfaction. This would ideally have been seen in a more regular and efficient bus system, but could also be improved through the impression that the bus company was investing in new equipment. Yet, the bus operator's accumulated knowledge and experience told him that the response of customers to changes in the bus service were typically some time after the changes had been affected - normally eighteen months after a change had taken place (a new route, improved bus conditions etc.). In terms of timescales, the network managers and the bus managers were working to two different time periods. The Network managers were expecting an immediate response from bus operators, which the operators (looking to customer responses) were thinking in terms of years, not months. From the beginning the bus operator was sceptical about the system's pervasive but invisible character, saying that BLISS amounted to 'priority by stealth'. The fact that the system works 'behind the scenes' robbed it of being a visible addition to the bus service.

Bus Drivers

The onboard equipment was installed such that it was not immediately visible or accessible to the bus driver. What information had been given to the drivers (and this was limited) emphasized the GPS aspect, which recorded the location of the bus and transmitted it to the instation. The advantages of the bus priority aspect were completely obscured. The bus driver's experience of the system was purely via the passenger's experience of the PIPs, which, as we have seen, were installed so that they could only be seen by the passengers and not the bus drivers. BLISS resulted in additional 'monitoring' by the passengers.

Bus drivers are largely self-managing. There are a small number of bus supervisors, who are, or who have been, drivers. The decisions that affect efficient running are made at street level. One of the clearest examples of street level control of buses comes in terms of managing 'headway' (the time between buses on the same route). As headway reduces and buses 'bunch', the decision will be made (either between bus drivers over radio communication, or by a supervisor) that the relatively empty bus overtakes the one running late. This results in a reordering of the buses on the route and hence the BLISS journey numbers. A consequence of which is that PIPs are listening for the wrong buses. The new technical arrangements effectively move control from the street-level to the operations management offices.

The issue of journey number brings us to an example that points up the social aspects of the BLISS computer system. While notionally part of existing bus driver practice, journey numbers had far less importance. Historically when a driver starts a route she or he must enter information into the ticket machine to bring up the correct 'fare table' (based upon the companies fare structuring). To do this all that the driver needs to do is enter the correct route, and the correct direction of travel. They do not have to enter the correct journey number. Given that at best the BLISS system is invisible to them, and at worst it is a new form of monitoring, it is unsurprising that this practice changed only slowly.

Passengers

Bus passengers bring a wealth of accumulated understandings and expectations. For example passengers on the Park and Ride service are used to driving to a car park and there being two or three buses waiting to take them into town. Passengers in the research focus groups expressed detailed knowledge of which particular bus stops to catch a bus at. Park and Ride passengers would go to the bus stop immediately before the one typically used by tourists (near to a city attraction), so as to be assured a seat. On other routes, experienced passengers shared a common understanding of when a particular service ran from their local stop. Infrequent passengers rely on 'journey frequency' information that mimics this understanding.

Intersecting Architectures

We have introduced the different architectures of BLISS and suggested how they relate to one another. Drawing on our ethnographic material, we identify three instances of 'breakdown', when these architectures interact or intersect, and use them to pursue a detailed analysis of the system.

In the Cab

The first moment of intersection and breakdown that we would like to consider was pivotal to the functioning of the total system. For some time a number of individual buses were not recognised by BLISS; hence both the bus priority and customer information aspects of the system did not function. The first explanation formulated for this was a technical one.

The existing ticket machine acts as an interface between the existing equipment on the bus and the new BLISS equipment. It transpired that the communication between them was problematic. As an issue of 'configuration' it was seen as a technical problem to be addressed by the system developers. It later transpired that the situation was more complicated. There were in fact two different types of ticket machines installed on the buses. The "Wayfairer 3" was installed on all normal routes and the "ERG" was installed on the newer Park and Ride buses (although at times the buses from one type of route could be sequestered for the other). The latter ticket machines were more likely to work with the new equipment, the former, older ticket machines did not work at all. The Park and Ride buses were the first to be installed with the BLISS equipment and so the true nature of the problem did not manifest itself until later in the installation process.

The "older" ticket machine became talked about in terms of redundancy, rather than a technical problem of the design of the communication between the ticket machine and the onboard controller, and the bus operator agree to upgrade the ticket machines on all the buses. However, the technology problem later turned to one of driver volition.

An alternative explanation was offered for the non-functioning system. Each progression of bus from one end of a route to the other has a unique 'journey number'. Without the correct identifying number the bus is invisible to BLISS. While an operational feature, journey numbers were new to the drivers. It was suggested that, either through inexperience or deliberate act, the drivers were inputting an incorrect number into the ticket machine at the beginning of a route. The BLISS developer suspected that it was deliberate: in other installations of GPS tracking systems there had been instances of vandalism premised upon driver concerns about monitoring.

We are not in a position to decide whether drivers were deliberately using an incorrect number, however the issue of monitoring draws out two conflicting meanings attributed to BLISS by the bus drivers and the bus operator. To the bus driver the GPS aspects of the system was indeed a form of behaviour monitoring; to the bus operator the system is a tool for attaining better management of the bus drivers from a distance. These interpretations speak to, and reinforce, the existing antipathy between drivers and management that is premised upon driver resistance to office-bound management control. The contrast is best seen in the manager's decision to place one of the first BLISS consoles (displaying bus movement over a map) in the canteen of the bus drivers, even before a place had been found in the offices of the operations team for the console (and the additional Busnetlive software). The Operations manager confided in an interview that this was a deliberate move to press on the drivers the monitoring features. As an aside, at the end of the research period there was not a console installed in the Operations office and no member of Operations staff had been trained to use the console and there was uncertainty about how it would be incorporated into work practices.

The imposition of sequential journey numbers by the BLISS system runs contrary to established on-the-ground management practices. Being able to re-order, and redirect buses allows on-street supervisors to respond to contingencies in real time such as 'bunching' that results naturally from the mobility dynamics of buses (as a bus is delayed it picks up more customers, which incurs greater 'loading time' and delays it further. The bus behind has fewer customers and proceeds quicker). BLISS threatens to undermine the role of the on-street supervisor that has mediated management power and control in the past.

In this case then the architecture of BLISS has the journey number as a 'software' component. This simple architectural feature however intersects with established architectures of practice and meaning and results in the potential of system failure and worsening management relations.

At the Bus Stop

This following instance is another intersection between the system architecture and the architecture of meaning and practice. This time it is the passenger at the bus stop who is centre stage. Our analysis revealed themes within the passengers' discussions about the PIPs, which we can understand in terms of the "ambiguity" and "accountability" of the interface content. The following photographs, taken by the researcher, depict a PIP over a period of 6 minutes (according to the clock in the bottom right hand corner of the display) and show inconsistencies in the predicted times.

At 15:46 the first bus is predicted to be nine minutes away, and the second fifteen. One minute later and the second bus is still fifteen minutes away and the first eight minutes.

One minute later and the second bus is apparently a minute closer, but the first bus is now going to arrive in six minutes.

Four minutes later and the first bus is 'DUE', yet the second bus is twelve minutes away. From being six minutes apart, there appears to be double that time now.

The reasons for the inconsistencies are technical. BLISS does not in fact calculate its output 'in real time': the position of the bus is only calculated and transmitted every thirty seconds. This positional information is then turned into an estimate of progression (and hence speed of the vehicle) through comparison with following position information. The accuracy is greater the number of instances plotted. This may take three, perhaps four instances, which introduces a delay of at least ninety seconds. In addition if a bus is stationary over consecutive positioning events (loading passengers at a bus stop for example) the process is restarted.

We have an instance in our data that highlights these features. A passenger information panel was fitted to a bus stop that was only a short distance from the beginning of the route. It took the bus perhaps two to three minutes to travel this distance and hence the PIP, if it displayed any information at all, moved directly to display 'Due'. This was in part because 'Due' is display when a bus is two or less minutes away, but mainly because the system did not have enough time to calculate the progress of the bus from its resting position at the beginning of the route. Added to this, the bus stop in question was positioned on a long straight road, which meant that a person standing at the bus stop could see the bus long before the PIP told them the bus was due. This particular instance is an intersection between the built environment, the system architecture and the architectures of meaning and practice.

A simple way to interpret the inconsistencies in the PIP display is to 'read into' the information a mundane cause: the second bus has been delayed and the first bus has sped up. This requires an understanding of 'congestions' and other traffic features that influences bus movement. However, and this is the key issue here, if 'predicted time' is hostage to the everyday contingencies of traffic movement, what use is it, and more importantly, how reliable is it? One suggestion made by the developers, is that the intervening time – denoted by the prediction – could be used to do other activities away from the stop. However if a person had read the display at 15:46 and come back eight minutes later, they would have missed their bus. It only takes a few instances of this happening for the passenger at the bus stop to not trust the information again.

In the Control Room

Our third example is based on a less tangible moment of intersection because it concerns the perspective of the transport network managers 'behind closed doors' and concerns a question of configuration. It serves as an example of the intersection between BLISS and the architecture of existing transport computer systems. It speaks to a breakdown in BLISS due to the software configuration that switches off the prioritization of BLISS buses at certain times.

One of the other systems in the UTMC is called 'SCOOT' (not an acronym). It exists in a sibling relationship to BLISS and has its own objective. SCOOT seeks to create and maintain a so-called 'greenwaving' effect - the steady flow of all traffic through successive traffic lights (see Figure 3).

When a BLISS enabled bus joins from a side street a corridor on which SCOOT is working, a decision has to be made by the central computer system (the UTMC) about which objective to pursue, either the 'greenwaving' objective, or the 'prioritisation' of the BLISS enabled bus. Should the main traffic's progress be interrupted by giving the bus priority, or should the bus wait? This question was a matter of ongoing discussion between the Network Management staff and the developers of the system (and had not been decided by the end of the study). The key point is

Figure 3. SCOOT's greenwaving objective – as described in interview with the network manager

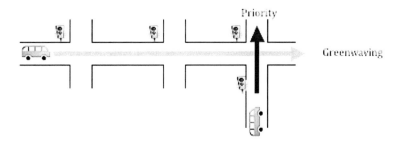

Figure 4.

that even with a fully working BLISS, it could be possible for the 'prioritization' request of BLISS equipped buses to fail.

This single issue of configuration highlights a number of interesting features. First we should note that SCOOT is not limited to one bus company and is 'owned' by the council's transport management office rather than the bus operator. It acts on bus movements, but is also seen as a way to control all traffic through the city. Interestingly buses are seen as 'thermometers' of all traffic movement. SCOOT effects changes in the total traffic movement, and hence exists at a higher level of traffic management than the bus based BLISS. It is likely (although only hinted at by the Network manager in interview) that SCOOT's objective will prevail. If this were to be the case, BLISS's prioritisation function would be undermined, and the system would fail to function in its objective of helping the buses run to time.

We have detailed three instances of 'breakdown': the first premised upon the interface between the system and the bus driver, the second between the system and the bus passenger, and third the system and the network manager. We will now use these readings of moments of intersection, interaction and breakdown to address our initial concerns about mobility and their wider implications for e-planning.

Discussion: BLISS, Mobility and Motility

Instead of fixed geographical regions the emerging field of Mobilities (Urry, 1999, Hannam *et al*, 2006) conceives the city in terms of movement and a 'set of flows' (people, transport, information etc.). At the same time this fluidity is afforded through 'dynamic ordering processes which have produced our chief sense of urban space and time [and is] produced through the design of mundane instruments of encounter' (Amin & Thrift, 2002, p. 83). Hannam *et al*. (2006) similarly note that 'mobilities' are reliant upon 'moorings' (p.3), a variety of geographically fixed features such as roads, traffic lights, mobile phone transmitters and the like. In terms of public transportation and the movement of people through the city then, bus management and information technologies can be conceived as forms of 'moorings' or 'mundane instruments of encounter'.

The mobilities approach is valuable here. Transportation is engendered by movement of people, equipment and information and the BLISS system is positioned as a means to augment and enhance such movement by speeding the flow of the bus through the city, easing the transition from one mode of movement (walking) to another (the bus) for passengers, through contingent and dynamic information provision and enabling the management of bus movement through the provision of detailed timing reports, that in turn enables iterative writing and rewriting of timing schedules.

There are at least two ways to conceptualise the relationships between the city, technology and human activity. Graham and Marvin (2001) analyse the city as constituted by multilayered 'landscapes' that include the technological infrastructures of utilities, information and transport and understand the city as a 'sociotechnical' object borne of the complex interconnections of these and other 'scapes'. Our use of the term architectures mirrors this idea of landscapes. Graham and Marvin assert that these urban arrangements of technology and space embody 'congealed social interests' (ibid. p. 11) and produce 'sociotechnical geometrics of power'. Further '[t]he construction of spaces of mobility and flow for some, however, always involves the construction of barriers for others' (ibid. p.11). Rather than providing for a static and taken for granted backdrop for human activities in the city, the various technological infrastructures result in a 'splintered urbanism' in which inequality is central. Different spatial arrangements and technologies are seen to afford 'different degrees of '*motility*' or 'potential for mobility' (Kaufman, 2002), with motility being a crucial dimension of unequal power relations' (Hannam *et al*, 2006, p.3).

Coutard and Guy (2007) in contrast present a less deterministic analysis of the technologised city, by emphasizing the 'ambivalence inherent to all technologies, the significant potential of contestation of, and resistance, (sic) to technology-supported forms of discrimination, and the deeply contingent nature of the process of appropriation of new technologies' (p. 713). Their argument has the active human working to incorporate and change the technologies appearing in the city into their own practices and understandings. Key here is the ideas of 'resistance' and 'contingent appropriation' that leads us to drill down to those moments of meaningful interaction between a person and the urban technology. Such approaches are part of a 'new kind of urbanism' that emphasises the ongoing and contingent nature of cities as emerging from the dynamic relationships between technologies and people and where the 'regulation of cities through technology is constantly contested and renegotiated' (Coutard and Guy, 2007, quoting Amin and Thrift, 2002). Accordingly 'each urban moment can spark performative improvisations which are unforeseen and unforeseeable' (Amin & Thrift, 2002:4). Motility here is not only afforded by and inherent in technology but is also a performative outcome of ongoing interaction with technology.

Our moments of breakdown not only reveal "system failures" but also contingent processes of meaning attribution, moments of resistance and appropriation. Rather than the technology determining particular structures of power, we have seen that the meaning of BLISS is dynamic. It is contested and such alternate meanings lead to instances of resistance, such as when bus drivers undermine the system through a single action. Rather than looking for increased mobility through the e-planning afforded by ITS, it is clear that we need to attend to the *differential motilities* that are enabled by the intersection of the architectures we have sketched out above. The uneven, contingent and so performative aspect of ITS explains how patterns of (uneven) motility rather than the production of uniformly improved mobility characterise the e-planned urban transport system. This is especially important of course for those the ITS implementation is supposed to serve, the passenger.

Passenger Motility and BLISS

Flamm and Kaufman (2006) set out a structure for understanding motility that makes the abstract concept amenable to analytic processes. They define motility as 'how an individual or group takes possession of the realm of possibilities for mobility and builds on it to develop personal projects' (p. 168). They identify three main factors involved in motility: (i) access rights, (ii) skill or aptitude for mobility and (iii) cognitive appropriation.

Access rights: The first of these they list as such factors as ownership of an automobile, access to reserved parking place, privately owned 'light' vehicles, public transportation passes, etc. To this list we might add in terms of bus transportation, an appropriately placed bus stop and well timed bus routes.

Skills: Travel is so common that we tend to forget that it is necessary for a person to possess specific skills and aptitudes. This may be as obvious as having a driving license but can also extend to the ability to walk short distances, or manoeuvre up a bus step. For bus travel a person needs skills in timing their travel, including the skilled use of a timetable schedule, they need to understand the customs and practices involved in taking a bus (and the ability to remain calm when the behaviour of others is troubling or the bus is running late), and they also need local knowledge about the area, and perhaps alternative routes and facilities. Flamm and Kaufman concludes that "learning to master a transportation mode is first and foremost a question of accumulating experience which therefore requires a medium to long-term learning process" (p. 168). When starting to accumulate this experience it is essential that there be an accommodating environment in which to learn new skills and fail if necessary. In terms of our use of architectures, the passengers' history of experience is an essential element of their motility, but also these experience structures set up interpretations, expectations and preferences.

Cognitive appropriation: This starts from the idea that to cope with the range of possibilities involved in transportation, people create simplified 'representations' of the opportunities and form evaluations of their different merits and usefulness. There are numerous factors involved in these representations including "rapidity, accessibility, independence, cost, safety, reliability, comfort, quality of services and environmental impact (energy efficiency, polluting emissions, noise" (p. 180). However, key, according to Flamm and Kaufman, are safety and reliability. Reliability is defined as "any means of travel that functions well and that we can trust to take us to the right destination and in a timely manner through all the day's activities" (p. 180). This implies that a means of travel is predictable and that the "aids for finding one's direction along the way are sufficient for first-time journeys, and the trip can be organised to suit one's activities" (p. 180). For planned journeys on unfamiliar routes 'punctuality' is key, while improvised trips require 'temporal flexibility' and 'spatial accessibility' are important.

Flamm and Kaufman highlight cognitive appropriation as "what people do with access and skills" (p. 169) in that an individual must consciously incorporate the technology into their lives based upon the resources to hand (in themselves, and in the world). What is interesting and relevant for our analysis is that Flamm and Kaufmann connect across the social, individual and perceptual domains and their prioritisation of cognitive appropriation fits well with our methodology that pursues wider issues of social, organisational and individual mobility through single moments of user perception. Our moments of intersection and interaction similarly combine these different aspects. Also in contrast to engineering notions of dependability and reliability, such notions prioritise the contingent interpretations of those attempting to use the technology.

In our discussion of the PIP interface we showed that 'the system' is judged in the context of experience and interpretation. The motility afforded is then highly contingent. The 'access rights' assigned are dependent upon the progressive and emergent interaction with the system. At one moment this might be stable – or perceived to be – when a person first looks at the display. Continued and conscious waiting and watching has, however, the potential *on every next turn* to undermine this. The cognitive appropriation is then continually in flux, as is the authenticity of the skilled passenger role. Only at some time later will a representation of the experience be engineered and added to the accumulated experiences to form an 'evaluation'. And hence only over some period (a quite extended period according to both Flamm and Kaufman and the bus operator) will the decision be formed.

So what does this say about the motility of BLISS? According to our analysis, it is difficult to see how motility can be improved at this stage.

One response might be to say that only when the ambiguities of the PIP can be countered – when the interface presents 'reliable' information in the sequence of predictions – can we imagine the information displays to impact upon passenger experience, public acceptance and non-bus-using public awareness of the benefits of the computer system. In regards to the prioritisation, again we could look to some future date when the ambiguities in the system are diminished. This would entail agreement over the relationships between BLISS and other computer objectives such as SCOOT and also when drivers are adequately informed and reassured about their place in the system and the benefits it will bring. In sociological terms we would be looking toward a form of 'stabilization' (Kline & Pinch, 1996).

The key point is that stabilization does not rely purely on technical progression. Any technology implementation is entwined with social and physical aspects. And more, the consequences of a technology are the outcome of a combination of 'human' and 'machinic' agency (Pickering, 1995) and do not rest purely on the technological design. The best that we can say is that technology affords a range of outcomes, that are specified in line with social interpretations and physical contexts, which are themselves premised upon a range of more or less established constraints and expectations. All of which does not leave us powerless. We can still criticize aspects of the design and make recommendations for future systems.

We might for example criticise the form of the information display as part of the overall design of the system. Perhaps 'predicted arrival time' is the wrong information to show, or perhaps presenting a prediction with no means to appreciate how this information was calculated is the problem. The information is meant to effect a sense of reliability when it never will, even if the technical aspects are changed and the prediction is presented in a way that does not open the possibility of questioning (as a picture of a bus on a map for example). Perhaps the problem is that it is assumed that sequential time expressed in numerals is the answer, this being a technical answer to a social and experiential concern. It imposes a technical frame on the history of experience of the bus passengers, instead of understanding the accepted ambiguities of catching a bus.

With regard the priority system of BLISS, we have seen instances of when the intersections between the understandings of the system can undermine its working, and we have begun to appreciate the contested nature of these purposes. We might then speak to a broader metric for a 'working system', one not based solely on the technical readout, and numerical output of the system itself. There is an issue here of circularity. The technical judge of whether BLISS works as a means to effects efficiency is BLISS itself, expressed as a numerical percentage of the number of buses running within five minutes of their registered timing points. The BLISS system does not mean that journey times will be shorter since its underlying objective is to bring buses in line with the schedules. Given what has already been said about the nature of the technical and social temporalities within the system, 'success', in terms of efficiency, is as contested and variable as any other aspect of the system.

FUTURE RESEARCH DIRECTIONS

The study above was based upon a particular instance of the implementation of ICTs into public transport management and provision. It represents then a specific example, and the analysis is specific to this example. Our concerns have been local rather than at a more national or long term level. This means that it is difficult for us to make predictions about the future trends of these kinds of interventions. What we are able to do is talk to the immediate aftermath of this particular implementation and from it make generalized claims about the nature and likely character of future implementations.

BLISS became a platform on which to build the service of the bus company situated within the ICT structure of the city. Whether this connection was planned in advance is unknown, yet the relationship between BLISS and the next stage of development was central to these developments. In other words the connection was contingent and highly consequential. It is quite likely that similarly contingent development happens in other urban contexts given the iterative nature of the development and adoption of innovation systems.

Despite such contingency, what we can see is the emergence of complementary informational infrastructures – such as new street-based information kiosks – that exploit yet add to elements of the transport system, and in turn a recursive development as bus operators go on to refine their fleet to exploit what are seen to be economies of scale and saving on labor costs. However, the *social* relationship between the bus and its passenger constrained how far such changes could go. Elongated 'bendy' buses and a particular novel arrangement of bus driver and ticket provision such that passengers were not able to interact with the driver to buy a ticket but instead had to use an onboard ticketing machine. It was found that this caused problems for passenger 'loading' times, especially when the person mounting the bus was elderly. Journey times were doubled, and the timetable's accuracy threatened. The bus company eventually removed the machines and returned to a conductor (renamed a 'customer services representative'). It is certainly ironic that the attempted modernization actually resulted in more traditional arrangements.

These two aspects of the study – contingent innovation across different actors coupled with a workability that requires continued reference to the social relationships of the service – will be likely to shape future trends in this form of e-transport in years to come.

CONCLUSION

Mobility is better understood in terms of 'motility', the potential for mobility. We sought to reveal the motility of an intelligent transport system called BLISS by conceptualising the system in its interpretive, physical and machinic locales by understanding different architectures and looking to see how these architectures relate to one another. We understood these as moments of intersection and interaction and looked to reveal these 'assemblages' of socio-technical features through specific instances. As those moments when the social technical arrangements become visible to interpretation, they are also moments when there is an empirical warrant for approaching them analytically. In ethno-methodological terms they are moments of trouble, and hence become contingent matters of 'ethno-methods' for making sense of the world. The BLISS implementation is one that has taken considerable 'labour' across all the social actors discussed above. As May and Finch (2009) have argued elsewhere, all social processes are 'complex and emergent' and 'can be transformed by the proximity of, or interaction with, some other process or event. Implementation processes are no different. …This means that predictions of the outcome of a complex and emergent process are a problem. After all, social processes transform as they are produced and reproduced. (p. 548)

This conceptual line leads to an appreciation of the complexity of BLISS understood through a social and urban frame of reference, which contrasted with the simplistic engineering frame that equated such features as passenger reliability with numerical efficiency, and imposed a rationale that requires the elimination of ambiguities. The social frame instead would embrace these ambiguities and so inform future e-planning for improved motility and so genuine mobility.

REFERENCES

Amin, A., & Thrift, N. J. (2002). *Cities: Reimagining the Urban*. Cambridge, UK: Polity Press.

(2008). *Chambers Dictionary* (11th ed.). Edinburgh, UK: Chambers.

Coutard, O., & Guy, S. (2007). STS and the City: Politics and Practices of Hope. *Science, Technology & Human Values*, *32*(6), 713–734. doi:10.1177/0162243907303600

Department for Transport (DFT). (2005). *The development of bus real-time systems in the UK*. London: Department for Transport.

Dewsbury, J.-D. (2000). Performativity and the event: enacting a philosophy of difference. *Environment and Planning. D, Society & Space*, *18*(4), 473–497. doi:10.1068/d200t

Flamm, M., & Kaufmann, V. (2006). Operationalising the Concept of Motility: A Qualitative Study. *Mobilities*, *1*(2), 167–189. doi:10.1080/17450100600726563

Garfinkel, H. (1967). *Studies in Ethnomethodology*. Cambridge, UK: Polity Press.

Graham, S., & Marvin, S. (2001). *Splintering Urbanism: Networked Infrastructures, Technological Mobilites and the Urban Condition*. London: Routledge. doi:10.4324/9780203452202

Hannam, K., Sheller, M., & Urry, J. (2006). Editorial: Mobilities, Immobilities and Moorings. *Mobilities*, *1*(1), 1–22. doi:10.1080/17450100500489189

Heidegger, M. (1962). *Being and Time*. London: SCM.

Jacucci, C., Jacucci, G., Wagner, I., & Psik, T. (2005). A manifesto for the performative development of ubiquitous media. In *CC '05: Proceedings of the 4th decennial conference on critical computing*, Aarhus, Denmark. New York: ACM.

Kaufmann, V. (2002). *Re-thinking Mobility: Contemporary Sociology*. Aldershot, UK: Ashgate.

Kline, R., & Pinch, T. (1996). Users as Agents of Technological Change: The Social Construction of the Automobile in the Rural United States. *Technology and Culture*, *37*, 763–795. doi:10.2307/3107097

Latour, B. (1987). *Science in Action*. Cambridge, MA: Harvard University Press.

Latour, B. (2005). *Reassembling the Social: An Introduction To Actor-Network-Theory (Clarendon Lectures in Management Studies)*. Oxford, UK: Oxford University Press.

May, C., & Finch, T. (2009). Implementing, Embedding and Integrating Practices: An Outline of Normalisation Profess Theory. *Sociology*, *43*, 535–545. doi:10.1177/0038038509103208

Michael, M. (2000). Futures of the Present: From Performativity to Prehension. In Brown, N. (Eds.), *Contested Futures: A Sociology of Prospective Techno-Science* (pp. 21–39). Aldershot, UK: Ashgate.

Pickering, A. (1995). *The Mangle of Practice: Time, Agency and Science*. Chicago, IL: Chicago University Press.

Suchman, L. (2002). Practice-based design of information systems: Notes from the hyperdeveloped World. *The Information Society*, *18*, 1–6. doi:10.1080/01972240290075066

Urry, J. (1999). *Sociology Beyond Societies: Mobilities for the Twenty First Century (International Library of Sociology)*. London: Routledge.

Urry, J. (2007). *Mobilities*. Cambridge, UK: Polity Press.

Weiser, M. (1993). *Ubiquitous Computing*. Washington, DC: IEEE Computer.

Wright, P., & Monk, A. (1991). A cost-effective evaluation method for use by designers. *International Journal of Man-Machine Studies, 35*, 891–912. doi:10.1016/S0020-7373(05)80167-1

ADDITIONAL READING

Avineri, E. (2004). A cumulative prospect theory approach to passengers behvior modeling: Waiting time paradox revisited. *Intelligent Transportation Systems, 8*, 195–204.

Berg, M. (1998). The politics of technology: On bringing social theory into technological design. *Science, Technology & Human Values, 23*(4), 456–490. doi:10.1177/016224399802300406

Bijker, W. E., & Law, J. (1994). *Shaping technology/building society: Studies in sociotechnical change (inside technology)*. Cambridge, MA: The MIT Press.

Bonsall, P. (2000). Legislating for modal shift: Background to the UK's new transport act. *Transport Policy, 7*(3), 179–184. doi:10.1016/S0967-070X(00)00028-7

Campbell, S., & Fainstein, S. S. (2003). *Readings in planning theory*. Oxford, UK: Blackwell Publishers.

Department for Transport. (2002). *Flexible Transport Services*.

Flyvbjerg, B. (2002). Bringing power to planning research: One researcher's praxis story. *Journal of Planning Education and Research, 21*(4), 353. doi:10.1177/0739456X0202100401

Graham, S., & Marvin, S. (1999). Planning cybercities? Integrating telecommunications into urban planning. *The Town Planning Review, 70*(1), 89–114.

Graham, S., & Marvin, S. (2000). Urban planning and the technological future of cities. Cities in the Telecommunications Age: The Fracturing of Geographies.

Hajer, M., & Zonneveld, W. (2000). Spatial planning in the network society-rethinking the principles of planning in the Netherlands. *European Planning Studies, 8*(3), 337–355.

Hall, P. G. (2002). *Cities of tomorrow: An intellectual history of urban planning and design in the twentieth century*. Oxford, UK: Blackwell Publishers.

Healey, P. (2003). The communicative turn in planning theory and its implications for spatial strategy formation. *Readings in Planning Theory*, 237-255.

Hellawell, S. (2001). *Beyond Access: ICT and social inclusion*. London: Fabian Society.

Hine, J. (2007). Travel demand management and social exclusion. *Mobilities, 2*(1), 109–120. doi:10.1080/17450100601106450

Laurini, R. (2001). *Information systems for urban planning: A hypermedia co-operative approach*. New York: Taylor & Francis.

Lyons, G., Harman, R., Austin, J., & Duff, A. (2001). Traveler information systems research: A review and recommendations for transport direct. *Traveler Information Systems Research: A Review and Recommendations for Transport Direct*.

Martens, M. H. (2000). Assessing road sign perception: A methodological review. *Transportation Human Factors, 2*(4), 347–357. doi:10.1207/STHF2-4_4

Perry, D. C. (2003). Making space: Planning as a mode of thought. In Campbell, S., & Fainstein, S. S. (Eds.), *Readings in planning theory*. Oxford, UK: Blackwell Publishers.

Rajé, F. (2007). The lived experience of transport structure: An exploration of transport's role in people's lives. *Mobilities*, *2*(1), 51–74. doi:10.1080/17450100601106260

Reed, D. (2006, February 3). *Experiencing BLISS in between HCI and STS*. Talk given at Department of Informatics, Sussex University

Reed, D. J., & Wright, P. (2006a). Experiencing BLISS when becoming a bus passenger. In *DIS '06: Proceedings of the 6th ACM conference on designing interactive systems,* University Park, PA, USA. New York: ACM Press.

Reed, D. J., & Wright, P. (2006b). Place and the experience of BLISS. In HCI '2006: Engage. London: University of London.

Stradling, S. G., Meadows, M. L., & Beatty, S. (2000). Helping drivers out of their cars. Integrating transport policy and social psychology for sustainable change. *Transport Policy*, *7*(3), 207–215. doi:10.1016/S0967-070X(00)00026-3

Thrift, N., & Dewsbury, J. (2000). Dead geographies - and how to make them live. *Environment and Planning - Part D*, *18*(4), 411-433.

Webster, A., & Reed, D. (2007). Use of ICT in enhancing public transport mobility: From impaired mobility to motility. In Transport for the mobility impaired. CSAU seminar, department for transport.

KEY TERMS AND DEFINITIONS

Architecture: Utilizing the definition of architecture as 'the realization and reification of structures of relevance' (Chambers Dictionary, 2008) allows us to extend architecture as a metaphor to include people and places as components, and the dynamic relationships of people to the system over time in terms of their attributions of meanings and values.

Engineering Frame: An analytical approach premised on the assumption that the implementation of new technology is principally a technical problem.

Mobility: The movement of people in urban settings as enabled by both the urban landscape and transport infrastructure.

Motility: Motility, denotes the potential for mobility in a system. Given that mobility is unequally distributed and that therefore some are afforded greater mobility than others, the term also implies a form of power distribution and relationships.

Social Frame: An analytical approach to understanding the ways in which the social and the technological interact and are co-constructed over time by heterogeneous social actors and so shape the patterns of technical implementation and its meaning and utility.

APPENDIX: BLISS COMPONENTS

On Bus Equipment (1 Per Vehicle)

CR120 Power Supply and Radio Unit
DT110 DeltaTrak on board Computer
Panorama Dual-band whip, or LP VHF and GPS antennae.
SC100 Low Power Traffic Signal Priority Transmitter

PIP Installation Kit (1 Per PIP)

Red LED 3 line 30 character display with 29mm character height, with 30 character spaces per line, fitted with DT102 DeltaTrak board and Radio programmed with network file.
Panorama whip, or LP antenna as appropriate

Or

Pole mounting LCD display, fitted with DT102 DeltaTrak board and Radio programmed with network file.
Panorama whip, or LP antenna as appropriate

Base Station Kit

1 Key KF510 –20A Base Stations
1 Key PS114 Power Supply
1 19" Rack assembly
1 Antenna

In-station Kit

1 19" Rack Assembly
1 Dell PowerEdge 1650 computer
1 Dell PowerEdge 1650 computer Web Server)

Bus Operator Workstation

Dell Optiplex 240 - 17" LCD - specification:
Fast Intel. Pentium 4 processors up to 2.2 GHz, with 400MHz system bus
256KB Advanced Transfer Cache
Up to 1,024MB 133MHz SDRAM Memory
Enhanced Quietkey (Spacesaver) keyboard with 3 programmable hotkeys
Microsoft Intellimouse 2 Button Wheel PS/2 Mouse (+#5)
'SD' chassis stand for tower orientation

Microsoft. Windows. 2000 Professional (SP2) (Excluding Docs. & Recovery CD)
NTFS Hard Drive File Format System

Traffic Signal Junctions (1 Kit Per Junction)

SC135 TSPU Assembly
SC160 PSU

Chapter 20
RFID in Urban Planning

Les Pang
University of Maryland University College, USA

Vanessa Morgan-Morris
Constellation Energy and Community College of Baltimore County, USA

Angela Howell
Department of Defense, USA

ABSTRACT

Radio Frequency Identification (RFID) is a significant emerging technology that enables the automation of numerous applications globally. Professions, businesses and industries have integrated this technology into their procedures and it has resulted in great advances in the accuracy of data, operational efficiencies, logistical enhancements and other process improvements. This chapter discusses the application of RFID technology to support the needs and requirements within the realm of urban planning. First, the historic and technical background behind RFID is reviewed. Illustrative examples of its use are presented. Next, the technology's potential is explored in terms of a practical tool for urban planners. Consequently, issues and challenges associated with RFID are identified and considerations to be made when applying the technology are offered. Finally, the outlook for RFID technology is examined as an instrument in urban development and the expected exponential growth of the technology is discussed.

INTRODUCTION

Urban planners face major challenges in the collection of data towards analyzing the constantly changing social, economic and environmental conditions in cities. This data is instrumental in the development of accurate models used to plan urban communities -- many of which are facing critical issues and challenges.

Current methodologies include the use of remote sensing technology in the form of satellite data collection to determine land use trends in urban as well as rural areas. State-of-the-art data collection techniques also include surveys and interviews, traffic monitoring devices and ethnographic observations.

RFID technology offers a new and innovative approach to data collection that will revolutionize urban design and planning. Through the use of these small monitoring devices, data accuracy will be significantly improved thus resulting in planning

DOI: 10.4018/978-1-61520-929-3.ch020

models with higher precision. These devices can detect trends at the pedestrian and vehicular level as opposed to the remote sensing satellite techniques at the atmospheric level. However, as with all technologies, the use of RFID presents challenges to the urban planner that must be considered and addressed accordingly. Also, the outlook of this technology needs to be examined to see where it will take the urban planning profession in the future.

This chapter starts with the background of RFID in terms of its technical underpinnings and its history and provides illustrative applications. Next, it examines the relevance of RFID to the urban planning profession, namely as a planning tool for data collection. This chapter then identifies the key issues and challenges associated with RFID that are pertinent to its application in the urban design and planning environment. Based on the identified issues and challenges, considerations when implementing RFID technology are explored. Finally, the outlook for RFID technology is examined in terms of its use as an instrument in urban development and projections of the expected growth and expansion of the RFID technology are presented.

BACKGROUND

RFID Technology

RFID is one of many technologies known as auto-identification (Auto-ID) technology. Auto-ID is a very broad term that relates to several technologies that assist machines in identifying objects with automatic data capture. Besides RFID devices, common forms of Auto-ID technology are barcodes, smart cards, voice recognition, biometrics, optical character recognition (OCR) and others. This automation approach is used to identify items, capture item information and transport the data to a computer without manual entry. These systems increase efficiency, reduce data entry error and free people up to perform more productive jobs and functions. (RFID4U, 2006, pgs. 1-2, 1-3)

RFID utilizes radio frequency waves to transfer data between a reader and movable or stationary objects called tags. This data is used to identify, categorize and track objects as they move through a controlled environment. This technology consists of a combination of technology components:

- **Tags** - a microchip packaged with an antenna used to identify objects. Radio Frequency is used to transfer data between the tag and a reader.
- **Readers** - Reads data from a tag ("interrogates" a tag) and writes data to the tag. It communicates with the tag through an antenna that emits radio waves. Some readers can provide power to tags without their own battery. Readers are also called scanners.
- **Antennas** - Two types of antennas in an RFID system, one is in the tag and the other is connected to the reader. Usually it is referred to the one that is connected to the RFID reader.
- **Networking** - Transports data and commands.
- **Software** - Helps manage business processes; device drivers, filters, databases, middleware and user applications such as warehouse and supply chain management systems.

In basic terms, the RFID information flow begins when a reader and tag communicate with one another using radio frequency. The carrier signal is generated by the reader on request from the host application. This host manages the reader and issues commands. The carrier signal is sent out through the reader's antenna and hits the tag. The tag receives and modifies the carrier signal and reflects back a modulated signal. The reader antenna receives the modulated signal and then sends it to the reader. The reader will decode the signal into data in a digital format that is then returned to the host application. (RFID4U, 2006, pgs. 1-6, 1-7)

There are three basic types of RFID tags; passive, semi-passive and active. Table 1 shows the U.S. Government Accountability Office's summary of tag characteristics and several ways to distinguish between RFID tags.

Passive tags are tags that do not contain a power source such as a battery or radio transmitter. They work by using the reader's energy, radio frequency (RF) waves, to power their integrated circuits and transmit the digital reply back to the reader. By not having a battery, the passive tags are less expensive but they can only communicate when inside the read zone and energized by the reader's RF waves. Examples of passive tags are access control cards and badges and car keys. Cards and badges utilize readers mounted on the wall for individuals to bring their card or badge within four to six feet to gain access. The tag, powered by the reader's magnetic field energy, responds to the reader with a digital signal with a unique code to the card and that particular employee. This information is then sent over a network to a computer systems that will allow access or a door to be unlocked if the tag's code is in the facility access control database. Passive tags are primarily used in applications requiring low cost, simple and lightweight tags. The advantages of passive tags are its small size and weight, low cost and longer life -- 20 or more years. The disadvantages are a lower read range and high power readers are needed.

Semi-passive tags generally have batteries but may only be powering the circuit while it is active and then operating at a very low level while waiting for a reader's signal. Semi-passive tags generally live longer than active tags but have shorter lives than passive tags. The advantages of a semi-passive tag are that it can store more data on the tag due to the fact one does not need to power the tag from the reader. This tag can operate under longer read ranges -- up to 100 feet. One can couple the tag with environmental sensors and capture and record information such as temperature, pressure and relative humidity. Compared with passive tags, the disadvantages include higher cost, larger and heavier tag and limited life constrained by the battery capacity. (RFID4U, 2006, pgs. 2-8)

Active tags have a battery and radio transmitter and are used for longer range applications -- normally 300 to 750 feet. Typical applications are in warehouses to track pallets or large shipping cases. These active tags often contain environmental sensors. For example, refrigerated trucks can be equipped with active RFID tags and temperature sensors. When these are placed in the truck they can continuously monitor the inside temperature of the truck and then the tags can be interrogated and identify the temperature profile inside the truck. It is very important to ensure the merchandise has been kept within the appropriate temperature range while in the truck. Individuals can also change

Table 1. Typical characteristics of RFID tags

	Passive Tags	Semi-Passive Tags	Active Tags
Power Supply	External (from reader)	Internal battery	Internal battery
Read Range	Up to 20 feet	Up to 100 feet	Up to 750 feet
Type of Memory	Mostly read-only	Read-write	Read-write
Cost	$0.15 to $0.20	$2 to $10	$20 or more
Life of Tag	Up to 20 years	2 to 7 years	5 to 10 years

(United States Government Accountability Office, 2005, pg. 8)

the temperature settings within trucks based on the temperature readings on the tag. Active tags are highest in costs, larger, heavier and have a limited life due to the battery. (RFID4U, 2006, pgs. 2-8, 2-9)

As an example of the technology behind RFID, in January 2008, Fujitsu developed the first 64-kilobyte high-capacity Ferroelectric Random Access Memory (FRAM) RFID tags for the aviation industry. These tags have the highest capacity for handling the storage of large volumes of aircraft parts and maintenance history while still supporting traceability worldwide. The aviation industry has been driven to raise the quality and efficiency of aircraft maintenance by being able to track parts not just by identification codes but also by product and parts maintenance history. These chips support:

- Traceability of various products.
- Traceability of maintenance information for parts exchange between companies and across nations of the world with the compatibility with different frequencies used around the world.
- High-speed data writing capability and high durability.
- State-of-the-art security functions; password management for each part of the memory area.
- This technology allows for aviation industry and its customers to realize improved supply chain management through lower costs and shorter cycle times worldwide (Fujitsu, 2008).

Tag and readers operate on a protocol that is set and defined by the manufacturers or various standards-setting organizations. EPCglobal is a non-profit organization that is working to standardize electronic product codes and RFID technology. In December 2004 EPCglobal released the electronic product code Generation 2 protocol. Generation 2 specifications were based on input from over 60 technology, manufacturing and retail companies to detail the signaling layer of the communication link between the tag and reader. (RFID4U, 2006, pg. 2-19)

Although Generation 2 brought new capabilities, it still needs to overcome issues with the alignment with the International Standards Organization and it is not compatible with existing pre-Gen 2 tags and equipment. (Kleist, 2005, pg. 45) These protocols, whether proprietary or open, must be the same between the tag and reader to communicate. Readers can be handheld, mobile or stationary and have the capability of reading single or multi-protocols and frequencies. The selection criteria for choosing the type of reader are as follows:

- **Operating Frequency** – Must match tag requirements.
- **Multi-Protocol** – Desirable if a variety of different tags need to be read which may have utilized different protocols.
- **Encode** – If the reader is expected to write data to the tags this capability is required.
- **Meets Local Regulations** – United States and Europe have different power output requirements.
- **Memory** – Necessary for tag buffering and tag list management.
- **Networking Capability** – Capability to network several readers together and communicate with a host computer; and the ability of the reader to be configurable and upgradeable through the network.

Other selection criteria to be considered are location of reader, surrounding environment, weight, size, ruggedness and price. The reader's antennas are the most sensitive component of an RFID system. They are usually housed in an enclosure that can be easily mounted. Readers must be placed in a position where powering the tag and receiving the data can be at its optimum. (Kleist, 2005, pgs. 62-63)

The read range or the distance at which a tag can be read depends on the following: the frequency, the antenna gain (effectiveness of an antenna in a given direction), the orientation and polarization of the reader antenna and the tag antenna, as well as the placement of the tag on the object to be identified. (Technovelgy, 2008) Also environmental factors may affect its performance (this will be discussed later).

RFID tags can come in many different form factors. These include contactless smart cards, smart labels (barcode label with an embedded chip), tickets, ear tags (livestock), glass bead, encapsulated in plastic and rigid casing. (RFID Soup, 2008)

For human tracking purposes, RFID tags can be configured in various ways. Tags can be placed on wristbands and serve as a replacement for, say, identification bracelets on hospital patients. Attached to laces on a shoe, one type of RFID tag is called a ChampionChip which has the capability of tracking a large number of runners along a course. Tags can be embedded into human subjects. This technique is done currently using the VeriChip technology. The tag is roughly the size of a rice grain and it contains a unique verification number. Using a local anesthetic, it is painlessly implanted in an inconspicuous location on the body such as in the triceps area between the elbow and the shoulder of the right arm. (FuturePundit, 2003)

All in all, RFID technology is a tool that enables business to improve data quality, capture data real-time, reduce collection time, human intervention and cost. Each organization has many different reasons for implementing RFID:

- Cost Savings – No need to scan or open boxes, crates and other containers
- Time Savings – Faster check-in/checkout procedures
- Automated real-time inventory of information
- Automated real-time tracking and tracing
- Verify origin of items
- Reduce inventory loss

Information can be attached to item to be processed at a later time; history and environmental data can be captured and retained

Opens possibilities to further automate or fine-tune business processes

RFID History and Applications

RFID is not a new techology. The technology was first documented by Harry Stockman with the publication of his 1948 report "Communication by Means of Reflective Power." (Brach, 1997) During World War II, the British placed transponders on planes flying over France to determine if the aircraft were their own, "friendly," or an enemy's, "foe". The transponder in the plane had to appropriately respond to an interrogating signal to be considered "friendly," if not, they knew it was their enemy. This "IFF" or "Identify: Friend or Foe" system is still utilized for the current aviation traffic control for commercial and private planes. (Goodman, pg. 4) Patents for the technology began in the 1950's.

Over the decades, one can see how RFID applications have evolved in diverse environments. In the 1960's and 1970's there was a strong need for a secure method for the surveillance of sensitive materials, particularly during the transport of nuclear material. Therefore, developments in RFID tagging were accellerated for tracking this type of sensitive material.

In the late 1970's and 1980's RFID was utilized in animal tracking. At the University of Chicago, transponders were placed in the back of a dairy cow. This transponder contained the animals identification and tracked its temperature. The thought was that the animals health, ovulation and feeding habits could be tracked by identifying the cow and its temperature as well as tracking its feeding patterns. (Eagles Nest, 2002, p. 1-3)

Today the United States Department of Agriculture is considering implanting RFID tags in every cow in the United States to better track diseases, such as mad cow, as well the specific farm location history of each cow. (RFID Gazette, 2004)

In the 1990's, the railroad industry quickly moved to RFID after numerous challenges in attempting to implement barcode technologies to track rolling stock. On significant issue with barcode technologies is that the readers require line-of-sight so it is impossible to read through containers or walls to identify the cargo -- all the containers would have to be unloaded or the truck opened. With RFID the radio waves can penetrate the walls of a truck or container so they do not have to be unloaded. Also, barcode scanners utilize visual light or infrared to read the barcodes. RFID has longer read distances, do not require line of sight and it has the ability to read through various and extreme weather conditions required in the industry such as rain, fog, snow, dirt, oil and sun. (Eagles Nest, 2002) Based on the tracking of materials on railroad cars, many other industries took this type of application and modified it to their unique business requirements. These applications included identifying fleet vehicles (such as tractors, trailers and containerized cargo), toll collection on highways, personnel badges and car remote keyless entry. (Landt, 2006)

In the early 2000's, several large retailers (namely, Wal-mart, Target and Best Buy), the United States Department of Defense and others published their supply chain requirements pertaining the use of RFID and placed demands and timelines on all their vendors worldwide to place tags on all shipments of pallets and cases. (Brewin, 2007) This mandate drove thousands of supply chain companies worldwide to make RFID investments. No matter what their size they had to implement and revise their way of doing business in order to comply and develop their own RFID systems, otherwise, they would lose a profitable business partner. Wal-mart, since 2005, has required its top 100 suppliers to place RFID tags on all of its cases and pallets. (Kleist, 2004, pgs. 138-147)

The Department of Defense has been using active tags for identification of large containers. It is now the largest operational tagging system in the world. If the United States Department of Defense's Defense Logistics Agency (DLA) were a commercial enterprise it would be the second largest distributor in the world. The DLA's mission is to support rapid deployment and sustain warfare anywhere in the world at a moment's notice. Even if there is no infrastructure or set location, DLA must meet the demands and needs of supply chain requirements. RFID has proven to be a scalable technology that can address these requirements. (Kleist, 2004, pgs. 166-167)

Presently, the Los Angeles County Police Department plans to launch a pilot program to track 10 percent of their 18,000 prison inmates using RFID devices. RFID use on the incarcerated is not new, but this appears to be a first in the United States. Inmates will be required to wear a RFID bracelet which emits a beacon every two seconds; officers and staff will carry a similar device on their belts. RFID readers will transfer the data back to a central computer to track such unacceptable behavior as entering restricted areas and escaping containment. (Engadget, 2005)

There are numerous other examples of RFID use. Here is a growing list of areas where RFID technology has been applied:

- Supply chain
- Shipping and receiving
- Warehousing
- Retail outlet
- Inventory management
- Pharmaceutical
- Pedigree management
- Product authentication
- Documents for the Food and Drug Administration
- Healthcare
- Patient tracking

- Equipment tracking
- Service tracking
- Library and video stores
- Cashless payment - credit or debit cards
- Hospitality - gambling chips
- Amusement parks
- Document management
- Lawyer's office
- Hospitals
- Government Offices - passports, mail security and tracking
- Transportation management
- Rail cars and trucks tracking
- Toll and other payment collection - E-Z Pass® and SmarTrip®
- Vehicle theft detection - keyless entry
- Vehicle speed tracking
- Automobile tires
- Animal tracking – pets and livestock
- Airlines
- Baggage tracking
- Supply management – parts

RFID and Urban Planning

RFID promises to take the urban planner profession to a new level. As an automatic identification technology, RFID has the potential to significantly improve the quality, accuracy and efficiencies of data collection for urban planning and development purposes. RFID is the highest level of technology to help the urban planner and the final result is better planned cities, particularly those currently beleaguered with urban issues.

This section looks at actual implementations of RFID to support urban planning and then extrapolates these implementations to potential applications for the future. For these extrapolations, it will assume the economic viability of the widespread distribution of tags and readers, privacy and security issues are not a concern and that technical issues such as interference and the limited range of RFID are resolved. Suggestions for addressing these implementation issues and concerns are discussed later.

RFID as an Urban Planning Tool

Pedestrian Movement

In the Netherlands, RFID technology has been used recently to analyze the space utilization of a building by its occupants. Participants were provided a tag and the RFID system was able to record their movements with high precision. Thirty seven employees working on a floor agreed to wear a tag for three months. Sixteen readers were installed on the floor and positioned so that it could track employee movement throughout the floor. During the study period, a total of 360,000 events were recorded representing employee movements among different zones on the floor. (Tabak, 2008)

Using a similar RFID concept but on a wider scale geographically, data on person movements can be collected within an urban environment. The relatively low cost of the RFID tags makes this possible and therefore a large sample size can be attained. In addition, combining RFID, Global Positioning Systems (GPS) and geographic mapping systems, people can be accurately tracked real time on an urban regional map.

For an urban planner, a potential application of this nature can assist in the identification of land use and transportation requirements for urban locations such as a central business district, major shopping malls and schools and colleges. For example, equipping a sample of people with RFID tags and strategically placing the readers, one can readily identify their movements throughout the city. Compared to current surveys and interview, this information gained from RFID data will provide much more accurate data on land use characteristics, traffic patterns and congestion issues that an urban planner could use.

A specific example of how RFID technology can be used is based on an application in Denmark. In 2004, an amusement park placed RFID wristbands on children in order that they could be quickly located by a distributed network of readers in the park if they were lost or abducted. Using similar technology, children can be fitted with these RFID wristbands to track their movements in the central business district. Their movements could be plotted and special security requirements could be implemented in urban corridors frequented by the children.

An urban planner could use this RFID data to identify locations for satellite police stations, security kiosks and/or special monitoring stations within these corridors. On a broader scale, locations for new schools can be identified by studying cluster diagrams which specify the origins of the trips made by the school children. This is just one example where RFID can be use to determine land use requirements.

If this approach is expanded to more citizens, RFID technology could be also be used for demand forecasting of urban facilities such as municipal buildings, libraries and museums by monitoring data over time. As a result, projections can be made on whether these facilities need to be expanded, reduced in size or even relocated.

A province in Indonesia is expected to pass a law that would require people infected with HIV/AIDS to be implanted with an RFID chip. The chip would used to track their movements and discourage these individuals from infecting healthy individuals. (Reuters, 2008) In addition, various hospitals are using RFID systems to track a patient's location using RFID wristbands. It also provides real-time tracking of the location of doctors and nurses in the hospital. Expanding this concept regionally, RFID information can be used to assist the urban planner in locating hospitals, satellite medical centers and outpatient medical clinics.

Transit Passenger Behavior

The Washington, D.C. Metrorail was the first U.S. urban mass transit system to use RFID technology when it introduced the SmarTrip® card in 1999. Currently, the microchip-embedded card is used as a contact-less stored value fare card by passengers to facilitate paying for transit trips and parking lot fees. This card system can determine payment required based on the stations entered and exited. (SmarTrip, 2008)

The RFID technology behind the SmarTrip® card has the potential to be used to monitor origin and destination transit passenger flow patterns, locate congested areas throughout the system and identify passenger traffic behavior during rush hour periods.

Urban planners would find this information gained from RFID-based cards to be very useful. It could identify where new transit routes are needed. For example, most Metrorail transit lines in the Washington, D.C. area extend out from the downtown area in a radial manner similar to spokes of a wheel. RFID technology could establish transit demand patterns that would identify needed routes that would connect stations in a more circumferential configuration in the Metrorail transit network.

In addition, the RFID data could be used to identify congestion patterns which would lead to determining if and where transit stations and adjacent parking areas need to be addressed. Specifically, transit platforms may need to be expanded, additional parking space and structures required and supplemental equipment such as fare machines added.

For a transit planner, RFID data can be used to determine if trains need to increase or decrease its arrival frequency. The data could also be applied to determine optimal routing of trains in the system particularly during the rush hour periods.

RFID can also be used to support the requirements of other modes of transportation. The Chicago Transit Authority has offered the Chicago Card and the Chicago Card Plus® for bus payments since 2005. (Card Plus, 2008) This technology can be used to track bus passenger movements. This information can be used to determine the optimized locations of bus terminals and storage facilities identify the most effective bus routing with the urban area and establish bus service requirements such as arrival frequency.

Vehicular Movement

A key aspect of urban planning is the design of transport facilities including the layout of streets and highways, bridges and tunnels within the urban environment. RFID technology addresses this aspect by providing an improved technique for the collection of data related to vehicular movements.

A consortium of researchers, college representatives and private industry in India is investigating RFID technology to improve traffic management and reduce pollution in Canada, India and globally as part of a $2.5 million initiative. This consortium is to develop a technology solution to power a concept called an "intelligent vehicle-highway system" (IVHS) framework. IVHS is frequently referred to as "smart cars" and "smart highways" and incorporates advances in electronics, communications and information processing which includes RFID technology. (Weiner, 1999)

RFID can capture data related to traffic use and capacity without significantly increasing in the investment in road infrastructure. According the consortium's project manager:

"Pacing traffic to flow more evenly can reduce commuting time, fossil fuel use and harmful exhaust emissions. Managing existing road capacity more efficiently through intelligent transportation systems is also much more affordable than undertaking large infrastructure projects." (McMaster University, 2008)

There are several RFID techniques to support the concept of IVHS. Non-destructive RFID passive tags can be embedded into the road in the middle of every lane at a predetermined spacing. Serving as beacons, these tags can be used to precisely track vehicles on the road.

Another approach involves the use of the electronic number plate or e-Plate which is a license plate with an embedded broadcast tag. With a range of over 100 meters, it is an active tag with its own built-in battery having a life of up to 10 years. (Hawley, 2004)

In addition, EZPass® is a relatively established RFID technology used on the east coast of the United States to facilitate the payment of bridge and highway tolls. EZPass® tags are mounted onto cars, trucks and other vehicles. (Ashton, 2008) The RFID readers are permanently installed in toll plazas but, as an alternative, researchers at Rensselaer Polytechnic Institute and the New York State Department of Transportation (DOT) are testing solar-powered mobile RFID readers. This portable device can be deployed in areas and under conditions for which a permanent RFID installation would be too expensive or impractical. (Swedberg, 2006)

RFID systems using vehicular-mounted tags can be used to identify congested routes and traffic patterns within an urban area. An urban planner can establish origin and destination behavior, identify bottlenecks and monitor rush hour behavior. This data will lead to the better design of transportation networks and facilities.

For example, RFID systems can be used in an urban area to determine parking activity and determine related projected land use within a given community. Researchers at MIT have developed an ad hoc sensor network based on RFID technology that can help drivers in urban areas find available parking spots on the street. (Roberti, 2006)

Using a similar system, a planner can use this sensor network to analyze and assess the parking demand and capacity characteristics then plan and design an effective strategy which could involve the addition of parking spaces, new parking regulations, additional parking lots and structures and other actions.

Using this RFID parking data, an urban planner can consider the application of the "congestion pricing" concept, a novel strategy of requiring drivers entering the central business district to pay a surcharge during the rush hour period in order to reduce traffic congestion. This approach has been used in London, Singapore, Milan and Stockholm and is being contemplated in New York City. (Transportation Alternatives, 2008)

Urban planners are often responsible for establishing an optimal bicycle network in the city. Bicycle lanes and paths can be determined by equipping bicyclists with RFID tags and monitoring their travel patterns.

Other Applications

RFID tags have used for many years in tagging merchandise in retail stores. For clothing stores, they are often removed when customers leave the store whereas the tags are frequently left on for other establishments. Having operating RFID tags on merchandise after leaving the retail store can be used to study the traffic patterns of the customers. The resultant data can be useful for land use projections, specifically, identifying areas needing retail stores, shopping centers and even regional malls.

A number of libraries are presently utilizing RFID tags to quickly check out and check in books and videos. Urban planners can use these tags to learn the geographic dispersion of library patrons. It could help locate communities where a library, satellite library or bookmobile is needed.

RFID tags have been used to track airport luggage. Through a coordinated system, one can track the movement of airline passengers throughout the network of airports. It can then be used to estimate passenger demand for each airport which will influence the growth of the airport terminal facilities.

RFID Issues and Challenges

Human Implantation and Privacy

One of most interesting, but also very controversal uses, is the implantation of RFID tags in humans. On October 14, 2004, the Federal Drug Administration approved the practice of injecting humans with tracking devices for medical purposes. Using a special scanner hospital staff can retrieve information from the impanted chips, such as patients identity, their blood type and details of their condition all in hopes to speed treatment. It is not an intention to store any medical history or private information. (Gilbert, 2004) This technology targets patients suffering from Alzheimer's disease, diabetes, cardiovascular disease and other conditions requiring complex treatement. Many countries such as Mexico and Italy are testing and utilizing this technology, not only in the medical field but also as an authentication tool for use in building access security and to complete financial transactions. The attorney general of Mexico and approximately 200 people on his staff were implanted with a chip to control access to areas where confidential documents are kept. (RFID Gazette, 2004) Companies such as Advanced Digital Solutions have announced embedded RFID cash and credit card technology. The idea is to walk into a store, pick up the merchandise and leave the store without interacting with another human. (Future Pundit, 2003)

In October 2006, Digital Angel, a company that develops GPS and RFID products for consumer, commercial and government sectors, was granted a patent for an implantable RFID microchip that can determine glucose levels in the bodies of animals and humans. They already had a temperature-sensing implantable RFID microchip that can quickly and easily read the temperature of surrounding body tempature. These chips are sold mainly in the animal market and especially in the horse industry. The other potential bio-sensing application they are researching is monitoring pulse, oxygen levels and blood pressure. (Digital Angel, 2008)

Another company in Canada, Tribal Expressions, advertised itself as the first company in the world to professionally offer subdermal RFID placements as a service. Implants would provide access control to doors, lighting, alarm system, computers and even one's car. A smart house would be controlled by a tag implanted between the thumb and finger. (Kennedy, 2006)

In July 2007, CityWatcher.com, a small Cincinnati-based company that provides surveillance equipment, implanted two of its employees with RFID tags. They did this to restrict access to vaults that held sensitive data and images for police departments. They said they required a more sophisticated technique to protect high-end data and their chief executive likened the implanted chips to retina scans or fingerprints. Some people considered this as a significant assault on the employees' rights and privacy. (Dallas Morning News, 2007)

In June 2007, California passed a bill that outlawed the forced subdermal RFID tagging of humans, Senate Bill No. 362. (California State Senate, 2008) The forced RFID tagging was considered by many as the ultimate invasion of privacy. (Anderson, 2007) Both Wisconsin and North Dakota already have similar laws on the books. (Jardin, 2007) California legislature for the past several years has focused on RFID privacy and passed several bills that mandate privacy and security requirements for technology. There is also a draft report from the Department Homeland Security on the use of RFID for human identification to identify and track individuals. The Department did not recommend the use of the technology for tracking humans because it offered little to no benefits when compared to the consequences relating to privacy and data integrity. If and when it is used, specific security and privacy safeguards measures need to be implemented. (Department of Homeland Security, 2008)

Even when human embedding of RFID is not involved, there has been significant privacy concerns over the use of the technology. For example, there has been significant concern over the tracking of a person's tagged merchandise from a retail store.

Existing solutions to prevent privacy breach either places the burden on the consumer or depends on the limited range of the tags. Until now the "kill" command that can be applied to a tag appears to be the solution with the greatest potential. However, there is no confirmation of successful disablement made to the customer.

In certain schools in California, RFID wristbands were considered for use in tracking attendance when students walk through the door. The plan also entails placing readers around schools to monitor student movement throughout the day, but the initiative has not been implemented because parents have rallied against the system due to privacy concerns. Legislation is still pending in the California legislature to impose restrictions on ID technologies that communicate via radio frequency waves.

Security

RFID is emerging as a key technology enabler for commercial supply chains which span the globe. However, there is the possibility of competitors, industrial spies and disgruntled employees to intercept RFID data for nefarious purposes. There are a number of safeguards such as encryption and shielding, but not all industries and business have implemented these security controls.

Military applications of RFID may also be vulnerable to data interception by foreign adversaries. The Department of Defense needs to be concerned that RFID can be used as a way to identify of military personnel and their geographical location.

RFID backend systems such as the EPCglobal network can be targets of denial of service attacks by hackers.

Technological Limitations

Depending on the type, RFID tags have a limited range, so the number and distribution of the corresponding reader network becomes critical. Although the cost of the individual tags may be minimal, the price of multiple readers may be prohibitive even although technological progress has been made in reducing this cost.

High humidity environments and the presence of water (such as in bottled water, humans and cattle) can dampen RFID radio frequencies. Active rather than passive tags are needed within these challenging situations.

A significant challenge is that metal, such as those in bullets, guns and armor, used by the Department of Defense, can interfere with radio signals. Radio signals can also trigger certain types of ordnance. (Kleist, 2004, pgs 166-167)

Standards and International Compatibility

Globally, RFID data can not be easily exchanged among different systems and proprietary system components. One reason is that the metadata or record structure of the RFID data can vary and another is that the frequencies may be incompatible. As mentioned earlier, EPCglobal's Generation 2 protocol is not compatible with the International Standards Organization specifications nor with existing pre-Gen 2 tags and equipment. In the United States, the RFID frequencies used do not match those in Japan or Europe. Unlike barcodes which have a universal standard, no such standard is emerging among the major RFID players.

Perhaps this is a role that a United Nations Committee on RFID technology can assume. Besides developing a truly global RFID standard, the committee could address the major controversies surrounding the safety and security of this technology and enable the world to fully embrace and utilize RFID technology.

Common Vision

One key challenge for all decision-makers is to create a shared vision and a common set of goals and objective on how RFID technology can keep urban cities more innovative and competitive in the global economy.

RFID Implementation Considerations

Clearly the issues and challenges that face RFID technologies need to be addressed when implementing an urban planning effort.

- **Human Implantation** – The legal, moral and social implications of human implantation of RFID needs to be well- understood before embarking on this approach.
- **Privacy** – The use and purpose of the RFID must be fully disclosed to all participating users. Also, all personal data collected through the RFID system need to be protected including the provision of backup data storage.
- **Security** – Encryption, network controls, physical safeguards and other techniques are needed to prevent unauthorized access to RFID data.
- **Technological Limitations** – It is important to avoid radio-frequency challenged environments; otherwise, the use of active tags may be required. For a geographically dispersed survey, more readers will be required and placed at strategic locations. This may be cost prohibitive so a detailed cost analysis is required.
- **Standards and International Compatibility** – The standards that facilitate data exchange and equipment compatibility should be selected. If other countries involved, there is a need to investigate their applicable RFID standards for compatibility.
- **Common Vision** – All stakeholders need to meet and agree on the purpose of the proposed RFID system and the desired outcomes. RFID is a tool and an enabler of improved processes and not an end to itself.

FUTURE RESEARCH DIRECTIONS

RFID as an Instrument of Urban Development

South Korea is building a large "ubiquitous city" or what they term a "U-city" using extensive government investments. RFID technology is playing a major role in its development. This city, New Songdom, is located on a man-made island about 40 miles from Seoul and is being built based on the concept of ubiquitous computing from the ground up. Everything is in an environment that is connected to a computer and RFID tags and sensors. Data is transmitted through wireless connections linking the home with life in and around the city. (Campbell, 2005)

Songdo is being built by Songdo IBD which is a Joint Venture Partnership between Gale International, one of the largest United States real estate developers and POSCO E&C, a subsidiary of POSCO Steel, the second largest steel company in the world. John Kim, one of the leaders in developing this city, has termed this new way of life to be the "U-life" with massive implimentation of RFID, smart cards and sensor based devices. (O'Connel, 2005) The city is not scheduled to be completed until 2014 but already people are already applying to live there. (Gale International, 2008)

This is one example of the role RFID technology will take in this new city:

"Imagine public recycling bins that use radio-frequency identification technology to credit recyclers every time they toss in a bottle; pressure-sensitive floors in the homes of older people that can detect the impact of a fall and immediately contact help; cell phones that store health records and can be used to pay for prescriptions." These are among the services dreamed up by industrial-design students at California State University, Long Beach, for possible use in New Songdo City. (Campbell, 2005)

Growth and Expansion of RFID

The RFID market is expected to grow exponentially. There were 600 million RFID tags sold in 2005 alone, and the value of the market, including hardware, systems and services, is expected to increase by a factor of ten between 2006 and 2016. The number of tags delivered in 2016 will be over 450 times the number delivered in 2006. (Das, 2008)

Studies show that industrial applications using RFID -- particularly transport and logistics, access control, real-time location, supply chain management, manufacturing and processing, agriculture, medicine and pharmaceuticals -- are expected to grow strongly. But RFID devices will also influence Government (e.g., eGovernment, national defense and security and military logistics) and consumer sectors (e.g., personal safety, sports and leisure, smart homes and smart cities).

Counter to intuition, RFID will not replace barcodes for many years except in certain sectors. The reason for this prolonged coexistence of the two technologies is primarily cost. RFID tags costs range from $0.15 to $20 or even more whereas printed barcodes are less than a penny each. (United States Government Accountability Office, 2005, pg. 8) However, costs are expected to drop as demand increases and technology advances.

Although not yet operational, chipless RFID allows for identifying tags without the use of an integrated circuit. Instead the tag uses threaded fibers that reflect a portion of the reader's signal back. This unique return signal can be used as an identifier of the object. This will allow tags to be printed on the object itself thereby lowering costs. (Technovelgy, 2008)

CONCLUSION

Clearly, RFID is an exploding technology that has and will impact society in many ways. The urban planner needs to examine and assess this critical technology and evaluate its application to his or her practice accordingly. As part of this process, the urban planner should be aware of its limitations and drawbacks to prevent unwanted controversy. This technology needs to be harnessed for the good of the profession.

REFERENCES

Anderson, N. (2007). California Outlaws the Forced Subdermal RFID Tagging on Humans. *Arstechnica.* Retrieved November 11, 2008, from http://arstechnica.com/news.ars/post/20070904-california-outlaws-forced-rfid-tagging-of-humans.html

Ashton, K. (2008). RFID Power to the People. *RFID Journal.* Retrieved November 11, 2008, from http://www.rfidjournal.com/article/articleview/4417/1/82/

Bracht, R., Miller, E. K., & Kuckertz, T. (1997). Using an Impedance-Modulated Reflector for Passive Communications. *IEEE Xplore.* Retrieved on November 1, 2008, from http://ieeexplore.ieee.org/Xplore/login.jsp?url=/iel3/4924/13698/00631742.pdf?arnumber=631742

Brewin, B. (2007). Spread of RFID in Defense Slower than Expected. *Government Executive.* Retrieved November 1, 2008 from http://www.govexec.com/story_page_pf.cfm?articleid=37434&printerfriendlyvers=1

California State Senate. (2008). Senate Bill No. 362. Retrieved November 11, 2008 from http://info.sen.ca.gov/pub/07-08/bill/sen/sb_0351-0400/sb_362_bill_20070627_amended_asm_v95.pdf

Campbell, A. (2005). Korea's Ubicomp Vision Offers Opportunities for the West. *RFID Weblog.* Retrieved October 17, 2008, from http://www.rfid-weblog.com/50226711/koreas_ubicomp_vision_offers_opportunity_for_the_west.php

Card Plus. (2008). CTA Smart Card Sales Start. *Chicago Transit Authority.* Retrieved December 1, 2008, from http://www.transitchicago.com/news/archpress.wu?action=displayarticledetail&articleid=111804

Dallas Morning News. (2007). *'Chipping' Away at Privacy?* Retrieved November 11, 2008 from http://www.dallasnews.com/sharedcontent/dws/news/healthscience/stories/072207dnnatchip.350ef26.html

Das, R. (2008). RFID Market $2.77Bn in 2006 to $12.35Bn in 2010. *IDTechEX.* Retrieved November 3, 2008 from http://www.idtechex.com/research/articles/rfid_market_2_77bn_in_2006_to_12_35bn_in_2010_00000409.asp

Department of Homeland Security. (2008). *The Use of RFID for Human Identification.* Retrieved November 2, 2008, from http://www.dhs.gov/xlibrary/assets/privacy/privacy_advcom_rpt_rfid_draft.pdf

Digital Angel. (2008). *Miraculous Medical Potential.* Retrieved November 11, 2008, from http://www.digitalangel.com/biosensor.aspx

Eagles Nest. (2002). *RFID: The Early Years 1980-1990.* Retrieved November 1, 2008, from http://members.surfbest.net/eaglesnest/rfidhist.htm

Engadget. (2005). RFID to be used for inmate tracking in L.A. Retrieved December 1, 2008, from http://engadget.com/search/?q=RFID+use&invocationType=wl-gadget

Fujitsu. (2008). *Fujitsu Develops World's First 64Kbyte High-Capacity FRAM RFID Tag for Aviation Applications.* Retrieved November 5, 2008 from http://www.fujitsu.com/global/news/pr/archives/month/2008/20080109-01.html

FuturePundit. (2003). *Human Subdermal Credit Card Announced.* Retrieved November 11, 2008, from http://www.futurepundit.com/archives/001824.html

Gale International. (2008). *Gateway to Northeast Asia*. Retrieved October 18, 2008, from http://www.songdo.com/default.aspx

Gilbert, A. (2004). Protecting your ID: RFID chips in humans gets green light. *Silicon.com*. Retrieved November 9, 2008, from http://www.silicon.com/research/specialreports/protectingid/0,3800002220,39124983,00.htm

Goodman, B. (2008). WWII Story: This Is it! Part-IX. *All Aviation Flight Online*. Retrieved November 10, 2008, from http://aafo.com/library/history/B-17/b17part9.htm

Hawley, A. (2004 October). RFID technology using Active or Passive Tags - A Very Hot Issue. *Traffic Engineering + Control, 45*(9), 324-326.

Jardin, X. (2007). CA Bill Would Ban Forced Subdermal RFID-tagging of Humans. *Boingboing*. Retrieved November 11, 2008 from http://www.boingboing.net/2007/09/04/ca-bill-bans-forced.html

Kennedy, K. (2006). Subdermal Keys/Tribal Expressions. *Tribal Expressions*. Retrieved November 11, 2008 from http://tribalexpression.blogspot.com

Kleist, R. A., Chapman, T. A., Sakai, D. A., & Jarvis, B. S. (2005). *RFID Labeling* (2nd ed.). Irvine, CA: Printronix.

Landt, J., & Catlin, B. (2001). Shrouds of Time, The history of RFID, An AIM Publication. *RFID Consultation*. Retrieved October 29, 2008 from http://www.rfidconsultation.eu/docs/ficheiros/shrouds_of_time.pdf

McMaster University. (2008). Canada-India RFID project looks to Improve Traffic Flow, Reduce Pollution. *Physorg.com*. Retrieved November 1, 2008, from http://www.physorg.com/news133112758.html

O'Connel, P. L. (2005). Korea's High-Tech Utopia, Where Everything is Observed. *The New York Times*.

Reuters. (2008). *Indonesia's Papua plans to tag AIDS sufferers*. Retrieved December 1, 2008 from http://blog.t1production.com/indonesias-papua-plans-to-tag-aids-sufferers

RFID4U. (2006). *RFID Workshop Student Handbook*. Irvine, CA: RFID4U.

RFID Gazette (2004). *The Future is Here: A Beginner's Guide to RFID*. Retrieved November 2, 2008, from http://www.rfidgazette.org/2004/06/rfid_101.html

RFID Soup (2008). *Form Factor*. Retrieved December 1, 2008 from http://rfidsoup.pbwiki.com/Form+Factor

Roberti, M. (2006). *Ending the Hunt for Parking*. RFID Journal.

SmarTrip. (2008). SmarTrip More than a smart card. It's pure genius. *Washington Metropolitan Area Transit Authority*. Retrieved December 1, 2008 from http://www.wmata.com/riding/SmarTrip.cfm

Swedberg, C. (2006). *Solar-Powered RFID Reader Measures Road Traffic*. RFID Journal.

Tabak, V., de Vries, B., & Dijkstra, J. (2008). *RFID Technology Applied for Validation of an Office Simulation Model*. Eindhoven University of Technology Faculty of Architecture, Building, and Planning.

Technovelgy. (2008). *Chipless RFID Tag*. Retrieved December 1, 2008 from the Technovelgy website: http://www.technovelgy.com/ct/Technology-Article.asp?ArtNum=28

Technovelgy. (2008). *RFID Reader*. Retrieved December 1, 2008 from the Technovelgy web site: http://www.technovelgy.com/ct/Technology-Article.asp?ArtNum=54

Transportation Alternatives. (2008). *Congestion Pricing*. Retrieved December 1, 2008, from http://www.transalt.org/campaigns/congestion

United States Government Accountability Office. (2005). *Information Security: Radio Frequency Identification Technology in the Federal Government, Publication No. GAO-05-551*.

KEY TERMS AND DEFINITIONS

Active Tags: Tags that have a battery and radio transmitter. They are used for longer range applications, normally 300 to 750 feet.

Antennas: Devices used to communicate between the tag and the reader. There are two types of antennas in an RFID system, one is in the tag and the other is connected to the reader. Usually it is referred to the one that is connected to the RFID reader.

EPCglobal Networks: A non-profit organization that is working to standardize electronic product codes and RFID technology.

Passive Tag: Tags that do not contain a power source such as a battery.

Radio Frequency Identification (RFID): An auto-identification technology which utilizes radio frequency waves to transfer data between a reader and a tag attached or embedded in movable or stationary objects.

Readers: A device that reads data from a tag ("interrogates" a tag) as well as writes data to the tag. It communicates with the tag through an antenna that emits radio waves. Readers are also called scanners.

RFID Software: Programs which help manage business processes. These include device drivers, filters, databases, middleware and user applications such as warehouse and supply chain management systems.

Semi-Passive Tags: Tags that generally have batteries but may only be powering the circuit while it is active and then operating at a very low level while waiting for a reader's signal.

Tags: A microchip with an integrated circuit packaged with an antenna used to identify objects. Radio Frequency is used to transfer data between the tag and a reader.

Chapter 21
E-Planning:
Information Security Risks and Management Implications

Stephen Kwamena Aikins
University of South Florida, USA

ABSTRACT

This chapter discusses the security risks and management implications for the use of information technology to manage urban and regional planning and development processes. The advancement in GIS technology and planning support systems has provided the opportunity for planning agencies to adopt innovative processes to aid and improve decision-making. Although studies show that a number of impediments to the widespread adoption these technologies exist, emerging trends point to opportunities for the integration of planning supporting systems with various models to help estimate urban growth, environmental, economic and social impact, as well as to facilitate participatory planning. At the same time, information security infrastructure and security preparedness of most public agencies lag behind vulnerabilities. Drawing on the literature on planning, e-planning and information security, the author argues that the emergence of e-planning as an efficient approach to urban planning and development also poses enormous security challenges that need to be managed to ensure integrity, confidentiality and availability of critical planning information for decision-making.

INTRODUCTION AND BACKGROUND

The purpose of this chapter is to discuss the security risks and management implications for the use of information technology (IT) in managing urban planning and development. Numerous societal problems are explored and addressed in urban and regional planning agencies, including urban growth, unemployment and economic revitalization, transportation, environmental degradation and protection, neighbourhood decline and redevelopment, conservation of land and natural resources, provision of open space, parks and recreational facilities, etc. Planning is therefore a future-oriented activity, strongly conditioned by the past and present. It links "scientific and technical knowledge to actions in

DOI: 10.4018/978-1-61520-929-3.ch021

the public domain" (Friedmann 1987, 38). Ideally, planning happens via public discourse between all groups and individuals interested in and/or affected by urban development and management activities pursued by the public and or private sector, although such comprehensive sharing of information and decision making is rarely found in practice (Nedovic'-Budic' 2000).

E-planning is the use of IT-based systems such as geographical information system (GIS), database management system (DBMS) and planning support system (PSS) for managing urban and regional planning and development processes. The advancement in IT and related hardware, particularly in relation to computer aided design (CAD), GIS, DBMS and PSS has provided the opportunity for local government authorities to adopt innovative and effective technologies to aid and improve the management and decision-making in urban development process. With CAD, maps and plans can now be prepared digitally. DBMS allows all the maps, plans and other data to be properly kept and easily retrieved. Using GIS, digital data represented on the maps and plans can then be retrieved and spatially analysed for decision-making purposes.

IT-based systems can also overcome the problems of paper-based systems. Smooth and swift flow of data and information between various stages of urban development enhances promptness and accuracy of decision-making and facilitates effective and efficient management. However, prompt and accurate decision-making can only be achieved if sufficient and accurate data and information are readily available to decision-makers. With a paper based system, files and folders that are physically moved from one office or department to another can only be accessed by one officer at a time, resulting in potential loss of documents and delay in decision making. In addition, a paper-based system is error prone and requires large storage space. If properly designed and implemented, an IT-based system can help resolve the problems of paper-based system by storing data in digital format to reduce storage space and minimize loss of critical information, and by keeping data centrally to facilitate access and enhance data security.

Effective automation of the planning and development processes, however, requires the design and development of multi-department systems that require integration of key components with various security implications that if unaddressed, could compromise the integrity, confidentiality and availability of critical information used for decision making. These components include workflow applications, data model for information sharing, the agency-wide network, as well as hardware and software to support the applications. Although several studies have been conducted on the use of information technology to aid urban planning and development, (e.g. French & Wiggins 1990, French & Skiles 1996, Warnecke et al. 1998, Yaakup et al. 2004) few studies are yet to be performed on the information security risks of e-planning and potential solutions. Drawing on the literature on planning, e-planning and information security, this chapter aims at filling the existing research gap by arguing that the emergence of e-planning as an efficient approach to urban planning and development provides enormous technological opportunities and security challenges for the planning profession. Therefore comprehensive information security management solutions are needed to ensure the integrity, confidentiality and availability of planning information for decision-making.

E-PLANNING EFFICIENCY AND NETWORK SECURITY CONTROLS

In planning analysis, information is derived from printed maps, field surveys, aerial photographs and satellite images. GIS systems enable data from a wide variety of sources and data formats to be integrated together in a common scheme of geographical referencing, thereby providing up-to-date information (Coulson & Bromley 1990) and enhancing operational efficiency. GIS is part of a planning support system (PSS) which alone can support decision-making and urban problem solving to a considerable extent. However, planners will have to adopt existing GIS tools to support their needs. PSS facilitate the process of planning via integrated developments usually based on multiple technologies and common interface (Nedovic'-Budic'2000). The integration of multiple technologies exposes planning data and information to security risks that needs to be addressed through a security management framework to ensure effective information security controls over e-planning resources such as aerial photographs, plans, field survey data, outputs of analytical models, satellite images, etc.

A good security management framework addresses monitoring and control of network security issues related to security policy compliance, technologies, and actions based on decisions made by humans. Additionally, it has the objective of ensuring availability, confidentiality and integrity of the information within an organization (Hentea 2007). Availability refers to the information system running for the required amount of time to meet service level agreements. Confidentiality is the protection against unauthorized disclosure of sensitive information. Integrity refers to the protection against unauthorized modification of data. With good security controls, PSS can contribute to secure data management, analysis, problem solving, design, decision-making, and communication activities.

Klosterman (1997) describes a useful PSS as one that allows the user to select the appropriate tool from a planning toolbox; link it to the appropriate database; perform the required calculations based on user assumptions of current and future conditions and quickly present the results in the form of charts, maps, and other media. Thus, with the advancement of GIS and other visualization technologies, planning decision tools have been substantially improved and applied in a variety of situations (Mandelbaum 1996; Brail & Klosterman 2001). However, adjusting generic multipurpose technology such as GIS and related software to specific applications and methodological requirements still presents a significant challenge. Vonk, Geertman & Schot (2005) identified a number of impediments to the widespread adoption of planning support systems, including lack of awareness of such systems, lack of experience with them, and lack of recognition of their value. The lack of proficiency in the use of PSS and GIS raises security concerns regarding unauthorized modification and disclosure of planning data on agency network.

Planning agencies can implement proficiency training and controls to install auditing/logging software to detect, log and store any adverse activity, in addition to access restriction to specific PSS applications. Access restriction includes having a data center with access cards to prevent the bad "guys from having unrestricted physical access to systems" (Microsoft Corporation, 2003), and having other physical devices such as routers hopefully located within a wiring closet. The agencies can also use "internal routers" to segment their networks into smaller networks for security reasons, such as isolating networks from each other, and for performance and system availability reasons such as increasing available routes for data to travel and increase bandwidth for users. In addition, "border routers" could also be used to connect to the local government's Internet Service Provider (Pastore, 2003). Additionally, the agencies should maintain up-to-date network configuration documentation, and ensure logical configurations within the routers, firewalls and servers are not open to attack and possibly compromised. This can be achieved by keeping up-to-date with vendor patches, ensuring there are complex administrative passwords and settings, as hackers usually scan and probe networks for insecure factory default passwords and settings (McClure et al., 2002; Ruth and Hudson and Microsoft Corporation 2003).

A more recent development, which has motivated the development of PSS, is the view of planning as "a process for articulation and negotiation among stakeholders, consensus building and dispute resolution" (Susskind and Cruikshank 1987; Leung 2003;p. 22). Taking advantage of ubiquitous data, neighbourhoods and environmental groups have adopted GIS to support advocacy activities (Carver, Evans, Kingston & Thurston 2001; Sieber 2006). As argued by Forester (1989) and Innes (1996), the key to claiming validity of negotiated conclusions and planning actions are institutions and the personnel that facilitate inclusive and sincere conversations. As a key component of PSS, the simplest forms of participation GIS (PGIS) can make data available to neighborhood groups, while the most sophisticated ones can solicit input from participation about conditions, plans, and proposals (Talen 2000), and link these comments to map locations when appropriate. In this regard, the use of tools and technology such as PSS help participants in the planning process to visualize alternative scenarios and their impacts, and to actively make and examine assumptions in real time.

PGIS provides the framework for collaborative decision making (Shiffer 1992). For example, the National Neighbourhood Indicators Partnership (NNIP) aims to capture grassroots information on neighbourhood conditions and make it available to neighborhood groups for use in community development activities. This democratization of information has empowered a number of non-traditional groups by providing them with GIS data analysis capabilities (Drummond & French2008). In the area of land use planning, the United States Department of Interior's bureau of land management (BLM) has launched a new program that creates more efficient business practices and encourages an open and collaborative land use planning process through leveraging information technology.

BLM administers 261 million acres of public lands in a manner that sustains the health, diversity, and productivity of these lands for the use and enjoyment of future generations. As part of the e-planning project, BLM has partnered with ESRI to build a common planning data model and core land management tools for the BLM enterprise. Powered by ESRI software programs ArcIMS and ArcSDE, the project is run from the Planning, Assessment and Community Support Group in the BLM's Washington office, the BLM National Science and Technology Center in Denver, Colorado, and the BLM Alaska state office in Anchorage, Alaska (Zulick 2003). BLM also uses Documentum's content management system that centralizes the storage, classification and security of enterprise content. The geographic dispersion of the e-planning system, however, requires information system security controls to help secure the Wide Area Network of BLM. This is especially important, considering the risks posed by the exponential growth of the Internet and organizational data networks, as well as convergence of Internet, wireless multi-media applications and services (Miller 2001).

Planning agencies can use firewalls to filter the traffic between their network and the Internet to address the numerous threats resulting from Web related attacks, unauthorized access and privacy of stakeholders. Studies by the Federal Trade Commission (FTC 2004) revealed that 53% of all reported fraud were Internet related (e.g. spam, phishing, spyware, and other malware threats). Another study by the National Association of State Chief Information Officers (NASCIO 2006a) shows nearly 5% of Americans have been victims of identity theft many of which were Internet related, within the past five years, resulting in estimated financial losses of $48 billion per year. It is generally agreed that a properly configured firewall would be set to use a "default-deny rule" (Pastore, 2003; Internet Security Systems, 2002). What this means is that the agency would lock down all ports, and then open any desired ports as needed, to enable the firewall to block any unwanted traffic. Multi-department planning agencies can consider two firewalls – a border or perimeter firewall to the external world, and an internal firewall to protect various entity servers from internal users in order to reduce or prevent internal data theft (Cisco Systems, 2002; Frederick, 2001).

Content filtering can also be used by planning agencies to block access to specific types of web content, which may be deemed inappropriate for the entities' Internet users (N2H2, 2003; Frederick, 2001). A solution for content filtering can be either appliance based or software based. An appliance is an all-in-one "black box" that contains its own software and databases. A software solution would need a separate server running a mainstream operating system. What may be applicable in the planning agency environment can be determined by the size of the network, bandwidth availability, vendor support, degree of testing, etc. If implemented properly, content filtering systems have the potential to yield significant benefits to planning agencies.

A key benefit of e-planning through the use of PSS is that it provides the framework for an integrated workflow process across governmental organizations for creating, enhancing and updating agency databases that can be easily shared both within and between the agencies and organizations. The integrated software can work together to handle the entire workflow from data creation to information distribution in an environment that supports information technology standards and interoperability with existing systems. In the management of the urban development process, an electronic development review process can be implemented to ensure that plans for development adhere to inter-governmental and regional requirements and also protect citizens from environmental or public safety hazards and support economic development. For example, a web-based multi-department land use approval system that encompasses workflow applications, data model for information sharing, local area and wide area networks, as well as hardware and software to support the applications can help various planning agencies to overcome problems of paper-based systems, and ensure smooth and swift flow of data and information between various stages of urban development for prompt and accurate decision-making. Such a system can also help address the problems of paper-based system by storing data in digital format to reduce storage space and minimize loss of critical information, and also keep data centrally to facilitate access and enhance data security.

One concern that every e-planning agency data network needs to address is the issue of anti-virus protection. Over the past couple of years, some of the most damages done to many networks have been virus attacks. For any anti-virus network software development, a planning agency can focus on a comprehensive solution that addresses the desktop, server, gateway, firewall and email servers for example. This defense-in-depth approach is essential, as the agency cannot trust the desktop solution 100%. Desktops must be kept up-to-date and protected at all times with the anti-virus product that is managed centrally for control. As part of its comprehensive security management program, the State of Michigan installed software that filters spam and serves as anti-virus solution for incoming and outgoing email. Through this solution the state stopped monthly averages of 74,519 viruses in 2003, 657,271 viruses in 2004 and 181,238 viruses in 2005 (NASCIO, 2006b).

Virtual Private Network (VPN) access can also be granted to remote users in an effort to protect the agency's data. To minimize the risk of unauthorized intrusion, the VPN concentrator could be located outside the firewall. This is necessary to avoid compromise of network security through tunnel connection. In addition, an Intrusion detection system (IDS), or an Intrusion Prevention System (IPS) may be considered in the efforts to provide defense-in-depth security for the data network. For large planning agencies with complex networks such as the BLM, it is a good idea to have intrusion detection mechanism in place quickly to detect serious probing or threat activities (Frederick 2001). In addition to the agency network, appropriate controls should be put in place to secure the e-planning database.

Securing E-Planning Database

Data collection and storage take up a considerable proportion of planning resources. As argued by Arbeit (1993), the effort put in database development is sometimes so immense that little time is left for analysis and for creative activities in designing plans and/or policies. To address this problem, urban planning and development related databases must be based on clear understanding of planning problems, process, and context (leClercq 1990). Furthermore, since planning databases are usually derived by compiling data from multiple sources and varying quality and scales, it is necessary to apply the rules of interoperability and integration (Devogele et al. 1998). Therefore, the integration of readily available data sets is one way to reduce database development and maintenance time (Nevodic'Budic' 2000).

In addition to collecting primary data, planners draw on numerous secondary data sources, including other agencies and government census, to acquire and integrate data into a useful database. This implies as part of planning support systems, databases on properties, streets, utilities, and detailed neighborhood characteristics have to be developed and secured, either internally or externally by other government agencies, with planners having access to the database. Most of these systems will be employed in local government settings where the nature, intensity, and impact of urban development and re-development are contested daily between various public and private stakeholders and decision-makers (Nevovic'-Budic' 2000). Regardless of the type of database system used for e-planning decision-making, the topologies for developing the system should enable critical business intelligence functionality (Bontempo & Zagelow 1998). They should be built within appropriate time frames and budgets, and with flexibility needed to meet the agency's ever-evolving requirements.

A key feature of a good database system is data integrity, which refers to the protection of data from defacement and replacement. Planners have the responsibility for ensuring that all documents on the GIS, PSS and related systems are unalterable by untrusted sources. To accomplish this, file integrity should be maintained and examined on a regular basis. The number of trusted sources with the ability to publish content should be limited to the fewest possible, and the publishing mechanism should be separate from the delivery mechanism (Herzog et al., 2001). To enhance integrity, access to the agency's data should follow a least privilege model. All public access should be restricted and enforced at multiple layers. For example, separate user IDs and passwords can be required when logging onto the network, and onto specific workflow and database applications. Additionally, concurrency control should be implemented to allow multi-task access without undermining data integrity.

As noted by Panda and Giordano (1999), protection of the system from unauthorized users, detection of any hostile activities and recovery from any damage caused by intruders are the three major activities in building a secure system. Data integrity rules are the recommended approach for enforcing data quality in operational databases (Lee et al. 2004). These rules are well-grounded in relational database theory (Codd 1970, 1990) and are widely used. Despite established theory and history of the practical use of integrity rules, data quality problems persist in organizations (Becker 1998; Brodie 1980). The failure to link integrity rules to organizational changes and the conventional practice of viewing the application of data integrity as a one-time static process are among the reasons that data quality problems plague organizations. Lee et al. (2004) therefore propose that the application of data integrity be viewed as a dynamic, continuous process embedded in an overall data quality improvement process.

As stated by Amman, Jajodia, McCollum and Blaustein (1997), the damage to an entity's data can spread to other parts of the database through legitimate users as they update fresh data after reading any damaged data. Therefore, timely and faster recovery is required to stop further spreading of damage. To help minimize unauthorized access and damage to e-planning data, and to detect, track and response effectively to all computer incidents and alerts, a planning agency can implement a computer and human reporting system. The computer system tracks all alerts as they occur, and classifies them initially as high, medium or low priority. Alerts such as an automated scan of the file transfer protocol (FTP) port would generate a low priority, whereas a Denial of Service signature would generate a high priority alert (Frederick, 2001). To ensure effectiveness, all alerts should be reviewed by daily security monitoring personnel, depending on the size of the agency and the complexity of the network, in order to investigate, take appropriate action, and document the entire process for late evidentiary and reference usage.

Management Challenges and Implications

Implementing the necessary controls to protect a planning agency's network and database is a proactive security management approach that minimizes vulnerabilities. Unfortunately, information security infrastructure of most organizations, including planning agencies, is a reactive-based approach such as detection of vulnerabilities and applying software updates (Cardoso & Freire 2005) as opposed to a proactive approach (Gordon, Loeb & Lucyshyn 2003) that is based on planning. In addition to technical security controls such as firewalls, passwords, intrusion detection, etc., a planning agency's security includes issues that are typically process and people issues such as policies, training, habits, awareness, procedures, and a variety of other less technical and non-technical issues (Heimerl & Voight 2005) such as social engineering. However, security education and awareness have been lagging behind the rapid and widespread use of the new digital infrastructure (Tassabehji 2005). All these factors make security a process which is based on interdisciplinary techniques (Maiwald 2004; Mena 2004).

Efficient e-planning security management implies the need for an event management approach with enhanced real-time capabilities, adaptation, and generalization to predict possible attacks and to support human actions. Dowd & McHenry (1998) argue "network security must be better understood and embraced" and recommend strategies such as knowing the potential attacker, the value of protected assets, and understanding the sources of risk such as poorly administered system, social engineering, external or internal intrusion. To deliver protection against the latest generation of cyber threats, the rules of pre-emptive protection have to meet criteria for effectiveness, performance, and protection. Therefore, effectiveness of e-planning security management system is determined by the intelligence of the system, defined as the ability to detect unknown attacks with accuracy, along with enough time to strategically take action against intruders (Wang 2005).

Effective e-planning can only be achieved if reliable and accurate information from a variety of sources is organized, stored and utilized by decision-makers. Typically, a well organized e-planning system has information on sensing, cartography, surveying, spatial statistics and other pertinent data residing on systems like GIS and PSS, and such information needs to be protected against unauthorized attack. This implies urban and regional planning agencies and their information security managers have the responsibility for containing damage from compromised systems, preventing internally and Internet launched attacks, providing systems for logging and intrusion detection, and building frameworks for administrators to securely manage agency networks. Unfortunately, information technology security initiatives often must compete with other information technology resource demands that appear to provide more tangible and immediate business value. The funding and resource constraints facing many state and local governments make it imperative for urban and regional planners to design and implement network and database security models that take into account the necessary steps and control measures that provide basic information security at the most cost-effective means.

Implementing a cost-effective security model implies going beyond the obvious items such as physical security, routers, firewalls and anti-virus, and looking at several other important issues, which includes confidentiality, data integrity, content filtering and incidence response. A cost-effective means of securing a planning agency's network and database begins with the documentation of an information security program that reflects the goals of the agency, a realistic assessment of the risks faced by the agency and identification of the resources (manpower, hardware, budget) that are available (Oxenhanlder, 2003). The implementation of the program begins with the risk assessment and the development of an enterprise-wide information security plan. Managing e-planning security implies understanding planners' environments and conducting IT risk assessment to determine their security needs and formulate plans that include strategic goals to protect critical infrastructure and citizen privacy. Once the plan is in place an information security program that embodies security management structure and comprehensive policy, related standards and procedural guidelines should be developed to equip the planners and their information security managers with the tools needed to implement the policy (GAO, 2001). This includes guidelines and standards for Authorization and Authentication, Logging, Auditing and Monitoring, Vulnerability and Patch Management, and Virus Management.

Aligning security policy with management controls, operational controls and technical controls is useful in ensuring effective vulnerability assessment mapping. Providing visible and regular security awareness training, including email updates and newsletters, is critical to effective enforcement of the program. To ensure adequate understanding of the planning agency's security posture, the security program should provide for a central management of incident handling. The program should also provide for compliance reviews and enforcement of the policies, procedures and standards (OMB, 2000). This could be accomplished through self-assessment and reporting, vulnerability and penetration testing, site security reviews, and a tool that queries agents on every desktop to give the security configuration status (NSA, 2002). A planning agency can thus manage its network and database risks by balancing the need for security and cost-effectiveness through information security decisions that restrict access to its network and databases, protect against virus, control network traffic, ensure policy compliance and provide information assurance within the confines of available funding structure.

FUTURE RESEARCH DIRECTIONS

Despite their potential to revolutionize the planning profession, studies conclude that with the exception of metropolitan transportation planning applications, the use of many e-planning tools like GIS and PPS is limited because of their complexity and voracious appetite for input data, resulting in basic mismatch between their functions and what planners really do (Drummond & French 2008). Although several researchers have used GIS tools and other software for the purpose of various type of impact assessment (e.g. Lindsey & Nguyen 2004; Shadewald et al. 2001) the practical attempts to integrate development impact methodology with decision tools and GIS have been slow. This is in spite of the need for closer consideration of development impacts relative to a variety of interests expressed by public and private stakeholders in increasingly complex decision settings (Appleton & Lovett 2005). A number of authors have therefore argued for the development of planning support systems that would integrate GIS, urban and environmental models and visualization (Brail & Klosterman 2001). As suggested by Vonk, Geertman and Schot (2005), the impediments to the widespread adoption of PSS include the lack of awareness of such systems, lack of experience with them, and lack of recognition of their value. It follows therefore that GIS-based planning support systems that are user friendly and capable of supporting economic, demographic and land use forecasting, environmental modelling and transportation planning are needed.

GIS has traditionally contributed to planning in the area of analytic thinking by easing the task of data collection and manipulation of geospatial data. Although techniques such as overlays, buffers, routing, and gravity models are now a routine part of the planner's tool kits, there is still a significant room for improvement in modelling and planning as most urban models have focused on forecasting the amount and location of urban growth (Drummond & French 2008). An emerging trend is the development of a class of models that estimate the consequences of urban growth in addition to the traditional capabilities. For example, Index and CommunityViz are two GIS-based software tools that have the capability to estimate land consumption, traffic, and environmental impacts of plans and proposed development projects. Another GIS-based model developed by the Federal Emergency Management Agency (FEMA) that can be useful to planners is HAZUS-MH, which is capable of estimating the physical damage and the resulting social and economic impacts of floods, hurricanes and earthquakes.

A potential for robust planning support systems that could enable planners and stakeholders to better understand the consequences of alternative causes of action is the integration of GIS with various models like traffic impact models, stormwater models, as well as economic, fiscal impact and social impact models. Although the vision for having these systems remains elusive at the moment, significant progress appears to have been made in the efforts to produce them. Previously, GIS had been unsuccessful in supporting design but with the recent ArchSketch extension for ArcGIS, we may see some integration of informal drawing into GIS whereby a sketch-friendly GIS design interface allows planners to explore alternative designs and model the fiscal, traffic, and environmental impacts of each alternative (Drummond & French 2008). In the arena of participatory planning, PGIS has the potential to move beyond solicitation of inputs in the plan making process when linked to the distributed platform of Internet GIS, thereby supporting unprecedented levels of citizen participation.

The above-mentioned trends, especially in the area of modelling and functional integration, have impact on efficient information security management from the standpoint of both data and network vulnerability. Many security technologies are not integrated and each technology provides information in its own format and meaning. Many devises and systems generate hundreds of events, and report various problems or symptoms. They may all come at different times and from different vendors with different reporting and management capabilities and different update schedules. Hentea (2007) argues these technologies lack features of aggregation and analysis of the data collected, and vendors may provide little or no consistent characterization of events that represent the same symptom. Thus, security analysts must choose how best to select observations, isolating aspects of interest. Additionally, the limitations of each security technology combined with the growth of attacks impact the efficiency of information security management and increase the activities to be performed by network administrators (Hentea 2007).

Consequently, comprehensive solutions are needed to include attack detection and filtering, attack source trace back and identification, and attack prevention and pre-emption (Chang 2002) in order to securely integrate GIS with other planning models. Also needed is the increase of automated auditing and intelligent reporting mechanisms that support security assessment and threat management. With systems becoming too complex for even the most skilled system integrators to install, configure, optimize and maintain, (Kephart & Chess 2003) a proposed solution is autonomic computing systems that can manage themselves given high-level objectives from administrators. According to Hentea (2007), these systems require capabilities for self-configuration, self-optimization, self-healing and self-protection but unfortunately, successful autonomic computing is still in the future, many years away.

Contrary to autonomic systems, an emerging trend is the focus on systems that are based on human-agent effective interaction. For example, security policies can control agent execution and communicate with a human to ensure that agent behaviour conforms to desired constraints and objectives of the security policies (Bhatt, Bertino, Ghafoor & Joshi 2004; Bradshaw, Cabri & Montanari 2003). Thus, for effective and secure integration of e-planning software such as linking GIS to various models or PGIS to Internet GIS, security event management solutions are needed to integrate data from various security and network products to discard false alarms, correlate events from multiple sources, and identify significant events to reduce the unmanaged risks and improve the operational security efficiency.

As discussed, the funding and resource constraints facing many state and local governments make it imperative for urban and regional planners to design and implement network security models that take into account the necessary steps and control measures that provide information security at the most cost-effective means. The comprehensive security management solutions needed to manage integrated e-planning systems require an economic perspective that recognizes that while investment in information security is good, more investment beyond a certain level is not always worth the cost. Given that public sector organizations possess limited resources, the adoption of an information security framework for considering decisions regarding the allocation of scarce information security dollars may be appropriate. In a study that considers how the vulnerability of information and the loss associated with such vulnerability affect the optimal level of resources that should be devoted to securing information, Gordon & Loeb (2002) concluded that for a broad class of security breach probability functions, the optimal amount to spend on information security is an increasing function of the level of vulnerability of such information.

The study also found that for a second class of breach probability functions, the optimal amount increases, but ultimately decreases with the level of vulnerability, suggesting that managers allocating information security budget should normally focus on information that falls into the mid-range vulnerability to security breaches. Hence a meaningful endeavor for e-planning security managers may be to partition information sets into low, middle, and high levels of security breach vulnerability. The analysis by Gordon & Loeb (2002) shows that for two broad of classes of security breach probability functions, the optimal amount to spend on information security never exceeds 37% of the expected loss resulting from a security breach. Hence the optimal amount to spend may be far less than even the expected loss from a security breach. Naturally, information security vendors and consultants will focus on huge potential losses from security breaches in order to sell their products and service. The information security manager of the planning agency should be astute enough to be aware that the expected losses are typically an order of magnitude smaller than such potential losses. The implication here is that the planning agency and its security manager must carefully balance the agency's security needs against resource constraints and other priorities.

CONCLUSION

The information age has changed the way organizations like urban and regional planning agencies operate. The advancement of GIS, PPS and other visualization tools has substantially improved planning decisions in a variety of situations. Although system complexity and other factors have limited extensive uses of planning models, PSS can provide the framework for an integrated workflow process across governmental organizations for creating, enhancing and updating agency databases that can be easily shared both within and between the agencies and organizations. Likewise, although GIS has traditionally contributed to planning in the area of analytic thinking by easing the task of data collection and manipulation of geospatial data, the potential for integration with other models to expand planning capabilities and participatory planning exists in the not too distant future. This poses security challenges for planners, and requires security management solutions that document and ensure compliance with security policies, standards and guidelines, contain damage from compromised systems, prevent internally and Internet launched attacks, provide systems for logging and intrusion detection, and build frameworks for administrators to securely manage agency data and networks. Such solutions need to be adopted from an economic perspective that balances the planning agency's security needs against resource constraints and other priorities with the goal of enhancing availability, confidentiality and integrity of planning information resources.

REFERENCES

Amman, P., Jagodia, S., McCullum, C., & Blaustein, B. (1997). Surviving information warfare attacks on databases. In *Proceedings of the IEEE Symposium on Security and Privacy* (pp. 164-174)

Arbeit, D. (1993). Resolving the data problem: A spatial information infrastructure for planning support. In *Proceedings of Third International Conference on Computers in Urban Planning and Urban Management*, Atlanta, GA, July 23-25.

Becker, S. (1998). A practical perspective on data quality issues. *Journal of Database Management*, *35*, 35–37.

Bhatti, R., Bertino, E., Ghafoor, A., & Joshi, J. B. D. (2004). XML-based specification for web services document security. *IEEE Computer*, *37*(4), 41–49.

Bontempo, C., & Zageclow, G. (1998). The IBM data warehouse architecture. *Communications of the ACM*, *4*(9), 38–48. doi:10.1145/285070.285078

Bradshaw, J. M., Cabri, J., & Montanari, R. (2003). Taking bask cyberspace. *IEEE Computer*, *36*(7), 89–92.

Brail, R., & Klosterman, R. (Eds.). (2001). *Planning support systems: Integrating geographic information systems, models and visualization tools*. Redlands, CA: ESRI Press.

Brodie, M. I. (1980). Data quality in information systems. *Information & Management*, (3): 245–258. doi:10.1016/0378-7206(80)90035-X

Cardoso, R. C., & Freire, M. M. (2005). Security vulnerabilities and exposures in Internet systems and services. In Pagani, M. (Ed.), *Encyclopedia of multimedia technology and networking* (pp. 910–916). Hershey, PA: Idea Group Reference.

Carver, S., Evans, A., Kingston, R., & Thurston, R. (2001). Public participation, GIS and cyber-democracy: Evaluating on-line spatial decision support systems. *Environment and Planning B, 28*(6), 907–921. doi:10.1068/b2751t

Chang, R. K. C. (2002). Defending against flooding-based distributed denial-of-service attacks: A tutorial. *IEEE Communications Magazine, 40*(10), 42–51. doi:10.1109/MCOM.2002.1039856

Cisco Systems. (2002). *Design guide Cisco IOS firewall*. Retrieved February 27, 2005, from http://www.cisco.com/warp/public/cc/pd/iosw/prodlit/firew_dg.htm

Codd, E. F. (1970). A relational model of data for large shared data banks. *Communications of the ACM, 13*(6), 377–387. doi:10.1145/362384.362685

Coulson, M., & Bromley, R. (1990). The assessment of the users needs for corporate GIS: The example of Swansea Council. In *Proceedings of the European Conference on Geographic Information Systems* (pp. 209-217). Amsterdam: EGIS Foundation.

Devogele, T., Parent, C., & Spaccapietra, S. (1998). On spatial database integration. *International Journal of Geographical Information Science, 12*(4), 335–352. doi:10.1080/136588198241824

Dowd, P. W., & McHenry, J. T. (1998). Network security: It's time to take it seriously. *IEEE Computer, 31*(9), 24–28.

Drummond, W. J., & French, S. P. (2008). The future of GIS. *Journal of the American Planning Association. American Planning Association, 74*(2), 161–176. doi:10.1080/01944360801982146

Forester, J. (1989). *Planning in the face of power*. Los Angeles: University of California Press.

Frederick, K. (2001). Network monitoring for intrusion detection. *Security Focus*. Retrieved March 3, 2005, from http://online.securityfocus.com/infocus/1220

French, S. P., & Skiles, A. E. (1996). Organizational structures for GIS Implementation. In M. J. Salling (Ed.), *URISA '96 Conference Proceedings*, Salt Lake City, Utah (pp. 280-293).

French, S. P., & Wiggins, L. I. (1990). California planning agency experience with automated mapping and geographic information systems. *Environment and Planning B, 17*(4), 441–450. doi:10.1068/b170441

Friedman, J. (1987). *Planning in the public domain: From knowledge on action*. Princeton, NJ: Princeton University Press.

Gordon, L. A., & Leob, M. P. (2002). The economics of information security investment. *ACM Transactions on Information and System Security, 5*(4), 438–457. doi:10.1145/581271.581274

Gordon, L. A., Loeb, M. P., & Lucyshyn, W. (2003). Information security expenditures and real options: A wait-and-see approach. *Computer Security Journal, 19*(2), 1–7.

Heimerl, J. L., & Voight, H. (2005). Management: The foundation of security program design and management. *Computer Security Journal, 21*(2), 1–20.

Hentea, M. (2007). Intelligent system for information security management: Architecture and design issues. *Issues in Information Science and Technology, 4*, 29–43.

Herzorg, P., et al. (2001). The open source security testing manual, v1.5. *IdeaHamster*. Retrieved May 6, 2005, from http://www.ideahamster.org/osstmm.htm

Hwang, M. S., Tzeng, S. F., & Tsai, C. S. (2003). A new secure generalization of threshold signature scheme. In Proceedings of International Technology for Research and Education (pp. 282-285).

Innes, J. E. (1996). Planning through consensus building: A new view of the comprehensive planning ideal. *Journal of the American Planning Association. American Planning Association, 62*(4), 460–472. doi:10.1080/01944369608975712

Innes, J. E. (1998). Information in communicative planning. *Journal of the American Planning Association. American Planning Association, 64*(11), 52–63. doi:10.1080/01944369808975956

Internet Security Systems. (2002). *Real Secure network protection.* Retrieved January 19, 2004, from http://www.iss.net/products_services/enterprise_protection/rsnetwork/index.php

Kephart, J. O., & Chess, D. M. (2003). The vision of automatic computing. *IEEE Computer, 36*(1), 41–50.

Klosterman, R. E. (1997). Planning support systems: A new perspective on computer-aided planning. *Journal of Planning Education and Research, 17*(1), 45–54. doi:10.1177/0739456X9701700105

leClercq, F. (1990). Information supply to strategic planning. *Environment and Planning. B, Planning & Design, 17*, 429–440. doi:10.1068/b170429

Lee, Y. W., Pipino, L., Strong, D. M., & Wang, R. Y. (2004). Process Embedded Data Integrity. *Journal of Database Management, 15*(1), 87–103.

Leung, H. J. (2003). *Land use planning made plain* (2nd ed.). Toronto, Canada: University of Toronto Press.

Lindsy, G., & Nguyen, D. B. L. (2004). Use of greenway trails in Indiana. *Journal of Urban Planning and Development, 130*(4), 213–217. doi:10.1061/(ASCE)0733-9488(2004)130:4(213)

Maiwald, E. (2004). *Fundamentals of network security.* New York: McGraw-Hill/Technology Education.

Mandelbaum, S. J. (1996). Making and breaking planning tools. *Computers, Environment and Urban Systems, 20*(2), 71–84. doi:10.1016/S0198-9715(96)00001-4

McClure, S., Scambray, J., & Kurtz, G. (2002). *Hacking exposed: Network security secrets & solutions* (4th ed.). New York: McGraw-Hill Osborne Media.

Mena, J. (2004). Homeland security connecting the DOTS. *Software Development, 12*(5), 34–41.

Microsoft Corporation. (2003). The ten immutable laws of security. *Microsoft Security Response Center.* Retrieved November 21, 2005, from http://www.microsoft.com/technet/treeview/default.asp?url=/technet/columns/security/essays/10imlaws.asap

Miller, S. K. (2001). Facing the challenge of wireless security. *IEEE Computer, 34*(7), 16–18.

National Association of State Chief Information Officers (NASCIO). (2006a). *The IT security business case: Sustainable funding to manage the risk.* Retrieved March 4, 2007, from http://www.nascio.org/publications/documents/NASCIOIT_Security_Business_Case.pdf

National Association of State Chief Information Officers (NASCIO). (2006b). Appendix F: Securing the state of Michigan information technology resources. In *Findings from NASCIO's strategic cyber security survey.* Retrieved December 17, 2008, from http://www.michigan.gov/documents/AppendixF_149547_7.pdf

National Institute of Standards and Technology. (2002). *Risk management guide for information technology systems.* Retrieved November 26, 2005, from http://csrc.nist.gov/publications/nistpubs/800-30/sp800-30.pdf

Nedovic-Budic, Z. (2000). Geographic information science implications for urban and regional planning. *URISA Journal, 12*(2), 81–93.

Office of Management and Budget. (2000, November 28). *Management of federal information resources. OM Circular A-130*. Retrieved December 15, 2005, from http://www.whitehouse.gov/omb/circulars/a130/a130trans4.html

Oxlenhandler, D. (2003). *Designing a secure local area network*. SANS Institute.

Panda, B., & Giordano, J. (1999). Reconstructing the database after electronic attacks. In Jajodia, S. (Ed.), *Database security XII: State and prospects* (pp. 143–156). Boston, MA: Kluwer Academic Publishers.

Pastore, M. (2003). Infrastructure and connectivity. In Security + study guide. Boston: Sybex.

Ruth, A., & Hudson, K., & Microsoft Corporation. (2003). Network Infrastructure Security. In *Security + certification training kit*. Redmond, WA: Microsoft Press.

Shiffer, M. J. (1992). Towards a collaborative planning system. *Environment and Planning B*, *19*(6), 709–722. doi:10.1068/b190709

Sieber, R. (2006). Public participation geographic information systems: A literature review and framework. *Annals of the American Association of Geographers*, *96*(3), 491–507. doi:10.1111/j.1467-8306.2006.00702.x

Susskind, L., & Cruikshank, J. (1987). *Breaking the impasse: Consensual approaches to resolving public disputes*. New York: Basic Books.

Talen, E. (2000). Bottom-up GIS: A new a tool for individual and group expression in participatory planning. *Journal of the American Planning Association. American Planning Association*, *66*(3), 491–807.

Tassabehji, R. (2005). Information security threats. In Pagani, M. (Ed.), *Encyclopedia of multimedia technology and networking* (pp. 404–410). Hershey, PA: Idea Group.

United States General Accounting Office. (2001). *Federal information system control audit manual*. GAO/AIMD-12.19.6. Retrieved March 7, 2006, from http://www.gao.gov/special.pubs/ail2.19.6.pdf

Vonk, G., Geertman, S., & Schot, R. (2005). Bottlenecks blocking the widespread usage of planning support systems. *Environment & Planning A*, *37*(5), 909–924. doi:10.1068/a3712

Warnecke, L., Beattie, J., Cheryl, K., Lyday, W., & French, S. (1998). *Geographic information technology in cities and counties: A nationwide assessment*. Washington, DC: American Forests.

Yaakup, A. B., Abu Bakar, Y., & Sulaiman, S. (2004). Web-based GIS for collaborative planning and public participation toward better governance. In *Proceedings of the 7th International seminar on GIS for Developing Countries*, Johor Bahru, May 10-12.

Zulick, C. (2003). New Era in Land Use Planning. *ESRI, Government Matters*. Retrieved December 17, 2008, from http://www.esri.com/news/arcuser/0404/eplanning.html

ADDITIONAL READING

Andreas, E. F. (2002). *On the necessity of management of information security: The standard ISO17799 as international basis*. Retrieved January 27, 2007, from http://noweco.com/wp_iso17799e.htm

Arbaugh, W. A. (2004). A patch in nine saves time. *IEEE Computer*, 82-83.

Austin, R. D., & Darby, C. A. R. (2003). The myth of secure computing. *Harvard Business Review*, *81*(6), 120–126.

(2003). *Authentication Policy for Federal Agencies*. Indianapolis, IN: Cisco Press.

Bhagyavati, & Hicks, G. (2003). A basic security plan for a generic organization. *Journal of Computing Sciences in Colleges, 19*(1), 248-256.

Birman, K. P. (1996). Software for reliable networks. *Scientific American, 274*(5), 64–69. doi:10.1038/scientificamerican0596-64

Blunk, I., & Vollbrecht, J. (1998). PPP extensible authentication protocol (EAP). *IETF RFC 2284.*

Braun, T., & Diot, C. (1995). Protocol Implementation using integrated layer processing. In *Proceedings of SIGCOMM-95.*

Buzzard, K. (2003). Adequate Security – What exactly do you mean? *Computer Law & Security Report, 19*(5), 406–410. doi:10.1016/S0267-3649(03)00510-7

Carver, S., Evans, A., Kingston, R., & Thurston, R. (2001). Public participation, GIS and cyberdemocracy: Evaluating on-line spatial decision support systems. *Environment and Planning B, 28*(6), 907–921. doi:10.1068/b2751t

Chen, D., Lin, T., Huang, T., Lee, C., & Hiaso, N. (2009). Experimental e-deliberation in Taiwan: A comparison of online and face-to-face citizens' conferences in Beitou, Taipei. In Reddick, C. (Ed.), *Handbook of Research on Strategies for Local E-Government and Implementation: Comparative Studies* (pp. 323–347). Hershey, PA: IGI Global.

Cogan, A., Sharpe, S., & Hertzberg, J. (1986). Citizen participation. In So, F. S., Hand, I., & Madowell, B. D. (Eds.), *The Practice of State and Regional Planning.* Chicago, IL: American Planning Association.

Conroy, M. M., & Evans-Cowery, J. (2004). Informing and interacting: The use of e-government for citizen participation in planning. *Journal of E-Government, 1*(3).

Craig, W., Weiner, H., & Harris, T. (2002). *Community empowerment, public participation and Geographic Information Science.* Philadelphia, PA: Taylor & Frances.

Davide, J. (2002). Policy Enforcement in the Workplace. *Computers & Security, 21*(6), 506–513. doi:10.1016/S0167-4048(02)01006-4

French, S. P., & Skiles, A. E. (1996). Organizational structures for GIS Implementation. In M. J. Salling (Ed.), *URISA '96 Conference Proceedings* (pp. 280-293).

Gross, T. (2002). E-democracy and community networks: Political visions, technological opportunities and social reality. In Gronlund, A. (Ed.), *Electronic government: Design, applications and management* (pp. 249–266). Hershey, PA: Idea Group Publishing.

Horton, M., & Mugge, C. (2003). *Network security – Portable Reference.* New York: McGraw-Hill/Osborne.

Howard, T. L. J., & Gaborit, N. (2007). Using virtual environment technology to improve public participation in urban planning process. *Journal of Urban Planning and Development, 133*(4), 233–241. doi:10.1061/(ASCE)0733-9488(2007)133:4(233)

Hudson-Smith, A., Batty, E. S., & Batty, S. (2003). *Online participation: The Woodberry down experiment.* CASA Working Paper 60. London: CASA.

ITSEC. (1991). *Information Technology Security Evaluation Criteria, Provisional Harmonized Criteria. Commission of the European Communities.* DG XIII.

Malik, S. (2003). *Network security principles and practices.* Indianapolis, IN: Cisco Press.

Mason, A. G. (2002). *Cisco secure virtual private networks.* Indianapolis, IN: Cisco Press.

Meadows, C. (2000, October 24-26). *A framework for denial of service analysis*. Presented at Third Information Survivability Workshop – ISW-2000.

National Security Agency. (2002). *Router security configuration guide by system and network attack center*. SNAC.

NIST. (2002). *Procedures for handling security patches*. Retrieved February 6, 2007, from http://csrc.nist.gov/publications/nistpubs/800-40/sp800-40.pdf

Nyanchama, M., & Stafania, M. (2003 May). Analyzing enterprise network vulnerabilities. *Information Systems Security*, 44-49.

VeriSign. (2004). *Securing wireless local area networks, understanding network vulnerabilities*. Retrieved February 19, 2007, from http://www.verisign.com/resources/wp/index.html#networksSecurity

KEY TERMS AND DEFINITIONS

Database Management: A collection of program that enables the user to create, maintain and query databases for data and generate reports.

Database: A collection of data stored in meaningful extraction for a user. It contains information stored in such a way that enables a computer program to select desired portions of data.

E-Planning: The use of IT-based systems such as geographical information system (GIS), database management system (DBMS) and planning support system (PSS) for managing urban planning and development processes.

Geographic Information System (GIS): A system enables data from a wide variety of sources and data formats to be integrated together in a common scheme of geographical referencing, thereby providing up-to-date information.

Information Security Management: A framework for ensuring the effectiveness of information security controls over information resources such as aerial photographs, plans, field survey data, outputs of analytical models, satellite images, etc. It addresses monitoring and control of security issues related to security policy compliance, technologies, and actions based on decisions made by a human.

Planning Support System (PSS): A system that facilitates the process of planning via integrated developments usually based on multiple technologies and common interface. PSS contributes to data management, analysis, problem solving, and design, decision-making, and communication activities.

Vulnerability: A weakness in a network computer system security procedures, administrative controls, system design, implementation, internal controls, etc. that could be exploited by a threat to gain an unauthorized access to information, disrupt critical processing, or violate a system security policy.

Chapter 22
E-Planning Applications in Turkish Local Governments

Koray Velibeyoglu
Izmir Institute of Technology, Turkey

ABSTRACT

This chapter examines the pivotal relationship between e-planning applications and their organizational context. It employs various evaluation frameworks by searching explicit and implicit structures behind the implementation process. The study is largely based on the statement that 'the organizational and user dimension of implementation factors more than technical ones, constitute the main obstacles to the improvement of e-planning tools in urban planning agencies'. The empirical part of the study scrutinizes the personal and situational factors of users in the process of implementation, benefits and constraints of an e-planning implementation and planning practitioners' perception of new technologies on urban planning practice and debate. Using a case study research in Turkish local governments, the findings of this study reveal that the organizational and human aspects of high order information systems are still the biggest obstacle in the implementation process.

INTRODUCTION

An increasingly developmental role beyond the traditional role of service provision pushes city governments to be more proactive and inventive in the application of information and communication technologies (ICTs). For the purposes of city governments the role of ICTs encompasses a number of fields and actions like catalyzing economic development, inhibiting social inequality, controlling and managing urban development and supporting accessibility to local information and services. In urban development, ICTs contribute to management and monitoring of urban development through various e-applications and spatial information systems like geographical information systems (GIS). ICTs contribution to urban development, however, is closely aligned with the soft organizational realities (e.g., cultural, structural, political, personal factors) that are highly influential in the implementation of ICT-based systems and policies. Therefore, in order to understand the impacts of ICTs in organizations one should carefully look into the computer-aided work practice and everyday use of sound technological systems.

The chapter heavily draws on a comprehensive field study conducted in a sample of local government urban planning agencies of Turkey. The major question that lies behind this study is how organizational and human aspects of technological applications affect success and failure of e-planning. The study, then, addresses the soft organizational realities of e-planning applications.

The term e-planning applications used in this chapter refer to the use of ICTs and information systems to advance urban planning and management. Technological applications on which e-planning are based could be grouped into three categories (Budthimedhee et al., 2002). The first group of technologies refers to technologies about data management and distribution. The second is related with mapping and processing of spatial data that is vital for urban planning and management. The last group is composed of interface technologies, which are concerned with creating more effective information environments via representation and modeling.

The basis of the study is largely determined by the human factor. Therefore, e-planning applications of selected planning agencies is measured by using qualitative methods (through surveys and interviews), and the supporting documents based on respondents' perceptions. The user surveys sketch the profile of a planning practitioner in three ways: (1) personal and technological background (e.g. age, sex, education, computer literacy, job title, duration of work), (2) technical knowledge on e-planning applications (3) attitude toward new technologies and their future role in planning practice and debate. Interviews also provide background information on the obstacles and benefits derived from e-planning applications in case study organizations.

This study has three parts. Firstly, as a background, the chapter includes a historical overview of ICTs and e-planning in Turkish case. Then, soft organizational realities of e-planning are exemplified in the selected Turkish local governments. An empirical study reveals the current state of e-planning applications in respondent local governments. It also scrutinizes planning practitioner's commitment and dependence on e-planning applications in practice. The chapter concludes with lessons learnt from e-planning in the case study organizations and with a prospective research agenda on the organizational dimension of ICTs for urban planning agencies.

BACKGROUND

Development of Urban Technologies in Turkish Local Governments

According to data provided from the 2000 General Population Census, 32% of the population live in 16 cities with metropolitan municipality organizations, and 22 million people live in municipal jurisdictions which comprise 41% of Turkey's population (DIE, 2000). The model of metropolitan municipality has been redesigned in recent years and the new municipal legislation (Law 5216), enforced the use of spatial information systems (e.g., GIS) for inter-municipal activities. Although such legal provisions about GIS can be considered a positive reform, they created new implementation problems for the metropolitan municipalities. As growing IS literature indicates, setting up and running a high order information system is not just a mere technical process but a process that, to a large extent, is affected by many complex factors such as human resources, organizational and socio-political structure (Cavric, 2002; Klosterman, 2001; Vonk 2003).

Municipalities in Turkey are responsible for delivering basic urban services (e.g., infrastructure development, fire department operations, garbage collection, planning services etc.) to the public that require collecting accurate information about their environment and efficient use of these data to perform their tasks. Municipalities collect and manage both spatial data (e.g., master plan and application plans, land-use maps, cadastre maps) and non-spatial data (e.g., water system revenues, environmental taxation, and building permits) in their operations. Almost 80% of total data exploited by municipalities is spatial data. Municipalities are increasingly using GIS, spatial decision support systems (SDSS), management information systems (MIS), and the Internet to carry out administrative, auditing, and planning tasks more efficiently. Among these technologies, GIS is the key for municipal operations with its enormous strength to process spatial data.

Plan making, programming and decision-making are largely dependent on the collection, storing, preservation and management of spatial information and information related to urban land use. Therefore, public works and urban planning are the major issues of Municipal GIS operations. In the Turkish case, Eskisehir, Istanbul, and Bursa are the leading cities, in which implementation process of GIS projects is still in progress (Velibeyoglu & Yigitcanlar, 2008). Since municipalities are the primary customers for the GIS market in Turkey the term urban information system (UIS) has become more popular in recent years (Velibeyoglu, 2005). By combining many of the municipal services into UIS, it is aimed that many municipal services will be integrated with local e-government applications. Thus the system will be shared with other local governments and central public organizations in the city, and thus service integration will be obtained.

This trend encompasses a strategy that aims to increase sharing of information both within and among institutions (Cavric, 2002). However, compared to the types of departmental and individual system implementation, this approach needs accumulation of knowledge, which means local governments should cope with more problems within the current organizational structure. Problems in running and using UIS in Turkish metropolitan municipalities can be summarized in three major points:

Software: When we look into the current state of municipal e-planning applications, it is seen that technology investments are primarily focused on CAD-based systems. The rationale behind this tendency is the complexity of the GIS packages and their requirements for having a good command of a foreign language and advanced knowledge of programming and computer literacy. For this reason, many municipalities turned to CAD-based applications instead of costly GIS investments.

Staffing and Management Structures: There are major bottlenecks in recruiting trained staff in terms of obtaining and processing spatial data. This is largely because of the problem of public sector employment policy that neither computer skills nor individual productivity is encouraged and rewarded by the administrative system.

Data Issues: Digitalization of spatial data is in its initial stages, and lacks quality and quantity. For this reason, the concept of UIS in Turkey is frequently considered for digital map production. In terms of data sharing, municipal organizations cannot competently interact with the public and other public initiatives on-line. In recent years, however, some promising implementations have been done. Telecommunications infrastructure of Izmir Metropolitan Municipality called "IzmirNET" has provided data exchange between district municipalities and public institutions in the city of Izmir (Velibeyoglu & Yigitcanlar, 2008).

History of E-Planning in Turkey

Turkey was first confronted with computer systems in the early 1960s. In the same period, started state planning and, within this process, the concept of urban planning gained prevalence. After the 1960s, land use models and transportation models applied contemporary planning techniques, in

'Metropolitan Planning Offices' set up in Istanbul, Ankara and Izmir. But considering the fact that mainframe computers were less in number and costly in prices and run only by computer experts, it can be said that scientific methods for urban planning and systems approach could not find true place in the practice of Turkish urban planning in this decade, a period in which the impacts of rapid urbanization were more severe (Velibeyoglu, 2004).

The 1980s witnessed technological transformation as much as it did economic and political changes. Parallel to the liberal economic policies implemented, there was a rapid development in the telecommunications sector and microcomputers (PCs) entered the Turkish market very quickly (Wolcott & Cagiltay, 2001). Local governments began to gain financial and administrative autonomy, and the reconstruction plans delegated to them after 1985 (with the Reconstruction Law 3194) pave the way to important developments in the professional practice. In the same period, metropolitan municipalities were established by Law 3030. The first applications in the area of information systems in planning were carried out by metropolitan municipalities. During this period, the Istanbul Metropolitan Municipality opened tender for the production of digital maps covering its jurisdictions. Since then, many local governments adopted institutional GIS systems (Velibeyoglu, 2004).

In recent years, spatial information systems in urban planning, especially GIS, have been mentioned frequently but they have not been fully adopted yet. In the metropolitan municipalities of big Turkish cities such as Istanbul, Ankara, Izmir and Bursa, and in some metropolitan district municipalities (e.g., Kadıköy-Istanbul, Nilüfer-Bursa), GIS has been in use. Despite a number of opportunities such as design, visualization, participation, decision making and modeling provided by information technologies and systems, very few of them have been used in planning agencies. As the early adopters of these systems local government institutions in Turkey have gained important achievements in the automation of basic urban services. However, no innovative use of various e-planning applications such as remote sensing, spatial modeling, simulation and forecasting, for example, in the institutional practice of urban planning and management has been observed so far (Velibeyoglu & Saygin, 2005). Beginning in the pre-implementation stage there are serious obstacles in this process. The following section examines the problems and the opportunities faced by planning organizations.

E-Planning in Practice: The Case of Turkish Metropolitan Planning Agencies

The empirical research was based on the assumption that organizational and user dimensions, more than technical factors, constitute the main obstacles faced by urban planning agencies in the utilization of e-planning applications (Velibeyoglu, 2004). The implementation environment incorporates not just technological but also social and organizational factors. The role of user values and organizational environment is even more important in developing countries. In order to test this argument, planning departments of three Turkish metropolitan municipalities (Ankara, Izmir, and Bursa) were selected as a major unit of analysis according to criteria below:

- awareness and operation of various e-planning applications;
- existence of a planning department as a single, defined unit within the organization;
- easy access to richer financial resources and larger size of planning staff;
- availability of different planning issues in different operational levels that creates further opportunities to observe intra/inter organizational context.

Figure 1. The variables and their relationships with the objectives of the survey

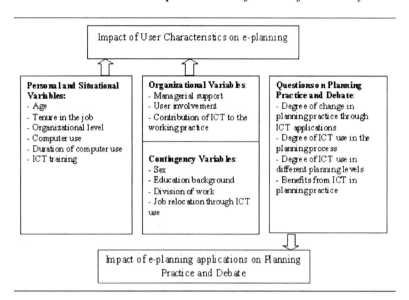

A user survey was conducted in the planning departments of selected Metropolitan Municipalities - Ankara, Izmir, Bursa - in 2004 (Velibeyoglu, 2004). The user survey was applied to all staff members in the planning departments (n=47). The participants were asked to fill out all the 17 structured questions within three major parts even if they were not familiar with the e-planning applications. This survey had three major aims: (1) to draw a profile of planning practitioners working in the respondent municipal organizations (e.g. age, sex, education, computer literacy, job title, duration of work), (2) to measure the degree of their knowledge of planning technologies and their perception of implementation of e-planning applications in the department, (3) to identify their attitude toward new technologies and their future role in planning practice and debate. Variables that were used in this survey are listed in Figure 1.

Additionally, interviews were conducted with managers in the case study organizations to learn more about implementation process. This step consisted of six structured qualitative interviews and ten complementary interviews with other responsible interviewees in and out of respondent organizations. The interviewees were selected based on their influence on the development of e-planning applications in the department and professional reputation –preferably the chief executive of the department. All interviews were carried out at the interviewee's place of work. To verify and supplement the interview, other complementary data was also used.

User Characteristics

Personal and situational characteristics may influence one's perception of information and the way one processes such information. Individuals with distinct characteristics and backgrounds are expected to have unique ways of interacting with information systems and, consequently, have different attitudes and behavior toward such systems (Khalil & Elkordy 2001). In this context, the first part of the survey investigated the relationship of user's age, sex, education background, division of work, tenure in the job, organizational level, ICT training and duration of computer use (Tables 1 and 2).

Table 1. Frequency distribution of personal and situational variables

	Frequencies	Percent
Age:		
(1) 20-25	2	4.3
(2) 26-31	15	31.9
(3) 32-37	17	36.2
(4) 38-43	9	19.1
(5) 44-49	3	6.4
(6) 50+	1	2.1
Sex:		
Female	31	66
Male	16	34
Education background:		
Undergraduate Degree	40	85.1
Master Degree	7	14.9
Tenure in the job:		
(1) Less than one year	1	2.1
(2) 1-2 years	2	4.3
(3) 3-5 years	5	10.6
(4) 6-10 years	18	38.3
(5) 11-15 years	14	29.8
(6) 16 years or more	7	14.9
Organizational Level:		
(1) City planner and Other Staff	37 (+7)	93.6
(2) Department manager	2	4.3
(3) Division Manager	1	2.1
Computer Training:		
(1) Formal Computer Training	26	55.3
(2) Informal or no computer training	21	44.7

Correlation Analysis

In this part of the study, the selected variables of personal and situational characteristics are compared to the uses of e-planning applications expressed by the staff in respondent organizations. All possible e-planning applications were grouped according to Piracha and Kammeier's (2002) classification of planners' knowledge namely, "hands-on use" and "advanced level use":

- Hands-on use includes basic CAD, basic GIS, Office Programs and Internet-based tools.
- Advanced level use requires some additional knowledge including Planning Support Systems (PSS), MIS and Project Management tools (PM), Statistical and Modeling programs, and knowledge of computer programming (Table 3). The results of correlation analysis have been obtained as follows:

Age: A person's willingness to accept a new technology or a change may differ with the person's age. Since younger users generally display a more positive attitude toward the information system, they are more ready to accept the change. A negative correlation between hands-on software use and age ($r=-0.39$, $p<0.01$) has been found.

Tenure in the Job: User attitude and behavior toward e-planning applications are expected to vary with the user's work experience, measured as tenure in the industry, organization, or job. A

Table 2. Mean rankings of users

	Mean (years)	S.D.
Age	2.98 (32-37)	1.07
Tenure in the job	4.34 (6-10)	1.12
Duration of computer use	5.11	2.88

Table 3. Frequency distribution of planners' knowledge of e-planning applications (n=47)

	Knowledge on E-Planning Applications							
	Don't know (1)		Heard of/read about (2)		Worked with (3)		Mean	S.D.
Hands-on Software	Freq.	%	Freq.	%	Freq.	%		
CAD	5	10.6	13	27.7	29	61.7	2.51	.69
GIS	9	19.1	23	48.9	15	31.9	2.13	.71
Office	1	2.1	8	17	38	80.9	2.79	.46
Internet	2	4.3	6	12.8	39	83	2.79	.50
Advance Level Software Applications								
Programming Tools	24	51.1	21	44.7	2	4.3	1.53	.58
PSS	40	85.1	7	14.9	0	0	1.15	.36
MIS/ PM	32	68.1	15	31.9	0	0	1.32	.47
Spreadsheet	28	59.6	16	34	3	6.4	1.47	.62

negative correlation also found between 'hands-on software use' and 'tenure in the job' (r=-0.53, p<0.01) on sampling data.

Organizational Level: The organizational level of the user's job determines his/her responsibilities and decisions and, consequently, his/her informational needs. Therefore, computer use may vary at different managerial levels. A negative correlation found between hands-on software use and organizational level (r=-0.33, p<0.05). This indicates that people at the managerial positions do not need to improve their knowledge in terms of hands-on software use. Besides, findings also support the results in the "tenure in the job" category. However, since the proportion of the people at managerial position is low (6.4%) and the study does not focus directly on the use of ICT by managers, it may be difficult to reach a generalization from these findings.

Computer use: Of the people working in the planning departments surveyed, 72% of them (37/47) use computers on their daily work practice. All the ones who do not use computers do not use any type of ICT software either.

Duration of Computer use: Lengthy use of e-planning applications may strengthen the user's belief in its usefulness, which, consequently, may increase his/her use of the system. In this survey, the positive correlation between hands-on software use and duration of computer use (r=0.41, p<0.01) supports this statement. Lengthy use of e-planning applications by planning practitioners (mean= 5.11 years) has had a positive impact on hands-on software use.

Computer Training: It was found that there was no significant correlation between computer training and hands-on software use. It was also found that although 55.3% of the people (26/47) have received formal computer training, they do not use e-planning applications related to their everyday tasks at work; in practice, it may be one of the likely explanations for this inconsistency. Another explanation is that even though 44.7% of the people (21/47) have not received formal computer training, they may be able to use software packages in practice with the help of the knowledge gained during informal processes.

Other Findings

- Education background, gender, and division of work were found to have no effect on users' attitudes toward hands-on software use.

- No correlation has been encountered between software requiring advance level knowledge (e.g., planning support systems, electronic spreadsheet, or hypermedia tools) and personal and situational variables. Advance level software applications have not yet been utilized in the respondent organizations since the planning practitioners' knowledge that has related to them is very limited. There is a positive correlation encountered, although small, between "software requiring advance knowledge" and "duration of computer use" in terms of programming (r=0.35, p<0.05) and electronic spreadsheet (r=0.29, p<00.5) software. On the other hand, the ratio of hands-on software use is relatively high (63.8%) within the respondent organizations that provide an opportunity in terms of awareness rising for advance level e-planning applications.
- Another set of variables affecting the usage of e-planning applications are organizational variables namely "user involvement" and "management support". User involvement is defined as a psychological state, which refers to the importance and personal relevance of a system to its users. Management support is defined as the extent to which the upper managerial levels provide an appropriate amount of support for information technologies and systems (Seliem et al., 2003). It is expected that both user involvement and management support lead to an increase in the system usage. When scrutinizing hands-on software usage there was no significant relationship found in terms of management support and user involvement. However, management support seems to have positive influence on user involvement (r=0.35, p<0.05). More than half of the respondents (31/47) say that they are encouraged to use e-planning applications within their organizations.

The majority of the respondents (89.4%) also need to use e-planning applications in their daily planning tasks. While the ratio of computer use is 72.4% and hands-on software use 63.8%, the need to use e-planning applications is far above these numbers. This situation confirms respondents' positive attitude towards e-planning applications without being influenced by the organizational environment. As a matter of fact, 93.7% of the respondents found e-planning applications useful in their daily work. Another result obtained, in terms of organizational determinants, is that the respondents' ability to use e-planning applications has no positive impact on their organizational levels. Although the managers claims that capability to use ICT-based software was one of the most important criteria to be considered during the recruitment process, as survey results suggest, using e-planning applications does not play an important role in personal carrier development at work.

Regression Analysis

Following the results of correlation analysis, the total numbers of independent variables were condensed into four personal-situational dimensions and two organizational dimensions respectively. A stepwise regression analysis was performed on the data set. The objective was to use the several

Figure 2. Relations of independent variables and dependent variable

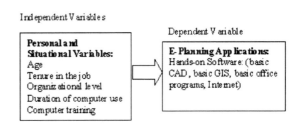

independent variables whose values were known to predict the single dependent value. This technique was repeated until it was determined that all significant predictor variables had been included (Figure 2). The given three models have predicted the independent variable (Hands-on Software use) in various degrees (Table 4).

Planners' Perception of E-Planning Applications

The second part of the survey is devoted to planning practitioners' perception of new technologies on urban planning practice and debate. The findings of this part can be summarized as follows:

Firstly, respondents were asked whether widespread use of e-planning applications make significant changes in urban planning practice or not. The majority of the planners (38/47) who have answered this question think that using e-planning applications will have significant transformational effects on the nature of urban planning practice. This underlines the fact that technological developments have a stronger impact on the respondents than the view of "new tools to perform old tasks" (see Klosterman, 2001).

Secondly, respondents were asked to rank the frequency of using e-planning applications during urban planning process from "the most" to "the least" on a 5-point Likert scale. Findings suggested that data collecting, issuing and evaluating features of e-planning applications were pointed as the most important (mean=1.58) whereas influence of these applications as an enabler on the participation process in planning was evaluated as the least important (mean=4.17). E-planning applications were not considered as a new participation means by the respondents. This result also confirms the common idea that new technologies support the techno-rational side of planning (see Wheeler, 2001).

Thirdly, respondents were asked to rank 'at which level of the planning process were e-planning applications used most effectively' on a 4-point Likert scale from the "the most" to "the least". These applications were said to be used the most frequently at the 'urban design' and 'community planning' levels. Considering that the majority of software packages used in urban planning practice is CAD-based and mostly effective on architectural scale, one may find the results reasonable. Findings alternatively indicate the need for the establishment of an urban information system that would be appropriate at planning levels such as regional or metropolitan planning.

Lastly, respondents were asked to rank the benefits of using e-planning applications during urban planning process from "the most" to "the least" on a 5-point Likert scale. They gave more importance to data collection and processing (mean=1.36), considering that these applications can also substitute routine planning tasks (mean=2.38). Here, too, functions such as decision-support, participation and data-sharing capability of e-planning applications seemed to have been neglected by respondents. It would be

Table 4. Stepwise regression: (n=47) ((1) Dependent Variable: hands-on software use)

Model	R	R Square	Adj. R Square	Std. Error of the Estimate	F
1	.442 [a]	.196	173	.380	8.51
2	.608 [b]	.370	.332	.341	9.96
3	.675 [c]	.456	.407	.321	9.22

a. Predictors: tenure in the job
b. Predictors: tenure in the job, duration of computer use
c. Predictors: tenure in the job, duration of computer use, computer training

possible to have improvements in these areas if a complete urban information system with all its functions working efficiently was established and if the participants' knowledge and skills to use the software was substantially improved.

Obstacles and Benefits about Implementation Process

Interviews with managerial staff in respondent organizations indicated some obstacles and benefits derived from e-planning applications. It is obvious that despite increasing use of e-planning applications in planning departments, it will take some time to realize the benefits of their implementation. In fact, even though the planning departments of respondent municipal organizations may have more advantages in particular areas compared to the other small scale municipalities, these selected organizations have a very short history of implementation. Therefore, while the obstacles were mentioned in detail by the interviewees, the benefits derived from implementation remained relatively peripheral for them. Besides, since the interviewees were planning heads or professional staff members, views of elected officials, other decision-making bodies, and citizens on obstacles and benefits from e-planning applications were reflected only indirectly.

In order to form a systematic framework to summarize implementation hurdles, a 'pentagon prism model' was adopted (see Vonk 2003):

Software-Hardware

- Software and hardware systems are not adjusted to perform daily planning tasks and they are incompatible with current planning rules and legislations.

Humanware

- Personnel who do not follow up technological innovations (quality);
- Shortage of staff which is open to ICT-based work processes (quantity).

Dataware

- Insufficient and unreliable supply of base maps, cadastre maps and other spatial data that would be essential for urban planning functions;
- Problem of updating digital data periodically;
- Insufficient planning information shared with the public that reduces the opportunity of participatory practices;
- Digital data repetitions due to miscommunication between organizations.

Orgware

- Existing ICT technologies which do not match the idea of e-local government;
- Canceling investments on e-planning applications as a result of constant changes in the managerial levels;
- Authorities approve technical reports not having full comprehension of them and give limited support for their implementation.

Planning Task & Work Process

- There is an urgent need to form teams that would work on contemporary planning of good quality rather than on the quantity of plan making;
- Although e-planning applications have been established, most of the analysis essential for urban planning is still done through traditional methods and in a two-dimensional environment;
- Production of digital data (both spatial and attribute data) has not been fully adjusted to the needs of urban planning tasks;
- There is considerable waste of time during auditing and approval processes as a result

of not having office automation and digital archive that consists of information such as planning decisions and planning approval dates.

External World (e.g., Legal Rules, Finance, Politics)

- Digitally-produced plans have not legal validity;
- It is not possible to encourage the personnel to use e-planning applications within the existing legislation;
- Establishing legal procedures related to selling, sharing and copywriting of plans created in a digital environment is still lacking in terms of organizational bodies that would deal with such problems.

Possible benefits derived from implementation of e-planning applications can be summarized in two main categories. Increase in the possible benefits derived from implementation depends on the elements such as diversification in areas of use, reorganization of work process and increase in duration of utilization.

Improvement in Decision-Making

- Accuracy in technical information collected through e-planning applications encourages decision-makers towards techno-rational accuracy;
- Carrying planning tasks into a digital environment provides an opportunity for planning departments to communicate with other departmental units and politicians more easily and to convey the problems to them more quickly and accurately;
- During quantitative plan audits the risk of wasting time gets smaller and the process of decision making and implementation gets easier.

Improvement in Work Process

- There is efficient use of time through use of planning software that is in compliance with plan making procedure and consequently the efficiency in daily routine tasks,
- There is considerable reduction in errors during routine planning tasks and hence improvement in planning accuracy;
- It is possible to facilitate control among different planning scales, and there is an opportunity for design and presentation at any required scale;
- Analysis and synthesis in planning studies are faster and accurate since various data layers may be superimposed easily in a computerized environment.

FUTURE RESEARCH DIRECTIONS

Studies of Obermeyer and Pinto (2008) bring forth a body of literature that draws attention to the different kinds of ICT research. To them, ICT research would follow a pattern that begins with technological problems, proceeds through financial aspects, continues with institutional issues, and finally culminates in examinations of the effect of the technology on society. In the future, ICT research in developing countries should give priority to social and organizational dimensions contrary to what happened until now, as the case of Turkey seems to suggest.

It is also necessary to undertake future studies on the measurement of tangible and intangible benefits of e-planning projects. Cost-benefit analysis on tangible benefits can be used as an effective tool. Intangible benefits (e.g. better working morale, better communication within the organization, responsibility, organizational effectiveness) bear great importance in the implementation of information systems (see Pick, 2005). Strategic tools such as 'stakeholders analysis' and 'SWOT analysis' could provide important clues for institutions to be aware of their own characteristics and to be able to design strategies to overcome implementation problems.

CONCLUSION

This chapter revealed a high level of awareness about e-planning applications among planning practitioners. The major findings of this study confirm that the organizational factors are important obstacles to the improvement of e-planning applications in urban planning agencies. Especially the organizational aspects of higher order information systems, like GIS, are among the most important barriers in the implementation process. The key points identified in the user survey and interviews in the case study organizations can be summed up as follows:

- Potential benefits expected from e-planning applications are closely related with the elements such as the diversification of the areas of use, re-organization of the work process, and the extension of the implementation period.
- In the case study organizations, it has been seen that the most important obstacles that could affect the functioning of planning technologies are orgware and dataware. One of the interesting results is that humanware does not constitute an important obstacle for implementation. In the interviews, staff quality was defined by the head of planning departments as one of the most important strengths of the institution. The fact that the mean age of the planners in these institutions is relatively low (mean=2.98 equivalent of 32-37) and that duration of computer use is high (mean=5.11 years), and users involvement is quite high (90% of total respondents) is effective.
- Another important point is environmental instability and change. It is also known that the effect of external macro factors is highly important during the process of implementation especially in developing countries.

Like many things in the knowledge era, it seems inevitable that there must be some changes in the role of planners. The results of the research indicate that the planners working in respondent municipal organizations are young (70% age between 26-37), have high job tenures (6-10 years) and more than half have taken formal computer training (55.3%). More than half (78%) of the practitioners have experience of 5 years in computer use. The three variables namely "tenure on the job", "duration of computer use", and "computer training" were able to account for approximately 40% of the total causes of hands on software usage.

In conclusion, e-planning technologies are not value-neutral and these tools have the possibility to shape the goals and the agenda of the profession (Wheeler, 2001). In the case study organizations, an important part of urban planners think that applying e-planning technologies would have an important role in the future planning practice. Even e-planning technologies have not replaced most of the routine planning tasks, and a great majority of respondents found these technologies to be an important contribution to their work practice. Since planners' jobs mostly cover other than non-routine tasks, unlike clerical staff, they are feared about their job satisfaction in a lesser extent. Although using e-planning applications does not constitute justification for job ranks within the organization being equipped with high level of computer competency gave ample opportunity for autonomy, variety and feedback in work experience, and also provides collaborations with other non-planner professionals such as computer experts. Eventually, this study illustrates that, in spite of the optimism, a considerable amount of time is still needed for Turkish public planning agencies in their quest for computer-aided planning to be fully tailored into daily work practice.

REFERENCES

Budthimedhee, K., Li, J., & George, V. (2002). ePlanning: A Snapshot of the Literature on Using the World Wide Web in Urban Planning. *Journal of Planning Literature*, *17*(2), 227–246. doi:10.1177/088541202762475964

Cavric, B. (2002). *Human and Organisational Aspects of GIS Development in Botswana*. Paper Presented at GSDI 6 Conference, Budapest, Romania.

DIE. (2000). *Devlet Istatistik Enstitusu* [Turkish Statistical Institute]. Ankara, Turkey: Nufus Sayimi Sonuclari. [Results of the 2000 General Population Census]

Khalil, O. E. M., & Elkordy, M. M. (2001). The Relationship of Some Personal and Situational Factors to IS Effectiveness: Empirical Evidence from Egypt. In Shaw, N. G. (Ed.), *Strategies for Managing Computer Software Upgrades*. London: Idea Group Inc.

Klosterman, R. E. (2001). Planning Support Systems: A new perspective on Computer-aided Planning. In Brail, R., & Klosterman, R. (Eds.), *Planning Support Systems*. New York: ESRI, Inc.

Nedovic-Budic, Z. (2008). Afterword: Planning for Creative Urban Regions. In Yigitcanlar, T., Velibeyoglu, K., & Baum, S. (Eds.), *Creative urban regions: harnessing urban technologies to support knowledge city initiatives* (pp. 312–318). Hershey, PA: IGI Global.

Obermeyer, N. J., & Pinto, J. K. (2008). *Managing Geographic Information Systems* (2nd ed.). New York: The Guilford Press.

Pick, J. (2005). Costs and benefits of GIS in Business. In Pick, J. (Ed.), *Geographic Information Systems in Business*. Hershey, PA: IGI Publishing.

Piracha, A. L., & Kammeier, D. H. (2002). Planning-support systems using an innovative blend of computer tools. *IDPR*, *24*(2), 203–221.

Seliem, A., Ashour, A., Khalil, O., & Millar, S. (2003). The Relationship of Some Organizational Factors to Information Systems Effectiveness: A Contingency Analysis of Egyptian Data. *Journal of Global Information Management*, *11*(1).

Velibeyoglu, K. (2004). *Institutional Use of Information Technologies in City Planning Agencies: Implications from Turkish Metropolitan Municipalities*. Unpublished doctoral dissertation, Izmir Institute of Technology, Izmir, Turkey.

Velibeyoglu, K. (2005). Urban Information Systems in Turkish Local Governments. In Marshall, S., Taylor, W., & Yu, X. (Eds.), *Encyclopedia of developing regional communities with information and communication technology* (pp. 709–714). Hershey, PA: Information Science Reference.

Velibeyoglu, K., & Saygin, O. (2005). *Spatial Information systems in Turkish Local Government: implications from recent practices*. Paper presented at the CUPUM 05: Ninth International Conference on Computers in Urban Planning and Urban Management, CASA, London.

Velibeyoglu, K., & Yigitcanlar, T. (2008). Understanding the supply side: ICT experience of Marmara, Turkey. In Yigitcanlar, T., Velibeyoglu, K., & Baum, S. (Eds.), *Creative Urban Regions: Harnessing Urban Technologies to Support Knowledge City Initiatives* (pp. 245–262). Hershey, PA: Information Science Reference.

Vonk, G. (2003). *Opportunities for Participatory Planning Support Systems (PPSS) in spatial planning. UNetworks in the Delta: ProjectsU*. The Netherlands: Utrecht University.

Wheeler, S. (2001). Technology and Planning: A Note of Caution. *Berkeley Planning Journal, 15*, 85–89.

Wolcott, P., & Çagiltay, K. (2001). Telecommunications, Liberalization, and the Growth of the Internet in Turkey. *The Information Society, 17*, 133–141. doi:10.1080/019722401750175685

ADDITIONAL READING

Cackowski, J. M. (2002). Commentary on Information Technology and Planning. *Journal of Planning Literature, 17*(2), 187–188. doi:10.1177/088541202762475937

Campbell, B. R., & McGrath, G. M. (2003). The Information System Within the Organization: A Case Study. In Cano, J. (Ed.), *Critical Reflections on Information Systems: A Systematic ApproachU*. Hershey, PA: Idea Group Inc.

Carr, T. R. (2003). Geographic Information Systems in the Public Sector. In Garson, G. D. (Ed.), *UPublic Information Technology: Policy and Management Issues. UHershey*. PA: Idea Group Inc.

Geertman, S. (2002). Participatory planning and GIS: a PSS to bridge the gap. *Environment and Planning. B, Planning & Design, 29*, 21–35. doi:10.1068/b2760

Gilfoyle, I., & Thorpe, P. (2004). *Geographic Information Management in Local Government*. Boca Raton, FL: CRC Press. doi:10.1201/9780203484920

Hee, C., & Bae, C. (2002). Information technology for planners: the gmforum. *Environment and Planning. B, Planning & Design, 29*, 883–894. doi:10.1068/b12823t

Heeks, R. (1999). *Reinventing government in the information age: international practice in IT-enabled public sector reform* (Heeks, R., Ed.). London: Routledge. doi:10.4324/9780203204962

Heeks, R. (2002). Information Systems and Developing Countries: Failure, Success, and Local Improvisations. *The Information Society, 18*, 101–112. doi:10.1080/01972240290075039

Mennecke, B. E., & West, L. (2001). Geographic Information Systems in Developing Countries: Issues in Data Collection, Implementation and Management. *Journal of Global Information Management, 9*(4), 44–54.

Nedovic-Budic, Z. (2002). Local Government Applications. In Bossler, J. (Ed.), *Manual of Geospatial Science and Technology* (pp. 563–574). London: Taylor and Francis.

Nedovic-Budic, Z., & Pinto, J. K. (2001). Organizational (Soft) GIS Interoperability: Lessons From the U.S. *International Journal of Applied Earth Observation and Geoinformation, 3*(3), 290–298. doi:10.1016/S0303-2434(01)85035-2

Pitkin, B. (2001). A Historical Perspective of Technology and Planning. *Berkeley Planning Journal, 15*, 32–55.

Ramasubramanian, L. (1999). GIS Implementation in Developing Countries: Learning from Organisational Theory and Reflective Practice. *Transactions in GIS, 3*(4), 359–380. doi:10.1111/1467-9671.00028

Sims, I. M., & Standing, C. (2003). Impact on Society: The Missing Dimension in Evaluating the Benefits of IT in the Public Sector. In Kamel, S. (Ed.), *UManaging Globally with Information Technology*. London: IRM Press.

Tecim, V. (2001). *Cografi Bilgi Sistemleri Tabanli Valilik Bilisim Sistemi* [A GIS based Governorship Information System]. Istanbul: Paper Presented at Cografi Bilgi Sistemleri Bilisim Günleri / GIS Days in Turkey, Fatih University.

Vonk, G. A., Geertman, S., & Schot, P. (2006). New Technologies Stuck in Old Hierarchies: The Diffusion of Geo-Information Technologies in Dutch Public Organizations. *Public Administration Review, 67*(4), 745–756. doi:10.1111/j.1540-6210.2007.00757.x

KEY TERMS AND DEFINITIONS

Local e-Government: Refers to information, services or transactions that local governments provide online to citizens using Internet and Web sites.

Planning Support System (PSS): Is an integrative system in urban planning consisting of a combination of geographic information system, a broad range of computer-based models and a variety of visualization tools for presenting the results of the models.

Pentagon-Prism Model: An assessment methodology in order to identify critical success factors in urban policy. It includes a systematic investigation into five necessary conditions: hardware (the tangible investments), software (the investment in knowledge), orgware (the organizational structure), finware (the financial aspects) and ecoware (the effects on the ecology).

Spatial Data: Any information about the location and shape of, and relationships among, geographic features, usually stored as coordinates and topology.

Spatial Decision Support System (SDSS): Is a spatial information system supporting decision making process related to complex spatial problems such as determining the optimal location of public services such as educational institutions or public parks.

SWOT Analysis: Is a scan of the internal and external environment as an important part of the strategic planning process. Environmental factors internal to the organization usually can be categorized as strengths (S) or weaknesses (W), and those external to the organization can be classified as opportunities (O) or threats (T).

Urban Information System (UIS): Is a powerful means for governments in meeting long-term strategic planning and management challenges. It provides a heightened awareness of the interdependency among environmental, social, and economic health and the impact of decisions made by neighboring jurisdictions, government agencies, and private business.

Chapter 23
GIS Implementation in Malaysian Statutory Development Plan System

Muhammad Faris Abdullah
International Islamic University Malaysia, Malaysia

Alias Abdullah
International Islamic University Malaysia, Malaysia

Rustam Khairi Zahari
International Islamic University Malaysia, Malaysia

ABSTRACT

The chapter presents the current state of GIS implementation in Malaysian development plan system. It offers an overview of GIS implementation worldwide, touching briefly on the history of GIS, planners' early acceptance of the system, factors that promote GIS implementation, level of usage among planners, and factors that impede successful GIS implementation. At the end, the chapter provides a comparison between the state of GIS implementation in Malaysian statutory development plan system with the state of GIS implementation worldwide. The evidence was derived from three main sources: literature, empirical observation of GIS implementation in Malaysia, and a survey conducted in 2008.

INTRODUCTION

After over two decades since its introduction into the planning fields, geographical information systems (GIS) has become one of the important tools-of-the-trade for planners (Ceccato & Snickars, 2000; Drummond & French, 2008; Gocmen, 2009). Despite planners' early resentment towards GIS, they have now become one of the most frequent users of the systems (Budic, 2000; Ceccato & Snickars, 2000; Geertman, 2002; Gilfoyle & Wong, 1998; Gocmen, 2009). In describing planners' early resentment towards GIS, Klosterman (1997) points out that this was largely due to past failures of efforts to computerize planning, such as the failure of large-scale urban modelling. Early GIS implementation was also expensive, and the software was highly complicated for planners liking.

Beginning in the middle of 1980s, the adoption of GIS among local governments in the United

DOI: 10.4018/978-1-61520-929-3.ch023

States of America began to increase slowly, and then sharply in the 1990s (Drummond & French, 2008). Researchers attributed this change in GIS adoption rate to better and cheaper hardware and software availability, as well as better GIS data accessibility (Drummond & French, 2008; Geertman, 2002; Gilfoyle & Wong, 1998; Klosterman, 1999).

Key actions by governments also helped to accelerate GIS adoption in the 1990s. For instance, the publication of Chorley Report in 1987 has helped to increase GIS awareness and provided fundamental directions for GIS development in Britain (Gilfoyle & Wong, 1998). In Wisconsin, U.S.A., the state-funded Wisconsin Land Information Program and the enactment of the Comprehensive Planning Law of 1999 have significantly contributed to increase GIS adoption in the state (Gocmen, 2009).

Since the 1990s, the use of GIS among planners has been widespread. Planners began to adopt this 'new' method in the course of their work, especially in terms of map-making and land data storing. GIS-based information systems were developed and deployed to allow planners and stakeholders better access to information and data (Craglia & Signoretta, 2000; Gilfoyle & Wong, 1998; Heeks, 2002). However, widespread implementation of GIS in the planning field does not translate into full utilization of GIS application. Research indicates that planners' regular use of GIS has been largely limited to the basic functions of the systems, such as mapping and accessing land information, for routine operational and management tasks including permit processing, land data storing and map presentation (Budic, 2000; Gill, Higgs, & Nevitt, 1999; Klosterman, 1997; Mennecke & West, 2001). Even at present, prevalent use of GIS among planners continues to centre on rudimentary applications while advanced applications, such as spatial analysis and modelling, remain underutilized (Gocmen, 2009).

While GIS application among planners continues to be underutilized, the fate of GIS-based information systems is more difficult to assess, mainly because such assessment is highly subjective and timing-dependent (Heeks, 2002). What is considered a success to one person may be a failure to another, and what is considered a success today may be a failure tomorrow. The difficulty in assessing the success and failure of GIS-based information systems is made worst by the propensity of system developers to only report success story. However, several authors suggest that there are many cases of failed GIS-based information systems (Abdullah et al, 2002; Heeks, 2002; Lee & Ahmad, 2000). For instance, Heeks (2002) says that,

On the basis of... these surveys, one may estimate that something like one-fifth to one-quarter of industrialized-country IS (i.e. information system) projects fall into the category of total failure category; something like one-third to three-fifths fall into the partial failure category; and only a minority fall into success category. (p. 102)

Nevertheless, researchers tend to agree that organisational factors are more important than technological ones in ensuring a successful implementation of an information system (Abdullah et al., 2002; Heeks, 2002; Innes & Simpson, 1993; Ramasubramanian, 1999). Organisational mission to implement and support information system, retaining key and trained staff involved in the development and implementation of the system, data availability, and system designers' understanding of users' needs are among the organisational factors identified as key ingredients for a successful information system.

BACKGROUND

Malaysian Statutory Development Plan System

Town and country planning system in Malaysia is governed by the Town and Country Planning Act

Figure 1. Malaysian statutory development plan hierarchy

1976[1] (Johar, 2003; Abdullah & Ariffin, 2003). Central to this Act are the requirements for planning authorities in Malaysia to formulate planning and development policies through preparation of statutory development plans and to regulate development through development control process. The statutory development plans required under the Act are national physical plan, state structure plan, local plan and special area plan.

The administration of this Act is carried out at three different governmental levels which are the Federal, the State and the Local Government (Ahmad Sarkawi, 2006). Planning authorities are established at each level of the government to undertake planning of different scale and functions.

The Town and Country Planning Act 1976 was amended in 2001 to make provision for an improved system of statutory development plans in Malaysia. The 2001 amendment introduced additional set of plans into the system, which aims at introducing a more integrated and holistic approach to planning, and enhancing the control and regulation of planning in Malaysia (Ahmad Sarkawi, 2006). Prior to the amendment, the Malaysian statutory development plan system was a two-tier system consisted of structure plans and local plans. Following the amendment, the system now consists of a hierarchy of development plans ranging from national to local levels. At the top of the hierarchy is the national physical plan. This is followed by the state structure plan, the local plan and the special area plan (Figure 1).

The national physical plan sits at the top of the hierarchy of statutory development plans in Malaysia. Prepared by the Federal Department of Town and Country Planning (FDTCP), the plan provides strategic development framework for the whole of West Malaysia. The plan was approved by the National Physical Planning Council in 2005 and shall be in effect until 2020. It is subject to revision every five years.

The state structure plan is prepared by the State Department of Town and Country Planning (State DTCP). The plan translates the development framework of national physical plan to state level, and provides planning policies and proposals for the whole area within the state boundary. The form and functions of state structure plan remain similar to the 'old' structure plan prior to the 2001 amendment of the Town and Country Planning Act 1976. The only difference is that the current state structure plan is a state-wide plan, while the previous structure plan was prepared for a part of an area within state boundary. At present, eight out of eleven states in West Malaysia already had their state structure plans published in the official gazette. Once published, the plan becomes a legal document that guides the planning and development within the state.

The 2001 amendment also affects the local plan. Although the form and functions of local plan remain the same as before the amendment, the plan is now a district-wide plan. Previously, it covered only a part of an area under a local planning authority's jurisdiction. Thus, the present local plan is commonly referred to as the 'district local plan' in order to distinguish it from the 'old' local plan.

Local plan is prepared by local planning authority to set out detail policies and proposals pertaining to development and use of land in the district. The plan also translates strategic policies and proposals of state structure plan into details to be implemented at the district level. Usually this involves detailing the exact location, components and the cost of development. At present, a total of 108 local plans have been published throughout West Malaysia.

Similar to the local plan, the special area plan is also prepared by local planning authority. Nevertheless, special area plan covers a smaller area in comparison to the local plan and contains proposals that are more detailed in nature. Special area plan is prepared for an area that requires special planning because of its unique character. At present, special area plans have been prepared for areas that require special conservation efforts because of their historical or natural environmental value, and for areas that are subjected to intense development pressure.

In general, the contents of these statutory development plans consist of written statements and maps. The written statements present the planning policies and development proposals for the planned area, while the maps act as supporting documents that clarify the policies and proposals through graphical representations. The national physical plan and the state structure plan are subject to revision every five years. On the other hand, the revision of a local plan can be performed by the local planning authority whenever necessary.

Preparations of all statutory development plans in Malaysia are outsourced to private planning consultants. A team of planning consultants is appointed for each statutory development plan preparation project and is responsible for collecting data, performing analyses and writing reports. Meanwhile, the planning authorities administer the preparation process by providing supervision and funds, as well as handling the publicity and the gazette process of the plan. The planning consultants, based on analyses of the collected data, propose planning policies and development proposals for the planned area for commentary by the planning authorities. Once finalised, the plan will be put on public display before being published in the government's gazette. Figure 2 presents the flow chart of a typical statutory development plan preparation process in Malaysia.

Although there are other types of development plans prepared by planning authorities in Malaysia, such as regional plans and rural growth centre plans, their preparations are not bounded by the requirements in the Town and Country Planning Act 1976. Thus, these plans are sometimes termed as non-statutory development plans. For the purpose of this chapter, any reference to development plans in the following discussions will refer to the statutory development plans.

History of GIS Implementation in Malaysian Statutory Development Plan System

GIS was introduced in Malaysia in the early 1980s. Early record shows that GIS was first used by Malaysian Department of Agriculture in 1981 (Idrus & Harman Shah, 2006). It was only over a decade later that GIS was officially used in planning in Malaysia.

The year 1993 marked the first GIS use in Malaysian planning when Putrajaya, the new Malaysian Federal Government Centre, was being planned (Abu Bakar, 2004). In the process, GIS (MapInfo version 1) was used, but only for map presentation. In that same year, the Local Plan for Parit Buntar and Bagan Serai area was also being prepared. This presented another opportunity for FDTCP to try to use GIS in planning. Thus, the first executive information system for a development plan, known as SMaRT, short for *Sistem Maklumat Rancangan Tempatan*, or Local Plan Information System, was developed for that local plan (Abu Bakar, 2004).

Beginning with these two GIS initiatives, GIS use in Malaysian planning gradually increased. Early widespread implementation of GIS centred on its implementation in the preparation of devel-

Figure 2. Typical development plan preparation process and integration with GIS (Adapted from Jabatan Perancangan Bandar dan Desa Semenanjung Malaysia, 2001 and Jabatan Perancangan Bandar dan Desa Semenanjung Malaysia, 2007)

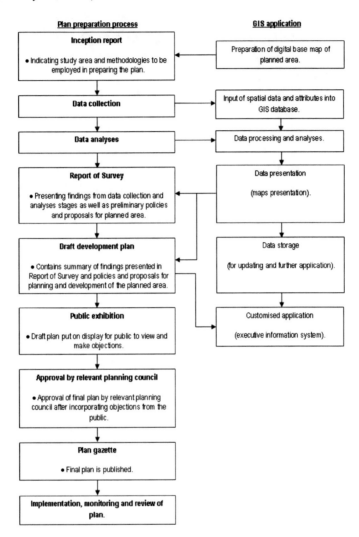

opment plans. By 1997, GIS was already embedded into the preparation process of development plans in Malaysia (Ibrahim et al., 2004). From the initial focus on development plan preparation, GIS was then implemented in other activities of planning including development control and land use change monitoring, as well as being the platform on which tailor-made e-planning systems were developed (Abdullah, 2004).

Today, GIS is a common knowledge among planners in Malaysia. Formal GIS training is now offered to all planning graduates at all planning schools in the country (Ibrahim et al., 2004). On-demand GIS trainings are also offered by various parties, including universities and private GIS companies, to public and private planners. GIS research and development are being pursued actively with universities playing significant roles in developing GIS model and exploring new areas where GIS can add value in the planning process (Abdullah, Abdullah, & Ibrahim, 2009; Samat, 2006; Tan, 2005).

Information System Planning Masterplan

In 1996, the Malaysian Government launched e-government initiatives as part of its efforts to employ information and communication technology (ICT) in order to reinvent the way it operates. E-government was one of the seven flagship applications initiated under the Multimedia Super Corridor programme (Lee & Ahmad, 2000). Under the programme, government departments were required to formulate and implement proposals for integration of ICT in their work in order to achieve speedier and better service delivery. In line with the Malaysian Government's e-government aspiration, in 1997, FDTCP introduced its Information System Planning Masterplan. This masterplan outlines the department's approach towards integration of ICT into planning in Malaysia. Nevertheless, many of the ICT systems proposed to be developed under the masterplan have some GIS elements in them. According to Mohd. Nazri Abdullah (2004), based on the detail explanation in the masterplan, it can be concluded that 80% of the proposed systems would lead to some GIS applications in planning. Among the ICT systems proposed by the masterplan are:

- Development and Planning System;
- Planning Monitoring System;
- Planning Approval System;
- Counter Service System;
- Appeal Board System;
- State Planning Council Decision Support System; and
- Data Warehousing and Information System.

The masterplan recognizes the benefits of GIS to planning, and thus outlines measures to integrate GIS into planning in a more coordinated approach. Among the key thrusts of the masterplan pertaining to GIS are to increase GIS awareness and knowledge among FDTCP officers, and to promote GIS use in planning, especially in the development plan preparation process.

GIS Implementation in Statutory Development Plan Preparation Process

The Information System Planning Masterplan's proposal to promote GIS use in development plan preparation process is perfectly appropriate for several reasons. Firstly, development plan preparation involves the production of a significant number of maps. For instance, Terengganu State Structure Plan contains 62 maps and Perak State Structure Plan contains 52 maps (Table 1). Nevertheless, it has to be noted that these are just the number of maps included in the published plan. The number of maps prepared throughout the entire plan preparation process is notably higher since there are also other reports like the inception report and the report of survey (Figure 2), which have to be produced in the process and these reports contain a sizeable number of maps as well. In fact, report of survey usually contains more maps than the published plan itself. Therefore, even though the number of maps included in Pahang State Structure Plan is only 15 (Figure 2), the total number of maps produced during the Plan's entire preparation process is considerably higher. Due to the high number of maps produced during development plan preparation process, many opportunities to use GIS arise in that process. Maps, which were usually prepared using computer-aided design (CAD) software (and in some cases using Microsoft PowerPoint) previously, can now be prepared using GIS software.

Secondly, statutory development plan forms the basis for development control for the planned area. The plan would determine the type of development allowable in an area, or land parcel, as well as any restrictions (such as development density, plot ratio and building height) associated to that land. The use of GIS would development

Table 1. Number of maps in published state structure plans

No.	State Structure Plan	No. of maps	No. of policies
1.	Terengganu State Structure Plan	69	158
2.	Perak State Structure Plan	52	49
3.	Melaka State Structure Plan	29	165
4.	Pulau Pinang State Structure Plan	29	79
5.	Selangor State Structure Plan	26	100
6.	Negeri Sembilan State Structure Plan	25	111
7.	Pahang State Structure Plan	15	29
8.	Johor State Structure Plan	n/a	n/a

of an information system and its usage in the development control process. Such information system would allow users, especially council planners, to quickly search for the land parcel affected by a planning application using computer and to ascertain the planning policies and the development control requirements related to that land. Such system would also help speed up the development control process considerably as compared to manual search and reference tasks, especially when a plan may consist of more than one hundred policies (refer Table 1). The less time taken to process a planning application would fit very well into the e-government initiatives, which seek for speedier service delivery.

Problems Faced During Early GIS Implementation Period

Following the introduction of the Information System Planning Masterplan, FDTCP began its promotion of GIS. Officers from the department were sent to GIS trainings and road shows were conducted at various town planning departments in the country. FDTCP also conducted several workshops to discuss in detail the areas where GIS can be used in the development plan preparation process and, later on, introduced GIS to private planning consultants who were involved in development plan preparation projects (Ibrahim et al., 2004).

Early attempts to integrate GIS into Malaysian statutory development plan preparation process were not without problems. When initially introduced, GIS received lukewarm reaction from FDTCP officers, created confusion among private planning consultants and failed to achieve what it was intended to do. Mohd Ali Abu Bakar (2004) reports that despite the promotion and the road shows, GIS was not initially well received by many officers from the professional category, such as planners. Many of them perceived GIS merely as a mapping tool and thus only useful to sub-professional officers, such as technical assistants and technicians.

When GIS was introduced to private planning consultants, the reception was not any better as well. To many of them, GIS was completely a new thing and most were unclear about how GIS was to be integrated into the development plan preparation process. When further clarification was sought, FDTCP's response suggested that what the department wanted was for digital maps to be used in the preparation of the plans. However, as stated by Mansor, et al. (2004), initially many consultants still decided not to use GIS. As digital plans and maps could be produced easily using CAD software, not all consultants decided to venture into using GIS software in their work. Instead, many of them opted to continue using CAD software, which they were already using prior to the GIS introduction.

Early implementation of GIS in the development plan preparation process also faced the problem of standardisation, or the lack of it (Bahagian Teknologi Maklumat, 2006). At that time, FDTCP did not provide any detail guidelines on land use colour classification, database structure, metadata structure and the like. Lack of detailed guidelines led to non-standardised GIS outputs prepared by the few consultants who decided to use GIS in the plan preparation.

The absence of detailed guidelines from FDTCP and lack of GIS knowledge among consultants also resulted in limited use of GIS in the development plan preparation process. GIS was viewed as a map drawing tool and thus, used as one. Maps were drawn using GIS software but without data attributes. Spatial analyses were still conducted manually despite the availability of GIS software and digital maps. However, the results of these analyses were then drawn onto the maps using GIS.

Quality Control of GIS Outputs

Having learnt lessons from the problems faced during the early period of GIS implementation in the development plan preparation process, in 1999, FDTCP introduced measures to control the quality of GIS outputs from the process. A guideline on GIS application for local plan preparation process was produced (Abu Bakar, 2004). Although brief in nature, the guideline was rather useful as foundation towards standardisation of GIS outputs. The guideline was tested on several local plan preparation projects before it was finally improved. In 2001, it was officially included in the revised FDTCP's Manual for Local Plan Preparation (Jabatan Perancangan Bandar dan Desa Semenanjung Malaysia, 2001; Abu Bakar, 2004).

Similarly, the FDTCP's Manual for State Structure Plan Preparation was also revised in 2003 to include guidelines for GIS application in the plan preparation process. Other guidelines concerning map preparation, metadata and database structure were also produced. These guidelines are constantly revised and improved. Today, the level of details provided in these guidelines is quite remarkable. For instance, the latest version of FDTCP's *Manual Penyediaan Pelan GIS* (or GIS Plan Preparation Manual) goes to the extent of providing step-by-step instructions on how to prepare land use data layers (Bahagian Teknologi Maklumat, 2007b). The guideline for *Format Metadata dan Struktur Pengkalan Data Sistem Maklumat Geografi (GIS) Kajian Rancangan Tempatan* (or Format for Metadata and GIS Database Systems for Local Plan) is similarly detailed in contents. The 78-page guideline includes samples of metadata format for each land use layer and pre-determines the name of each of the layers (Bahagian Teknologi Maklumat, 2007a).

Another measure taken by FDTCP to ensure standardisation of GIS outputs is the appointment of GIS consultant into the team of consultants that are involved in preparing a development plan. When GIS was initially introduced into the development plan preparation process, the responsibility to prepare GIS maps was on each of the planning consultant in the team. Such arrangement lacked the necessary coordination to ensure that all the maps produced were uniformly formatted. With GIS consultant in the team, the responsibility to coordinate maps preparation is on the GIS consultant. Because of their advanced knowledge in GIS technologies, GIS consultants are also tasked to conduct GIS analyses, develop GIS database and develop executive information system for the plan (Jabatan Perancangan Bandar dan Desa Semenanjung Malaysia, 2007). In terms of map preparation *per se*, the responsibility remains with individual consultants in the team. However, increasingly, in many development plan preparation projects, GIS consultants are also given the task to prepare maps on behalf of the planning consultants. In such cases, the planning consultants still provide the required information for GIS maps preparation to GIS consultants..

Level of GIS Application in Statutory Development Plan Preparation

Despite having GIS consultants' expertise at disposal, present GIS use in most statutory development plan preparation projects is still limited to map preparation and presentation with minimum GIS analyses being performed (Ibrahim et al., 2004; Nor Salehi Kassim, personal communication, December 11, 2008). In most instances, GIS analyses performed are restricted to basic analyses like overlays and catchment (buffer) analyses (G. L. Chua, personal communication, December 12, 2008). Nevertheless, although rare, there were cases of advanced GIS analyses performed in development plan preparation, such as in the preparation of the national physical plan and the Kulim District Local Plan (Abdullah, Abdullah, & Shahbuddin et al., 2009). In both cases, GIS was coupled with decision support systems to conduct analysis and generate maps for land suitability (Amandus Jr, 2006; Kassim & Islam, 2009).

In explaining the reason behind the limited use of GIS analyses in the development plan preparation process, Ibrahim, et al. (2004) states that:

...the current limitations in terms of GIS use in development plan preparation arisen due to inadequate comprehension among private planning consultants on the analyses capability of GIS and also on how GIS can help to ease their work. In addition, the preparation of development plan does not solely involve analyses of spatial data, in many instances, analyses of aspatial data, or a combination of both, are equally important. Marrying the spatial analyses with aspatial ones might be a little on the complicated side to many planning consultants. (p. 3)

Although planning consultants who are involved in the development plan preparation are aware of GIS functions, they are not aware of GIS full capability for analysis. In particular, they do not know whether any of the GIS analyses can be used to replace what they have been doing manually. In the same time, GIS consultants also tend to overlook their responsibility to advice planning consultants on GIS analyses because they are too preoccupied with developing GIS database and the executive information system, which are deemed as the core of their work. Additionally, they receive no request for GIS analyses to be performed from planning consultants (G.L. Chua, personal communication, December 12, 2008).

The superficial knowledge of GIS among planning consultants, as described above, can be attributed to the lack of GIS training received. It should be noted that many of the planning consultants involved in preparing development plans are seniors in their fields. Many never received any formal GIS training during their university days. Although commercial GIS training is available at present, senior consultants prefer to not attend the training sessions. Instead, younger staff are sent for the training. A common idea shared by many senior planning consultants is that GIS (and ICT) suits younger people better.

Executive Information System for Statutory Development Plan

Executive information system (EIS) is another aspect of GIS application in the statutory development plan preparation process. The requirement to produce EIS for development plan is now mandatory in all development plan preparation projects (Jabatan Perancangan Bandar dan Desa Semenanjung Malaysia, 2007). EIS for development plan is basically a computer system that stores all GIS maps prepared for a development plan. Users can then access these maps through the EIS and perform some GIS functions using the toolbar on the system's graphical user interface. Usually, the toolbar includes only basic GIS functions such as calculating distance, accessing information of land parcels and displaying selected layers of information. There are also EIS that stores the digital version of the published development plan as well. As such, the system is frequently used by the planning authorities' officers as a substitute for the hardcopy version of the plan.

The development of EIS for development plan is advancing. Earlier, EIS was developed mainly as a stand-alone system to store GIS maps and reports produced during the development plan preparation process. The database of the system was fixed and only updated when revision was done to the plan. However, recently, EIS began to be developed as a multi-user system and deployed over local area network. The system also now includes features of a charting[2] system, where the database can be easily updated through the system's interface. With these features, planners are able to update the land information in the database without having to actually use GIS software.

EIS for development plan targets three types of users, which are:

- Executives, including council planners, who make decisions about planning applications;
- System administrators who are usually planners and technical staff of planning authorities;
- The general public, including developers and land owners.

Access to information stored in EIS database is restricted according to the type of users. Executive users have access to all information including the digital version of both the report of survey and the published development plan. However, they do not have access to update the GIS database. Public users neither have access to the reports nor to update GIS database. Instead, they can only access information regarding the proposed land use of a particular land parcel and the development restrictions attached to that land. System administrators, obviously, have access to all information and to update the GIS database.

Besides being used as a medium for quick access to planning information, EIS is also used in development control, counter services and public exhibition of development plan. FDTCP has also begun to encourage the development of internet-based EIS (Jabatan Perancangan Bandar dan Desa Semenanjung Malaysia, 2007). It is hoped that by being internet-based, future EIS can be accessed more easily by users, especially the public. With the present system, public users can only access the system at planning authorities' offices, either at the information kiosk or service counter.

EIS Implementation Problems

While EIS development is mandatory for all statutory development plans in Malaysia, the success of its implementation is uncertain due to lack of research on this issue. Nonetheless, it seems that some planning authorities are more successful in implementing EIS than others. EIS implementation at planning authorities such as at Melaka State DTCP, Petaling Jaya City Council and Seberang Perai Municipal Council are often referred to, by those in the planning fraternity, as examples of successful EIS (G. L. Chua, personal communication, December 12, 2008; Mohd Ramzi Mohd Hussain, personal communication, December 10, 2008).

As mentioned previously, EIS for development plan is developed during the plan preparation process. The cost to prepare the plan, including EIS, is jointly funded by FDTCP and the planning authority whose area is being planned for. In most cases, the former contributes towards a major portion of the funding. Once EIS development is completed, the GIS consultant will create an installation disc of the system and hand it over to the planning authority. A manual on how to use the system is also prepared and printed out by the GIS consultant for the planning authority's use. It is then up to the planning authority to install, maintain and upgrade the system. Yet, it is common for planning authorities to fail to maintain the system. Often times, the reasons cited for such failure are lack of personnel and financial capacities.

In hindsight, the EIS for development plan is actually forced upon planning authorities by FDTCP. When EIS became a mandatory requirement in development plan preparation, planning authorities had to accept the system when it is handed over to them although they may not have the resources to maintain the system. Unfortunately, many of the planning authorities do not have the resources to maintain the system. The problem of not having sufficient personnel, especially trained personnel, and funds to maintain EIS is quite significant among many planning authorities, especially those operating in the rural area. In terms of personnel, some planning authorities in the rural area do not even have a planner on their payroll. In cases like this, it is usually the technical assistant who will act as the planner for the area. There are also cases where a planning authority has only one staff, either a planner or a technical assistant, to handle planning matters of the area. To worsen the matter, some of these planning authorities are not even equipped with a computer in which the system can be installed (Aishah Abdullah, personal communication, November 18, 2008; Muhammad Hakimi Mohd. Hussain, personal communication, November 18, 2008). Their existing computers are obsolete that they do not have the right specifications to run EIS applications.

GIS-Based Statutory Development Plan Monitoring System

The monitoring of statutory development plan implementation has not been given sufficient attention by planning authorities in Malaysia, even at present. In fact, many planning authorities do not monitor the implementation of their development plans. In the event where monitoring of development plan is undertaken, the focus is solely on land use change. Meanwhile, other aspects of the plans, such as protection of environmental resources, management of traffic and provision of public facilities, are not included in the monitoring exercise.

Although the monitoring of development plan implementation is not widely practiced, the need for it is clearly identified, especially for the state structure plan. The Manual for Preparation of Structure Plan proposes a procedure that can be employed by States DTCP to monitor their state structure plans (Jabatan Perancangan Bandar dan Desa Semenanjung Malaysia, 2007). Additionally, all published state structure plans include a section that outlines the monitoring programme that the States DTCP intend to undertake. Interestingly, while the manual does not specifically propose the use of any ICT application in state structure plan monitoring, the plans describe the possibility of GIS being used for monitoring purposes.

The only GIS-based monitoring system for state structure plan in Malaysia is the one developed by Negeri Sembilan State DTCP. Known as GIS9, the system was developed for the purpose of monitoring land use change in the state and evaluating the change in relation to the land use allocation proposed in Negeri Sembilan State Structure Plan. The system is placed on a server located at the Negeri Sembilan State DTCP office. The system is also connected to all local planning authorities in the state in order to enable them to update the system's database when change of land use occurs in their area. The updated data will then be used to determine the actual land use allocation in the state. This information will then be compared with the land use allocation proposed by the state structure plan. Thereafter, the difference between actual land use allocation and the proposed one will be calculated and the results will be presented in terms of percentage (UGisP, 2006).

GIS-Based Monitoring System Implementation Problems

Despite being potentially useful, the implementation of GIS9 is rather unsuccessful. This is due to the failure of local planning authorities in updating the data for their area and poor system's infrastructure. As aforementioned, the system is very dependent on local planning authorities updating the land use data for their area whenever there is a change of land use. However, the local planning authorities failed to do this. As in the case of EIS previously, lack of personnel was often cited as one of the reasons for their failure to update the data (Anuar Maidin, personal communication, November 11, 2008). As a result, the land use data in the system database is outdated and does not reflect the actual land use allocation in the state. Thus, the data cannot be used to provide accurate monitoring results.

The system is also plagued by poor infrastructure set up. The server experiences too many problems that lead to numerous down time, which at times extending well over two months (Haibernarisal Baijuri, personal communication, November 16, 2008). The long down time leads to data updating exercise being put on hold and discourages users from using the system.

Impacts of GIS Implementation in Statutory Development Plan System

The implementation of GIS in statutory development plan system has impacted Malaysian planning scene in a number of ways. Perhaps, the biggest impact is how it has contributed towards the increase of GIS critical mass in the country. Although Rogers' diffusion of innovation theory (as cited in Ramasubramanian, 1999) suggests that critical mass precedes adoption of technology, in the case of GIS implementation in Malaysia, the reverse may have happened. When FDTCP introduced GIS in the development plan preparation process, few planners were aware of what GIS was all about. Nevertheless, FDTCP had already decided to make GIS implementation in the plan preparation process mandatory. This created the demand for planners with knowledge on GIS.

Warnecke, Beatie & Lyday (as cited in Drummond & French, 2008) traced the beginning of large-scale introduction of GIS into planning back to 1984. So, while it was already 13 years that planners elsewhere in the world have been using GIS in their work, many Malaysian planners still found GIS a completely new invention when it was officially introduced into Malaysian development plan preparation process in 1997 (Mansor Ibrahim et al., 2004). However, the result of a survey conducted among planners in Malaysia revealed that by 2004, already more than 50% of the planners have used GIS software. A similar survey in 2007 showed that the figure had risen to almost 70% (Abdullah, Abdullah, & Ibrahim et al., 2005). Looking from this angle, one can see the profound impact of GIS implementation in development plan preparation process in creating GIS critical mass in Malaysian planning. It took only 10 years since GIS was officially introduced into the development plan preparation process for it to become widespread among Malaysian planners.

Based on our observation, we would like to offer two main reasons for the popularity of GIS among Malaysian planners after its use was made mandatory in the development plan preparation process. Firstly, in the late 1990s, Malaysia and several other countries in the Asian region were faced with economic recession due to the currency crisis. Prior to the recession, Malaysia was developing rapidly and many private planners were involved as consultants in the construction industry, especially in terms of formulating masterplans for new townships and submitting planning applications for new development. However, during the recession, the construction industry in Malaysia was badly hit that it almost halted. Suddenly, many of the private planners found that there was no job in the construction industry. Around the same time, FDTCP began commissioning a new batch of development plan preparation projects and was looking for private planners to be appointed as consultants in order to prepare these plans. Since planning jobs were scarce, many private planners had no choice but to get involved in those development plan preparation projects. It was through their involvement in those projects that they began to obtain exposure to GIS and GIS software.

Planners from the public sector also involved in development plan preparation projects, although not as consultants, but usually as a project manager, as a technical committee member or as owner of the plan. Thus, similar to private planners, their involvement in the projects helped to expose them to GIS. Additionally, many of them had the benefit of attending formal GIS training. FDTCP, for instance, had established a programme under which GIS trainings are offered to planners from the public sector. The programme, known as Geotechnical Spatial Analysis Research and Development, was established in 2000 as a platform for FDTCP to undertake research and training on spatial planning and decision-making tools relevant to planning such as GIS, multi-criteria evaluation and spatial decision support systems (Jabatan Perancangan Bandar dan Desa Semenanjung Malaysia, 2008).

Secondly, the implementation of GIS in the development plan preparation process has also led to planning schools in the country to begin to offer, or place greater emphasis on, GIS subjects with the aim of producing more marketable graduates. Graduates are not only trained through GIS-specific subjects, but also through heavy GIS integration, both theoretically and practically, within other subjects such as design studios, urban design, planning theory and practice, environmental planning and so on (Ibrahim et al., 2004). In fact, during the late 1990s and early 2000s, having the knowledge to operate GIS software became one of the, if not the most, important criteria for planning graduates to secure employment at planning firms and authorities. Today, although the importance of GIS knowledge in securing employment has reduced slightly (but still important nonetheless), planning schools continue to offer GIS-specific subjects and the integration of GIS in other related subjects remains high. These ensure that planning graduates in the country continue to get exposure to GIS, at least into the foreseeable future. In turn, it helps to retain the critical mass needed to support continued use of GIS in Malaysian planning.

FUTURE RESEARCH DIRECTIONS IN GIS APPLICATION IN MALAYSIAN STATUTORY DEVELOPMENT PLAN SYSTEM

Recent developments indicate that the use of GIS in Malaysian statutory development plan system will continue to flourish. Two areas of great opportunity lie in the integration of GIS with decision support systems in the process of development plan preparation and the development of GIS-based systems for monitoring development plan implementation.

Within the past few years, efforts to institutionalise decision support systems into Malaysian planning have been on-going (Abdullah, Abdullah, & Ibrahim, 2009). Experimental projects in which GIS was coupled with decision support systems in order to provide better decision making during the development plan preparation process were undertaken. Whether those projects have added value to the plan preparation process is, however, unclear at the moment. Presently, it seems that there is lack of research on this issue. Nevertheless, those projects had contributed towards creating among Malaysian planners on the integration of GIS with decision support systems and had laid the foundation for further application of GIS and decision support systems. FDTCP is also working on a guideline on decision support systems implementation in the development plan preparation process at the moment. How this guideline will turn out remains to be seen.

Recently, the development plan monitoring is getting increasing attention from planning authorities. FDTCP is currently working to develop a GIS-based development plan monitoring model (Nor Salehi Kassim, personal communication, November 4, 2008). There are also efforts to develop GIS-based systems for development plan monitoring at other planning authorities. For instance, Selangor DTCP will begin developing their own GIS-based systems soon and Kuala Lumpur City Hall is also looking into the idea of having GIS-based systems to help them monitor their local plan.

CONCLUSION

Looking at GIS implementation in Malaysian statutory development plan system, one can identify several common problems in relation to GIS implementation in planning worldwide.

These are:

- The use of GIS is largely limited to basic functions.
- The implementation failure is due to organisational factors rather than technical.
- Data availability limits the success of GIS-based information systems.

The implementation of GIS in Malaysian development plan preparation process faces several persistent problems. For instance, advanced GIS applications in the preparation of development plans are still underutilized. Despite increased GIS awareness among planners, GIS application in the development plan preparation process continues to be rudimentary in nature. This is most likely caused by insufficient GIS knowledge among senior planning consultants who are involved in the plan preparation. To mitigate the problem, in the short term, it may be necessary for FDTCP to produce a guideline that details out the types of GIS analyses required in the plan preparation process. At present, the same measure is already undertaken by FDTCP in order to promote the use of planning analyses in development plan preparation process. In its Manual for Preparation of Structure Plan, FDTCP has identified a list of required planning analyses that need to be undertaken during the state structure plan preparation process. These analyses include, for example, consistency analysis between plans' objectives and policies, policy impact analysis, land requirement analysis and cost benefit analysis (Jabatan Perancangan Bandar dan Desa Semenanjung Malaysia, 2007). Similar approach may be taken by FDTCP for GIS analyses.

In the long run, the number of GIS analyses used in the development plan preparation may increase as younger planners with GIS technical know-how become planning consultants in the future. Since GIS is relatively new in Malaysia, so are GIS courses at planning schools in the country. While the senior planners may not have the opportunity to learn GIS during their university days, younger planners have had formal GIS training at university.

Many researchers point out that strong organisational support is important to ensure successful GIS implementation (Abdullah, A. et al., 2002; Innes & Simpson, 1993; Ramasubramanian, 1999). Thus, relevant parties must continue the drive towards advanced application of GIS in statutory development plan system. More GIS research should be encouraged to discover the areas in which advanced GIS application can add value to the development plan preparation process, and funds must be made available for such research.

The need for strong organisational support is more visible in the case of GIS-based information systems implementation. EIS for development plan could prove useful for planning authorities in implementing the policies and proposals of their development plans, and as a medium to convey planning information to the general public. However, its implementation, especially at local authority level, is plagued by the problem of inadequate financial and personnel resources. Although FDTCP has been funding the development cost of EIS, it is one-off in nature and this proved to be insufficient. As Alias Abdullah, Muhammad Faris Abdullah & Fauzan Nordin

(2003) point out, information system like EIS is a 'living' system and thus:

...require maintenance and system capacity building from time to time. Accordingly, sufficient funding must be made available. Funding is required for manpower to operate and maintain the system, and for capacity building, which include hardware and software upgrades as well as the overall improvement of the system. (p. 79)

For planning authorities that face financial restriction in maintaining their EIS, regular funding from FDTCP may be necessary to subsidise part of the maintenance cost. Additionally, the planning authorities must be committed to EIS and not use the subsidy for other purposes, except for maintaining the EIS.

The effort to utilize GIS in monitoring the implementation of development plan has been severely hampered by the problem of data unavailability. This is hardly surprising because researchers have found that data unavailability continues to be one of the most prominent factors that limit successful GIS implementation in planning (Gilfoyle & Wong, 1998; Mennecke & West, 2001). Therefore, to overcome this problem, solutions have to be focused towards ensuring data availability. Very often data available are generally inaccurate and incomplete, but these are also the best available data (Klosterman & Abdullah, A., 2008). Thus, it is important that any GIS-based systems for monitoring development plan implementation do not require extensive data sets that are difficult to obtain in order for it to work. The system must also be flexible enough to accommodate the best available data.

Another interesting development pertaining to GIS implementation in Malaysian statutory development plan preparation process is the effort to institutionalise decision support systems in the process. These efforts are welcomed but it has to be implemented with care. For planners who are unfamiliar with, or have little knowledge of, decision support systems, it could be a daunting task to apply it in their work. If its implementation is rushed, it would only lead to planners finding it too difficult and consequently losing interest to continue to use decision support systems. Researchers have suggested that one of the important characteristics of successful technological implementation is that it must allow for incremental and small trials (Innes & Simpson, 1993; Ramasubramanian, 1999). Planners must be exposed slowly to decision support systems. The decision support technology and models adopted must be able to sustain planners' interest in the system and relevant to the planning tasks. We must not adopt complex models which are only of pedagogic value but not useful for plan making purposes (Batty, 2004).

REFERENCES

Abdullah, A., Abdullah, M. F., & Nordin, M. F. (2002). *Spatial information system as a tool in managing urban development process*. Paper presented at the 18th EAROPH World Planning Congress.

Abdullah, A., Abdullah, M. F., & Nordin, M. F. (2003). Managing urban development process by using spatial information system. *Planning Malaysia, 1*, 71–92.

Abdullah, A., Abdullah, M. F., Shahbuddin, M. N. A., & Klosterman, R. E. (2009). Modelling the reality in planning: SPDSS experience in Malaysia. *Malaysian Journal of Environmental Management, 10*(1), 119–132.

Abdullah, M. F., Abdullah, A., & Ibrahim, M. (2009). Institutionalisation of spatial planning and decision support systems for planning and governance in Malaysia. In Davide, G., & Abdullah, A. (Eds.), *Spatial decision support for urban and environmental planning: a collection of case studies*. Shah Alam, Malaysia: Arah Publications.

Abdullah, M. F., Abdullah, A., Ibrahim, M., & Abdul Samad, D. (2005). A study on the demand of spatial planning and decision support system in Malaysia. *Planning Malaysia, 3*, 47–60.

Abdullah, M. F., & Ariffin, I. (2003). Incorporating sustainable development objectives into development plans through strategic environmental assessment. *Planning Malaysia, 1*, 1–18.

Abdullah, M. N. (2004). Pelaksanaan sistem maklumat geografi di Jabatan Perancangan Bandar dan Desa Negeri: dilema dan potensi. *Malaysian Townplan, 2*(1), 29–37.

Abu Bakar, M. A. (2004). Sistem maklumat geografi di Jabatan Perancangan Bandar dan Desa Semenanjung Malaysia. *Malaysian Townplan, 2*(1), 19–23.

Ahmad Sarkawi, A. (2006). The legal considerations of neighbouring lands in development planning: the Malaysian context. Newcastle-upon-Tyne, United Kingdom.

Amandus, P. Jr. (2006). MCDM and land suitability analysis for spatial development: case study Kulim local plan. In Shamsuddin, K. (Ed.), *Implementing Multi Criteria Decision Making in Malaysian Town Planning*. Kuala Lumpur: Federal Department of Town and Country Planning Peninsular Malaysia.

Bahagian Teknologi Maklumat. (2007a). Format Metadata dan Struktur Pengkalan Data Sistem Maklumat Geografi Kajian Rancangan Tempatan. Kuala Lumpur: Jabatan Perancangan Bandar dan Desa Semenanjung Malaysia.

Bahagian Teknologi Maklumat. (2007b). Manual Penyediaan Pelan GIS. Kuala Lumpur: Jabatan Perancangan Bandar dan Desa Semenanjung Malaysia.

Batty, M. (2004). Dissecting the streams of planning history: technology versus policy through models. *Environment and Planning. B, Planning & Design, 31*, 326–330. doi:10.1068/b3103ed

Budic, D. Z. (2000). Geographic information science implications for urban and regional planning. *URISA, 12*(2), 81–93.

Ceccato, V. A., & Snickars, F. (2000). Adapting GIS technology to the needs of local planning. *Environment and Planning. B, Planning & Design, 27*, 923–937. doi:10.1068/b26103

Craglia, M., & Signoretta, P. (2000). From global to local: the development of local geographic information strategies in the United Kingdom. *Environment and Planning. B, Planning & Design, 27*, 777–788. doi:10.1068/b2651

Drummond, W. J., & French, S. P. (2008). The Future of GIS in Planning: Converging Technologies and Diverging Interests. *Journal of the American Planning Association. American Planning Association, 74*(2), 161–174. doi:10.1080/01944360801982146

Geertman, S. (2002). Participatory planning and GIS: a PSS to bridge the gap. *Environment and Planning. B, Planning & Design, 29*, 21–35. doi:10.1068/b2760

Gilfoyle, I., & Wong, C. (1998). Computer applications in planning: Twenty years'. *Planning Practice and Research, 13*(2), 191–197. doi:10.1080/02697459816193

Gill, S., Higgs, G., & Nevitt, P. (1999). GIS in planning departments: Preliminary results from a survey of local planning authorities. *Planning Practice and Research, 14*(3), 341–361. doi:10.1080/02697459915643

Gocmen, Z. A. (2009). *GIS use for planning in Wisconsin's public agencies*. Paper presented at the 11th International Conference on Computers in Urban Planning and Urban Management.

Heeks, R. (2002). Information Systems and Developing Countries: Failure, Success, and Local Improvisations. *The Information Society, 18*(2), 101–112. doi:10.1080/01972240290075039

Ibrahim, M., Abdullah, M. F., & Abdul Samad, D. (2004). *Geographical information system (GIS) in Malaysian planning education.* Paper presented at the 7th International Seminar on GIS for Developing Countries (GISDECO).

Idrus, S., & Harman Shah, A. H. (2006). 'Metabolisme' GIS di Malaysia. *Buletin Geospatial Sektor Awam, 1,* 13–19.

Innes, J. E., & Simpson, D. M. (1993). Implementing GIS for planning. *Journal of the American Planning Association. American Planning Association, 59*(2), 230. doi:10.1080/01944369308975872

Jabatan Perancangan Bandar dan Desa Semenanjung Malaysia. (2001). *Manual Penyediaan Rancangan Tempatan (Pindaan 2001).* Kuala Lumpur: Jabatan Perancangan Bandar dan Desa Semenanjung Malaysia.

Jabatan Perancangan Bandar dan Desa Semenanjung Malaysia. (2007). *Manual Penyediaan Rancangan Struktur.* Kuala Lumpur: Jabatan Perancangan Bandar dan Desa Semenanjung Malaysia.

Jabatan Perancangan Bandar dan Desa Semenanjung Malaysia. (2008). Geotechnical spatial analysis research and development: fokus dan agenda. Retrieved December 12, 2008, from http://www.townplan.gov.my/penyelidikan_gsard.php

Johar, F. (2003). Environmental concern in local planning. *Planning Malaysia, 1,* 19–34.

Kassim, N. S., & Islam, R. (2009). Spatial AHP for the National Physical Plan of Malaysia. In Davide, G., & Abdullah, A. (Eds.), *Spatial decision support for urban and environmental planning: a collection of case studies.* Shah Alam, Malaysia: Arah Publications.

Klosterman, R. E. (1997). Planning support systems: a new perspective on computer-aided planning. *Journal of Planning Education and Research, 17*(1), 45–54. doi:10.1177/0739456X9701700105

Klosterman, R. E. (1999). New perspectives on planning support systems. *Environment and Planning. B, Planning & Design, 26,* 317–320. doi:10.1068/b260393

Klosterman, R. E., & Abdullah, A. (2008). *Computer technologies in urban planning.* Paper presented at the Medina Forum Seminar on Developing and Deploying Planning Support Systems.

Lee, L. M., & Ahmad, M. J. (2000). *Local authority networked development approval system.* Paper presented at the Planning Digital Conference.

Maklumat, B. T. (2006). Sistem pengurusan data JPBD Semenanjung Malaysia. *Buletin Geospatial Sektor Awam, 1/2006,* 26–30.

Mennecke, B. E., & West, L. A. J. (2001). Geographic Information Systems in Developing Countries: Issues in Data Collection, Implementation and Management. (Technology Information). *Journal of Global Information Management, 9*(4), 44–54.

Ramasubramanian, L. (1999). GIS Implementation in Developing Countries: Learning from Organisational Theory and Reflective Practice. *Transactions in GIS, 3*(4), 359–380. doi:10.1111/1467-9671.00028

Samat, N. (2006). *Applications of geographic information systems in urban land use planning in Malaysia.* Paper presented at the 4th Taipei International Conference on Digital Earth.

Tan, T. S. (2005). Electronic local authority management system. *Planning Malaysia, 3,* 27–46.

UGisP. (2006). GIS9: Pembangunan Pengkalan Data Sistem Maklumat Geografi Negeri Sembilan Darul Khusus - Laporan Akhir Fasa III.

ADDITIONAL READING

Abdullah, A. Kamalruddin Shamsuddin, & Abdullah, M. F. (Eds.). (2004). *Applications of planning and decision support systems*. Kuala Lumpur: Bureau of Consultancy & Entrepreneurship, International Islamic University Malaysia.

Abdullah, M. F., Abdullah, A., & Rustam Khairi Zahari. (2009). *GIS application in monitoring development plan implementation in Malaysia: weaknesses and potential*. Paper presented at the 11th International Conference on Computers in Urban Planning and Urban Management.

Brail, R. K., & Klosterman, R. E. (Eds.). (2001). *Planning support systems: integrating geographic information systems, models, and visualization tools*. Redlands, CA: ESRI Press.

Bruton, M. J. (2007). *Malaysia: the planning of a nation*. Kuala Lumpur: PERSADA.

Budic, D. Z. (1993). GIS use among southeastern local governments. *Journal of Urban and Regional Information Systems Association*, *6*(6), 529–546.

Geertman, S., & Stillwell, J. (Eds.). (2003). *Planning support systems in practice*. Heidelberg, Germany: Springer.

Goh, B. L. (1990). *Urban planning in Malaysia*. Petaling Jaya, Malaysia: Tempo Publishing.

Huxhold, W. (1991). *An introduction to urban geographic information systems*. New York: Oxford University Press.

Johar, F. (2006). Environmental sustainability in selected local plans in Malaysia. In Kadouf, H. A., & Al-Junid, S. Z. (Eds.), *Land use planning and environmental sustainability in Malaysia: policies and trends* (pp. 255–274). Kuala Lumpur: Research Centre, International Islamic University Malaysia.

Kadouf, H. A., & Al-Junid, S. Z. (Eds.). (2006). *Land use planning and environmental sustainability in Malaysia: policies and trends*. Kuala Lumpur: Research Centre, International Islamic University Malaysia.

Kadouf, H. A., & Maidin, A. J. (2006). Theory and practice in land use planning in Malaysia: an overview. In Kadouf, H. A., & Al-Junid, S. Z. (Eds.), *Land use planning and environmental sustainability in Malaysia: policies and trends* (pp. 77–113). Kuala Lumpur: Research Centre, International Islamic University Malaysia.

Kaniclides, A., & Kimble, C. (1995). *A development framework for executive information system*. Paper presented at the GRONICS '95, Groningen, The Netherlands.

Kasmin, A. M. (1999). *Kerangka pelaksanaan dasar rancangan struktur*. Shah Alam, Malaysia: Pusat R&D, Pejabat Setiausaha Kerajaan Negeri Selangor.

Kassim, N. S. (2003). Konsep dan proses e-submission dalam permohonan perancangan. *Malaysian Townplan*, *1*(1), 19–25.

Matori, A. N. (2007). Development of a user interface application for information delivery - a case study of Ipoh City Council's Department of Town Planning. *Buletin Geospatial Sektor Awam*, *1*, 24–29.

Muhammad, Z. (2000). Integrated rural planning and development approach - the Malaysian experience. *Habitat Malaysia*, *3*, 1–34.

Munir, A. B., & Zulhuda, S. (2007). *Becoming e-cities: legal issues and challenges*. Paper presented at the 2nd International Symposium on Knowledge Cities, Shah Alam.

Nedovic-Budic, Z. (1998). The impact of GIS technology. *Environment and Planning. B, Planning & Design*, *25*, 681–692. doi:10.1068/b250681

Rogers, E. (1983). *Diffusion of Innovations* (3rd ed.). New York: Free Press.

Shamsuddin, K. (Ed.). (2006). *Implementing multi-criteria decision making in Malaysian town planning*. Kuala Lumpur: Federal Department of Town and Country Planning Peninsular Malaysia.

Stillwell, J., Geertman, S., & Openshaw, S. (Eds.). (1999). *Geographical information and planning*. Heidelberg, Germany: Springer.

Vonk, G., Geertman, S., & Schot, P. (2005). Bottlenecks blocking widespread usage of planning support systems. *Environment & Planning A*, *37*, 909–924. doi:10.1068/a3712

Wan Ismail, W. H. (2007). *Pematuhan perancangan (teori)*. Paper presented at the Mesyuarat Pegawai Kanan Perancang Bandar dan Desa Malaysia ke 21.

Wegener, M. (1998). GIS and spatial planning. *Environment and Planning. B, Planning & Design*, *25*, 48–52.

Yaakup, A. (2004). *GIS as a tool for development planning and monitoring*. Paper presented at the NGIS 2004.

Yaakup, A., Abu Bakar, S. Z., & Sulaiman, S. (2009). Decision support systems for urban sustainability in Malaysia. *Malaysian Journal of Environmental Management*, *10*(1), 101–118.

Yaakup, A., Johar, F., Maidin, M. A., & Ahmad, E. F. (2004). *GIS and decision support systems for Malaysian development plan studies*. Paper presented at the International Seminar on SPDSS 2004.

KEY TERMS AND DEFINITIONS

Decision Support Systems: Computerized systems that can be used to help users in making decision over complex and ill-structured problems.

Development Plan: A plan which is usually prepared by planners to guide or promote development in an area. It usually contains maps of the area, and planning policies and proposals for the area.

Executive Information Systems: Computerized systems that are developed to provide important information in a concise manner to senior executives. This information can be used by the executives in making decision over a business problem.

Federal Department of Town and Country Planning: Is the planning authority at the Federal level that is responsible in charting the general direction of planning and development for the nation. Its functions also include promoting town and country planning in Malaysia and providing advisory services to the Federal Government and planning authorities at State and Local levels. It also provides technical and monetary supports to State and Local level planning authorities in the preparation of their development plans.

Geographical Information Systems (GIS): Computerized systems that can be used to store, analyze and display spatially-referenced data.

GIS-Based Systems: Computerized systems which are developed using GIS software as the platform.

Local Planning Authority: Local planning authority of an area is usually the local authority of the area, such as city hall, the municipal council or the district council. It is responsible in regulating, controlling and determining the planning and development direction of its area.

State Department of Town and Country Planning: Each state in the Peninsular Malaysia has its own Town and Country Planning Department which oversees the planning and development matters within the State boundary and provides advisory services to the State government. State Town and Country Planning Department also acts as local planning authority, handling development control affairs for areas within the state, which are not under the jurisdiction of any local authority.

Statutory Development Plan: Development plan which preparation is required by law.

ENDNOTES

[1] States in East Malaysia (e.g., Sabah and Sarawak), and the three Federal Territories (e.g., Kuala Lumpur, Putrajaya and Labuan) in Malaysia are not bounded by the Town and Country Planning Act 1976. They, however, have their own planning acts which are almost similar to the Act.

[2] Charting is a process where information about planning applications is 'drawn' onto standard cadastral map. Traditionally, this was done by hand where draughtsman would draw on the map the boundary of land parcel where planning applications were submitted, as well as other basic information about the applications. Many planning authorities have now replaced their manual charting with GIS.

Compilation of References

Aaltonen, M. (2007). Chronotope Space – Managing Time and the Properties of Strategic Landscape. *Foresight*, 9(4), 58–62. doi:10.1108/14636680710773830

Aaltonen, M. (2007). The Return to Multi-Causality. *Journal of Futures Studies*, 12(1), 81–86.

Aaltonen, M. (2007). *The Third Lens. Multi-ontology Sense-making and Strategic Decision-making*. Aldershot, UK: Ashgate.

Aaltonen, M., & Sanders, T.I. (2006). Identifying Systems" New Initial Conditions as Influence Points for Future. *Foresight: Journal of futures studies, strategic thinking and policy*, 8(3), 28-35.

Abbot, J., Chambers, R., Dunn, C., Harris, T., de Merode, E., & Porter, G. (1998). Participatory GIS: opportunity or oxymoron? *IIED PLA Notes*, 33, 27–33.

Abbott, J. (2001). Democracy @ Internet.asia? The challenges to the emancipatory potential of the net: Lessons from China and Malaysia. *Third World Quarterly*, 22(1), 99–114. doi:10.1080/01436590020022600

Abdullah, A., Abdullah, M. F., & Nordin, M. F. (2002). *Spatial information system as a tool in managing urban development process*. Paper presented at the 18th EAROPH World Planning Congress.

Abdullah, A., Abdullah, M. F., & Nordin, M. F. (2003). Managing urban development process by using spatial information system. *Planning Malaysia*, 1, 71–92.

Abdullah, A., Abdullah, M. F., Shahbuddin, M. N. A., & Klosterman, R. E. (2009). Modelling the reality in planning: SPDSS experience in Malaysia. *Malaysian Journal of Environmental Management*, 10(1), 119–132.

Abdullah, M. F., & Ariffin, I. (2003). Incorporating sustainable development objectives into development plans through strategic environmental assessment. *Planning Malaysia*, 1, 1–18.

Abdullah, M. F., Abdullah, A., & Ibrahim, M. (2009). Institutionalisation of spatial planning and decision support systems for planning and governance in Malaysia. In Davide, G., & Abdullah, A. (Eds.), *Spatial decision support for urban and environmental planning: a collection of case studies*. Shah Alam, Malaysia: Arah Publications.

Abdullah, M. F., Abdullah, A., Ibrahim, M., & Abdul Samad, D. (2005). A study on the demand of spatial planning and decision support system in Malaysia. *Planning Malaysia*, 3, 47–60.

Abdullah, M. N. (2004). Pelaksanaan sistem maklumat geografi di Jabatan Perancangan Bandar dan Desa Negeri: dilema dan potensi. *Malaysian Townplan*, 2(1), 29–37.

Abramowitz, M. (1995). The origins of the postwar catch-up and governance boom. In Fagerberg, J., Versbangen, B., & von Tunzelmann, N. (Eds.), *The dynamics of technology, trade and growth*. Brookfield: Edward Elgar.

Abu Bakar, M. A. (2004). Sistem maklumat geografi di Jabatan Perancangan Bandar dan Desa Semenanjung Malaysia. *Malaysian Townplan*, 2(1), 19–23.

Adams, B. (1993). Sustainable development and the greening of development theory. In Schuurman, F. J. (Ed.), *Beyond the Impasse – New direction in development theory* (pp. 207–222). London: Zed Books.

Adams, W. M. (2001). *Green Development: environment and sustainability in the Third World*. London: Routledge.

ADEME. (2006). *Réussir un projet d'urbanisme durable*. Paris: Le Moniteur.

Agenda 21: Earth Summit - The United Nations Programme of Action from Rio. *(1993)*.

Ahmad Sarkawi, A. (2006). The legal considerations of neighbouring lands in development planning: the Malaysian context. Newcastle-upon-Tyne, United Kingdom.

Ahn, H. J., & Lee, H. (2004). An agent-based dynamic information network for supply chain management. *BT Technology Journal*, *22*, 18–27. doi:10.1023/B:BTTJ.0000033467.83300.c0

Aichholzer, G., & Kozeluh, U. (2007). eParticipation and Democracy: Evaluation dimensions and approaches. In B. Lippa (Ed.), DEMO-net: Research workshop report – Frameworks and methods for evaluating e-participation (pp. 31-43). Bremen, Germany: IST Network of Excellence Project, Akrich, M. (1995). The de-scription of technical objects. In W. E. Bijker & J. Law (Eds.), Shaping Technology/Building Society: Studies in Sociotechnical Change (pp. 205-224). Cambridge, MA: MIT Press.

Aichholzer, G., & Westholm, H. (2009). Evaluating eParticipation Projects. Practical Examples and Outline of an Evaluation Framework. *European Journal of ePractice*. Retrieved from http://www.epracticejournal.eu/document/5511

Aikins, S. K. (2008). Practical measures for securing government networks. In Garson, G. D., & Khosrow-Pour, M. (Eds.), *Handbook of Research on Public Information Technology* (pp. 386–394). New York: Information Science Reference.

Aitken, S. (2002). Public Participation, Technological Discourses, and the Scale of GIS. In Craig, W., Harris, T., & Weiner, D. (Eds.), *Community Participation and Geographic cal Information Systems*. London: Taylor & Francis. doi:10.1201/9780203469484.ch27

Albrechts, L., & Denayer, W. (2001). Communicative Planning, Emancipatory Politics and Postmodernism. In Paddison, R. (Ed.), *Handbook of Urban Studies*. Thousand Oaks, CA: SAGE Publications.

Alexander, C. (1964). *Notes on the Synthesis of Form*. Cambridge, MA: Harvard University Press.

Alexander, C. (1988). A City is Not a Tree. In Thackara, J. (Ed.), *Design After Modernism*. London: Thames and Hudson.

Alexander, C., Ishikawa, S., Silverstein, M., Jacobson, M., Fiksdahl-King, I., & Angel, S. (1977). *A Pattern Language*. New York: Oxford University Press.

Alford, J. (2002). Defining the Client in the Public Sector: a social exchange perspective. *Public Administration Review*, *62*(3), 337–346. doi:10.1111/1540-6210.00183

Alfredsson, B., & Wiman, J. (2001). Planning in Sweden - Fundamentals Outlined. In Christoferson, I. (Ed.), *Swedish Planning - In Times of Diversity*. Gävle, Sverige: The Swedish Society for Town and Country Planning.

Al-Kodmany, K. (1999). Using visualization techniques for enhancing public participation in planning and design. *Landscape and Urban Planning*, *45*, 37–45. doi:10.1016/S0169-2046(99)00024-9

Al-Kodmany, K. (2000). Extending geographic information systems to meet neighborhood planning needs: The case of three Chicago communities. *Journal of the Urban and Regional Information Systems Association*, *12*(3), 19–37.

Al-Kodmany, K. (2000). GIS in the Urban Landscape: reconfiguring neighborhood planning and design processes. *Landscape Research*, *25*(1), 5–28. doi:10.1080/014263900113145

Al-Kodmany, K. (2001). Supporting imageability on the World Wide Web: Lynch's five elements of the city in community planning. *Environment and Planning. B, Planning & Design*, *28*(6), 805–832. doi:10.1068/b2746t

Al-Kodmany, K. (2002). GIS and the artist: shaping the image of a neighbourhood in participatory environmental design. In Weiner, D., Harris, T. M., & Craig, W. J. (Eds.), *Community participation and geographic information systems* (pp. 320–329). London: Taylor and Francis. doi:10.1201/9780203469484.ch24

Al-Kodmany, K. (2002). Visualization tools and methods in community planning: From freehand sketches to virtual reality. *Journal of Planning Literature*, *17*, 189–211. doi:10.1177/088541202762475946

Al-Kodmany, K. (2007). Creative approaches for augmenting two-way spatial communication and GIS. *GIS Development*, *3*(12), 1–9.

Allen, B., Juillet, L., Paquet, G., & Roy, J. (2001). E-governance and government online in Canada: Partnerships, people and prospects. *Government Information Quarterly*, *18*, 93–104. doi:10.1016/S0740-624X(01)00063-6

Allen, E. (2001). INDEX: software for community indicators. In Brail, K. R., & Klosterman, R. E. (Eds.), *Planning support systems* (pp. 229–261). Redlands, CA: ESRI Press.

Allmendinger, P. (2001). Planning. In *Postmodern Times*. London: Routledge.

Almendinger, P. (2002). *Planning theory*. Basingstoke, UK: Palgrave Macmillan.

Al-Qahtaani, A. (2002). *A Manual on the Rites of Hajj* (2nd ed.). London: Invitation to Islam Press.

Alty, R., & Darke, R. (1987). A city centre for people: involving the community in planning for Sheffield's central area. *Planning Practice and Research*, *2*(3), 7–12. doi:10.1080/02697458708722679

Amandus, P. Jr. (2006). MCDM and land suitability analysis for spatial development: case study Kulim local plan. In Shamsuddin, K. (Ed.), *Implementing Multi Criteria Decision Making in Malaysian Town Planning*. Kuala Lumpur: Federal Department of Town and Country Planning Peninsular Malaysia.

Amin, A., & Thrift, N. J. (2002). *Cities: Reimagining the Urban*. Cambridge, UK: Polity Press.

Amman, P., Jagodia, S., McCullum, C., & Blaustein, B. (1997). Surviving information warfare attacks on databases. In *Proceedings of the IEEE Symposium on Security and Privacy* (pp. 164-174)

An, L., Lindsernman, M., Qi, J., Shortridge, A., & Lui, J. (2005). Exploring complexity in a human-environment system: an agent based spatial model for multidisciplinary and multiscale integration. *Annals of the Association of American Geographers. Association of American Geographers*, *95*(1), 54–79. doi:10.1111/j.1467-8306.2005.00450.x

Anderson, C. A. (2004). An update on the effects of playing violent video games. *Journal of Adolescence*, *27*, 113–122. doi:10.1016/j.adolescence.2003.10.009

Anderson, L. W., & Krathwohl, J. (Eds.). (2001). *A taxonomy for learning, teaching, and assessing: A revision of Bloom's taxonomy of educational objectives*. New York: Longman.

Anderson, N. (2007). California Outlaws the Forced Subdermal RFID Tagging on Humans. *Arstechnica*. Retrieved November 11, 2008, from http://arstechnica.com/news.ars/post/20070904-california-outlaws-forced-rfid-tagging-of-humans.html

Andrienkoa, G., Andrienkoa, N., Vossa, H., & Carter, J. (1999). Internet mapping for dissemination of statistical information. *Computers, Environment and Urban Systems*, *23*(1), 425–441. doi:10.1016/S0198-9715(99)00044-7

Anttiroiko, A.-V. (2003). eGovernment. eGovernment-alan tutkimuksen ja opetuksen kehittäminen Tampereen yliopistossa [Development of the research and teaching of eGovernment at the University of Tampere]. Retrieved December 3, 2008, from http://www.uta.fi/laitokset/ISI/julkaisut/eGovernment-raportti_1-2002.html#Luku1

Anttiroiko, A.-V., & Malkia, M. (Eds.). (2007). *Encyclopedia of Digital Government*. Hershey, PA: Idea Group Publishing.

Aoki, Y. (1983). An empirical study on the appraisals of landscape types by residential groups – Tsukuba Science City. *Landscape Planning*, *10*, 109–130. doi:10.1016/0304-3924(83)90055-2

Aoki, Y. (1999). Trends in the study of the psychological evaluation of landscape. *Landscape Research*, *24*(1), 85–94. doi:10.1080/01426399908706552

Aouad, G., Lee, A., & Wu, S. (2006). *Constructing the Future: nD Modelling*. New York: Taylor & Francis.

Arbeit, D. (1993). Resolving the data problem: A spatial information infrastructure for planning support. In *Proceedings of Third International Conference on Computers in Urban Planning and Urban Management*, Atlanta, GA, July 23-25.

Arendt, R. G. (1996). *Conservation Design for Subdivisions: A Practical Guide for Creating Open Space Networks*. Washington, DC: Island Press.

Arias, E. (1996). Bottom-up Neighborhood Revitalisation: A Language Approach for Participatory Decision Support. *Urban Studies (Edinburgh, Scotland)*, *33*(10), 1831–1848. doi:10.1080/0042098966402

Armstrong, M. (1994). Requirements for the development of GIS-based group decision-support systems. *Journal of the American Society for Information Science American Society for Information Science*, *45*(9), 669–677. doi:10.1002/(SICI)1097-4571(199410)45:9<669::AID-ASI4>3.0.CO;2-P

Arnberger, A., & Haider, W. (2007). Would you displace? It depends! A multivariate visual approach to intended displacement from an urban forest trail. *Journal of Leisure Research*, *39*(2), 345–365.

Arnberger, A., Eder, R., Brandenburg, C., & Reichhart, T. (2007). Assessing landscape preferences of urban population for terraced areas. In Faculty of Forestry (Ed.), Landscape Assessment - From Theory to Practice: Applications in Planning and Design (pp. 111-119). Beograd, Srbija: Planeta Print.

Arnesen, R. R., & Danielsson, J. (2007). Protecting citizen privacy in digital government. In Anttiroiko, A.-V., & Malkia, M. (Eds.), *Encyclopedia of Digital Government* (pp. 1358–1363). Hershey, PA: Idea Group Publishing.

Arnstein, S. R. (1969). A Ladder of Citizen Participation. *Journal of the American Institute of Planners*, *35*, 216–224.

Arnstein, S. R. (1971). A ladder of citizen participation. *Journal of the American Planning Association. American Planning Association*, *35*(4), 216–224. doi:10.1080/01944366908977225

Arrow, K. J. (1963). *Social Choice and Individual Values*. New Haven, CT: Yale University Press.

Asgarkhani, M. (2007). The Reality of Social Inclusion through Digital Government. *Journal of Technology in Human Services*, *25*(1/2), 127–146. doi:10.1300/J017v25n01_09

Ashton, K. (2008). RFID Power to the People. *RFID Journal*. Retrieved November 11, 2008, from http://www.rfidjournal.com/article/articleview/4417/1/82/

Åström, J. (2004). *Mot en digital demokrati? Teknik, politik och institutionell förändring*. Örebro Studies in Political Science 9. Örebro, Sverige: Örebro University.

Åström, J., & Granberg, M. (2007). Urban Planners, Wired for Change? Understanding Elite Support for E-participation. *Journal of Information Technology & Politics*, *4*(2), 63–77.

Aurigi, A. (2005). *Making the Digital City: The Early Shaping of Urban Internet Space*. Aldershot, UK: Ashgate Publishing.

Aurigi, A., & de Cindio, F. (Eds.). (2008). *Augmented Urban Spaces. Articulating the Physical and Electronic City*. Aldershot, UK: Ashgate.

Axelrod, R. (1997). *The complexity of cooperation: Agent-based models of competition and collaboration*. Princeton, NJ: Princeton University Press.

Bahagian Teknologi Maklumat. (2007). Format Metadata dan Struktur Pengkalan Data Sistem Maklumat Geografi Kajian Rancangan Tempatan. Kuala Lumpur: Jabatan Perancangan Bandar dan Desa Semenanjung Malaysia.

Bahagian Teknologi Maklumat. (2007). Manual Penyediaan Pelan GIS. Kuala Lumpur: Jabatan Perancangan Bandar dan Desa Semenanjung Malaysia.

Bains, S. (2002 April). Wired cities. *Communications International*, 21-25.

Baker, S. (2006). *Sustainable development*. New York: Routledge.

Balram, S., & Dragićević, S. (2006). *Collaborative Geographic Information Systems*. Hershey, PA: Idea Group Inc.

Barbanente, A., Camarda, D., Grassini, L., & Khakee, A. (2007). Visioning the regional future: Globalisation and regional transformation of Rabat/Casablanca. *Technological Forecasting and Social Change, 74*, 763–778. doi:10.1016/j.techfore.2006.05.019

Barber, B. (1984). *Strong Democracy. Participatory Democracy for a New Age*. Berkeley, CA: University of California Press.

Barber, B. (1998). *A place for US: How to make Society Civil and Democracy Strong*. New York: Hill and Wang.

Barnard, E., Cloete, L., & Patel, H. (2004). Language and technology literacy barriers to accessing government services. In Traunmüller, R. (Ed.), *Electronic Government* (pp. 37–42). Berlin: Springer.

Barton, H., Grant, M., & Guise, R. (Eds.). (2003). *Shaping Neighbourhoods: A Guide for Health, Sustainability and Vitality*. London: Spon Press.

Batty, M. (1995). Planning Support Systems and the New Logic of Computation. *Regional Development Dialogue, 16*(1), 1–17.

Batty, M. (1996). Planning, late-20th-century style. *Environment and Planning. B, Planning & Design, 23*(1), 1–2.

Batty, M. (1998). Digital planning. In Sikdar, K., & Rao, K. (Eds.), *Computers in urban planning and urban management* (pp. 13–30). New Delhi: Narosa.

Batty, M. (2004). Dissecting the streams of planning history: technology versus policy through models. *Environment and Planning. B, Planning & Design, 31*, 326–330. doi:10.1068/b3103ed

Baugesetzbuch (German Federal Building Code). (2006). Retrieved November 8, 2008, from http://www.gesetze-im-internet.de/bbaug/BJNR003410960.html

Becker, S. (1998). A practical perspective on data quality issues. *Journal of Database Management, 35*, 35–37.

Becker, S. A. (2005). E-Government Usability for Older Adults. *Communications of the ACM, 48*(2), 102–104. doi:10.1145/1042091.1042127

Bell, D. (1973). *The Coming of Post-industrial Society*. New York: Basic Books.

Bell, S. (1986). Information systems planning and operation in less developed countries, part 1: planning and operational concerns. *Journal of Information Science, 12*(5), 231–245. doi:10.1177/016555158601200503

Bellamy, C., & Taylor, J. (1998). *Governing in the Information Age*. Buckingham, UK: Open University Press.

Benyon-Davies, P. (2007). Models for e-government. *Transforming Government: People. Process and Policy, 1*(1), 7–28.

Berghäll, E., & Koncitz, J. (1997). Urbanisation and sustainability. In OECD (Ed.), Sustainable development OECD policy approaches for the 21st Century (pp. 117-127). Paris: OECD Publications.

Berke, P. R., Godschalk, D. R., & Kaiser, E. J. (2006). *Urban land use planning*. Urbana, IL: University of Illinois Press.

Bertelsmann Stiftung. (2004). *Kommunaler Bürgerhaushalt: Ein Leitfaden für die Praxis*. Retrieved November 22, 2008, from http://www.buergerhaushalt.org/wp-content/plugins/wp-publications-archive/openfile.php?action=open&file=5

Bertin, J. (1982). *Graphische Darstellungen – Graphische Verarbeitung von Informationen, übersetzt und bearbeitet von Wolfgang Scharfe*. Berlin: Walter de Gruyter.

Bertuglia, C. S., Giuliano, B., & Mela, A. (1998). Introduction. In Bertuglia, C. S., Giuliano, B., & Mela, A. (Eds.), *The city and its Sciences* (pp. 1–92). Heidelberg, Germany: Physica-Verlag.

Betty, M., & Longley, P. (1994). *Fractal Cities*. London: Academic Press.

Beyer, H., & Holtzblatt, K. (1998). *Contextual Design: Defining Customer-Centered Systems*. New York: Academic Press.

Bhatti, R., Bertino, E., Ghafoor, A., & Joshi, J. B. D. (2004). XML-based specification for web services document security. *IEEE Computer, 37*(4), 41–49.

Birch, A. H. (2001). *The Concepts and Theories of Modern Democracy* (2nd ed.). London: Routledge.

Bishop, I. D., Barry, M., McPherson, E., Nascarella, J., Urquhart, K., & Escobar, F. (2002). Meeting the Need for GIS Skills in Developing Countries: The Case of Informal Settlements. *Transactions in GIS, 6*(3), 311-326.

Bishop, I. D., Ye, W., & Kardaglis, C. (2001). Experiential approaches to perception response in virtual worlds. *Landscape and Urban Planning, 54*(1-4), 117–125. doi:10.1016/S0169-2046(01)00130-X

Blaug, R. (2002). Engineering Democracy. *Political Studies, 50*, 102–116. doi:10.1111/1467-9248.00361

Blumstengel, A. (1998). *Entwicklung hypermedialer Lernsysteme* (Dissertation). Berlin: Wissenschaftlicher Verlag.

Boerwinkel, H. W., & Jansen, W. J. A. (1994). CAD-Modelle online bewertet. In: Garten+Landschaft 10/1994

Bohm, K. (1985). *Med- och motborgare i stadsplanering. En historia om medborgardeltagandets förutsättningar.* Stockholm: Liber.

Bolay, J.-C. (1998). Habitat des pauvres en Amérique Latine. In Rossel, P., Bassand, M., & Roy, M.-A. (Eds.), *Au-delà du laboratoire, les nouvelles technologies à l'épreuve de l'usage.* Lausanne, France: PPUR.

Bolay, J.-C. (2006). Slums and urban development: Questions on society and globalization. *European Journal of Development Research, 18*(2), 284–298. doi:10.1080/09578810600709492

Bolay, J.-C., & Cissé, G. (2001). Urban environmental management: new tools for urban players. In KFPE (Ed.), Enhancing research capacity in developing and transition countries (pp. 175-182). Bern, Switzerland: Geographica Bernensia.

Bontempo, C., & Zageclow, G. (1998). The IBM data warehouse architecture. *Communications of the ACM, 4*(9), 38–48. doi:10.1145/285070.285078

Booher, D., & Innes, J. E. (2002). Network Power in Collaborative Planning. *Journal of Planning Education and Research, 21*(3), 221–236. doi:10.1177/0739456X0202100301

Boon, J. A. (1992). Information and development: some reasons for failure. *The Information Society, 8*(3), 227–241.

Borja, J., & Castells, M. (1997). *Local and global: management of cities in the information age.* London: Earthscan.

Borning, A., Friedman, B., Davis, J., & Lin, P. (2005). Informing public deliberation: Value sensitive design of indicators for a large-scale urban simulation. In *Proc. ECSCW 2005* (pp. 449-468.)

Borning, A., Friedman, B., Davis, J., Gill, B., Kahn, P., Kriplean, T., & Lin, P. (2009). Public participation and value advocacy in information design and sharing: Laying the foundations in advance of wide-scale public deployment. *Information Polity, 14*(1-2), 61–74.

Borning, A., Ševčíková, H., & Waddell, P. (2008b). A Domain-Specific Language for Urban Simulation Variables. In *Proceedings of the 9th Annual International Conference on Digital Government Research*, Montréal, Canada, May 2008.

Borning, A., Waddell, P., & Förster, R. (2008a). UrbanSim: using simulation to inform public deliberation and decision-making. In Chen, H. (Eds.), *Digital Government: Advanced Research and Case Studies.* Berlin: Springer-Verlag.

Borri, D., & Camarda, D. (2006). Visualizing space-based interactions among distributed agents: Environmental planning at the inner-city scale. *Lecture Notes in Computer Science, 4101*, 182–191. doi:10.1007/11863649_23

Borri, D., Camarda, D., & De Liddo, A. (2004). Envisioning environmental futures: Multi-agent knowledge generation, frame problem, cognitive mapping. *Lecture Notes in Computer Science, 3190*, 230–237.

Borri, D., Camarda, D., & De Liddo, A. (2008). Multi-agent environmental planning: A forum-based case-study in Italy. *Planning Practice and Research, 23*, 211–228. doi:10.1080/02697450802327156

Boschma, R. (2003). *The competitiveness of regions from an evolutionary perspective.* Paper presented at the Conference of Regional Studies Association, Pisa, Italy, 12–15 April 2002.

Bossell, H. (1999). *Indicators for Sustainable Development: Theory, Method, Applications. A Report to the Balaton Group*. Winnipeg, Manitoba, Canada: International Institute for Sustainable Development.

Bossell, H. (2007). *Systems and Models Complexity, Dynamics, Evolution, Sustainability*. Books on Demand GmbH.

Bourdakis, V. (1997). Virtual Reality: A Communication Tool for Urban Planning. In Asanowicz, A., & Jakimowitz, A. (Eds.), *CAAD-Towards New Design Conventions* (pp. 45–59). Bialystok, Poland: Technical University of Bialystok.

Bourdakis, V. (1998). Navigation in Large VR Urban Models. In Heudin, J. C. (Ed.), *Virtual Worlds Lecture Notes in Artificial Intelligence* (pp. 345–356). Berlin, Heidelberg: Springer-Verlag.

Bourdakis, V. (2004). Developing VR Tools for an Urban Planning Public Participation ICT Curriculum; The PICT Approach. In B. Rudiger, B. Tournay & H. Orbaek (Eds.), Architecture in the Network Society: eCAADe2004 Proceedings (pp. 601-607). Copenhagen: eCAADe.

Bourdakis, V. (2008). Low Tech Approach to 3D Urban Modelling. In M. Muylle (Ed.), Architecture. 'in computro', eCAADe 26 Proceedings (pp. 959-964). Antwerpen: eCAADe.

Bousquet, F., & Le Page, C. (2004). Multi-agent simulations and ecosystem management: A review. *Ecological Modelling*, *176*, 313–332. doi:10.1016/j.ecolmodel.2004.01.011

Bouton, N. (2009). *Designing for the Community Experience*. Retrieved May 28, 2009 from http://www.slideshare.net/nickbouton/designing-for-the-community-experience-vanue-may-2609-1497974

Bracht, R., Miller, E. K., & Kuckertz, T. (1997). Using an Impedance-Modulated Reflector for Passive Communications. *IEEE Xplore*. Retrieved on November 1, 2008, from http://ieeexplore.ieee.org/Xplore/login.jsp?url=/iel3/4924/13698/00631742.pdf?arnumber=631742

Bradford, N. (2005). *Place-based Public Policy: Towards a New Urban and Community Agenda for Canada*. Research Report F/51. Family Network.

Bradshaw, J. M., Cabri, J., & Montanari, R. (2003). Taking bask cyberspace. *IEEE Computer*, *36*(7), 89–92.

Brail, R. K. (2001). Planning Support Systems: A new perspective on computer-aided planning. In Brail, R. K., & Klosterman, R. E. (Eds.), *Planning Support Systems: Integrating Geographic Information System, Models, and Visualization Tools*. New York: ESRI Press.

Brail, R., & Klosterman, R. (Eds.). (2001). *Planning support systems: Integrating geographic information systems, models and visualization tools*. Redlands, CA: ESRI Press.

Brandt Commission. (1980). *North-south: A programme for survival*. London: Pan Books.

Bravo, J., Hervás, R., Chavira, G., & Nava, S. (2006). Mosaics of visualization: An approach to embedded interaction through identification process. *Lecture Notes in Computer Science*, *4101*, 41–48. doi:10.1007/11863649_6

Breitenstrom, C., Eckert, K., & von Lucke, J. (2008). *The EU-Services Directive – Point of Single Contact – Framework Architecture and Technical Solutions - White Paper Version 1.0*. Berlin: Fraunhofer-Institute for Open Communication Systems FOKUS. Retrieved September 15. 2008, from http://www.fokus.fraunhofer.de/en/elan/publikationen/Infomaterial/white_paper/DLR_1_0/index.html

Brennan, R./ LGMB [Local Government Management Board] (1991). The Equal Opportunities Guide. Luton: LGMB.

Brewin, B. (2007). Spread of RFID in Defense Slower than Expected. *Government Executive*. Retrieved November 1, 2008 from http://www.govexec.com/story_page_pf.cfm?articleid=37434&printerfriendlyvers=1

Brodie, M. I. (1980). Data quality in information systems. *Information & Management*, (3): 245–258. doi:10.1016/0378-7206(80)90035-X

Brody, S. D., Godschalk, D. R., & Burby, R. J. (2004). Mandating citizen participation in Plan Making: Six Strategic Planning Choices. *Journal of the American Planning Association. American Planning Association, 69*(3), 245–264. doi:10.1080/01944360308978018

Brown, I. (1999). Developing a Virtual Reality User Interface for Geographic Information Retrieval on the Internet. *Transactions in GIS, 3*(3), 207–220. doi:10.1111/1467-9671.00018

Brown, L. D. (1991). Bridging organizations and sustainable development. *Human Relations, 44*(8), 807–831. doi:10.1177/001872679104400804

Brunner, K. (2001). Kartengestaltung für elektronische Bildanzeigen. *Kartographische Bausteine, 19.*

Brunner-Friedrich, B. (2004). *InMuKIS – Konzept eines benutzergruppenangepassten interaktiven multimedialen kartographischen Informationssystems für die Schule zur Präsentation raumbezogener Informationen.* Dissertation; Technische Universität Wien.

Bryson, V. (2007). *Gender and the Politics of Time. Feminist theory and contemporary debates.* Bristol, UK: Policy Press.

Buchanan, E. A. (2004). *Readings in virtual research ethics. Issues and controversies.* Hershey, PA: Information Science Publishing.

Buchsbaum, T. (2008). *E-democracy and E-Parliament. Thoughts and Standard-setting of the Council of Europe.* Retrieved December 3, 2008 from http://www.bmeia.gv.at/fileadmin/user_upload/bmeia/media/AOes/e-Democracy/CAHDE_2008/Sofia_IPAIT_-_Buchsbaum_080610_final.pdf

Budic, D. Z. (2000). Geographic information science implications for urban and regional planning. *URISA, 12*(2), 81–93.

Budic, Z. (1993). GIS Use Among South-eastern Local Governments - 1990/1991 Mail Survey Results. *Journal of Urban and Regional Information Systems Association, 5*(1), 4–17.

Budic, Z. (1994). Effectiveness of Geographic Information Systems in Local Planning. *Journal of the American Planning Association. American Planning Association, 60*(2), 244–263. doi:10.1080/01944369408975579

Budthimedhee, K., Li, J., & George, R. V. (2002). e-Planning: a snapshot of the literature on using the World Wide Web in urban planning. *Journal of Planning Literature, 17*(2), 227–246. doi:10.1177/088541202762475964

Bunch, M. J. (2000). *An Adaptive Ecosystem Approach to Rehabilitation and Management of the Cooum River Environmental System in Chennai, India.* PhD Thesis, University of Waterloo, Waterloo, Canada.

Buzin, R. (2001). *Multimedia-Kartographie – Eine Untersuchung zur Nutzer-Orientierung kartomedialer Atlanten.* Diss., TU Dresden, Der Andere Verlag, Osnabrück.

California State Senate. (2008). Senate Bill No. 362. Retrieved November 11, 2008 from http://info.sen.ca.gov/pub/07-08/bill/sen/sb_0351-0400/sb_362_bill_20070627_amended_asm_v95.pdf

Caloundra City. (2009). *Strategic City Plan of the Caloundra City.* Retrieved on March 13, 2009, from http://maproom.caloundra.qld.gov.au

Camagni, R. (2002). *On the concept of territorial competitiveness: sound or misleading?* Paper presented at the 42nd Congress of the European Regional Science Association (ERSA). Dortmund Germany, 27–31 August 2002.

Camble, E. (1994). The information environment of rural development workers in Borno State, Nigeria. *African Journal of Library Archives and Information Sciences, 4*(2), 99–106.

Camhis, M. (1979). *Planning Theory and Philosophy.* London: Tavistock Publications.

Campagna, M. (2007). *Citizens Participation in Urban Planning.* Summary Note for the ESF-LiU Conference, Electronic Democracy - Achievements and Challenges. Vadstena, Sweden, 21-25 November 2007. Retrieved December 13, 2008, from http://www.docs.ifib.de/esf-conference07/conf_programme.html

Campagna, M., & Deplano, G. (2004). Evaluating geographic information provision within public administration websites. *Environment and Planning. B, Planning & Design, 31*, 21–37. doi:10.1068/b12966

Campbell, A. (2005). Korea's Ubicomp Vision Offers Opportunities for the West. *RFID Weblog.* Retrieved October 17, 2008, from http://www.rfid-weblog.com/50226711/koreas_ubicomp_vision_offers_opportunity_for_the_west.php

Caperna, A. (2005). *ICT per un progetto urbano sostenibile*. Milan: Tesionline.

Caperna, A. (2007). *Linee per un Piano d'Azione per lo Sviluppo Sostenibile del territorio del Parco Regionale dei Monti Lucretili*. Paper presented at the meeting of a.d.a. on the Environmental State Report, Rome, Italy.

Caperna, A., & Salingaros, N. A. (2006). *La complessità come nuovo approccio metodologico alla progettazione urbana ed architettonica*. Paper presented at the meeting of Istituto Nazionale di Urbanistica (INU) on Urban Environment and Complexity, Genoa 22-23 June 2006 (Italy).

Capra, F., Steindl-Rast, D., & Matus, T. (1992). *Belonging to the Universe: New Thinking About God and Nature*. New York: Penguin Press Science.

Card Plus. (2008). CTA Smart Card Sales Start. *Chicago Transit Authority.* Retrieved December 1, 2008, from http://www.transitchicago.com/news/archpress.wu?action=displayarticledetail&articleid=111804

Carden, F. (2008, October). *Ordinary Word, Extraordinary Confusion: Cause in development evaluation*. Paper presented at the Conference of the European Evaluation Society, Lisbon.

Cardoso, R. C., & Freire, M. M. (2005). Security vulnerabilities and exposures in Internet systems and services. In Pagani, M. (Ed.), *Encyclopedia of multimedia technology and networking* (pp. 910–916). Hershey, PA: Idea Group Reference.

Carmona, M., & Burgess, R. (2001). *Strategic planning and urban projets*. Delft, The Netherlands: Delft University Press.

Carney, D. (1998). Michigan Communities Launch Largest Regional GIS. *Newsletter of Regional Geographic Information Systems.* An Agency of Grand Valley Metropolitan Council. Retrieved July 20, 2009, from http://www.gvmc-regis.org/news/news4.html

Carson, R. (2002, October 21). The Art of Planning and Politics. *The Planetizen.* Retrieved July 20, 2009, from http://www.planetizen.com/node/67

Carver, S. (2001). Public participation using web-based GIS. *Environment and Planning. B, Planning & Design, 28*(1), 803–804. doi:10.1068/b2806ed

Carver, S. (2003). The Future of Participatory Approaches Using Geographic Information: Developing a Research Agenda for the 21st Century. *URISA Journal, 15*(1), 61–71.

Carver, S., Evans, A., Kingston, R., & Thurston, R. (2001). Public participation, GIS and cyberdemocracy: Evaluating on-line spatial decision support systems. *Environment and Planning B, 28*(6), 907–921. doi:10.1068/b2751t

Caso, O. (1999). The city, the elderly, and telematic. Design aspects of telematic applications in a residential neighbourhood. In *Transformation, 2.* Delft University Press.

Castells, M. (1992). The world has changed: can planning change? *Landscape and Urban Planning, 22*, 73–78. doi:10.1016/0169-2046(92)90009-O

Castells, M. (1996). *Rise of the Network Society: The Information Age: Economy, Society and Culture*. Cambridge, MA: Blackwell Publishers, Inc.

Castells, M. (1997). The education of city planners in the information age. Berkeley Planning Journal, No. 12.

Castells, M., & Hall, P. (1994). *Technopoles of the World: The Making of Twenty-first-century Industrial Complexes*. London: Routledge.

Cavric, B. (2002). *Human and Organisational Aspects of GIS Development in Botswana*. Paper Presented at GSDI 6 Conference, Budapest, Romania.

Ceccato, V. A., & Snickars, F. (2000). Adapting GIS technology to the needs of local planning. *Environment and Planning. B, Planning & Design, 27*, 923–937. doi:10.1068/b26103

Cernea, M. (1987). Farmer organizations and institution building. *Regional Development Dialogue, 8*(2), 1–24.

Cessford, G. R. (2003). Perception and reality of conflict: Walkers and mountain bikes on the Queen Charlotte Track in New Zealand. *Journal for Nature Conservation, 11*(4), 310–316. doi:10.1078/1617-1381-00062

Chadwick, A., & May, C. (2003). Interaction between States and Citizens in the Age of the Internet: E-Government in the United States, Britain, and the European Union. *Governance, 16*(2), 271-300. Retrieved December 3, 2008 from http://www.bmeia.gv.at/fileadmin/user_upload/bmeia/media/AOes/e-Democracy/CAHDE_2008/Sofia_IPAIT_-_Buchsbaum_080610_final.pdf

Chambers, R. (1994). Participatory rural appraisal: challenges, potentials and paradigm. *World Development, 22*(10), 1437–1454. doi:10.1016/0305-750X(94)90030-2

Chambers, R. (1997). *Whose reality counts: putting the first last*. London: ITDG Publishing.

Chang, R. K. C. (2002). Defending against flooding-based distributed denial-of-service attacks: A tutorial. *IEEE Communications Magazine, 40*(10), 42–51. doi:10.1109/MCOM.2002.1039856

Chapman, R., & Slaymaker, T. (2002). *ICTs and rural development: review of literature, current interventions and opportunities for action*. ODI Working Paper 192. London: Overseas Development Institute.

Charitos, D. (2008). Precedents for the design of Locative Media as hybrid spatial communication interfaces for social interaction within the urban context. In Saariluoma, P., Isomaki, H. M., & Isomaki, H. (Eds.), *Future Interaction Design II*. New York: Springer Verlag.

Checkland, P. (1981). *Systems Thinking, Systems Practice*. Chichester, UK: Wiley.

Checkland, P. (1999). *Soft Systems Methodology: a 30-year Retrospective*. Chichester, UK: John Wiley & Sons.

Checkland, P., & Holwell, S. (1998). *Information, Systems and Information System - Making Sense of the Field*. Chichester, UK: John Wiley & Sons.

Chen, X., Bishop, I. D., & Abdul Hamid, A. R. (2002). Community exploration of changing landscape values: the role of the virtual environment. In D. Suter & A. Bab-Habiashar (Eds.), *Proceedings of the sixth digital image computing - techniques and applications conference - DICTA 2002* (pp. 273-278). Melbourne, Australia: Australian Pattern Recognition Society.

Chen, Y. (2007). *Modelling And Matching: A Methodology For Complex ePlanning Systems Development*. PhD thesis, University of Salford, UK.

Chen, Y., Kutar, M., & Hamilton, A. (2007). Modelling and Matching: A Methodology for ePlanning System Development to Address the Requirements of Multiple User Groups. In D. Schuler (Ed.), Online Communities and Social Computing, HCII 2007 (LNCS 4564, pp. 41–49). Berlin: Springer-Verlag.

Chess, C., & Purcell, K. (1999). Public Participation and the Environment: Do We Know What Works? *Environmental Science & Technology, 33*(16), 2685–2692. doi:10.1021/es980500g

Cho, G. (2005). *Mastering geographic information science: technology, applications and management*. New York: John Wiley and Sons.

Cisco Systems. (2002). *Design guide Cisco IOS firewall*. Retrieved February 27, 2005, from http://www.cisco.com/warp/public/cc/pd/iosw/prodlit/firew_dg.htm

CNU. (1998). *Charter of the New Urbanism*. San Francisco: Congress for the New Urbanism.

Coburn, J. (2003). Bringing Local Knowledge into Environmental Decision Making: Improving Urban Planning for Communities at Risk. *Journal of Planning Education and Research, 22*(4), 420–433. doi:10.1177/0739456X03022004008

Codd, E. F. (1970). A relational model of data for large shared data banks. *Communications of the ACM, 13*(6), 377–387. doi:10.1145/362384.362685

CoE. (2003). *European Charter on the Participation of Young People in Local and Regional Life*. Strasbourg, France: Council of Europe.

Coglianese, C. (2004). Information Technology and Regulatory Policy: New Directions for Digital Government Research. *Social Science Computer Review, 22*(1), 85–91. doi:10.1177/0894439303259890

Cohen, G. (2003). *Urban future and ICT policy: perceptions of urban decision makers.* PhD Dissertation, Free University, Amsterdam.

Cohen, G., van Geenhuizen, M., & Nijkamp, P. (2001). *Urban Planning and Information Communication Technology.* Amsterdam: Timbergen Institute.

Cohen, J. E. (2006). Citizen satisfaction with contacting government on the internet. *Information Polity, 11*, 51–65.

Colditz, C., Coconu, L., Deussen, O., & Hege, H. C. (2005). Real-Time Rendering of Complex Photorealistic Landscapes Using Hybrid Level-of- Detail Approaches. In Trends in Real-Time Landscape Visualisation and Participation – Proceedings at Anhalt University of Applied Sciences. Heidelberg, Germany: Herbert Wichmann Verlag.

Coleman, S., & Goetze, J. (2001). *Bowling Together. Online Public Engagement in Policy Deliberation.* London, UK: Hansard Society and BT. Retrieved from http://bowlingtogether.net

Coleman, S., Åström, J., Freschi, A. C., Mambrey, P., & Moss, G. (2008). *Making eParticipation Policy – A European Analysis* (Demo-net booklet D14.4.).

Congress of Local and Regional Authorities. (2000). *Guidelines for a policy on citizens' responsible participation in municipal and regional life. Resolution 91.* Strasbourg, France: Council of Europe.

Congress of Local and Regional Authorities. (2008). *Electronic democracy and deliberative consultation on urban projects. Resolution 267.* Strasbourg, France: Council of Europe.

Congress of Local and Regional Authorities. (2008). *Electronic democracy and deliberative consultation on urban projects. Recommendation 249.* Strasbourg, France: Council of Europe.

Conroy, M. M., & Berke, P. R. (2004). What Makes a Good Sustainable Development Plan? An analysis of factors that influence principles of sustainable development. *Environment & Planning A, 36*, 1381–1396. doi:10.1068/a367

Conroy, M. M., & Evans-Cowley, J. (2006). E-participation in planning: an analysis of cities adopting on-line citizen participation tools. *Environment and Planning. C, Government & Policy, 24*(3), 371–384. doi:10.1068/c1k

Conroy, M. M., & Gordon, S. I. (2004). Utility of Interactive Computer Based Materials for Enhancing Public Participation. *Environmental Planning and Management, 47*(1), 19–33. doi:10.1080/0964056042000189781

Cooke, P., Uranga, M., & Etxebarria, G. (1997). Regional Innovation Systems: Institutional and Organisational Dimensions. *Research Policy, 26*, 475–491. doi:10.1016/S0048-7333(97)00025-5

Cooke, P., Uranga, M., & Etxebarria, G. (1998). Regional Innovation Systems: An Evolutionary Perspective. *Environment & Planning A, 30*, 1563–1584. doi:10.1068/a301563

Cornford, J., & Pollock, N. (2003). *Putting the University Online: Information, Technology and Organizational Change.* Buckingham, UK: Open University Press.

Cornford, J., Wessels, B., Richardson, R., Gillespie, A., McLoughlin, I., & Kohannejad, J. (2004). *Local e-Government: Process Evaluation of Electronic Local Government in England.* London: ODPM.

Couclelis, H., & Monmonier, M. (1995). Using SUSS for resolve NIMBY: How spatial understanding support systems can help with the 'not on my back yard' syndrome. *Geographical Systems, 2*(1), 83–101.

Coulson, M., & Bromley, R. (1990). The assessment of the users needs for corporate GIS: The example of Swansea Council. In *Proceedings of the European Conference on Geographic Information Systems* (pp. 209-217). Amsterdam: EGIS Foundation.

Council of Ministers. (2001). *Recommendation to member states on the participation of citizens in local public life.*

Recommendation (2001)19. Strasbourg, France: Council of Europe.

Coutard, O., & Guy, S. (2007). STS and the City: Politics and Practices of Hope. *Science, Technology & Human Values, 32*(6), 713–734. doi:10.1177/0162243907303600

Coward, L. A., & Salingaros, N. A. (2004). The Information Architecture of Cities. *Journal of Information Science, 30*(1), 101–112.

Craglia, M., & Signoretta, P. (2000). From global to local: the development of local geographic information strategies in the United Kingdom. *Environment and Planning. B, Planning & Design, 27*, 777–788. doi:10.1068/b2651

Craig, W. (1998). The Internet Aids Community Participation in the Planning Process. *Computers, Environment and Urban Systems, 22*(4), 393–404. doi:10.1016/S0198-9715(98)00033-7

Craig, W. (2002). *Community Participation and Geographic Information Systems*. London: Taylor and Francis.

Craig, W. J., Harns, T. M., & Weiner, D. (Eds.). (2002). *Community Participation and Geographic Information Systems*. London: Taylor and Francis.

Craig, W., Harris, T., & Weiner, D. (1998, October 15-17). *Empowerment, Marginalization and Public Participation GIS*. Report of Varenius Workshop, Santa Barbara, California. Retrieved July 20, 2009 from http://www.ncgia.ucsb.edu/Publications/Varenius_Reports/PPGIS98.pdf

Creasy, S., Gavelin, K., Fisher, H., Holmes, L., & Desai, M. (2007). *Engage for Change: The Role of Public Engagement in Climate Change Policy. The result of research undertaken for the Sustainable Development Commission*. Retrieved December 11, 2008, from http://www.sd-commission.org.uk/publications.php?id=618

Cullingworth, B., & Nadin, V. (2006). *Town and Country Planning in the UK* (14th ed.). London: Routledge.

Curwell, S., & Hamilton, A. (2003). *The roadmap (final deliverable) for the INTELCITY one-year RTD roadmap project (2002-2003)*. Funded by the EU DG Research Information Society Technologies (IST) Programme (IST-2001-37373).

Dallas Morning News. (2007). *'Chipping' Away at Privacy?* Retrieved November 11, 2008 from http://www.dallasnews.com/sharedcontent/dws/news/healthscience/stories/072207dnnatchip.350ef26.html

Damanpour, F. (1996). Bureaucracy and Innovation Revisited: Effects of Contingency Factors, Industrial Sectors, and Innovation Characteristics. *The Journal of High Technology Management Research, 7*(2).

Daniel, T. C., & Meitner, M. M. (2001). Representational validity of landscape visualizations: The effects of graphical realism on perceived scenic beauty of forest vistas. *Journal of Environmental Psychology, 21*, 61–72. doi:10.1006/jevp.2000.0182

Danielson, P. (2007). Digital morality and Ethics. In Anttiroiko, A.-V., & Malkia, M. (Eds.), *Encyclopedia of Digital Government* (pp. 377–381). Hershey, PA: Idea Group Publishing.

Danziger, S., & Rafal, R. (2009). The effect of visual signals on spatial decision making. *Cognition, 110*, 182–197. doi:10.1016/j.cognition.2008.11.005

Darke, R. (2000). Public participation, equal opportunities, planning policies and decisions. In P. Allmendinger, A. Prior, & J. Raemaekers (Eds), (2000). Introduction to Planning Practice (pp. 385-412). Chichester: John Wiley and Sons.

Das, R. (2008). RFID Market $2.77Bn in 2006 to $12.35Bn in 2010. *IDTechEX*. Retrieved November 3, 2008 from http://www.idtechex.com/research/articles/rfid_market_2_77bn_in_2006_to_12_35bn_in_2010_00000409.asp

Day, D. (1997). Citizen participation in the planning process: An essentially contested concept? *Journal of Planning Literature, 11*(3), 421–434. doi:10.1177/088541229701100309

De Angelis, G. (1996). *I Monti Lucretili*. Roma: Provincia di Roma.

De Graaf, R., & Dewulf, G. (2002). Interactive urban planning, hype or reality. In *3rd International Conference on Decision Making in Urban and Civil Engineering*, London.

De la Barra, T. (2001). Integrated land us and transport modeling: the Tranus experience. In Brail, R. K., & Klosterman, R. E. (Eds.), *Planning support systems* (pp. 129–156). Redlands, CA: ESRI Press.

deLeon, P. (1992). The Democratization of the Policy Sciences. *Public Administration Review, 52*(2), 125–129. doi:10.2307/976465

DeMers, M. (2005). *Fundamentals of geographic information systems.* New York: John Wiley and Sons.

DEMO-net. (2007). Introducing eParticipation. DEMO-net booklet series No. 1. Brussels, Belgium: European Commission European Commission. (2003a). Directive 2003/4/EC. On public access to environmental information.

Densham, P. (1991). Spatial decision support system. In Maguire, J., & Rhind, D. (Eds.), *Geographical information systems: Principles and applications* (pp. 403–412). London: Longman.

Densham, P., Armstrong, M., & Kemp, K. (1995). *Collaborative spatial decision making, Scientific Report of the NCGIA Initiative 6 Specialist Meeting, NCGIA Technical Report 90-5.* National Center for Geographic Information and Analysis, UCSB., California.

Department for Transport (DFT). (2005). *The development of bus real-time systems in the UK.* London: Department for Transport.

Department of Homeland Security. (2008). *The Use of RFID for Human Identification.* Retrieved November 2, 2008, from http://www.dhs.gov/xlibrary/assets/privacy/privacy_advcom_rpt_rfid_draft.pdf

Development, G. I. S. (2002). *GIS Development History: Milestones of GIS.* Retrieved on May 2, 2002, from http://www.gisdevelopment.net/history/1960-1970.htm

Devogele, T., Parent, C., & Spaccapietra, S. (1998). On spatial database integration. *International Journal of Geographical Information Science, 12*(4), 335–352. doi:10.1080/136588198241824

Dewsbury, J.-D. (2000). Performativity and the event: enacting a philosophy of difference. *Environment and Planning. D, Society & Space, 18*(4), 473–497. doi:10.1068/d200t

Dialogue Project Report. (2002). Retrieved from http://www.beepknowledgesystem.org/Search/ShowCase.asp?CaseTitleID=182&CaseID=574

DIE. (2000). *Devlet Istatistik Enstitusu* [Turkish Statistical Institute]. Ankara, Turkey: Nufus Sayimi Sonuclari. [Results of the 2000 General Population Census]

Digital Angel. (2008). *Miraculous Medical Potential.* Retrieved November 11, 2008, from http://www.digitalangel.com/biosensor.aspx

DiMaggio, P., & Hargittai, E. (2001). *From the Digital Divide to Digital Inequality: Studying Internet Use as Penetration Increases.* Woodrow Wilson School, Centre for the Arts and Cultural Policy Studies, Princeton University.

Dimitrova, D. V., & Chen, Y.-C. (2006). Profiling the Adopters of E-Government Information and Services: The Influences of Psychological Characteristics, Civic Mindedness, and Information Channels. *Social Science Computer Review, 24*(2), 172–188. doi:10.1177/0894439305281517

Dittrich, Y., & Floyd, C. C., & Klischewski, R. (Eds.). (2002). Social Thinking, Software Practice. London: MIT.

Dittrich, Y., Eriksen, S., & Wessels, B. (2009). *From Knowledge Transfer to Situated Innovation: cultivating spaces for co-operation in innovation and design between academics, user-groups and ICT providers.* Karlskrona, Sweden: Blekinge Institute of Technology.

Dobson, J. E. Blyth., A. J. C., Chudge, J., & Strens, R. (1994). The ORDIT approach to organisational requirements. In M. Jirotka & J. Goguen (Eds.), Requirements Engineering: Social and Technical Issues (pp. 87-106). London: Academic Press.

Dodig-Crnkovic, G., & Horniak, V. (2007). Ethics and privacy of communications in the e-Polis. In Anttiroiko, A.-V., & Malkia, M. (Eds.), *Encyclopedia of Digital Government* (pp. 740–744). Hershey, PA: Idea Group Publishing.

Dokonal, W., & Martens, B. (2002). Round Table Session on 3D-City-Modeling. In *20th eCAADe Conference Proceedings* (pp. 610-613). Warsaw, Poland: eCAADe.

Dollner, J., Baumann, K., & Buchholz, H. (2006). Virtual 3D City Models as Foundation of Complex Urban Information Spaces. In the Proceeding of the 11th International Conference on Urban Planning and Regional Development in the Information Society. 13th – 16th February 2006, Vienna, Austria.

Döllner, J., Buchholz, H., Nienhaus, M., & Kirsch, K. (2005). Illustrative Visualization of 3D City Models. In *Proceedings of SPIE – Visualization and Data Analysis 2005 (VDA),* San Jose, CA, USA (pp. 42-51).

Dowd, P. W., & McHenry, J. T. (1998). Network security: It's time to take it seriously. *IEEE Computer, 31*(9), 24–28.

Doyle, S., Dodge, M., & Smith, A. (1998). The Potential of Web Based Mapping and Virtual Reality Technologies for Modelling Urban Environments. *Computers, Environment and Urban Systems, 22*(2), 137–155. doi:10.1016/S0198-9715(98)00014-3

Drewe, P. (1998). *In search of new spatial concepts, inspired by Information Technology.* Report for the Ministerie van Volkshuisvesting, Ruimtelijke Ordening en Milieubeheer.

Drewe, P. (2000). *ICT and Urban Form. Urban Planning and Design - Off the beaten track.* Report for the Ministerie van Volkshuisvesting, Ruimtelijke Ordening en Milieubeheer.

Drewe, P. (2002). The Network City – from Utopia to new paradigm. *Atlantis, 13*(5).

Drewe, P. (2003). *ICT and urban form. Old dogma, new tricks.* Delft: Delft University of Technology.

Drummond, W. J., & French, S. P. (2008). The Future of GIS in Planning: Converging Technologies and Diverging Interests. *Journal of the American Planning Association. American Planning Association, 74*(2), 161–174. doi:10.1080/01944360801982146

Dryzek, J. S. (2000). *Deliberative Democracy and Beyond.* Oxford, UK: Oxford University Press.

Dühr, S., Bates-Brkljac, N., & Counsell, J. (2005). Public Digital Collaboration in Planning. In *Cooperative Design, Visualization, and Engineering* (pp. 186-193). Retrieved August 20, 2008, from http://dx.doi.org/10.1007/11555223_20.

Dunleavy, P., Margetts, H., Bastow, S., Callaghan, R., & Yared, H. (2002). *Progress in implementing e-Government in Britain: Supporting Evidence for the National Audit Office Report Government on the Web II.* London: LSE Public Policy Group and School of Public Policy UCL.

Durand-Lasserve, A., & Royston, L. (2002). *Holding their ground. Secure land tenure for the urban poor in developing countries.* London: Earthscan.

Dutil, P. A., Howard, C., Langford, J., & Roy, J. (2007). Rethinking Government-Public Relationships in a Digital World: Customers, Clients, or Citizens? *Journal of Information Technology & Politics, 4*(1), 77–90. doi:10.1300/J516v04n01_06

Dutta, S., Lopez-Carlos, A., & Mia, I. (2006). *Global Information Technology Report, 2005-2006.* New York: Palgrave-Macmillian.

Dutton, W. H., & Shepherd, A. (2006). Trust in the Internet as an Experience Technology. *Information Communication and Society, 9*(4), 433–451. doi:10.1080/13691180600858606

Eagles Nest. (2002). *RFID: The Early Years 1980-1990.* Retrieved November 1, 2008, from http://members.surfbest.net/eaglesnest/rfidhist.htm

Economou, D., Coccossis, H., & Deffner, A. (2005). Athens, a Capital City Under the Shadow of the State: 'Too Many Cooks Spoil the Broth. In Hendriks, F., & van Stipdonk, V. (Eds.), *Urban-regional Governance in the European Union: Practices and Prospects* (pp. 83–99). The Hague: Elsevier.

ECTP [European Council of Town Planners] (1998/2003). *The New Charter of Athens 2003: The European Council of Town Planners' Vision for Cities in the 21st century.* Lisbon.

ECTP. (2003). *The New Charter of Athens*. Bruxells, Belgium: European Council of Town Planners.

Edquist, C. (1997). Systems of Innovation Approaches (Their Emergence and Characteristics). In Edquist, C. (Ed.), *Systems of Innovation: Technologies, Institutions and Organizations*. Washington, DC: Pinter Publishers.

Edwardes, A. J., & Purves, R. S. (2007). A theoretical grounding for semantic descriptions of place. M. Ware & G. Taylor (Eds.), *Proceedings of 7th Intl. Workshop on Web and Wireless GIS, W2GIS 2007* (pp. 106-120). Berlin: Springer.

Elliott, L. (2004). *The global politics of the environment*. New York: Palgrave Macmillan.

Ellis, S. R. (1993). *Pictorial Communication: Pictures and the Synthetic Universe, Pictorial Communication in Virtual and Real Environments*. London: Taylor & Francis.

Elwood, S. (2006). Participatory GIS and community planning: restructuring technologies, social processes and future research in PPGIS. In S. Balram & Suzana Dragićević (Eds.), Collaborative Geographic Information Systems (pp. 66-84). Hershey, PA: Idea Group Publishing.

Elwood, S., & Leitner, H. (1998). GIS and Community-based Planning: Exploring the Diversity of Neighbourhood Perspectives and Needs. *Cartography and Geographic Information Systems*, *25*(2), 77–88. doi:10.1559/152304098782594553

Engadget. (2005). RFID to be used for inmate tracking in L.A. Retrieved December 1, 2008, from http://engadget.com/search/?q=RFID+use&invocationType=wl-gadget

Engeström, Y. (2008). *From Teams to Knots. Activity-theoretical studies of collaboration and learning at work*. New York: Cambridge University Press. doi:10.1017/CBO9780511619847

Eräsaari, R. (2006). Objektiivisuus, asiantuntijat ja instituutiot. In J. Parviainen (Eds.), Kollektiivinen asiantuntijuus (pp. 19-54). Tampere, Finland: Tampereen yliopistopaino.

Erickson, B., Lloyd-Jones, M. T., Nice, S., & Roberts, M. (1999). Place and Space in the Networked City: Conceptualising the Integrated Metropolis. *Journal of Urban Design*, *4*(1).

ESCAP. (1992). Social development strategy for the escap region towards the year 2000 and beyond. Bangkok. United Nations, ST/ESCAP/1124.

Esnard, A. (2007, January). Interoperable Three-Dimensional GIS: Urban Modeling with ArcGIS 3D Analyst and SketchUp. *ArcUser*. Retrieved July 20, 2009 from http://www.esri.com/news/arcuser/0207/urban.html

ESRI. (2002). *Architecture of ArcIMS*. Retrieved on June 29, 2004, from http://www.esri.com/software/arcims/architecture.html

ESRI. (2003). ArcGIS 3D Analyst: Three Dimensional Visualization, Topographic Analysis, and Surface Creation. *ESRI White Paper*, 1-16.

ESRI. (2004). *ArcGIS 3D Analyst*. Retrieved on June 29, 2004, from http://www.esri.com/software/arcgis/arcgisxtensions/3danalyst

Etzioni, A. (1967). Mixed-scanning: A Third Approach to Decision-Making. *Public Administration Review*, *27*(5), 385–392. doi:10.2307/973394

European Commission. (1997). *Compendium of European planning systems. Regional Development Studies. Report 28*. Luxembourg: Office for Official Publications of the European Communities.

European Commission. (1999). *European Spatial Development Perspective: Towards Balanced and Sustainable Development of the Territory of the European Union*. Luxembourg: Office for Official Publications of the European Communities.

European Commission. (2003). *Directive 2003/35/EC*.

European Commission. (2006). *The Riga Ministerial Declaration on eInclusion*. Retrieved from http://ec.europa.eu/.../events/ict_riga_2006/doc/declaration_riga.pdf

European Commission. (2007). *Measuring progress in eInclusion, Riga Dashboard*. Retrieved from http://

ec.europa.eu/.../einclusion/docs/i2010_initiative/riga-dashboard.pdf

Evans, R. (2002). E-commerce, competitiveness and local and regional governance in Greater Manchester and Merseyide: a preliminary assessment. *Urban Studies (Edinburgh, Scotland)*, *39*(5-6), 947–975. doi:10.1080/00420980220128390

Evans-Cowley, J., & Conroy, M. M. (2004). E-governance: on-line citizen participation tools for planners. [Chicago, IL: American Planning Association.]. *PAS Reporter*, 525.

Evans-Cowley, J., & Conroy, M. M. (2009). Local Government Experiences with ICT for Participation. In Reddick, C. (Ed.), *Handbook of Research on Strategies for Local E-Government Adoption and Implementation: Comparative Studies* (pp. 268–286). Hershey, PA: Information Science Publishing.

Faber, B. (1995). Extending electronic meeting systems for collaborative spatial decision making. In M. Densham & K. Kemp (Eds.), *Collaborative spatial decision-making: Scientific Report for the Initiative 17 Specialist Meeting, NCGIA Technical Report 95-14*. National Center for Geographic Information and Analysis, UCSB, California.

Faber, B. (1997). Active response GIS: An architecture for interactive resource modelling. In *GIS'97 Annual Symposium on Geographic Information Systems*, Vancouver, BC, March.

Fainstein, N. I., & Fainstein, S. S. (1979). New debates in urban planning: the impact of Marxist theory within the United States. *International Journal of Urban and Regional Research*, *3*, 381–403. doi:10.1111/j.1468-2427.1979.tb00796.x

Fainstein, S. (2000). New Directions in Planning Theory. *Urban Affairs Review*, *35*(4), 451–478.

Falch, M. (2006). ICT and the future conditions for democratic governance. *Telematics and Informatics*, *23*, 134–156. doi:10.1016/j.tele.2005.06.001

Faludi, A. (1973). The rationale of planning theory. In Faludi, A. (Ed.), *A Reader in Planning Theory* (pp. 35–53). Oxford, UK: Pergamon Press.

Faludi, A. (1973). What is planning theory? In Faludi, A. (Ed.), *A Reader in Planning Theory* (pp. 1–10). Oxford, UK: Pergamon Press.

Faust, N. (1995). The virtual reality of GIS. *Environment and Planning. B, Planning & Design*, *22*(1), 257–268. doi:10.1068/b220257

Felleman, J. (1997). *Deep Information: The Role of Information Policy in Environmental Sustainability*. Greenwich, CT: Ablex Publishing Corporation.

Ferber, J. (1999). *Multi-agent systems*. London: Addison-Wesley.

Fernández-Maldonado, A. M. (2004). *ICT-related Transformations in Latin American Metropolises*. Delft: Delft University Press.

Fetterman, D. (2001). *Foundations of Empowerment Evaluation*. London: Sage.

Fischer, F. (1990). *Technocracy and the Politics of Expertise*. Newbury Park, CA: Sage.

Fischer, F. (2000). *Citizens, experts, and the environment: The politics of local knowledge*. Durham, NC: Duke University Press.

Fischer, F., & Forester, J. (Eds.). (1993). *The Argumentative Turn in Policy Analysis and Planning*. Durham, NC: Duke University Press.

Fischer, M., & Nijkamp, P. (1993). *Geographic information systems, spatial modelling and policy evaluation*. New York: Springer-Verlag.

Fischer, P., Kubitzki, J., Guter, S., & Frey, D. (2007). Virtual driving and risk taking: Do racing games increase risk-taking cognitions, affect, and behaviors? *Journal of Experimental Psychology*, *13*(1), 22–31.

Fischler, R. (2000). Communicative Planning Theory. A Foucaldian Assessment. *Journal of Planning Education and Research*, *19*(4), 358–368. doi:10.1177/0739456X0001900405

Flamm, M., & Kaufmann, V. (2006). Operationalising the Concept of Motility: A Qualitative Study. *Mobilities*, *1*(2), 167–189. doi:10.1080/17450100600726563

Flanagan, M., Howe, D., & Nissenbaum, H. (2005). Values at play: Design tradeoffs in socially-oriented game design. In *Proc. CHI 2005* (pp. 751-760). New York: ACM Press.

Florida, R. (1995). Toward the Learning Region. *Futures, 27*(5), 527–536. doi:10.1016/0016-3287(95)00021-N

Flyvbjerg, B. (1998). *Rationality and Power: Democracy in Practice*. Chicago, IL: The University of Chicago Press.

Fog, H., Bröchner, J., Törnqvist, A., & Åström, K. (1989). *Det kontrollerade byggandet*. Stockholm: Carlssons.

Forester, J. (1989). *Planning in the face of power*. Los Angeles: University of California Press.

Forester, J. (1999). *The deliberative practitioner: Encouraging participatory planning processes*. Cambridge, MA: MIT Press.

Forlano, L. (2009). Codespaces: Community wireless networks and the reconfiguration of cities. In Foth, M. (Ed.), *Handbook of Research on Urban Informatics: The Practice and Promise of the Real-Time City* (pp. 292–309). Hershey, PA: IGI Global.

Forrester, J. (1989). *Planning in the face of power*. Berkeley, CA: University of California Press.

Foth, M. (Ed.). (2009). *Handbook of Research on Urban Informatics: The Practice and Promise of the Real-Time City*. Hershey, PA: IGI Global.

Foth, M., Choi, J. H., Bilandzic, M., & Satchell, C. (2008). Collective and Network Sociality in an Urban Village. In Mäyrä, F., Lietsala, K., Franssila, H., & Lugmayr, A. (Eds.), *Proceedings MindTrek: 12th international conference on entertainment and media in the ubiquitous era* (pp. 179–183). Tampere, Finland.

Fox, S., & Vitak, J. (2008). *Degrees of Access (May 2008 data)*. Presentation from the Pew Internet & American Life Project. Retrieved from http://www.pewinternet.org/Presentations/2008/Degrees-of-Access-(May-2008-data).aspx

Fraga, E. (2002). Trends in e-Government: How to Plan, Design, and Measure e-Government. In *Government Management Information Sciences (GMIS) Conference*, June 17, Santa Fe, New Mexico, U.S.A.

Frankhauser, P. (1994). *La Fractalité des Structures Urbaines*. Paris: Anthropos.

Franklin, S., & Graesser, A. (1997). Is it an agent, or just a program? A taxonomy for autonomous agents. *Lecture Notes in Computer Science, 1193*, 21–35. doi:10.1007/BFb0013570

Frederick, K. (2001). Network monitoring for intrusion detection. *Security Focus*. Retrieved March 3, 2005, from http://online.securityfocus.com/infocus/1220

Fredman, P., & Hörnsten, L. (2001). *Perceived crowding, visitor satisfaction and trail design in Fulufjäll National Park Sweden. Report*. Östersund, Sweden: European Tourism Research Institute.

French, S. P., & Skiles, A. E. (1996). Organizational structures for GIS Implementation. In M. J. Salling (Ed.), *URISA '96 Conference Proceedings*, Salt Lake City, Utah (pp. 280-293).

French, S. P., & Wiggins, L. I. (1990). California planning agency experience with automated mapping and geographic information systems. *Environment and Planning B, 17*(4), 441–450. doi:10.1068/b170441

Freschi, A. C., Raffini, L., Balocchi, M., & Tizzi, G. (2007). White paper. In Lippa, B. (Ed.), *DEMO-net: Research workshop report – Frameworks and methods for evaluating e-participation* (pp. 76–85). Bremen, Germany: IST Network of Excellence Project.

Friedman, A. (2007). *Sustainable Residential Development: Planning and Design for Green Neighborhoods*. New York: McGraw-Hill Professional.

Friedman, B., Borning, A., Davis, J., Gill, B., Kahn, P., Kriplean, T., & Lin, P. (2008). Laying the Foundations for Public Participation and Value Advocacy: Interaction Design for a Large Scale Urban Simulation. In *Proceedings of the 9th Annual International Conference on Digital Government Research*, Montréal, Canada, May 2008.

Friedman, B., Kahn, P., & Borning, A. (2006a). Value Sensitive Design and information systems. In Zhang, P., & Galletta, D. (Eds.), *Human-Computer Interaction and Management Information Systems: Foundations* (pp. 348–372). New York: M.E. Sharpe.

Friedman, B., Kahn, P., Hagman, J., Severson, R., & Gill, B. (2006b). The watcher and the watched: Social judgments about privacy in a public place. *Human-Computer Interaction*, *21*(2), 233–269. doi:10.1207/s15327051hci2102_3

Friedman, J. (1987). *Planning in the public domain: From knowledge on action*. Princeton, NJ: Princeton University Press.

Friedmann, J. (1996). Two centuries of planning theory: An Overview. In Mandelbaum, S. J., Mazza, L., & Burchell, R. W. (Eds.), *Explorations in planning theory* (pp. 10–29). New Brunswick, CT: Centre for Urban Policy Research.

Friend, J. (2002). *Stradspan: New Horizons in Strategic Decision Support*. Retrieved on May 2, 2002, from http://www.btinternet.com/~stradspan/program.htm

Friend, J., & Hickling, A. (1997). *Planning under pressure: The strategic choice approach*. London: Butterworth-Heinemann.

Fritsch, M., Repetti, A., Vuillerat, C.-A., & Schmid, G. (2008). Integrated and participatory land management. In *11th Interpraevent Conference*, Dornbirn.

Frohmann, E. (1997). *Gestaltqualitäten in Landschaft und Freiraum*. Wien, Austria: Österreichischer Kunst und Kulturverlag.

Fujitsu. (2008). *Fujitsu Develops World's First 64Kbyte High-Capacity FRAM RFID Tag for Aviation Applications*. Retrieved November 5, 2008 from http://www.fujitsu.com/global/news/pr/archives/month/2008/20080109-01.html

Fung, A. (2006). Varieties of Participation in Complex Governance. *Public Administration Review*, *66*(1), 66–75. doi:10.1111/j.1540-6210.2006.00667.x

FuturePundit. (2003). *Human Subdermal Credit Card Announced*. Retrieved November 11, 2008, from http://www.futurepundit.com/archives/001824.html

Gaerling, T., Golledge, R. G., et al. (Eds.). (1993). Behavior and Environment: Psychological and Geographical Approaches (Advances in Psychology Vol. 96). Amsterdam: North-Holland.

Gale International. (2008). *Gateway to Northeast Asia*. Retrieved October 18, 2008, from http://www.songdo.com/default.aspx

Gallent, N., Juntti, M., Kidd, S., & Shaw, D. (2008). *Introduction to Rural Planning*. New York: Routledge.

Gallopín, G. (1997). Indicators and their use: Information for decision-making. In Moldan, B., & Billharz, S. (Eds.), *Sustainability Indicators: Report of the Project on Indicators of Sustainable Development* (pp. 13–27). New York: Wiley.

Garfinkel, H. (1967). *Studies in Ethnomethodology*. Cambridge, UK: Polity Press.

Garling, T., & Evans, G. W. (Eds.). (1991). *Environment, Cognition, and Action: An Integrated Approach*. New York: Oxford University Press.

Garson, G. D., & Khosrow-Pour, M. (Eds.). (2008). Handbook of Research on Public Information Technology (Vol. 1 & 2). New York: Information Science Reference.

Gaye, M. (1996). *Entrepreneurial cities*. Dakar, Senegal: ENDA.

Geertman, S. (2002). Participatory planning and GIS: a PSS to bridge the gap. *Environment and Planning. B, Planning & Design*, *29*, 21–35. doi:10.1068/b2760

Geertman, S., & Stillwell, J. (2003). Planning support systems: an introduction. In Geertman, S., & Stillwell, J. (Eds.), *Planning support systems in practice* (pp. 3–22). London: Springer.

Geertman, S., & Stillwell, J. (Eds.). (2009). *Planning support systems best practices and new methods*. London: Springer.

Ghose, R. (2001). Transforming Geographic Information Systems into Community Information Systems. In *Transactions in GIS* (Vol. 5, pp. 141–163). Use of Information Technology for Community Empowerment.

Ghose, R. (2001). Use of Information Technology for Community Empowerment: Transforming Geographic

Information Systems into Community Information Systems. *Transactions in GIS*, *5*, 141–163. doi:10.1111/1467-9671.00073

Ghose, R. (2007). Politics of Scale and Networks of Association in PPGIS. *Environment & Planning A*, *39*(1), 1961–1980. doi:10.1068/a38247

Gibin, M., Mateos, P., Petersen, J., & Atkinson, P. (2009). Google Maps Mashups for local public health service planning. In Geertman, S., & Stillwell, J. (Eds.), *Planning support systems best practices and new methods* (pp. 227–241). London: Springer. doi:10.1007/978-1-4020-8952-7_12

Gilbert, A. (2004). Protecting your ID: RFID chips in humans gets green light. *Silicon.com*. Retrieved November 9, 2008, from http://www.silicon.com/research/specialreports/protectingid/0,3800002220,39124983,00.htm

Gilbert, R., Stevenson, D., Girardet, H., & Stren, R. (1996). *Making Cities Work: The Role of Local Authorities in the Urban Environment*. London: Earthscan.

Gilfoyle, I., & Wong, C. (1998). Computer applications in planning: Twenty years'. *Planning Practice and Research*, *13*(2), 191–197. doi:10.1080/02697459816193

Gill, S., Higgs, G., & Nevitt, P. (1999). GIS in planning departments: Preliminary results from a survey of local planning authorities. *Planning Practice and Research*, *14*(3), 341–361. doi:10.1080/02697459915643

Glander, T., & Döllner, J. (2007). Cell-Based Generalization of 3D Building Groups with Outlier Management. In ACMGIS 2007. New York: ACM.

Glenn, J. C., & Gordon, T. J. (2003). *Futures Research Methodology – V2.0*. AC/UNU Millennium Project.

Gobster, P. H. (1995). Perception and use of a metropolitan greenway system for recreation. *Landscape and Urban Planning*, *33*, 401–413. doi:10.1016/0169-2046(94)02031-A

Gocmen, Z. A. (2009). *GIS use for planning in Wisconsin's public agencies*. Paper presented at the 11th International Conference on Computers in Urban Planning and Urban Management.

Gooch, A., & Gooch, B. (2001). *Non-Photorealistic Rendering*. Natick, MA: AK Peters Ltd.

Goodchild, M. (2007). *Citizens as Sensors: The World of Volunteered Geography*. Retrieved May 10, 2009, from http://www.ncgia.ucsb.edu/projects/vgi/docs/position/Goodchild_VGI2007.pdf

Goodchild, M., & Densham, P. (1995). *Spatial decision support systems*. Scientific Report of the NCGIA Initiative 17 Specialist Meeting, NCGIA Technical Report 95-14. National Center for Geographic Information and Analysis, UCSB, California.

Goodman, B. (2008). WWII Story: This Is it! Part-IX. *All Aviation Flight Online*. Retrieved November 10, 2008, from http://aafo.com/library/history/B-17/b17part9.htm

Goodman, N. (1951). *The Structure of Appearance*. Cambridge, UK: Harvard UP.

Goodspeed, R. (2008). *Citizen Participation and the Internet in Urban Planning*. Unpublished master's thesis. Retrieved in January 1, 2009, from http://goodspeedupdate.com/wp-content/uploads/2008/11/goodspeed-internetparticipation.pdf

Gordon, L. A., & Leob, M. P. (2002). The economics of information security investment. *ACM Transactions on Information and System Security*, *5*(4), 438–457. doi:10.1145/581271.581274

Gordon, L. A., Loeb, M. P., & Lucyshyn, W. (2003). Information security expenditures and real options: A wait-and-see approach. *Computer Security Journal*, *19*(2), 1–7.

Graham, H. (2007). *Seeking sustainability in an age of complexity*. Cambridge, UK: Cambridge University Press.

Graham, S. (1998). The end of geography or the explosion of space? Conceptualizing space, place and information technology. *Progress in Human Geography*, *2*, 165–185. doi:10.1191/030913298671334137

Graham, S. (2002). Bridging urban digital divides? Urban polarization and information and communications technologies. *Urban Studies (Edinburgh, Scotland), 39*(1), 33–56. doi:10.1080/00420980220099050

Graham, S., & Marvin, S. (1996). *Telecommunications and the City*. London: Routledge. doi:10.4324/9780203430453

Graham, S., & Marvin, S. (2001). *Splintering urbanism. Networked infrastructures, technological mobilities and the urban condition*. London: Routledge. doi:10.4324/9780203452202

Granberg, M. (2004). *Från lokal välfärdsstat till stadspolitik. Politiska processer mellan demokrati och effektivitet*. Örebro Studies in Political Science 11. Örebro, Sverige: Örebro University.

Grant, J., Manuel, P., & Joudrey, D. (1996). A framework for planning sustainable residential landscapes. *Journal of the American Planning Association. American Planning Association, 62*, 331–345. doi:10.1080/01944369608975698

Green Mountain Institute for Environmental Democracy. (2005). *Environmental Democracy - What's in it for me?* Montpellier, France. Retrieved December 11, 2008, from http://www.gmied.org/comment.htm

Griffin, T., Harris, R., & Williams, P. (2002). *Sustainable Tourism*. London: Butterworth-Heinemann.

Grimes, S. (2003). The digital economy challenge facing peripheral rural areas. *Progress in Human Geography, 27*(2), 174–193. doi:10.1191/0309132503ph421oa

Grönlund, Å. (2001). Democracy in an IT-Framed Society. *Communications of the ACM, 44*(1), 23–26. doi:10.1145/357489.357498

Großegger, B., & Zentner, M. (2008). *Computerspiele im Alltag Jugendlicher. Gamer-Segmente und Gamer-Kulturen in der Altersgruppe der 11- bis 18-Jährigen*. Wien, Austria: Institut für Jugendforschung.

Gupta, J. N. D., Sharma, K. S., & Hsu, J. (2003). An Overview of Knowledge management. In Gupta, J. N. D., & Sushil, K. S. (Eds.), *Creating Knowledge Based Organizations*. Hershey, PA: Idea Group Publishing.

Gurstein, M. (2007). *What is community informatics (and Why Does It Matter)?* Milano, Italia: Polimetrica.

Gustafsson, G. (1981). *KPP-projektet om kommunal demokrati och planering*. (Rapport/Byggforskningsrådet 1981:5). Stockholm: Statens råd för byggnadsforskning.

Habermas, J. (1979). *Communication and the Evolution of Society* (McCarthy, T., Trans.). Boston: Beacon Press.

Habermas, J. (1984). *The Theory of Communicative Action* (Vol. 1). (McCarthy, T., Trans.). Boston: Beacon Press.

Hagen, M. (2000). Digital Democracy and Political Systems. In Hacker, K., & van Dijk, J. (Eds.), *Digital Democracy: Issues of Theory & Practice* (pp. 54–69). London: Sage.

Hake, G., Grünreich, D., & Meng, L. (2002). *Kartographie – Visualisierung raum-zeitlicher Information*. Berlin: Walter de Gruyter.

Haklay, M. (2002) *Public Environmental Information Systems: Challenges and Perspectives*. PhD thesis, Department of Geography, UCL, University of London.

Haklay, M. (2002). Public Environmental Information - Understanding Requirements and Patterns of Likely Public Use. *Area, 34*(1), 17–28. doi:10.1111/1475-4762.00053

Haklay, M. E., & Tobon, C. (2003). Usability Evaluation and PPGIS: Towards a User-centered Design Approach. In *Geographical Information Science, 17*, 577–592. New York: Taylor & Francis Ltd.

Halder, M., & Willard, T. (2003). *Information Society and Sustainable Development: Exploring the Linkages*. Paper presented at the World Summit on the Information Society.

Hall, P. (2002). *Cities of tomorrow: An intellectual history of urban planning and design in the twentieth century*. Oxford, UK: Blackwell.

Hall, P., & Ward, C. (1998). *Sociable Cities: The Legacy of Ebenezer Howard*. Chichester, UK: John Wiley & Sons.

Hamilton A., Burns, M., Arayici, Y., Gamito, P., Marambio, A. E., Abajo, B., Pérez, J., & Rodríguez-Maribona, I. A. (2005 September). *Building Data Integration System*.

Final Project Deliverable in the Intelligent Cities (IntelCities) Integrated Project, IST – 2002-507860, Deliverable 5.4c. University of Salford, UK.

Hamilton, A., Curwell, S., & Davies, T. (1998). A Simulation of the Urban Environment in Salford. In *Proceedings of CIB World Building Congress,* Gåvle, Sweden, 7-12 June 1998 (pp. 1847-1855).

Hampton, W. A. (1977). Research into public participation on structure planning. In Coppock, J. T., & Sewell, W. R. D. (Eds.), *Public Participation in Planning* (pp. 27–42). New York: Wiley.

Hannam, K., Sheller, M., & Urry, J. (2006). Editorial: Mobilities, Immobilities and Moorings. *Mobilities, 1*(1), 1–22. doi:10.1080/17450100500489189

Hannemann, A. (2008). *Strategic Urban Planning and Municipal Governance.* Saarbrücken, Germany: Verlag Dr. Muller Aktiengesellschaft & Co.

Hanzl, M. (2007). Information Technology as a Tool for Public Participation in Urban Planning: a Review of Experiments and Potentials. *Design Studies, 28*(3), 289–307. doi:10.1016/j.destud.2007.02.003

Harder, C. (1998). *Serving Maps on the Internet.* Redlands, California: Environmental Systems Research Institute Inc.

Harris, B. (1989). Beyond Geographic Information Systems: Computers and the Planning Professional. *Journal of the American Planning Association. American Planning Association, 55*(1), 85–90. doi:10.1080/01944368908975408

Harris, B. (1991). Planning Theory and the Design of Planning Support Systems. In *Second International Conference on Computers in Planning and Management,* 6-8 July, 1991, Oxford.

Harris, B., & Batty, M. (1993). Locational Models, Geographical Information and Planning Support System. *Journal of Planning Education and Research, 12,* 184–198. doi:10.1177/0739456X9301200302

Harris, T., & Elmes, G. (1993). GIS Applications in Urban and Regional Planning: The North American Experience. *Applied Geography (Sevenoaks, England), 13*(1), 9–27. doi:10.1016/0143-6228(93)90077-E

Harrison, C., & Haklay, M. (2002). The potential of public participation geographic information systems in UK environmental planning: appraisals by active publics. *Journal of Environmental Planning and Management, 45*(6), 841–863. doi:10.1080/0964056022000024370

Harrison, J., & Wessels, B. (2005). A new public service communication environment? Public service broadcasting values in the reconfiguring media. *New Media & Society, 7*(6), 861–880. doi:10.1177/1461444805058172

Hartley, T., Trinkler, I., & Burgess, N. (2004). Geometric determinants of human spatial memory. *Cognition, 94,* 39–75.

Harvey, D. (1985). On planning the ideology of planning. In Harvey, D. (Ed.), *The Urbanization of capital* (pp. 165–184). Oxford, UK: Blackwell.

Harvey, L. A. (2007). Digging the Digital Well. *Urban, 10*(1), 24–26.

Hawley, A. (2004 October). RFID technology using Active or Passive Tags - A Very Hot Issue. *Traffic Engineering + Control, 45*(9), 324-326.

Hazel, G., & Parry, R. (2004). *Making Cities Work.* Chichester, UK: Wiley-Academy.

Healey and Baker Consultants. (2001). *European E-locations Monitor.* London: Healey and Baker Consultants.

Healey, P. (1997). *Collaborative planning. Shaping places in fragmented societies.* London: MacMillan.

Healey, P. (2003). Collaborative planning in perspective. *Planning theory, 2*(2), 101-123.

Healey, P., & Khakee, A. (1997). *Making Strategic Spatial Plans: Innovation in Europe.* London: UCL Press.

Heeks, R. (2002). Information Systems and Developing Countries: Failure, Success, and Local Improvisations. *The Information Society, 18*(2), 101–112. doi:10.1080/01972240290075039

Heeks, R. (2003). E-government in Africa: promise and practice. *Information Policy, 7*(2-3), 97–114.

Heft, H., & Nasar, J. N. (2000). Evaluating environmental scenes using dynamic versus static displays. *Environment and Behavior, 32*(3), 301–322.

Heidegger, M. (1962). *Being and Time*. London: SCM.

Heimerl, J. L., & Voight, H. (2005). Management: The foundation of security program design and management. *Computer Security Journal, 21*(2), 1–20.

Helbig, N., Gil-Garcia, J., & Ferro, E. (2009). Understanding the complexity of electronic Implications from the digital divide literature. *Government Information Quarterly, 26*, 89–97. doi:10.1016/j.giq.2008.05.004

Held, D. (1987). *Models of Democracy*. Cambridge, UK: Polity Press.

Henecke, B., & Khan, J. (2002). *Medborgardeltagande i den fysiska planeringen - en demokratiteoretisk analys av lagstiftning, retorik och praktik*. (Working Paper in Sociology. 2002:1/Report No. 36, November 2002). Lund, Sverige: Sociologiska institutionen/Avdelningen för miljö- och energisystem, Lunds universitet.

Hentea, M. (2007). Intelligent system for information security management: Architecture and design issues. *Issues in Information Science and Technology, 4*, 29–43.

Herwig, A., Kretzler, E., & Paar, P. (2005). Using games software for interactive landscape visualization. In Bishop, I. D., & Lange, E. (Eds.), *Visualization in Landscape and Environmental Planning – Technology and Applications* (pp. 62–67). New York: Taylor & Francis.

Herzele, A., & Woerkum, C. (2008). Local Knowledge in Visually Mediated Practice. *Journal of Planning Education and Research, 27*, 444–455. doi:10.1177/0739456X08315890

Herzorg, P., et al. (2001). The open source security testing manual, v1.5. *IdeaHamster*. Retrieved May 6, 2005, from http://www.ideahamster.org/osstmm.htm

Hester, U., Mesicek, R., & Schnepf, D. (2004). *Prerequisites for a Sustainable and Democratic Application of ICT*. Presentation at the World Summit on Information Society.

Hetherington, J., Daniel, T. C., & Brown, T. C. (2004). Is motion more important than it sounds? The medium of presentation in environmental perception research. *Journal of Environmental Psychology, 13*, 283–291. doi:10.1016/S0272-4944(05)80251-8

Hillier, B. (1996). *Space is the Machine*. Cambridge, UK: Cambridge University Press.

Hillier, J. (2008, September). *Are we there yet?* Key note speech at the 40th Anniversary Conference of the Centre for Urban and Regional Studies, Helsinki University of Technology.

Hillier, J., & Healey, P. (Eds.). (2008). *Critical essays in planning theory* (*Vol. 1-3*). Aldershot, UK: Ashgate.

Hirschheim, R. (1992). Information Systems Epistemology: An Historical Perspective. In Galliers, R. (Ed.), *Information Systems Research: Issues, Methods and Practical Guidelines* (pp. 28–60). Oxford: Blackwell.

Hirschheim, R., & Klein, H. (1989). Four Paradigms of Information Systems Development. *Communications of the ACM, 32*(10), 1199–1215. doi:10.1145/67933.67937

Hjorth, P., & Bagheri, A. (2006). *Navigating towards sustainable development: a system dynamic approach*. Amsterdam: Elsevier Ltd.

Hoch, C. (2007). Pragmatic Communicative Action Theory. *Journal of Planning Education and Research, 26*(3), 272–284. doi:10.1177/0739456X06295029

Hoff, J. (2004). Members of Parliaments' use of ICT in a Comparative European Perspective. *Information Polity, 9*, 5–16.

Holliday, I., & Kwok, R. C. W. (2004). Governance in the information age: building e-government in Hong Kong. *New Media & Society, 6*(4), 549–570. doi:10.1177/146144804044334

Holman, N. (2008). Community participation: using social network analysis to improve developmental benefits. *Environment and Planning. C, Government & Policy, 26*, 525–543. doi:10.1068/c0719p

Holzer, M., & Seang-Tae, K. (2008). *Digital Governance in Municipalities worldwide (2007). A longitudinal assessment of municipal websites throughout the world.* Newark, NJ: Rutgers University.

Honadle, G., & Van Sant, J. (1985). *Implementation for sustainability: Lessons from integrated rural development.* West Hartford, CT: Kumarian Press.

Honor Fagan, G., Newman, D. R., McCusker, P., & Murray, M. (2006). *E-consultation: evaluating appropriate technologies and processes for citizens' participation in public policy.* E-Consultation Research Project, Final Report, 14 July 2006. Retrieved December 11, 2008, from http://www.e-consultation.org

Hopkins, L. (2001). *Urban Development: The Logic of Making Plans.* New York: Island Press.

Hopkins, L. D., Twidale, M., & Pallathucheril, V. G. (2004, July). *Interface Devices and Public Participation.* Paper presented at the Public Participation GIS Conference, Madison, WS.

Horelli, L. (2002). A methodology of participatory planning. In Bechtel, R., & Churchman, A. (Eds.), *Handbook of Environmental Psychology* (pp. 607–628). New York: John Wiley & Sons.

Horelli, L. (2003). *Valittajista tekijöiksi (From complainers to agents, Adolescents on the arenas of empowerment).* Espoo, Finland: Helsinki University of Technology.

Horelli, L. (2006). A Learning-based network approach to urban planning with young people. In Spencer, C., & Blades, M. (Eds.), *Children and Their Environments: Learning, Using and Designing Spaces* (pp. 238–255). Cambridge, UK: Cambridge University Press. doi:10.1017/CBO9780511521232.015

Horelli, L. (2009). Network Evaluation from the Everyday Life Perspective. A Tool for Capacity-Building and Voice. *Evaluation, 15*(2), 205–223. doi:10.1177/1356389008101969

Horelli, L., & Wallin, S. (2006). Arjen ajan hallintaa, kokemuksia suomalaisesta aikasuunnittelusta [Mastering of everyday life, Experiences from time planning in Finland]. Helsinki, Finland: Helsingin kaupungin tietokeskus.

Horelli, L., & Wallin, S. (2009, June). *Gendered community informatics for sustaining a glocal everyday life.* Paper presented at the City Futures 09 Conference, Madrid.

Horelli, L., Wallin, S., Innamaa, I., & Jarenko, K. (2007). *Introduction to the user-sensitive Ubi-Helsinki – a report on the preconditions for the user-sensitive design and its evaluation in the context of commercial, public and community services.* Espoo, Finland: Helsinki University of Technology, Centre for Urban and Regional Studies.

Horst, M., Kuttschreutter, M., & Gutteling, J. M. (2007). Perceived usefulness, personal experiences, risk perception and trust as determinants of adoption of e-government services in The Netherlands. *Computers in Human Behavior, 23*, 1838–1852. doi:10.1016/j.chb.2005.11.003

Hosmer, D. W., & Lemeshow, S. (2000). *Applied Logistic Regression.* New York: Wiley. doi:10.1002/0471722146

Howard, E. (2001). *Garden Cities of To-morrow.* New York: Books for Business.

Huang, B., & Lin, H. (1999). GeoVR: a web-based tool for virtual reality presentation from 2D GIS data. *Computers & Geosciences, 25*(1), 1167–1175. doi:10.1016/S0098-3004(99)00073-4

Hwang, M. S., Tzeng, S. F., & Tsai, C. S. (2003). A new secure generalization of threshold signature scheme. In *Proceedings of International Technology for Research and Education* (pp. 282-285).

Ibrahim, K. F. (2007). *Newnes Guide to Television and Video Technology, The Guide for the Digital Age – from HDTV, DVD and flatscreen technologies to Multimedia Broadcasting, Mobile TV and Blue Ray* (4th ed.). Burlington, MA: Newnes.

Ibrahim, M., Abdullah, M. F., & Abdul Samad, D. (2004). *Geographical information system (GIS) in Malaysian planning education.* Paper presented at the 7th International Seminar on GIS for Developing Countries (GISDECO).

ICLEI. (1996). *The Local agenda 21 planning guide.* Toronto, Canada: ICLEI.

IDA. (2006). *E-government Customer Perception Survey Conducted in 2006.* Retrieved from http://www.ida.gov.sg, date accessed 20 July 2006

IDA. (2007). *Annual E-government Customer Perception Survey Conducted in 2007.* Retrieved from http://www.ida.gov.sg/Publications/20071001171301.aspx

IDA. (2008). Annual survey on Infocomm Usage in households and Individuals for 2007. Retrieved from http://www.ida.gov.sg/doc/Publications/Publications_Level2/20061205092557/ASInfocommUsageHseholds07.pdf

Idrus, S., & Harman Shah, A. H. (2006). 'Metabolisme' GIS di Malaysia. *Buletin Geospatial Sektor Awam, 1,* 13–19.

iGOV2010 Project Steering Committee. (2006). *From Integrated Service to Integrated Government.* Singapore: Ministry of Finance, Infocomm Development Authority of Singapore.

ILO. (2002). *Women and men in the informal economy: a statistical picture.* Geneva, Switzerland: International Labor Organization.

iN2015 Infocomm Infrastructure, Services and Technology Development Sub-Committee. (2006). *Totally Connected, Wired and Wireless.* Report by the Infocomm Infrastructure, Services and Technology Development Sub-Committee, Singapore: IDA.

Inayatullah, S. (2007). *Questioning the Future: methods and tools for organizational and societal change.* Tamsui, ROC: Tamkang University.

Initiative eParticipation. (2005). *Elektronische Bürgerbeteiligung in deutschen Großstädten. Zweites Web-Ranking der Initiative eParticipation.* (M. Bräuer & T. Biewendt, Eds.). Retrieved from http://www.initiative-eparticipation.de

Innes, J. E. (1995). Planning theory's emerging paradigm: communicative action and interactive practice. *Journal of Planning Education and Research, 14,* 183–189. doi:10.1177/0739456X9501400307

Innes, J. E. (1996). Planning through consensus building: A new view of the comprehensive planning ideal. *Journal of the American Planning Association. American Planning Association, 62*(4), 460–472. doi:10.1080/01944369608975712

Innes, J. E. (1998). Information in communicative planning. *Journal of the American Planning Association. American Planning Association, 64*(11), 52–63. doi:10.1080/01944369808975956

Innes, J. E., & Booher, D. E. (1999). Consensus building as role playing and bricolage: toward a theory of collaborative planning. *Journal of the American Planning Association. American Planning Association, 65,* 9–26. doi:10.1080/01944369908976031

Innes, J. E., & Booher, D. E. (2004). Reframing Public Participation: Strategies for the 21st Century. *Planning Theory & Practice, 5*(4), 419–436. doi:10.1080/1464935042000293170

Innes, J., & Simpson, D. (1993). Implementing GIS for planning. *Journal of the American Planning Association. American Planning Association, 59*(1), 230–236. doi:10.1080/01944369308975872

Internet Security Systems. (2002). *Real Secure network protection.* Retrieved January 19, 2004, from http://www.iss.net/products_services/enterprise_protection/rsnetwork/index.php

Islam, P. (2007). Citizen-centric E-Government: The Next Frontier. *The Kennedy School Review, 7,* 103–108.

Jabatan Perancangan Bandar dan Desa Semenanjung Malaysia. (2001). *Manual Penyediaan Rancangan Tempatan (Pindaan 2001).* Kuala Lumpur: Jabatan Perancangan Bandar dan Desa Semenanjung Malaysia.

Jabatan Perancangan Bandar dan Desa Semenanjung Malaysia. (2008). Geotechnical spatial analysis research and development: fokus dan agenda. Retrieved December 12, 2008, from http://www.townplan.gov.my/penyelidikan_gsard.php

Jacob, G.-R., & Schreyer, R. (1980). Conflict in outdoor recreation: a theoretical perspective. *Journal of Leisure Research, 12*(4), 368–380.

Jacobs, J. (1961). *The Death and Life of Great American Cities*. New York: Vintage Books.

Jacucci, C., Jacucci, G., Wagner, I., & Psik, T. (2005). A manifesto for the performative development of ubiquitous media. In *CC '05: Proceedings of the 4th decennial conference on critical computing*, Aarhus, Denmark. New York: ACM.

Jaeger, P. T., & Thompson, K. M. (2004). Social information behavior and the democratic process: Information poverty, normative behavior, and electronic government in the United States. *Library & Information Science Research, 26*, 94–107. doi:10.1016/j.lisr.2003.11.006

James, W. (1908). *Pragmatism: A New Name for Some Old Ways of Thinking: Popular Lectures on Philosophy*. New York: Longman. Retrieved July 20, 2009, from Project Gutenberg: http://www.gutenberg.org/dirs/etext04/prgmt10.txt

Janczewski, L., & Portougal, V. (2007). Managing security clearances within government institutions. In Anttiroiko, A.-V., & Malkia, M. (Eds.), *Encyclopedia of Digital Government* (pp. 1196–1202). Hershey, PA: Idea Group Publishing.

Jankowski, P. (2009). Towards participatory geographic information systems for community-based environmental decision making. *Journal of Environmental Management, 90*(6), 1966–1971. doi:10.1016/j.jenvman.2007.08.028

Jankowski, P., & Nyerges, T. (2001). *Geographic information systems for group decision-making: Towards a participatory geographic information science*. London: Taylor and Francis.

Jankowski, P., & Stasik, M. (1997). Design considerations for space and time distributed collaborative spatial decision making. *Journal of Geographic Information and Decision Analysis, 1*(1), 1–12.

Jardin, X. (2007). CA Bill Would Ban Forced Subdermal RFID-tagging of Humans. *Boingboing*. Retrieved November 11, 2008 from http://www.boingboing.net/2007/09/04/ca-bill-bans-forced.html

Jarupathirun, S., & Zahedi, F. M. (2007). Dialectic decision support systems: System design and empirical evaluation. *Decision Support Systems, 43*, 1553–1570. doi:10.1016/j.dss.2006.03.002

Jho, W. (2005). Challenges for e-governance: protests from civil society on the protection of privacy in E-government in Korea. *International Review of Administrative Sciences, 71*(1), 151–161. doi:10.1177/0020852305051690

Jiang, B., & Li, Z. (2005). Geovisualization: Design, Enhanced Visual Tools and Applications. *The Cartographic Journal, 42*(1), 3–4. doi:10.1179/000870405X52702

Jirotka, M., & Goguen, J. A. (Eds.). (1994). *Requirements Engineering: Social and Technical Issues*. London: Academic Press.

Joerin, F., Rey, M. C., Desthieux, G., & Nembrini, A. (2001). Information et participation pour l'aménagement du territoire. *Revue Internationale de Géomatique, 11*(3-4), 309–332.

Johar, F. (2003). Environmental concern in local planning. *Planning Malaysia, 1*, 19–34.

Jonassen, D. H., & Grabowski, B. L. (1993). *Handbook of Individual Differences; Learning and Instruction*. Hillsdale, NY: Lawrence Erlbaum Associates.

Jones, S., & Fox, S. (2009). *Pew Internet Project Data Memo: Generations Online in 2009*. Memo from the Pew Internet & American Life Project. Retrieved from http://www.pewinternet.org/~/media//Files/Reports/2009/PIP_Generations_2009.pdf

Jorgensen, A., Hitchmough, J., & Calvert, T. (2002). Woodland spaces and edges: their impact on perception of safety and preference. *Landscape and Urban Planning, 59*, 1–11.

Jovero, S., & Horelli, L. (2002). *Nuoret ja paikallisuuden merkitys? Nuorten ympäristösuhteen tarkastelua Vantaan Koivukylä-Havukosken alueella (Young people and the meaning of locality)*. Espoo: TKK/YTK Julkaisuja.

Juhl, G. (1993). Government Agencies Let Their Hair Down about GIS. *Geographical Information Systems, 3*(7), 20–26.

Jungk, R., & Mullert, N. (1996). *Future workshop: How to create desirable futures*. London: Institute for Social Inventions.

Junker, B., & Buchecker, M. (2008). Aesthetic preferences versus ecological objectives in river restorations. *Landscape and Urban Planning, 85*(3-4), 141–154.

Kahn, P., Friedman, B., Perez-Granados, D., & Freier, N. (2006). Robotic pets in the lives of preschool children. *Interaction Studies: Social Behaviour and Communication in Biological and Artificial Systems, 7*(3), 405–436. doi:10.1075/is.7.3.13kah

Kain, J., & Soderberg, H. (2008). Management of complex knowledge in planning for sustainable development: the use of multi-criteria decision aids. *Environmental Impact Assessment Review, 28*, 7–21. doi:10.1016/j.eiar.2007.03.007

Kalman, R. E. (1969). *Topics in mathematical system theory*. New York: McGraw-Hill.

Kalu, K. N. (2007). Capacity Building and IT Diffusion, A comparative assessment of e-government in Africa. *Social Science Computer Review, 25*(3), 358–371. doi:10.1177/0894439307296917

Kao, J. (2009). Tapping the World's Innovation Hot Spots. *Harvard Business Review, 87*(3), 109–114.

Karjalainen, E., & Tyrväinen, L. (2002). Visualization in forest landscape preference research: A Finnish perspective. *Landscape and Urban Planning, 59*, 13–28. doi:10.1016/S0169-2046(01)00244-4

Kassim, N. S., & Islam, R. (2009). Spatial AHP for the National Physical Plan of Malaysia. In Davide, G., & Abdullah, A. (Eds.), *Spatial decision support for urban and environmental planning: a collection of case studies*. Shah Alam, Malaysia: Arah Publications.

Kasvio, A., & Anttiroiko, A.-V. (Eds.). (2005). e-City. Analysing efforts to Generate Local Dynamism in the City of Tampere. Tampere, Finland: Tampere University Press.

Katok, A. B., & Hasselblatt, B. (1999). *Introduction to the Modern Theory of Dynamical Systems*. Cambridge, UK: Cambridge University Press.

Kaufmann, V. (2002). *Re-thinking Mobility: Contemporary Sociology*. Aldershot, UK: Ashgate.

Kaylor, C., Deshazo, R., & Van Eck, D. (2001). Gauging e-government: a report on implementing services among American cities. *Government Information Quarterly, 18*, 293–307. doi:10.1016/S0740-624X(01)00089-2

Keisler, J., & Sundell, R. (1997). Combining multi-attribute utility and geographic information for boundary decisions: an application to park planning. *Journal of Geographic Information and Decision Analysis, 1*(2), 101–118.

Keller, P. R., & Keller, M. M. (1993). *Visual Cues: Practical Data Visualization*. Los Alamitos, CA: IEEE Computer Society Press.

Kelly, E. D., & Becker, B. (2000). *Community Planning: An Introduction to the Comprehensive Plan*. Washington, DC: Island Press.

Kelly, K. L. (1998). A systems approach to identifying decisive information for sustainable development. *European Journal of Operational Research, 109*, 452–464. doi:10.1016/S0377-2217(98)00070-8

Kelly, S. (2004). *Community Planning: How to Solve Urban and Environmental Problems*. New York: Rowman and Littlefield Publishers.

Kennedy, K. (2006). Subdermal Keys/Tribal Expressions. *Tribal Expressions*. Retrieved November 11, 2008 from http://tribalexpression.blogspot.com

Kephart, J. O., & Chess, D. M. (2003). The vision of automatic computing. *IEEE Computer, 36*(1), 41–50.

Kersten, G. E., Mikolajuk, Z., & Gar-On Yeh, A. (1999). *Decision Support Systems for Sustainable Development: A Resource Book of Methods and Applications*. Boston: Kluwer Academic Publishers.

Keul, A. G., & Martens, B. (1995). Simulation – How Does it Shape the Message? In *The Future of Endoscopy – Proceedings of the 2nd European Architectural Endoscopy Association Conference*. Wien, Austria: Technische Universität Wien, ISIS-ISIS Publications.

Khakee, A. (1989). Kommunal planering i omvandling 1947-1987. (Gerum nr 13, Gerum). Umeå, Geografiska institutionen, Umeå universitet.

Khakee, A. (1999). Demokratin i samhällsplaneringen. In E. Amnå (Ed.), Medborgarnas erfarenheter. (Demokratiutredningens forskarvolym V. SOU 1999: 113). Stockholm: Fakta info direkt.

Khakee, A. (2000) Samhällsplanering. Lund, Sverige: Studentlitteratur.

Khakee, A., Barbanente, A., & Borri, D. (2000). Expert and experiential knowledge in planning. *The Journal of the Operational Research Society, 51*, 776–788.

Khakee, A., Barbanente, A., & Puglisi, M. (2002). Scenario building for Metropolitan Tunis. *Futures, 34*, 583–596. doi:10.1016/S0016-3287(02)00002-2

Khakee, A., Barbanente, A., Camarda, D., & Puglisi, M. (2002). With or without? Comparative study of preparing participatory scenarios using computer-aided and traditional brainstorming. *Journal of Future Research, 6*, 45–64.

Khalil, O. E. M., & Elkordy, M. M. (2001). The Relationship of Some Personal and Situational Factors to IS Effectiveness: Empirical Evidence from Egypt. In Shaw, N. G. (Ed.), *Strategies for Managing Computer Software Upgrades*. London: Idea Group Inc.

Khoshaflan, S., & Buckiewicz, M. (1995). *Introduction to groupware, workflow, and workgroup computing*. London: John Wiley and Sons.

Kickert, W. J. M. Klijn. E-H., & Koppenjan, J. F. M. (Eds.). (1997). Managing Complex Networks: Strategies for the Public Sector. London: Sage.

Kingston, R. (2007). Public Participation in Local Policy Decision-making: The Role of Web-based Mapping. *The Cartographic Journal, 44*(2), 138–144. doi:10.1179/000870407X213459

Kingston, R. (2008). *The role of participatory e-Planning in the new English Local Planning System*. Retrieved December 3, 2008, from http://www.ppgis.man.ac.uk/

Kingston, R., Carver, S., Evans, A., & Turton, I. (2000). Web-based public participation geographical information systems: an aid to local environmental decision-making. *Computers, Environment and Urban Systems, 24*(1), 109–125. doi:10.1016/S0198-9715(99)00049-6

Kiss, C., Poltimae, H., Struminska, M., & Ewing, M. (2006). *Environmental Democracy. An Assessment of Access to Information, Participation in Decision-making and Access to Justice in Environmental Matters in Selected European Countries*. The Access Initiative Europe and The EMLA Association. Retrieved December 11, 2008, from http://www.emla.hu/img_upload/0aa155da39c21509c55c587879f86484/TAI_1.pdf

Kleist, R. A., Chapman, T. A., Sakai, D. A., & Jarvis, B. S. (2005). *RFID Labeling* (2nd ed.). Irvine, CA: Printronix.

Klijn, E. H., & Koppenjan, J. F. M. (2000). Politicians and Interactive Decision Making: Institutional Spoilsports or Playmakers. *Public Administration, 78*(2), 365–387. doi:10.1111/1467-9299.00210

Kline, R., & Pinch, T. (1996). Users as Agents of Technological Change: The Social Construction of the Automobile in the Rural United States. *Technology and Culture, 37*, 763–795. doi:10.2307/3107097

Klosterman, R. (2008). Urban future strategies: concepts and tools for a new urban management. In 2008 International Seminar on Future City, October 24, 2008, National Hanbat University, Daejeon, Korea.

Klosterman, R. E. (1997). Planning support systems: a new perspective on computer-aided planning. *Journal of Planning Education and Research, 17*(1), 45–54. doi:10.1177/0739456X9701700105

Klosterman, R. E. (1999). New perspectives on planning support systems. *Environment and Planning. B, Planning & Design, 26*, 317–320. doi:10.1068/b260393

Klosterman, R. E., & Abdullah, A. (2008). *Computer technologies in urban planning*. Paper presented at the Medina Forum Seminar on Developing and Deploying Planning Support Systems.

Knapp, S., Chen, Y., Hamilton, A., & Coors, V. (2009). An ePlanning Case Study in Stuttgart Using OPPA 3D. In Reddick, C. G. (Ed.), *Handbook of Research on Strategies for Local E-Government Adoption and Implementation: Comparative Studies*. Hershey, PA: IGI Global.

Kofman, F., & Senge, P. M. (1993). Communities of commitment: the heart of the learning organizations. *Organizational Dynamics, 22*, 5–19. doi:10.1016/0090-2616(93)90050-B

Koivisto, J. (2008, October). *Relational evaluation: A Step forward or a lapse in evaluation practice?* Paper presented at the Conference of the European Evaluation Society, Lisbon.

Komito, L. (2007). Community and inclusion: the impact of new communications technologies. *Irish Journal of Sociology, 16*(2), 77–96.

Korten, D. (1990). *Getting to the 21st century: Voluntary action and the global agenda*. West Hartford, CT: Kumarian Press.

Kovaic, M. (2005). The impact of national culture on worldwide egovernment readiness. *Informing Science Journal, 8*, 143–158.

Kraemer, K. L., & Dutton, W. H. (1982). The Automation of Bias. In Danziger, J. N., Dutton, W. H., Kling, R., & Kraemer, K. L. (Eds.), *Computer and Politics: High Technology in American Local Governments*. New York: Columbia University Press.

Kraus, K. (2007). *Photogrammetry - Geometry from Images and Laser Scans* (2nd ed.). Berlin: Walter de Gruyter.

Kubicek, H. (2007). *Electronic Democracy and Deliberative Consultation on Urban Projects - Putting E-Democracy into Context*. Report for the Congress of Local and Regional Authorities. Retrieved September 15, 2008, from http://www.ifib.de/publikationsdateien/Creative_final.pdf

Kubicek, H. (2008). Electronic Democracy Achievements and Challenges. In *the ESF Research Conference in Vadstena*, Sweden, 21-25 November 2007. Retrieved from http://www.docs.ifib.de/esfconference07

Kubicek, H., Lippa, B., & Westholm, H. (2009). Medienmix in der lokalen Demokratie. Berlin, Germany: Ed. sigma Kubicek, H., & Westholm, W. (20010). Consensus Building by Blended Participation. The Case of the Public Stadium Swimming Pool in Bremen. In S. French & D. R. Insua (Eds.), E-Democracy: A Group Decision and Negotiation Oriented Approach. Berlin: Springer.

Kubicek, H., Millard, J., & Westholm, H. (2007). Back-Office integration for online services between organizations. In Anttiroiko, A.-V., & Malkia, M. (Eds.), *Encyclopedia of Digital Government* (pp. 123–130). Hershey, PA: Idea Group Publishing.

Kuhn, T. S. (1962). *The Structure of Scientific Revolutions*. Chicago, IL: University of Chicago Press.

Kuo, E., & Choi, A. Mahizhnan, A. Peng, L-W, Soh, C. (2002). Internet in Singapore: A Study on Usage and Impact. Singapore: Times Academic Press.

Kurki, H. (2005). Lähitalouden ymmärtäminen ja asuinalueiden kehitys. In Neloskierrettä kaupunginosiin. Kumppanuudet ja roolit alueiden kehittämisessä (pp. 69-80). Kulttuuriasiainkeskus, Helsingin kaupunki.

Kwan, M. (2002). Feminist Visualization: Re-envisioning GIS as a Method in Feminist Geographic Research. *Annals of the Association of American Geographers. Association of American Geographers, 92*(4), 645–661. doi:10.1111/1467-8306.00309

Kwan, M. P., & Weber, J. (2003). Individual accessibility revisited: implications for geographical analysis in the twenty-first century. *Geographical Analysis, 35*, 341–353. doi:10.1353/geo.2003.0015

La Porte, T. M. (2005). Being good and doing well: Organizational Openness and Government Effectiveness on the World Wide Web. *Bulletin of the American Society for Information Science and Technology February/March*, 23-27.

Laing, R., Davies, A.-M., & Scott, S. (2005). Combining visualization with choice experimentation in the built environment. In Bishop, I. D., & Lange, E. (Eds.), *Visualization in Landscape and Environmental Planning – Technology and Applications* (pp. 212–219). New York: Taylor & Francis.

Landorf, C. (2009). Social inclusion and sustainable urban environments: an analysis of the urban and regional planning literature. In *Proceedings of Second International Conference on Whole Life Urban Sustainability and its Assessment*, Loughborough, UK, 22-24 April 2009 (pp. 861 -878).

Landt, J., & Catlin, B. (2001). Shrouds of Time, The history of RFID, An AIM Publication. *RFID Consultation*. Retrieved October 29, 2008 from http://www.rfidconsultation.eu/docs/ficheiros/shrouds_of_time.pdf

Lange, E. (1999). *Realität und computergestützte visuelle Simulation*. Vdf Zürich.

Lange, E. (2001). The limits of realism: perceptions of virtual landscapes. *Landscape and Urban Planning, 54*, 163–182. doi:10.1016/S0169-2046(01)00134-7

Lange, E. (2005). Issues and questions for research in communicating with the public through visualisations. In Buhmann, E., Paar, P., Bishop, I., & Lange, E. (Eds.), *Trends in Real-Time Landscape Visualization and Participation* (pp. 16–26). Heidelberg, Germany: Herbert Wichmann Verlag.

Lange, E., & Hehl-Lange, S. (2005). Future scenarios of peri-urban green space. In Bishop, I. D., & Lange, E. (Eds.), *Visualization in Landscape and Environmental Planning – Technology and Applications* (pp. 195–202). New York: Taylor & Francis.

Larsen, J., Axhausen, K., & Urry, J. (2006). Geographies of Social Networks: Meetings, Travel and Communications. *Mobilities, 1*(2), 261–283. doi:10.1080/17450100600726654

Latouche, S. (2001 May). En finir une fois pour toute avec le développement. *Le Monde Diplomatique*, 6-7.

Latour, B. (1987). *Science in action: How to follow scientists and engineers through society*. Cambridge, MA: Harvard University Press.

Latour, B. (2005). *Reassembling the Social: An Introduction To Actor-Network-Theory (Clarendon Lectures in Management Studies)*. Oxford, UK: Oxford University Press.

Laurian, L. (2004). Public participation in environmental decision making: Findings from communities facing toxic waste cleanup. *Journal of the American Planning Association. American Planning Association, 70*(1), 53–66. doi:10.1080/01944360408976338

Laurini, R. (2001). *Information Systems for Urban Planning: A Hypermedia Co-operative Approach*. New York: Taylor & Francis.

Le Corbusier (1971). *La Charte d'Athènes*. Paris: Seuil.

Le Corbusier. (1943/1987). *The Athens Charter*. Athens: Ypsilon.

Le Galès, P. (1988). Regulation, territory and governance. *International Journal of Urban and Regional Research, 22*(3), 482–506. doi:10.1111/1468-2427.00153

Leat, P., Seltzer, K., & Stoker, G. (2002). *Towards Holistic Governance: The New Reform Agenda*. Basingstoke, UK: Palgrave.

leClercq, F. (1990). Information supply to strategic planning. *Environment and Planning. B, Planning & Design, 17*, 429–440. doi:10.1068/b170429

Lee, L. M., & Ahmad, M. J. (2000). *Local authority networked development approval system*. Paper presented at the Planning Digital Conference.

Lee, S. L., Tan, X., & Trimi, S. (2005). Current Practices of Leading E-Government Countries. *Communications of the ACM, 48*(10), 99–105. doi:10.1145/1089107.1089112

Lee, Y. W., Pipino, L., Strong, D. M., & Wang, R. Y. (2004). Process Embedded Data Integrity. *Journal of Database Management, 15*(1), 87–103.

Leitner, C. (2003). *E-Government in Europe: The State of Affairs*. Maastricht, The Netherlands: European Institute of Public Administration.

Lenk, K. (2007). *Bürokratieabbau durch E-Government - Handlungsempfehlungen zur Verwaltungsmodernisierung für Nordrhein-Westfalen auf der Grundlage von Entwicklungen und Erfahrungen in den Niederlanden*. Report for the Information office d-NRW. Retrieved

October 31, 2008, from http://egovernmentplattform.de/uploads/media/Lenk_Buerokratieabbau.pdf von Lucke, J. (2008). *Hochleistungsportale für die öffentliche Verwaltung*. Lohmar: Eul-Verlag.

Leung, H. J. (2003). *Land use planning made plain* (2nd ed.). Toronto, Canada: University of Toronto Press.

Li, S., Guo, X., Ma, X., & Chang, Z. (2007). Towards GIS-Enabled Virtual Public Meeting Space for Public Participation. *Photogrammetric Engineering and Remote Sensing, 73*(6), 641–649.

Liggett, R., Friedman, S., & Jepson, W. (1995). Interactive design/decision making in a virtual urban world: Visual simulation and GIS. In *1995 ESRI User Conference*, ESRI, California.

Lilienfeld, R. (1978). *The rise of system theory*. New York: John Wiley.

Lindblom, C. E. (1959). The Science of Muddling 'Through'. *Public Administration Review, 19*, 288–304. doi:10.2307/973677

Lindgaard, G. (1994). *Usability Testing and System Evaluation: A guide for designing useful computer systems*. Technical Communications (Publishing) Ltd.

Lindsy, G., & Nguyen, D. B. L. (2004). Use of greenway trails in Indiana. *Journal of Urban Planning and Development, 130*(4), 213–217. doi:10.1061/(ASCE)0733-9488(2004)130:4(213)

Lippa, B., Aichholzer, G., Allhutter, D., et al. (2008). eParticipation. Evaluation and Impact. *DEMO-net Booklet 13.3*. Retrieved from http://www.ifib.de/publikationsdateien/DEMOnet_booklet_13.3_eParticipation_evaluation.pdf

Lipsey, R. G. (1997). Globalisation and national government policies: An economist's view. In Dunning, J. H. (Ed.), *Governments, globalisation, and international business*. London: Oxford University Press.

Lloyd, P. (1994). *Groupware in the 21st century: Computer supported collaborative working toward the millennium*. New York: Greenwood.

Lococo, A., & Yen, D. (1998). Groupware: Computer Supported Collaboration. *Telematics and Informatics, 15*(1), 85–101. doi:10.1016/S0736-5853(98)00006-9

Lodge, J. (2003). Toward an e-commonwealth? A tool for peace and democracy? *The Round Table, 372*, 609–621. doi:10.1080/0035853032000150627

Loiterton, D., & Bishop, I. (2005). Virtual environments and location-based questioning for understanding visitor movement in urban parks and gardens. In Buhmann, E., Paar, P., Bishop, I., & Lange, E. (Eds.), *Trends in Real-Time Landscape Visualization and Participation* (pp. 144–154). Heidelberg, Germany: Herbert Wichmann Verlag.

Longley, P., Goodchild, M., Maguire, D., & Rhind, D. (2001). *Geographic Information Systems and Science*. New York: Wiley.

Louviere, J. J., Hensher, D. A., & Swait, J. D. (2000). *Stated choice methods – Analysis and application*. Cambridge, UK: University Press.

Lynch, K. (1960). *The Image of the City*. Cambridge, MA: The MIT Press.

MacEachren, A. M. (1995). *How Maps Work: Representation, Visualization and Design*. New York: Guilford Press.

Macintosh, A. (2003). Using Information and Communication Technologies to Enhance Citizen Engagement in the Policy Process. In Promises and Problems of E-Democracy: Challenge of Online Citizen Engagement. Paris: OECD.

Macintosh, A., & Coleman, S. (2006). Multidisciplinary Roadmap and Report from eParticipation Research. *DEMO-net Deliverable D 4.2*. Retrieved from http://itc.napier.ac.uk/ITC/documents/Demo-net%204_2_multidisciplinary_roadmap.pdf

Macintosh, A., & Whyte, A. (2006). Evaluating how e-participation changes local democracy. *eGovernment Workshop '06 (eGOV06)*, 11 September 2006, Brunel University, London.

Macintosh, A., & Whyte, A. (2007). Towards an evaluation framework for e-participation. In Lippa, B. (Ed.),

DEMO-net: Research workshop report – Frameworks and methods for evaluating e-participation (pp. 43–57). Bremen, Germany: IST Network of Excellence Project.

Macintosh, A., Coleman, S., & Lalljee, M. (2005). *E-Methods for Public Engagement: Helping Local Authorities communicate with citizens.* Bristol City Council for the Local eDemocracy National Project. Retrieved December 11, 2008, from http://itc.napier.ac.uk/ITC/Documents/eMethods_guide2005.pdf

Mackinlay, J. (1986). Automating the Design of Graphical Presentations of Relational Information. *ACM Transactions on Graphics, 5*(2), 111–141. doi:10.1145/22949.22950

Madanipour, A. (2001). Multiple Meanings of Space and the Need for a Dynamic Perspective. In Madanipour, A., Hull, A., & Healey, P. (Eds.), *The Governance of Place. Space and planning processes* (pp. 154–168). Aldershot, UK: Ashgate.

Madon, S., Reinhard, N., Roode, D., & Walsham, G. (2009). Digital Inclusion Projects in Developing Countries: Processes of Institutionalization. *Information Technology for Development, 15*(2), 95–107. doi:10.1002/itdj.20108

Maenpaa, O. (2004). *E-government: Effects on Civil Society, Transparency and Democracy.* Presented at International Institute of Administrative Sciences, 26th Congress of Administrative Sciences, Seoul.

Maiwald, E. (2004). *Fundamentals of network security.* New York: McGraw-Hill/Technology Education.

Maklumat, B. T. (2006). Sistem pengurusan data JPBD Semenanjung Malaysia. *Buletin Geospatial Sektor Awam, 1/2006*, 26–30.

Malecki, E. J. (2002). Hard and soft networks for urban competitiveness. *Urban Studies (Edinburgh, Scotland), 39*(5-6), 929–945. doi:10.1080/00420980220128381

Malić, B. (1998). *Physiologische und technische Aspekte kartographischer Bildschirmvisualisierung.* Dissertation; Schriftenreihe des Instituts für Kartographie und Topographie der Rheinischen Friedrich-Wilhelms-Universität Bonn.

Mandelbaum, S. J. (1996). Making and breaking planning tools. *Computers, Environment and Urban Systems, 20*(2), 71–84. doi:10.1016/S0198-9715(96)00001-4

Mannermaa, M. (2008). Jokuveli. Elämä ja vaikuttaminen ubiikkiyhteiskunnassa [Some Brother, Life and Participation in the Ubiquitous Society]. Helsinki: WSOYpro.

Manning, R. (2007). *Parks and carrying capacity.* Washington, DC: Island Press.

Manning, R. E., & Freimund, W. A. (2004). Use of visual research methods to measure standards of quality for parks and outdoor recreation. *Leisure Sciences, 36*(4), 557–579.

Mansell, R., & Steinmueller, W. (2000). *Mobilizing the Information Society: strategies for growth and opportunity.* Oxford, UK: Oxford University Press.

Mantero, D. (2000). *Escursioni nel Parco Naturale Regionale Monti Lucretili. Provincia di Roma.* Roma: Italia.

Mantero, D., & Giacopini, L. (1997). *Guida al Parco Regionale Naturale dei Monti lucretili.* Roma: Regione Lazio, Italia.

March, J. G., & Olsen, J. P. (1989). *Rediscovering Institutions: The Organizational Basis of Politics.* New York: Free Press.

Marcinkowski, F. (2007). Media system and political communication. In Klöti, U., Knoepfel, P., Kriesi, H., Linder, W., Papadopoulos, Y., & Sciarini, P. (Eds.), *Handbook of Swiss politics* (2nd ed., pp. 381–402). Zürich, Deutschland: NZZ Publishing.

Margerum, R. D. (2002). Collaborative Planning Building Consensus and Building a Distinct Model for Practice. *Journal of Planning Education and Research, 21*, 237–253. doi:10.1177/0739456X0202100302

Märker, O. (2008). E-Partizipation als Gesamtsystem. *Standort - Zeitschrift für angewandte Geographie, 32*(3), 80-83.

Märker, O., & Wehner, J. (2008). E-Partizipation - Bürgerbeteiligung in Stadt- und Regionalplanung. *Standort - Zeitschrift für angewandte Geographie, 32*(3), 84-89.

Märker, O., Hagedorn, H., & Trénel, M. (2002). *Internet-Based Public Consultation: Relevance – Moderation – Software*. Retrieved February 25, 2009 from http://www.ercim.org/publication/Ercim_News/enw48/maerker.html

Martin, B., & Byrne, J. (2003). *Implementing E-government: Widening the lens. Electronic Journal of E-Government, 1*(1).

Mason, R. O., & Mitroff, I. I. (1981). *Challenging Strategic Planning Assumptions*. New York: John Wiley.

Masuda, Y. (1981). *The information society as Post-industrial society*. New York: World Future Society.

Maxcy, S. (2003). Pragmatic Threads in Mixed Methods Research in the Social Sciences: The Search for Multiple Modes of Inquiry and the End of the Philosophy of Formalism. In Tashakkori, A., & Teddlie, C. (Eds.), *Handbook of Mixed Methods in Social Q4 and Behavioral Research*. Thousand Oaks, CA: Sage.

May, C., & Finch, T. (2009). Implementing, Embedding and Integrating Practices: An Outline of Normalisation Profess Theory. *Sociology, 43*, 535–545. doi:10.1177/0038038509103208

Mayer, H. (2000). *Einführung in die Wahrnehmungs-, Lern- und Werbe-Psychologie*. Oldenburg, Deutschland: Oldenburg Wissenschaftsverlag.

McCarthy, J., & Lloyd, G. (2007). *From Property to People? Partnership, Collaborative Planning and Urban Regeneration*. Aldershot, UK: Ashgate.

McCaughey, M., & Ayers, M. D. (Eds.). (2003). *Cyberactivism. Online Activism in Theory and Practice*. New York: Routledge.

McClure, S., Scambray, J., & Kurtz, G. (2002). *Hacking exposed: Network security secrets & solutions* (4th ed.). New York: McGraw-Hill Osborne Media.

McGinn, M. (2001). *Getting Involved in Planning*. Edinburgh, UK: Scottish Executive Development Department.

McMaster University. (2008). Canada-India RFID project looks to Improve Traffic Flow, Reduce Pollution. *Physorg.com*. Retrieved November 1, 2008, from http://www.physorg.com/news133112758.html

McNeal, R. S., Tolbert, C. J., Mossberger, K., & Dotterweich, L. J. (2003). Innovating in Digital Government in the American States. *Social Science Quarterly, 84*(1), 52–70. doi:10.1111/1540-6237.00140

Meadows, D. H., Meadows, D., Randers, J., & Behrens, W. W. III. (1972). *The limits of Growth*. New York: Universe Books.

Medaglia, R. (2007). The challenged identity of a field: The state of the art of eParticipation research. *Information Polity, 12*(3), 169–181.

Melville, R. (2007). Ethical dilemmas in online research. In Anttiroiko, A.-V., & Malkia, M. (Eds.), *Encyclopedia of Digital Government* (pp. 734–739). Hershey, PA: Idea Group Publishing.

Members of the High Level Group on the Information Society. (1994, May 26). *Europe and the Global Information Society: Recommendations to the European Council* (The Bangemann Report).

Mena, J. (2004). Homeland security connecting the DOTS. *Software Development, 12*(5), 34–41.

Mennecke, B. E., & West, L. A. J. (2001). Geographic Information Systems in Developing Countries: Issues in Data Collection, Implementation and Management. (Technology Information). *Journal of Global Information Management, 9*(4), 44–54.

Metaverse. (2007). *Metaverse roadmap foresight framework*. Retrieved October 15, 2008, from http://www.metaverseroadmap.org/inputs.html

Metaxiotis, K., & Psarras, J. (2004). E-government: new concept, big challenge, success stories, *Electronic Government, an International Journal, 1*(2), 141-151.

Michael, M. (2000). Futures of the Present: From Performativity to Prehension. In Brown, N. (Eds.), *Contested Futures: A Sociology of Prospective Techno-Science* (pp. 21–39). Aldershot, UK: Ashgate.

Microsoft Corporation. (2003). The ten immutable laws of security. *Microsoft Security Response Center*. Retrieved November 21, 2005, from http://www.microsoft.com/technet/treeview/default.asp?url=/technet/columns/security/essays/10imlaws.asap

Miller, J., Friedman, B., Jancke, G., & Gill, B. (2007). Value tensions in design: The value sensitive design, development, and appropriation of a corporation's groupware system. In *Proc. of GROUP 2007* (pp. 281-290).

Miller, S. K. (2001). Facing the challenge of wireless security. *IEEE Computer*, *34*(7), 16–18.

Miraftab, F. (2003). The Perils of Participatory Discourse: Housing Policy in Post-apartheid South Africa. *Journal of Planning Education and Research*, *22*, 226–239. doi:10.1177/0739456X02250305

Mitchell, W. J. (1998). *City of Bits: Space, place, and the infobahn*. Cambridge, MA: The MIT Press.

Mitleton-Kelly, E. (2003). Complexity Research – Approaches and Methods: The LSE Complexity Group Integrated Methodology. In A. Keskinen, M. Aaltonen, & E. Mitleton-Kelly (Eds.), Organisational Complexity (pp. 55-74). Turku: Finland Futures Research Centre, Turku School of Economics and Business Administration, Mossberger, Karen (2008). Digital citizenship: the internet, society, and participation. Cambridge, MA: MIT Press.

Mockler, R. J. (1989). *Knowledge-Based Systems for Management Decisions*. Upper Saddle River, NJ: Prentice-Hall.

Molina, A. H. (1995). Sociotechnical Constituencies as Processes of Alignment: the rise of a large-scale European Information Initiative. *Technology in Society*, *17*(4), 385–412. doi:10.1016/0160-791X(95)00016-K

Mossberger, K., Tolbert, C. J., & McNeal, R. S. (2008). *Digital Citizenship. The internet, society and participation*. Cambridge, MA: The MIT Press.

Mugumbu, W. (2000). *GIS into Planning*. Retrieved from http://www.hbp.usm.my/thesis/heritageGIS/master/research/GIS%20into%20Planning.htm

Muhar, A. (2001). Three-dimensional modelling and visualisation of vegetation for landscape simulation. *Landscape and Urban Planning*, *54*(1-4), 5–19. doi:10.1016/S0169-2046(01)00122-0

Mukhija, V. (2001). Enabling slum redevelopment in Mumbai: Policy paradox in practice. *Housing Studies*, *18*(4), 213–222.

Mumford, E. (2000). *The CIAM discourse on Urbanism, 1928-1960*. Cambridge, MA: The MIT Press.

Nasar, J. L. (1983). Adult viewers' preferences in residential scenes. A study of the relationships of environmental attributes to preference. *Environment and Behavior*, *15*(5), 589–614. doi:10.1177/0013916583155003

National Association of State Chief Information Officers (NASCIO). (2006). *The IT security business case: Sustainable funding to manage the risk*. Retrieved March 4, 2007, from http://www.nascio.org/publications/documents/NASCIOIT_Security_Business_Case.pdf

National Association of State Chief Information Officers (NASCIO). (2006). Appendix F: Securing the state of Michigan information technology resources. In *Findings from NASCIO's strategic cyber security survey*. Retrieved December 17, 2008, from http://www.michigan.gov/documents/AppendixF_149547_7.pdf

National Institute of Standards and Technology. (2002). *Risk management guide for information technology systems*. Retrieved November 26, 2005, from http://csrc.nist.gov/publications/nistpubs/800-30/sp800-30.pdf

Nawwab, I. (1992 July). The Journey of a Lifetime. *Saudi Aramco World*, 24–35. Retrieved from http://www.saudiaramcoworld.com/issue/199204/the.journey.of.a.life time.htm

Ndione, E., de Leener, P., Jacolin, P., Perier, J.-P., & Ndiaye, M. (1993). *La ressource humaine, avenir des terroirs*. Paris: Karthala.

Nedovic-Budic, Z. (2000). Geographic information science implications for urban and regional planning. *URISA Journal*, *12*(2), 81–93.

Nedovic-Budic, Z. (2008). Afterword: Planning for Creative Urban Regions. In Yigitcanlar, T., Velibeyoglu, K.,

& Baum, S. (Eds.), *Creative urban regions: harnessing urban technologies to support knowledge city initiatives* (pp. 312–318). Hershey, PA: IGI Global.

Nedovic-Budic, Z. (2008). ICTs to support urban planning strategies. In 2008 International Seminar on Future City, October 24, 2008, National Hanbat University, Daejeon, Korea.

Needham, M. D., & Rollins, R. B. (2005). Interest group standards for recreation and tourism impacts at ski areas in the summer. *Tourism Management, 26*, 1–13. doi:10.1016/j.tourman.2003.08.015

Neisser, U. (1996). *Kognition und Wirklichkeit*. Stuttgart, Deutschland: Klett-Cotta.

Netchaeva, I. (2002). E-Government and E-Democracy: A Comparison of Opportunities in the North and South. *Gazette: The International Journal for Communication Studies, 64*(5), 467–477. doi:10.1177/17480485020640050601

Neudeck, S. (2001). Zur Gestaltung topografischer Karten für die Bildschirmvisualisierung; Dissertation; Fakultät für Bauingenieur- und Vermessungswesen. In *Schriftenreihe des Studiengangs Geodäsie und Geoinformation* (Vol. 74). Neubiberg: Universität der Bundeswehr München.

Newman, P., Manning, R. E., Pilcher, E., Trevino, K., & Savidge, M. (2006). Understanding and managing soundscapes in national parks: Part 1- indicators of quality. In D. Siegrist, C. Clivaz, M. Hunziker & S. Iten (Eds.), *Exploring the Nature of Management. Proceedings of the Third International Conference on Monitoring and Management of Visitor Flows in Recreational and Protected Areas* (pp. 193-195). Rapperswil, Switzerland: University of Applied Sciences.

Nilsson, K. (2003). Planning in a Sustainable Direction - The Art of Conscious Choices. Stockholm: Institutionen för infrastruktur och samhällsplanering, Kungliga Tekniska Högskolan.

Noisette, P. (1996). Le marketing urbain: outils du MT. In Decoutère, S., Ruegg, J., & Joye, D. (Eds.), *Le partenariat public-privé* (pp. 261–281). Lausanne, France: PPUR.

Nonaka, I., & Takeuchi, H. (1995). *The Knowledge-Creating Company. How Japanese Companies Create the Dynamics of Innovation*. New York: Oxford University Press.

Nordicom. (2006). *Nordicom-Sveriges Internetbarometer 2005*. (Nordicom-Sverige MedieNotiser Nr 2, 2006). Göteborg, Sverige: Göteborgs universitet.

Norman, D. A., & Draper, S. W. (Eds.). (1986). *User-centered System Design: New Perspectives on Human-Computer Interaction*. Hillsdale, NJ: Lawrence Erlbaum Associates, Inc.

Norris, D. F., Fletcher, P. D., & Holden, S. H. (2001). Is Your Local Government Plugged. In *Highlights of the 2000 Electronic Government Survey. Baltimore County*. Baltimore, MD: University of Maryland.

Norris, P. (2001). *Digital Divide? Civic Engagement, Information Poverty and the Internet Worldwide*. Cambridge, UK: Cambridge University Press.

Norris, P. (2005). *Developments in Party Communications*. Washington, DC: National Democratic Institute for International Affairs.

Noveck, B. S. (2009). *Wiki Government. How technology can make government better, democracy stronger, and citizen more powerful*. Washington, DC: Brookings Institution Press.

NSSG (National Statistical Service of Greece). (n.d.). *2001Census*. Retrieved May 2009 from http://www.statistics.gr/gr_tables/S1101_SAP_1_TB_DC_01_03_Y.pdf PDF

Nyström, J. (2003). Planeringens grunder. En översikt (2nd Ed.). Lund, Sverige: Studentlitteratur.

O'Connel, P. L. (2005). Korea's High-Tech Utopia, Where Everything is Observed. *The New York Times*.

Obermeyer, N., & Pinto, J. (2007). *Managing Geographic Information Systems* (2nd ed.). New York: Guilford Press.

ODPM (Office of the Deputy Prime Minister). (2004). *Planning Website Survey 2004: Survey of Planning Websites in England and Wales*. Retrieved from http://www.pendleton-assoc.com/planningsurvey2004.html

OECD. (1997). *Webcasting and Convergence: Policy Implications*. Paris: OECD. Retrieved from www.oecd.org/dsti/sti/it/cm/prod/e_97221.htm

OECD. (2001). *Citizens as partners: Information, Consultation and Public Participation in Policy-Making*. Paris, France: OECD.

OECD. (2001). *Engaging Citizens in Policy-making: Information, Consultation and Public Participation. PUMA Policy Brief, 10*. Paris, France: OECD.

OECD. (2002). *Public Governance and Management. Definitions and Concepts: E-government*. Retrieved December 3, 2008 from http://www.oecd.org/EN/aboutfurther_page/0,EN-about_further_page-300-nodirectorate-no-no--11-no-no-1,FF.html

OECD. (2003). *Promise and Problems of E-Democracy: Challenges of Online Citizen Engagement*. Paris, France: OECD.

OECD. (2005). *Evaluating Public Participation in Policy Making*. Paris, France: OECD.

OECD. (2007). *Working Party on the Information Economy. Participative Web: User-created Content*. Paris, France: OECD.

OECD. (2008). Information Technology Outlook 2008. Paris: Organisation for Economic Co-operation and Development.

Office of Management and Budget (OMB). (2007). *FY 2006 Report to Congress on Implementation of The E-Government Act of 2002*. Washington, DC: Office of Management and Budget.

Office of Management and Budget. (2000, November 28). *Management of federal information resources. OM Circular A-130*. Retrieved December 15, 2005, from http://www.whitehouse.gov/omb/circulars/a130/a130trans4.html

Oh, K. (1994). A perceptual evaluation of computer-based landscape simulations. *Landscape and Urban Planning, 28*, 201–216. doi:10.1016/0169-2046(94)90008-6

Onsrud, H. J., & Craglia, M. (2003). Introduction to the Special issues on Access and participatory Approaches in Using Geographic Information. *URISA Journal, 15*, 5–7.

Orlikowski, W., & Hofman, D. (1997). An Improvisational Model of Change Management: The Case of Groupware Technologies. *Sloan Management Review*, 11–21.

Osimo, D. (2008). *Web 2.0 in Government. Why and How. JRC Scientific and Technical Reports*. Brussels, Belgium: Joint Research Centres European Commission.

Owen, G. (1995). *Game theory*. New York: Academic Press.

Oxlenhandler, D. (2003). *Designing a secure local area network*. SANS Institute.

Paar, P., & Rekittke, J. (2005). Lenné3D - Walk-through Visualization of Planned Landscapes. *Visualization in landscape and environmental planning*, 152-162.

Paivio, A. (1986). *Mental representations: a dual coding approach*. Oxford, UK: Oxford University Press.

Pamuk, A. (2006). *Mapping global cities: GIS methods in urban analysis*. Redlands, CA: ESRI Press.

Panda, B., & Giordano, J. (1999). Reconstructing the database after electronic attacks. In Jajodia, S. (Ed.), *Database security XII: State and prospects* (pp. 143–156). Boston, MA: Kluwer Academic Publishers.

Parker, S., & Heapy, J. (2006). *The Journey to the Interface. How public service design can connect users to reform*. London: Demos.

Parviainen, J. (2006). Kollektiivinen tiedonrakentaminen asiantuntijatyössä. In J. Parviainen (Ed.), Kollektiivinen asiantuntijuus [Collective expertise] (pp. 155-187). Tampere, Finland: Tampereen yliopistopaino.

Pastore, M. (2003). Infrastructure and connectivity. In Security + study guide. Boston: Sybex.

Pateman, C. (1970). *Participation and Democratic Theory*. Cambridge, UK: Cambridge University Press.

Paul, S. (2007). A case study of E-governance initiatives in India. *The International Information & Library Review, 39*, 176–184. doi:10.1016/j.iilr.2007.06.003

Peart, M. N., & Ramos Diaz, J. (2007). *Comparative Project on Local e-Democracy. Initiatives in Europe and North America*. Geneva, Switzerland: University of Geneva. Retrieved December 11, 2008, from http://edc.unige.ch/edcadmin/images/ESF%20-%20Local%20E-Democracy.pdf

Peet, R. (1998). *Modern geographical thought*. Oxford, UK: Blackwell.

Peng, Z., & Tsou, M. (2003). *Internet GIS: distributed GIS services for the internet and wireless networks*. New York: John Wiley and Sons.

Perry, A. (2001, June 4). Getting Out the Message. *Time Magazine, 157*(22). Retrieved August 29, 2002, from http://www.time.com/time/interactive/politics/changing_np.html

Perry, C. (2000). Neighbourhood Unit: From the Regional Survey of New York and its Environs: *Vol. VII. Neighbourhood and Community Planning*. London: Routledge. (Original work published 1929)

Pessala, H. (2008). Sähköisiä kohtaamisia: Suomalaisten yhteiskunnallinen osallistuminen internetissä Electrical meetings [Societal participation in the internet by Finns]. Helsinki: Helsingin yliopisto, Viestinnän tutkimuskeskus CRC.

Peterson, J. A. (2003). *The birth of city planning in the United States, 1840-1917*. London: The John Hopkins University Press.

Pew Internet & American Life Project (Pew). (2009). *Demographics of Internet Users*. Retrieved from http://pewinternet.org/Data-Tools/Download-Data/Trend-Data.aspx

Pfeifer, N. (2002). 3D Terrain Models on the Basis of a Triangulation. In Geowissenschaftliche Mitteilungen (Vol. 65).

Piaget, J. (2003). *Meine Theorie der geistigen Entwicklung* (Fatke, R., Ed.). Berlin: Beltz Verlag.

Pick, J. (2005). Costs and benefits of GIS in Business. In Pick, J. (Ed.), *Geographic Information Systems in Business*. Hershey, PA: IGI Publishing.

Pickering, A. (1995). *The Mangle of Practice: Time, Agency and Science*. Chicago, IL: Chicago University Press.

Pickles, J. (1995). *Ground truth: the social implications of geographical information systems*. New York: Guilford Press.

PICT (Planning Inclusion of Clients Through E-training). (2006) *A Guide to Good Practice: People, Planners and Participation. Can ICT help?* Retrieved July 2009 from http://vr.arch.uth.gr/pict-dvd/PDF/wp8_guide/GUIDE_ENGLISH.pdf

Pierson, J., Mante-Meijer, E., Loos, E., & Sapio, B. (Eds.). (2008). Innovating for and by users. COST Office: COST Action 298.

Piracha, A. L., & Kammeier, D. H. (2002). Planning-support systems using an innovative blend of computer tools. *IDPR, 24*(2), 203–221.

Plewe, B. (1997). *GIS On-Line: Information, retrieval, mapping and the Internet*. Santa Fe, NM: On Word Press.

Postman, N. (1993). *Technopoly: The Surrender of Culture to Technology*. New York: Vintage Books.

Potapchuk, W. R. (1996). Building sustainable community politics: synergizing participatory, institutional, and representative democracy. *National Civic Review, 85*(3), 54–60. doi:10.1002/ncr.4100850311

Pratchett, L. (1998). Technological Bias in the Information Age. ICT Policy Making in Local Government. In Snellen, I., & van de Donk, W. (Eds.), *Public Information in an Information Age*. Amsterdam: IOS Press.

Pratchett, L. (2006). Understanding e-democracy developments in Europe. Scoping Paper. In *Ad hoc Committee on e-democracy (CAHDE)*, Strasbourg, 18-19 September 2006. Retrieved December 12, 2008, from http://www.coe.int/t/e/integrated_projects/democracy/02-activities/002_e-democracy/CAHDE(2006)E_Scopingpapers.asp

Pratchett, L., et al. (2005). Barriers to e-Democracy. *Local e-Democracy National Project*. Retrieved May 27, 2005, from http://www.e-democracy.gov.uk

Pratchett, L., et al. (2009). Empowering Communities to Influence Local Decisoin_Making. A Systematic Review of the Evidence. Department for Communiuties and Local Government, United Kingdom, June 2009. Retrieved March 12, 2010, from http://www.communities.gov.uk/publications/localgovernment/localdecisionlessons

Prensky, M. (2001, October). Digital Natives, Digital Immigrants. *Horizon, 9*(5).

Prusinkiewicz, P., & Lindenmayer, A. (1990). *The Algorithmic Beauty of Plants*. New York: Springer Verlag.

Ramasubramanian, L. (1999). GIS Implementation in Developing Countries: Learning from Organisational Theory and Reflective Practice. *Transactions in GIS, 3*(4), 359–380. doi:10.1111/1467-9671.00028

Randall, D., Hughes, J., & Shapiro, D. (1994). Steps towards a partnership: Ethnography and system design. In Jirotka, M., & Goguen, J. (Eds.), *Requirements Engineering: Social and Technical Issues* (pp. 241–258). London: Academic Press.

Randolph, J. (2004). *Environmental land-use planning and management*. Washington, DC: Island Press.

Rantanen, H. & Kahila, M. (2008). The SoftGIS approach to local knowledge. *Journal of Environmental Management, 90*(2009), 1981-1990.

Rantanen, H. (2004). *Paikallisyhteisöt internetissä. Julkaisujärjestelmät ja kolmas sektori. Sitran raportteja 44*. Helsinki, Finland: Edita Prima Oy.

Rantanen, H., & Nummi, P. (2009). Alueella on tietoa. In A. Staffans & E. Väyrynen (Eds.), Oppiva kaupunkisuunnittelu [Learning-based urban planning] (pp. 29-78). Helsinki University of Technology, Department of Architecture, Research Publication 2009/98.

Reddel, T., & Woolcock, G. (2004). From consultation to participatory governance? A critical review of citizen engagement strategies in Queensland. *Australian Journal of Public Administration, 63*(3), 75–87. doi:10.1111/j.1467-8500.2004.00392.x

Reeve, D., Thommason, E., Scott, S., & Simpson, L. (2002). Engaging Citizens: The Bradford Community Statistics Project. In S. Wise, Y. Kim, & C. Openshaw (Eds.), *GISRUK: GIS Research UK 10th Annual Conference*, 3rd-5th April 2002, University of Sheffield, Sheffield, UK (pp. 49-51).

Reichhart, T., Arnberger, A., & Muhar, A. (2007). A comparison of still images and 3-D animations for assessing social trail use conditions. *Forest Snow and Landscape Research, 81*, 77–88.

Repetti, A., & Desthieux, G. (2006). A relational indicatorset model for urban land-use planning and management: methodological approach and application in two case studies. *Landscape and Urban Planning, 77*, 196–215. doi:10.1016/j.landurbplan.2005.02.006

Repetti, A., & Prélaz-Droux, R. (2003). An urban monitor as support for a participative management of developing cities. *Habitat International, 27*, 653–667. doi:10.1016/S0197-3975(03)00010-9

Repetti, A., Soutter, M., & Musy, A. (2006). Introducing SMURF: a software system for monitoring urban functionalities. *Computers, Environment and Urban Systems, 30*, 686–707. doi:10.1016/j.compenvurbsys.2005.06.001

Rettie, R. (2008). Mobile Phones as Network Capital: Facilitating Connections. *Mobilities, 3*(2), 291–311. doi:10.1080/17450100802095346

Reuters. (2008). *Indonesia's Papua plans to tag AIDS sufferers*. Retrieved December 1, 2008 from http://blog.t1production.com/indonesias-papua-plans-to-tag-aids-sufferers

RFID Gazette (2004). *The Future is Here: A Beginner's Guide to RFID*. Retrieved November 2, 2008, from http://www.rfidgazette.org/2004/06/rfid_101.html

RFID Soup (2008). *Form Factor*. Retrieved December 1, 2008 from http://rfidsoup.pbwiki.com/Form+Factor

RFID4U. (2006). *RFID Workshop Student Handbook*. Irvine, CA: RFID4U.

Richmond, B. (1993). Systems thinking: critical thinking skills for the 1990s and beyond. *System Dynamics Review, 9*(2), 113–133. doi:10.1002/sdr.4260090203

Richmond, B. (1994). System dynamics/systems thinking: let's just get on with it. In *International Systems Dynamics Conference*, Sterling, Scotland, 1994.

Richter, S., & Bloem, G. (2008). Ein Plan soll in die Region. *Standort - Zeitschrift für angewandte Geographie, 32*(3), 104-107.

Roberti, M. (2006). *Ending the Hunt for Parking*. RFID Journal.

Robinson, A., & Petchenik, B. (1976). *The Nature of Maps*. Chicago: The University of Chicago Press.

Rogers, R., & Power, A. (2000). *Cities for a Small Country*. London: Faber and Faber.

Rohrmann, B., & Bishop, I. (2002). Subjective responses to computer simulations of urban environments. *Journal of Environmental Psychology, 22*, 319–331. doi:10.1006/jevp.2001.0206

Roininen, J. (2009). *Alue- ja yhdyskuntasuunnittelun arvioinnin fragmentoitunut luonne ja eheyttäminen* [Fragmentation and defragmentation of evaluation in regional and urban planning]. Unpublished doctoral dissertation. Espoo, Finland: Teknillinen korkeakoulu.

Ron, S. (2005). *Cognition and multi-agent interaction: From cognitive modeling to social simulation*. New York: Cambridge University Press.

Rose, J., Lippa, B., & Rios, J. (2009). (accepted). Technology Support for Participatory Budgeting. *International Journal of Electronic Government*.

Rowe, D., McGibbon, S., & Bell, O. (2005). Shared Source and Open Solutions for e-Government. In Cunningham, P., & Cunningham, M. (Eds.), *Innovation and the Knowledge Economy: Issues, Applications, Case Studies* (Vol. 2, pp. 375–381). Amsterdam: IOS press.

Rowe, G., & Frewer, L. G. (2000). Public Participation Methods: A Framework for Evaluation. *Journal of Science. Technology & Human Values, 25*(1), 3. doi:10.1177/016224390002500101

Rowe, G., & Wright, G. (2001). Expert opinions in forecasting. Role of the Delphi technique. In Armstrong, J. S. (Ed.), *Principles of forecasting: A handbook of researchers and practitioners* (pp. 125–144). Boston, MA: Kluwer Academic Publishers.

Rowe, N. C. (2007). Cyber Attacks. In Anttiroiko, A.-V., & Malkia, M. (Eds.), *Encyclopedia of Digital Government* (pp. 271–276). Hershey, PA: Idea Group Publishing.

Royal Town Planning Institute (RTPI). (2007). Guidelines on Effective Community Involvement and Consultation. *RTPI Good Practice Note 1*. Retrieved December 11, 2008, from http://www.rtpi.org.uk/download/364/RTPI-GPN1-Consultation-v1-2006.pdf

Rubinstein-Montano, B. (2003). Virtual communities as role models for organizational knowledge management. In Gupta, J. N. D., & Sushil, K. S. (Eds.), *Creating Knowledge Based Organizations*. Hershey, PA: Idea Group Publishing.

Ruth, A., & Hudson, K., & Microsoft Corporation. (2003). Network Infrastructure Security. In *Security + certification training kit*. Redmond, WA: Microsoft Press.

Saad-Sulonen, J. (2005). *Mediaattori – Urban Mediator: a hybrid infrastructure for neighborhoods*. Master of Arts Thesis in New Media, University of Art and Design, Helsinki. Retrieved January 15, 2009 from http://www2.uiah.fi/~jsaadsu/thesis.html

Sachs, I. (1997). *L'écodéveloppement: stratégie pour le XXie siècle*. Paris: Syros.

Sacramento City. (2002). *Sacsites: Sacramento's Business and Development Resource*. Retrieved on May 2, 2002, from http://www.maps.cityofsacramento.org/website/sacramentoed/ed.htm

Salingaros, N. A. (1998). Theory of the urban web. *Journal of Urban Design, 3*, 53–71. doi:10.1080/13574809808724416

Samat, N. (2006). *Applications of geographic information systems in urban land use planning in Malaysia*. Paper presented at the 4th Taipei International Conference on Digital Earth.

Sandercock, L. (1998). *Towards Cosmopolis*. New York: John Wiley & Sons.

Sandercock, L. (2003). Out of the closet: The importance of stories and storytelling in planning practice. *Planning Theory & Practice, 4*(1), 11–28. doi:10.1080/1464935032000057209

Santandreu, A. (2001). Rapid visual diagnosis, a rapid, low cost, participatory methodology applied in Montevideo. *Urban Agriculture Magazine, 5*, 13–14.

Sapient. (2000). *Smart Places: Collaborative GIS Approach*. Retrieved on May 2, 2002, from http://www.saptek.com/smart

Sargent, F. O., Lusk, P., Rivera, J. A., & Varela, M. (1993). *Rural Environmental Planning for Sustainable Communities*. Washington, DC: Island Press.

Sarjakoski, T. (1998). Networked GIS for Public Participation – Emphasis on Utilizing Image Data. *Computers, Environment and Urban Systems, 22*(4), 381–392. doi:10.1016/S0198-9715(98)00031-3

Sassen, S. (2000). *Cities in a World Economy*. Thousand Oaks, CA: Pine Forge/Sage Press.

Sassen, S. (2001). *The Global city*. Princeton, NJ: Princeton University Press.

Scardamalia, M., & Bereiter, C. (2003). Knowledge Building. In Guthrie, J. W. (Ed.), *Encyclopedia of Education* (2nd ed., pp. 1370–1373). New York: Macmillan Reference.

Schienstock, G., & Hämäläinen, T. (2001). *Transformation of the Finnish innovation system. A network approach. Sitra Reports series 7*. Helsinki: Hakapaino.

Schmidt, S. (1986). Pionjärer, efterföljare och avvaktare. Lund, Sverige: Kommunfakta Förlag.

Schuler, D., & Namioka, A. (Eds.). (1993). *Participatory Design: Principles and Practices*. Hillsdale, NJ: Lawrence Erlbaum Associates, Inc.

Schultz, R. A. (2006). *Contemporary Issues in Ethics and Information Technology*. Hershey, PA: Information Science Publishing.

Schulze-Wolf, T., & Habekost, T. (2008). E-Partizipation in der Raumordnung. *Standort - Zeitschrift für angewandte Geographie, 32*(3), 97-103.

Schumpeter, J. A. (1950). *Capitalism, Socialism and Democracy*. London: Allen and Unwin.

Scott, A. J. (2000). *Regions and World Economy. The Coming Shape of Global Production, Competition and Political Order*. New York: Oxford University Press.

Scott, A. J., & Roweis, S. T. (1977). Urban planning in theory and practice: a reappraisal. *Environment & Planning A, 9*(10), 1097–1119. doi:10.1068/a091097

Scott, D., & Oelofse, C. (2005). Social and Environmental Justice in South African Cities: Including Invisible Stakeholders in Environmental Assessment Procedures. *Journal of Environmental Planning and Management, 48*(3), 445–467. doi:10.1080/09640560500067582

Sehested, K. (2001). *Investigating Urban Governance - From the Perspective of Policy Networks, Democracy and Planning*. (Research Paper no. 1/01). Roskilde, Sverige. The Department of Social Sciences. Roskilde University.

Seliem, A., Ashour, A., Khalil, O., & Millar, S. (2003). The Relationship of Some Organizational Factors to Information Systems Effectiveness: A Contingency Analysis of Egyptian Data. *Journal of Global Information Management, 11*(1).

Sengers, P., Boehner, K., David, S., & Kaye, J. (2005). Reflective design. In Proc. 4th Decennial Conference on Critical Computing: Between Sense and Sensibility (Aarhus, Denmark, Aug 2005) ACM Press, 49-58.

Sester, M., Bernard, L., & Paelke, V. (2009). *Advances in Giscience*. Berlin: Springer.

Shakun, M. (Ed.). (1996). *Negotiation processes: modeling frameworks and information technology*. Boston, MA: Kluwer.

Sharma, S. K. (2004). Assessing E-government Implementations. *Electronic Government Journal, 1*(2), 198–212. doi:10.1504/EG.2004.005178

Sharma, S. K. (2006). An E-Government Services Framework. In Khosrow-Pour, M. (Ed.), *Encyclopedia of Commerce, E-Government and Mobile Commerce* (pp. 373–378). Hershey, PA: Information Resources Management Association.

Sharma, S. K., & Gupta, J. N. D. (2003). Building Blocks of an E-government – A Framework. *Journal of Electronic Commerce in Organizations, 1*(4), 34–48.

Shiffer, M. J. (1992). Towards a collaborative planning system. *Environment and Planning B, 19*(6), 709–722. doi:10.1068/b190709

Shiode, N. (2000). Urban Planning, Information Technology, and Cyberspace. *Journal of Urban Technology, 7*(2), 105–126. doi:10.1080/713684111

Shulman, S., Schlosberg, D., Zavestovski, S., & Courard-Hauri, D. (2003). Electronic Rulemaking: A Public Participation Research Agenda for the Social Sciences. *Social Science Computer Review, 21*(2), 162–178. doi:10.1177/0894439303021002003

Siebenhüner, B., & Barth, V. (2004). The Role of Computer Modelling in Participatory Integrated Assessment. *Environmental Impact Assessment Review, 25*, 367–389. doi:10.1016/j.eiar.2004.10.002

Sieber, R. (2006). Public Participation Geographic Information Systems: A Literature Review and Framework. *Annals of the American Association of Geographers, 96*(3), 491–507. doi:10.1111/j.1467-8306.2006.00702.x

Sieber, R. (2007). Spatial Data Access by the Grassroots. *Cartography and Geographic Information Science, 34*(1), 47–62. doi:10.1559/152304007780279087

Siemens, G. (2006). *Knowing knowledge*. Retrieved June1, 2008 from http://www.knowingknowledge.com

Siitonen, P. (1995). Future of Endoscopy. In *The Future of Endoscopy – Proceedings of the 2nd European Architectural Endoscopy Association Conference*, Technische Universität Wien, ISIS-ISIS Publications.

Silcock, R. (2001). What is E-government? *Parliamentary Affairs, 54*, 88–101. doi:10.1093/pa/54.1.88

Silva, C. N. (1994). *Política Urbana em Lisboa, 1926-1974*. Lisbon: Livros Horizonte.

Silva, C. N. (2003). Urban utopias in the twentieth century. *Journal of Urban History, 29*(3), 327–332. doi:10.1177/0096144203029003011

Silva, C. N. (2005). Charter of Athens. In Caves, R. W. (Ed.), *Encyclopedia of the City* (pp. 52–53). London: Routledge.

Silva, C. N. (2005). City Beautiful. In Caves, R. W. (Ed.), *Encyclopedia of the City* (pp. 69–70). London: Routledge.

Silva, C. N. (2005). New Charter of Athens. In Caves, R. W. (Ed.), *Encyclopedia of the City* (pp. 328–329). London: Routledge.

Silva, C. N. (2007). e-Planning. In A.-V. Anttiroiko & M. Malkia (Eds.), Encyclopedia of Digital Government (pp. 703-707). Hershey, PA: Idea Group Publishing.

Silva, C. N. (2007). Urban Planning and Ethics. In Rabin, J., & Berman, E. M. (Eds.), *Encyclopedia of Public Administration and Public Policy* (2nd ed.). New York: CRC Press / Taylor & Francis Group. doi:10.1201/NOE1420052756.ch410

Silva, C. N. (2008). Research Ethics in e-Public Administration. In Garson, G. D., & Khosrow-Pour, M. (Eds.), *Handbook of Research on Public Information Technology* (Vol. 1, pp. 314–322). New York: Information Science Reference.

Silva, C. N., & Syrett, S. (2006). Governing Lisbon: Evolving forms of city governance. *International Journal of Urban and Regional Research, 30*(1), 98–119. doi:10.1111/j.1468-2427.2006.00646.x

Simon, H. (1973). Applying information technology to organizational design. *Public Administration, 33*, 269–270.

Singapore Department of Statistics. (2001). *Singapore Census of Population 2000: Statistical Release 2—Education, Language and Religion*. Singapore: Department of Statistics.

SmarTrip. (2008). SmarTrip More than a smart card. It's pure genius. *Washington Metropolitan Area Transit*

Authority. Retrieved December 1, 2008 from http://www.wmata.com/riding/SmarTrip.cfm

Smith, A., & Rainie, L. (2008). *The Internet and the 2008 election*. Report from the Pew Internet & American Life Project. Retrieved from http://www.pewinternet.org/Reports/2008/The-Internet-and-the-2008-Election.aspx

Snellen, I. (2001). ICTs, Bureaucracies, and the Future of Democracy. *Communications of the ACM, 44*(1), 45–48. doi:10.1145/357489.357504

Song, Y., Bogdahn, J., Hamilton, A., & Wang, H. (2009). Integrating BIM with Urban Spatial Applications: A VEPS perspective. In *Handbook of Research on Building Information Modeling and Construction Informatics: Concepts and Technologies*. Hershey, PA: IGI Global.

SOU. (2000). *En uthållig demokrati. Politik för folkstyrelse på 2000-talet. Demokratiutredningens betänkande (State Commission Report)*. Stockholm: Fritzes.

SOU. (2001). *Att vara med på riktigt - demokratiutveckling i kommuner och landsting. Bilagor till betänkande av kommundemokratikommittén (State Commission Report)*. Stockholm: Fritzes.

South Yorkshire Partnership. (2006). *Progress in South Yorkshire, Sheffield*. Sheffield, UK: Sheffield Regional Development.

Southern Tier. (2002). *Southern Tier Regional Planning and Development Board: Community GIS*. Retrieved on May 2, 2002, from http://www.southerntierwest.org/st/cgis/html/locgovgis1.htm

Soutter, M., & Repetti, A. (2009). Land Management with the SMURF Planning Support System. In Geertman, S., & Stillwell, J. (Eds.), *Planning support systems best practices and new methods* (pp. 369–388). London: Springer. doi:10.1007/978-1-4020-8952-7_18

Spiess, E., Baumgartner, U., Arn, S. & Vez, C. (2002). Topographsiche Karten, Kartengraphik und Generalisierung. *Schweizerische Gesellschaft für Kartographie, Kartographische Publikationsreihe, 16*.

Sprinkel, D. A., & Brown, K. D. (2008). Using Digital Technology in the Field. *UTA Geological Survey. Survey Notes, 40*(1), 1–2.

Staffans, A. (2004). Vaikuttavat asukkaat. Vuorovaikutus ja paikallinen tieto kaupunkisuunnittelun haasteina [Influencial inhabitants. Local knowledge and interaction challenging urban planning, in Finnish]. Helsinki University of Technology, Centre for Urban and Regional Studies, Publication A 29. Helsinki, Finland: Yliopistopaino.

Staffans, A., & Väyrynen, E. (Eds.). (2009). Oppiva kaupunkisuunnittelu (Learning-based Urban Planning, in Finnish). Helsinki University of Technology, Department of Architecture, Research Publication 2009/98. Espoo, Finland: Painotalo Casper Oy.

Statistik Austria. (2008). *Haushalte mit Computer, Internetzugang und Breitbandverbindung 2004-2008*. Retrieved December 3, 2008, from http://www.statistik.at/web_de/static/ergebnisse_im_ueberblick_haushalte_mit_computer_internetzugang_und_breitba_022206.pdf

Steinmann, R., Kerk, A., & Blaschke, T. (2004). Analysis of Online Public Participatory GIS Applications with Respect to the Difference between the US and Europe. In *UDMS - Urban Data Management Symposium*, Chioggia, Italy, 2004. Retrieved from http://www.salzburgresearch.at/research

Stewart, J. (2007). Local Experts in the Domestication of Information and Communication Technologies. *Communication. The Information Society, 10*(4), 547–569. doi:10.1080/13691180701560093

Stiftung Mitarbeit. (2007). *E-Partizipation*. Bonn, Deutschland: Beteiligungsprojekte im Internet.

Still, K. (2008). *Jamarat Bridge – Accidents*. Retrieved July 20, 2009, from http://www.crowddynamics.com/Disasters/jamarat_bridge.htm

Storper, M. (1997). *The Regional World: Territorial Development in a Global Economy*. New York: The Guilford Press.

Strauss, W.-C. (2006). *Öffentlichkeits- und Trägerbeteiligung in der Bauleitplanung im und über das Internet*.

Erste Erfahrungen aus den Kommunen. Fachtagung Bauleitplanung und Internet. Berlin, Germany: Deutsches Institut für Urbanistik.

Strömgren, A. (2007). Samordning, hyfs och reda. Stabilitet och förändring i svensk planpolitik 1945-2005. Uppsala, Sverige: Acta Universitatis Upsaliensis.

Strothotte, T., & Schlechtweg, S. (2002). *Non-Photorealistic Computer Graphics: Modeling, Rendering and Animation*. San Francisco: Morgan Kaufman.

Sturges, P., & Neill, R. (1998). *The quiet struggle: information and libraries for people of Africa*. London: Mansell.

Suchman, L. (1987). *Plans and Situated Actions. The Problem of Human-machine Communication*. New York: Cambridge University Press.

Suchman, L. (2000). Located Accountabilities in Technological Production. Retrieved October 31, 2008, from http://www.comp.lancs.ac.uk/ sociology/soc039ls.html

Suchman, L. (2002). Practice-based design of information systems: Notes from the hyperdeveloped World. *The Information Society, 18*, 1–6. doi:10.1080/01972240290075066

Sudhira, H. S., & Ramachandra, T. V. (2009). A spatial planning support system for managing Bangalore's urban sprawl. In Geertman, S., & Stillwell, J. (Eds.), *Planning support systems best practices and new methods* (pp. 175–190). London: Springer. doi:10.1007/978-1-4020-8952-7_9

Susskind, L., & Cruikshank, J. (1987). *Breaking the impasse: Consensual approaches to resolving public disputes*. New York: Basic Books.

Swedberg, C. (2006). *Solar-Powered RFID Reader Measures Road Traffic*. RFID Journal.

Tabak, V., de Vries, B., & Dijkstra, J. (2008). *RFID Technology Applied for Validation of an Office Simulation Model*. Eindhoven University of Technology Faculty of Architecture, Building, and Planning.

Tahvanainen, L., Tyrväinen, L., Ihalainen, M., Vuorela, N., & Kolehmainen, O. (2001). Forest management and public perceptions – visual versus verbal information. *Landscape and Urban Planning, 53*, 53–70. doi:10.1016/S0169-2046(00)00137-7

Talen, E. (2000). Bottom-up GIS: A new a tool for individual and group expression in participatory planning. *Journal of the American Planning Association. American Planning Association, 66*(3), 491–807.

Talen, E. (2005). *New Urbanism and American Planning: the conflict of cultures*. London: Routledge.

Talen, E., & Ellis, C. (2002). Beyond Relativism. Reclaiming the search for good city form. *Journal of Planning Education and Research, 22*(1), 36–49. doi:10.1177/0739456X0202200104

Talen, E., & Shah, S. (2007). Neighbourhood Evaluation Using GIS: An Exploratory Study. *Environment and Behavior, 39*(5), 583–615. doi:10.1177/0013916506292332

Talvitie, J. (2001). *Incorporating the Impact of ICT into Urban and Regional Planning* (p. 10). Stockholm: European Journal of Spatial Development.

Talvitie, J. (2003). *The Impact of Information and Communication Technology on Urban and Regional Planning*. Report for Helsinki University of Technology, Department of Surveying.

Tambini, D. (1999). New media and democracy: The civic networking movement. *New Media & Society, 10*, 305–329. doi:10.1177/14614449922225609

Tan, S.-C., & Tan, A.-L. (2006). Conversational analysis as an analytical tool for face-to-face and online conversations. *Educational Media International, 43*(4), 347–361. doi:10.1080/09523980600926374

Tan, T. S. (2005). Electronic local authority management system. *Planning Malaysia, 3*, 27–46.

Tassabehji, R. (2005). Information security threats. In Pagani, M. (Ed.), *Encyclopedia of multimedia technology and networking* (pp. 404–410). Hershey, PA: Idea Group.

Taylor, N. (1998). *Urban Planning Theory since 1945*. London: Sage Publications.

Technovelgy. (2008). *Chipless RFID Tag*. Retrieved December 1, 2008 from the Technovelgy website:

http://www.technovelgy.com/ct/Technology-Article.asp?ArtNum=28

Technovelgy. (2008). *RFID Reader*. Retrieved December 1, 2008 from the Technovelgy web site: http://www.technovelgy.com/ct/Technology-Article.asp?ArtNum=54

Tepe, I. (2005). *Le lotissement à la périphérie de Thiès*. Dakar, Senegal: Ecocité.

Thaler, R. H. & Sunstein, C., R. (2008): Nudge. Improving Decisions about Health, Wealth, and Happiness. Yale University Press 2008

The Planning Service. (2004). *Planning Service Website*. Retrieved from http://www.planningni.gov.uk/

Thesaurus. (2009). *Governance*. Retrieved April 7, 2009 from http://www.answers.com/governance&r=67#Thesaurus

Thomas, J. C., & Streib, G. (2003). The new face of government: citizen-initiated contact in the era of e-government. *Journal of Public Administration: Research and Theory, 13*(1), 83–102. doi:10.1093/jpart/mug010

Tiamiyu, M. A., & Ogunsola, K. (2008). Preparing for E-Government: some findings and lessons from government agencies in Oyo State, Nigeria. *South African Journal of Library & Information Science, 74*(1), 58–72.

Tlauka, M., Brolese, A., Pomeroy, D., & Hobbs, W. (2005). Gender differences in spatial knowledge acquired through simulated exploration of a virtual shopping centre. *Journal of Environmental Psychology, 25*, 111–118. doi:10.1016/j.jenvp.2004.12.002

Tolbert, C. J., & Mossberger, K. (2006). The Effects of E-Government on Trust and Confidence in Government. *Public Administration Review, 66*(3), 354–369. doi:10.1111/j.1540-6210.2006.00594.x

Tolbert, C., Mossberger, K., & McNeal, R. (2008). Innovation and Learning: Measuring E-government Performance in the American States 2000-2004. *Public Administration Review, 68*(3), 549–563. doi:10.1111/j.1540-6210.2008.00890.x

Torres, L., Pina, V., & Acerete, B. (2006). E-Governance Developments in EU Cities, Reshaping Government Relation to Citizens. *Governancy, 19*(2), 687–699.

Townsend, A. (2009). Foreword. In Foth, M. (Ed.), *Handbook of Research on Urban Informatics: The Practice and Promise of the Real-Time City* (pp. xxii–xxvi). Hershey, PA: IGI Global.

Transportation Alternatives. (2008). *Congestion Pricing*. Retrieved December 1, 2008, from http://www.transalt.org/campaigns/congestion

Tripathy, G. (2002). *Web-GIS Based Urban Planning and Information System for Municipal Corporations – A Distributed and Real-Time System for Public Utility and Town*. Retrieved on May 2, 2002, from http://www.gisdevelopment.net/application/urban/overview/urbano0028pf.htm

Tuomi, I. (1999). *Corporate knowledge. Theory and Practice of Intelligent Organizations*. Helsinki, Finland: Metaxis.

Tura, T., & Harmaakorpi, V. (2003). Social Capital in Building Regional Innovative Capability: A Theoretical and Conceptual Assessment. Conference Report. In *43rd Congress of the European Regional Science Association (ERSA)*, Jyväskylä, Finland, 27–30 August 2003.

Tura, T., & Harmaakorpi, V. (2005). Measuring Regional Innovative Capability. In *Conference Report, 45th Congress of the European Regional Science Association (ERSA)*, Amsterdam, The Netherlands, 27–30 August 2005.

Turner, J. F. C. (1976). *Housing by people*. New York: Pantheon Books.

Tversky, B., & Hard, B. M. (2009). Embodied and disembodied cognition: Spatial perspective-taking. *Cognition, 110*, 124–129. doi:10.1016/j.cognition.2008.10.008

UGisP. (2006). GIS9: Pembangunan Pengkalan Data Sistem Maklumat Geografi Negeri Sembilan Darul Khusus - Laporan Akhir Fasa III.

UN E-Government Readiness Survey. (2008). Retrieved November 30, 2008, from http://unpan1.un.org/intradoc/groups/public/documents/UN/UNPAN028607.pdf

UN Global E-government Readiness Report. (2005). *From E-government to E-inclusion, UNPAN/2005/14*. New York: United Nations.

UN. (1989). *Convention on the Rights of the Child*. New York: United Nations.

UN. (1998). *Convention on access to information, public participation in decision-making and access to justice in environmental matters (Aarhus Convention)*. New York: United Nations.

UNECE Economic Commission for Europe. (2005). *Synthesis Report on the Status of Implementation of the Convention*. Retrieved December 11, 2008, from http://www.unece.org/env/pp/reports%20implementation.htm

UNEP/UNCHS. (2008). *Building an Environmental Management Information System (EMIS). Handbook with Toolkit*. SCP Source Book Series.

UN-Habitat. (2007). *Global report on human settlements 2007: enhancing urban safety and security*. London: Earthscan.

UN-Habitat. (2008). *State of the world cities 2008-2009*. Nairobi, Kenya: UN-Habitat.

United Nations. (2005). *UN Global E-government Readiness Report 2005: From E-government to E-inclusion. Department of Economic and Social Affairs, Division of Public Administration and Development Management*. New York: United Nations.

United Nations. (2008). *United Nations e-government survey 2008: from e-government to connected governance*. New York: United Nations Publication.

United States General Accounting Office. (2001). *Federal information system control audit manual*. GAO/AIMD-12.19.6. Retrieved March 7, 2006, from http://www.gao.gov/special.pubs/ai12.19.6.pdf

United States Government Accountability Office. (2005). *Information Security: Radio Frequency Identification Technology in the Federal Government, Publication No. GAO-05-551*.

Urry, J. (1999). *Sociology Beyond Societies: Mobilities for the Twenty First Century (International Library of Sociology)*. London: Routledge.

Urry, J. (2007). *Mobilities*. Cambridge, UK: Polity Press.

Vallerie, W. A., Park, L. O. B., Hallo, J. C., Stanfield, R. E., & Manning, R. E. (2006). *Enhancing visual research with computer animation: A study of crowding-related standards of quality for the loop road at Acadia National Park*. Retrieved December 4, 2008, from http://www.issrm2006.rem.sfu.ca/abstractdsip_popup.php?id=629

Valovirta, V., & Stern, E. (2006, October). *The evaluation of new policy instruments. Complexity and governance*. Paper presented at the Conference of the European Evaluation Society, London, UK.

Van den Berg, A. E., Vlek, C. A. J., & Coeterier, F. J. (1998). Group differences in the aesthetic evaluation of nature development plans: A multilevel approach. *Journal of Environmental Psychology, 18*, 141–157. doi:10.1006/jevp.1998.0080

Van der Meer, A., & Van Winden, W. (2003). E-governance in cities: A comparison of urban ICT policies. *Regional Studies, 37*(4), 407–419. doi:10.1080/0034340032000074433

Van Dyke Parunak, H., Brueckner, S., Fleischer, M., & Odell, J. (2004). A design taxonomy of multi-agent interactions. *Lecture Notes in Computer Science, 2935*, 123–137.

Van Geenhuizen, M. (2001). ICT and Regional Policy: Experiences in The Netherlands. In Heitor, M. (Ed.), *Innovation and Regional Development*. London: Edward Elgar Publishing.

Van Herzele, A. (2004). Local Knowledge in Action. Valuing Nonprofessional Reasoning in the Planning Process. *Journal of Planning Education and Research, 24*, 197–212. doi:10.1177/0739456X04267723

Van Winden, W. (2003). *Essays on Urban ICT Policies*. Doctoral dissertation, Erasmus Universiteit, Rotterdam.

Vassilakis, C. (2004). Barriers to Electronic Service Delivery. *e-Service Journal, 4*(1), 41-63.

Vegh, S. (2003). Classifying Forms of Online Activism. The case of Cyberprotests against the World Bank. In McCaughey, M., & Ayers, M. D. (Eds.), *Cyberactivism. Online Activism in Theory and Practice*. New York: Routledge.

Velibeyoglu, K. (2004). *Institutional Use of Information Technologies in City Planning Agencies: Implications from Turkish Metropolitan Municipalities*. Unpublished doctoral dissertation, Izmir Institute of Technology, Izmir, Turkey.

Velibeyoglu, K. (2005). Urban Information Systems in Turkish Local Governments. In Marshall, S., Taylor, W., & Yu, X. (Eds.), *Encyclopedia of developing regional communities with information and communication technology* (pp. 709–714). Hershey, PA: Information Science Reference.

Velibeyoglu, K., & Saygin, O. (2005). *Spatial Information systems in Turkish Local Government: implications from recent practices*. Paper presented at the CUPUM 05: Ninth International Conference on Computers in Urban Planning and Urban Management, CASA, London.

Velibeyoglu, K., & Yigitcanlar, T. (2008). Understanding the supply side: ICT experience of Marmara, Turkey. In Yigitcanlar, T., Velibeyoglu, K., & Baum, S. (Eds.), *Creative Urban Regions: Harnessing Urban Technologies to Support Knowledge City Initiatives* (pp. 245–262). Hershey, PA: Information Science Reference.

Veloso, M. M., Patil, R., Rybski, P. E., & Kanade, T. (2004). People detection and tracking in high resolution panoramic video mosaic. In *IEEE/RSJ International Conference. Intelligent robots and systems (IROS 2004)* (Vol. 2, pp.1323-1328).

VEPs. (2006). *VEPs Project Home Page*. Retrieved from http://www.veps3d.org

Vigoda-Gadot, E. (2004). Collaborative Public Administration: Some Lessons from the Israeli Experience. *Managerial Auditing Journal*, *19*(6), 700–711. doi:10.1108/02686900410543831

Vilmin, T. (1999). *L'aménagement urbain en France: une approche systémique*. Paris: CERTU.

Vircavs, I. (2006). Development of e-services in Urban Development Issues in Riga City. In *Proceeding of the 11th International Conference on Urban Planning and Regional Development in the Information Society*, 13 – 16 February 2006, Vienna, Austria.

Von Haldenwang, C. (2004). Electronic Government (E-Government) and Development. *European Journal of Development Research*, *16*(2), 417–432. doi:10.1080/0957881042000220886

von Ranke, F., Puhler, M., Wolf, P., & Krcmar, H. (2005). Software Engineering for e-Government Applications. In Cunningham, P., & Cunningham, M. (Eds.), *Innovation and the Knowledge Economy: Issues, Applications, Case Studies* (Vol. 2, pp. 397–404). Amsterdam: IOS press.

Vonk, G. (2003). *Opportunities for Participatory Planning Support Systems (PPSS) in spatial planning. UNetworks in the Delta: ProjectsU*. The Netherlands: Utrecht University.

Vonk, G., Geertman, S., & Schot, R. (2005). Bottlenecks blocking the widespread usage of planning support systems. *Environment & Planning A*, *37*(5), 909–924. doi:10.1068/a3712

Vorwerk, V., Märker, O., & Wehner, J. (2008). Bürgerbeteiligung am Haushalt. *Standort - Zeitschrift für angewandte Geographie*, *32*(3), 114-119.

Waddell, P., & Borning, A. (2004, February). A case study in digital government: Developing and applying UrbanSim, a system for simulating urban land use, transportation, and environmental impacts. *Social Science Computer Review*, *22*(1), 37–51. doi:10.1177/0894439303259882

Wagenaar, H., & Hajer, M. A. (Eds.). (2003). *Deliberative policy analysis: Governance in the network society*. Cambridge, MA: Cambridge University Press.

Waldon, R. S. (2006). *Planners and Politics: Helping Communities Make Decisions*. Chicago, IL: Planners Press.

Wallin, S., & Horelli, L. (in press). Methodology of a user-sensitive service design within urban planning. *Environment and Planning, B*.

Wang, H., & Hamilton, A. (2009). Extending BIM into Urban Scale by Integrating Building Information with Geospatial Information. In *Handbook of Research on Building Information Modeling and Construction Informatics: Concepts and Technologies*. Hershey, PA: IGI Global.

Ward, S. V. (2004). *Planning and Urban Change*. London: Sage Publications.

Warnecke, L., Beattie, J., Cheryl, K., Lyday, W., & French, S. (1998). *Geographic information technology in cities and counties: A nationwide assessment*. Washington, DC: American Forests.

Warren, T., & Gibson, E. (2002). The influence of referential processing on sentence complexity. *Cognition, 85*, 79–112. doi:10.1016/S0010-0277(02)00087-2

Warschauer, M. (2003). Social capital and access. *Universal Access in the Information Society, 2*, 315–330. doi:10.1007/s10209-002-0040-8

Warschauer, M. (2003). *Technology and Social Inclusion: Rethinking the Digital Divide*. Cambridge, MA: MIT Press.

Wates, N. (2008). *The Community Planning Event Manual: How to Use Collaborative Planning and Urban Design Events to Improve Your Environment*. London: Earthscan.

Wates, N. (Ed.). (2000). *The Community Planning Handbook: How People Can Shape Their Cities, Towns and Villages in Any Part of the World*. London: Earthscan.

Watson, R. T., & Mundy, B. (2001). A Strategic Perspective of Electronic Democracy. *Communications of the ACM, 44*(1), 27–30. doi:10.1145/357489.357499

WCED (World Commission on Environment and Development). (1987). *Our common future*. Oxford, UK: Oxford University Press.

Web Accessibility Initiative (WAI). (n.d.). *WAI-ARIA Overview*. Retrieved from http://www.w3.org/WAI/intro/aria

Weber, L. M., Loumakis, A., & Berman, J. (2003). Who participates and why? An analysis of citizens on the Internet and the mass public. *Social Science Computer Review, 21*(1), 26–42. doi:10.1177/0894439302238969

Weiser, M. (1993). *Ubiquitous Computing*. Washington, DC: IEEE Computer.

Welzel, C., Falk, S., & Müller-Mordhorst, F. (2005). *Standortfaktor Verwaltung: E-Government und Kundenservice in Nordrhein-Westfalen*. Retrieved October 31, 2008, from http://de.sitestat.com/bertelsmann/stiftung-de/s?bst.Suche.nrw_standortfaktor.pdf&ns_type=pdf&ns_url=http://www.bertelsmann-stiftung.de/cps/rde/xbcr/SID-0A000F0A-E87D2A32/bst/nrw_standortfaktor.pdf

Wenger, E., White, N., Smith, D., & Rowe, K. (2005). *Technology for Communities*. CHEFRIO Book Chapter v. 5.2. Retrieved from http://technologyforcommunities.com/CEFRIO_Book_Chapter_v_5.2.pdf

Wessels, B. (2000). Telematics in the East End of London: New Media as a Cultural Form. *New Media & Society, 2*(4), 427–444. doi:10.1177/14614440022225896

Wessels, B. (2007). *Inside the Digital Revolution: policing and changing communication with the public*. Aldershot, UK: Ashgate.

Wessels, B. (2008). Creating a regional agency to foster eInclusion: the case of South Yorkshire, UK. *European Journal of ePractice*, (3), 3-13.

Wessels, B., & Craglia, M. (2007). Situated innovation of e-social science: Integrating infrastructure, collaboration, and knowledge in developing e-social science. *Journal of Computer-Mediated Communication, 12*(2). Retrieved from http://jcmc.indiana.edu/vol12/issue2/wessles.html. doi:10.1111/j.1083-6101.2007.00345.x

Wessels, B., Walsh, S., & Adam, E. (2008). Mediating Voices: Community Participation in the Design of E-Enabled Community Care Services. *The Information Society, 24*(1), 3–39. doi:10.1080/01972240701774683

West, D. (2007). *Global E-Government, 2007*. Providence, RI: Center for Public Policy, Brown University.

West, D. M. (2000). *Assessing E-Government: The Internet, Democracy, and Service Delivery by State and Federal Governments*. Washington, DC: World Bank.

Weyns, D., Omicini, A., & Odell, J. (2007). Environment as a first class abstraction in multiagent systems. *Autonomous Agents and Multi-Agent Systems, 14*, 5–30. doi:10.1007/s10458-006-0012-0

Wheeler, S. (2001). Technology and Planning: A Note of Caution. *Berkeley Planning Journal, 15*, 85–89.

Wiedemann, P. M., & Femers, S. (1993). Public Participation in Waste Management Decision-making. *Journal of Hazardous Materials, 33*(3), 355–368. doi:10.1016/0304-3894(93)85085-S

Wild, A., & Marshall, R. (1999). Participatory practice in the context of local agenda 21: a case study evaluation of experience in three English local authorities. *Sustainable Development, 7*, 151–162. doi:10.1002/(SICI)1099-1719(199908)7:3<151::AID-SD111>3.0.CO;2-O

Wildavsky, A. (1987). *Speaking Truth to Power: The Art and Craft of Policy Analysis*. New Brunswick, CT: Transaction Publishers.

Williams, A., Robles, E., & Dourish, P. (2009). Urbaning the City: Examining and refining the assumptions behind urban informatics. In Foth, M. (Ed.), *Handbook of Research on Urban Informatics: The Practice and Promise of the Real-Time City* (pp. 1–20). Hershey, PA: IGI Global.

Williams, C. (1999). Local economic development. In Allmendinger, P., & Chapman, M. (Eds.), *Planning Beyond 2000* (pp. 176–187). Chichester, UK: John Wiley and Sons.

Williams, K., Forda, R., Bishop, I., Loiterton, D., & Hickey, J. (2007). Realism and selectivity in data-driven visualisations: A process for developing viewer-oriented landscape surrogates. *Landscape and Urban Planning, 81*(3), 213–224. doi:10.1016/j.landurbplan.2006.11.008

Williams, R., Stewart, J., & Slack, R. (2005). *Social Learning in Technological Innovation: Experimenting with Information and Communication Technologies*. Cheltenham, UK: Edward Elgar.

Winarso, H., & Mattingly, M. (1999). *Local participation in Indonesia's urban infrastructure investment programming: sustainability through local government involvement?* Bandung, Indonesia: Bandung Institute of Technology.

Winden, W. V., & Woets, P. (2004). Urban broadband Internet policies in Europe: a critical review. *Urban Studies (Edinburgh, Scotland), 41*(10), 2043–2059. doi:10.1080/0042098042000256378

Wohlers, T. E. (2009). The Digital World of Local Government: A Comparative Analysis of the United States and Germany. *Journal of Information Technology & Politics, 6*, 111–126. doi:10.1080/19331680902821593

Wolcott, P., & Çagiltay, K. (2001). Telecommunications, Liberalization, and the Growth of the Internet in Turkey. *The Information Society, 17*, 133–141. doi:10.1080/019722401750175685

Wooldridge, M. (2002). *An introduction to multi-agent systems*. London: Wiley.

World Conservation Union IUCN. (1980). *World conservation strategy: Living resource conservation for sustainable development*. Gland, Switzerland: IUCN.

World Tourism Organization (WTO). (2004). *Committee on Sustainable Development of Tourism, 2004*.

Wright, D. W. (1996). Infrastructure planning and sustainable development. *Journal of Urban Planning and Development, 122*(4), 111–117. doi:10.1061/(ASCE)0733-9488(1996)122:4(111)

Wright, P., & Monk, A. (1991). A cost-effective evaluation method for use by designers. *International Journal of Man-Machine Studies, 35*, 891–912. doi:10.1016/S0020-7373(05)80167-1

Wyld, D. C. (2008). Blogging. In Garson, G. D., & Khosrow-Pour, M. (Eds.), *Handbook of Research on Public Information Technology* (pp. 81–93). New York: Information Science Reference.

Xie, B., & Jaeger, P. T. (2008). Older Adults and Political Participation on the Internet: A Cross-cultural Comparison of the USA and China. *Journal of Cross-Cultural Gerontology, 23*, 1–15. doi:10.1007/s10823-007-9050-6

Yaakup, A. B., Abu Bakar, Y., & Sulaiman, S. (2004). Web-based GIS for collaborative planning and public participation toward better governance. In *Proceedings of the 7th International seminar on GIS for Developing Countries*, Johor Bahru, May 10-12.

Yamato City. (2002). *Online Urban Master Plan of the Yamato City*. Retrieved on May 2, 2002, from http://www.city.yamato.kanagawa.jp/t-soumu/TMP/e/index.html

Yang, K. (2005). Public Administrators' Trust in Citizens: A Missing Link in Citizen Involvement Efforts. *Public Administration Review, 65*(3), 273–285. doi:10.1111/j.1540-6210.2005.00453.x

Yigitcanlar, T. (2002). Community Based Internet GIS: A Public Oriented Interactive Decision Support System. In S. Wise, Y. Kim, & C. Openshaw (Eds.), *GISRUK: GIS Research UK 10th Annual Conference*, 3rd-5th April 2002. University of Sheffield, Sheffield, UK (pp. 63-67).

Yigitcanlar, T. (2005). Is Australia Ready to Move Planning to an Online Mode? *Australian Planner, 42*(2), 42–51.

Yigitcanlar, T. (2006). Australian local governments' practice and prospects with online planning. *URISA Journal, 18*(2), 7–17.

Yigitcanlar, T. (2008). A public oriented interactive environmental decision support system. In Wise, S., & Craglia, M. (Eds.), *GIS and Evidence-Based Policy Making* (pp. 347–366). London: Taylor and Francis.

Yigitcanlar, T., & Gudes, O. (2008). Web-based public participatory GIS. In Adam, F. (Ed.), *Encyclopedia of Decision Making and Decision Support Technologies* (Vol. 2, pp. 969–976). London: Information Science Reference.

Yigitcanlar, T., & Okabe, A. (2002). *Building Online Participatory Systems: Towards Community Based Interactive Environmental Decision Support Systems*. Tokyo: United Nations University, Institute of Advanced Studies.

Yigitcanlar, T., Baum, S., & Stimson, R. (2003). *Analyzing the Patterns of ICT Utilization for Online Public Participatory Planning in Queensland* (pp. 5–21). Australia: Assessment Journal.

Yong, S. L. (Ed.). (2003). *E-government in Asia: Enabling Public Service Innovation in the 21st Century*. Singapore: Times Media P/L.

Zeisel, J. (1981). *Inquiry by Design. Tools for Environment-Behavior Research*. Monterey, CA: Brooks/Cole.

Zhu, D., Li, Y., Shi, J., Xu, Y., & Shen, W. (2009). (in press). A service-oriented city portal framework and collaborative development platform. *Information Sciences*. doi:10.1016/j.ins.2009.01.038

Ziegler, W. (1991, June). Envisioning the future. *Futures*, 516–552. doi:10.1016/0016-3287(91)90099-N

Zimbardo, P. G. (1997). *Psychologie (5. Aufl.)*. Heidelberg, Germany: Springer Verlag.

Zube, E. H., Sell, J. L., & Taylor, J. G. (1982). Landscape perception: Research, application and theory. *Landscape Planning, 9*, 1–33. doi:10.1016/0304-3924(82)90009-0

Zulick, C. (2003). New Era in Land Use Planning. *ESRI, Government Matters*. Retrieved December 17, 2008, from http://www.esri.com/news/arcuser/0404/eplanning.html

About the Contributors

Carlos Nunes Silva, PhD, is Professor Auxiliar in the Institute of Geography and Spatial Planning at the University of Lisbon, Portugal. He has a degree in Geography (University of Coimbra), a post-graduation in European Studies (University of Coimbra - Faculty of Law), a master degree in Human Geography: Regional and Local Planning (University of Lisbon), and a PhD in Geography: Regional and Local Planning (University of Lisbon). His research interests are mainly focused on local government policies, history and theory of urban planning, urban and metropolitan governance, urban planning ethics, research methods, e-government and e-planning.

* * *

Muhammad Faris Abdullah is a town planner by training. He began his career as a project officer at a consulting firm before moving on to academic. He is currently attached to the Department of Urban and Regional Planning, International Islamic University Malaysia, where he was formerly Head of that Department. As a project officer, he was involved in a number of planning projects including development plan preparation projects. He was among the first to use geographical information systems (GIS) in development plan preparation in Malaysia. His experience with GIS is not limited to development plan preparation alone, but also in development of planning and decision support systems for local and international clients. As a lecturer, his involvement in planning projects is significantly reduced, but he still consulting on strategic environmental assessment and sustainability appraisal of development plans for the Government of Malaysia. He also maintains his interest in development plans, GIS and decision support systems by undertaking research and writing academic articles for journals, books and conferences.

Alias Abdullah holds the post of Professor of Urban and Regional Planning at the International Islamic University Malaysia. His specialisation lies in the fields of urban and regional planning as well as geographic information systems (GIS) and spatial planning and decision support systems (SPDSS). Throughout his academic career, he has published numerous articles and books relating to urban planning, GIS and SPDSS. He has also been actively involved in research and consulting for local and international clients. Dr. Alias also sits on a number of Government advisory boards, the latest being his appointment as a member of the Public Hearing Panel for the Kuala Lumpur 2020 Local Plan. He is also a Director of IIUM Holdings Ltd. and the Executive Director of IIUM Entrepreneurship and Consultancies Ltd. He is a Registered Town Planner with the Malaysian Board of Town Planners and a Fellow Member of the Malaysian Institute of Planners and Commonwealth Association of Planners.

About the Contributors

Stephen Kwamena Aikins is a faculty member in the Department of Government and International Affairs. He holds graduate degrees in Public Administration, Information Systems Management and Business Administration. He is a Certified Public Accountant, a Certified Information Systems Auditor and a Certified Business Manager. He has published in the areas of information security and e-government. His research interests include risk management policy, government auditing and public economics.

Kheir Al-Kodmany's recent research agenda reflects on recent planning experience of Hajj during a two-year fellowship in Saudi Arabia. Hajj, the annual Muslims' pilgrimage to Makkah is the largest congregational event in the world. Multiple research themes have emerged including defining critical information for planning; crowed dynamic modeling; real-time monitoring systems; complex spatial modeling; and power and planning. New research builds on previous extensive studies he made on participatory planning and technology including GIS, visualization and the Internet. Al-Kodmany continues teaching in the area of computer applications for planners and is presently the Co-Director of the Urban Data Visualization Laboratory and the Director of the Graduate Studies of the Urban Planning and Policy Department.

Arne Arnberger (Priv.Doz. Dipl.-Ing. Dr.) holds a master degree in landscape planning and landscape architecture as well as in environmental engineering and a doctoral degree in landscape planning and resource management. He is an associate professor teaching and researching at the Institute of Landscape Development, Recreation and Conservation Planning of the University of Natural Resources and Applied Life Sciences Vienna. His professional and research interests include outdoor recreation in protected areas, forests as well as in urban settings, green space and stress reduction effects, landscape aesthetics, social and ecological carrying capacities, crowding, ecotourism, and environmental education. He was involved in several national and international projects in urban parks and suburban recreation and conservation areas.

Joachim Åström holds a PhD in political science. Åström is a member of the Centre for Urban and Regional Studies (CUReS) and the School of Humanities, Education and Social Sciences (HumES), Örebro University, Sweden. He is also a member of the Network of Excellence DEMO-net funded under the European Commission's sixth framework program.

Scott Baum is Associate Professor and research fellow in the Urban Research Program at Griffith University, Queensland Australia. Since completing his PhD he has developed a strong research interest in understanding questions of social disadvantage at a community and individual level. His research has included work on social polarisation in global cities, community social disadvantage in cities and regions of Australia, the social vulnerability of climate change and extreme weather events and issues of e-governance development in Singapore. He has published over 50 journal articles, 20 book chapters, 4 books and 3 edited books.

Jean-Claude Bolay got his PhD in Political Sciences from University of Lausanne in 1985, after having spent 2 years in Colegio de Mexico, Mexico City, and one year in University of California Berkeley, working on rural – urban migrations. He has worked 4 years as project manager for the Swiss Agency for Development and Cooperation, in an urban project of slum upgrading in Duala, Cameroun, before being contracted by Ecole Polytechnique Fédérale de Lausanne (EPFL) as urban researcher. He

About the Contributors

is currently professor of urban sociology at the Laboratory of Urban Sociology Faculty of Natural and Built Environment, EPFL. He is director of the scientific cooperation at EPFL, heading the Cooperation@epfl unit and the UNESCO Chair in Technologies for Development. He has more than 25 years of experience in urban planning, specialized in habitat, urban environment, social action, institutional issues, and participatory approach, doubled with a long practice of projects and programs' evaluation. He has published around 80 papers and book chapters on the subject.

Alan Borning is Professor in the Department of Computer Science & Engineering, and Adjunct Professor in the Information School, at the University of Washington. He received a BA in Mathematics from Reed College, and a PhD in Computer Science from Stanford University, and is a Fellow of the Association for Computing Machinery. His current research is primarily in the areas of integrated urban simulation of land use, transportation, and environmental impacts; and in design methodologies for human values (in particular Value Sensitive Design). He has also done research in constraint-based languages and systems, object-oriented programming, and human computer interaction more broadly, and has published in all of these areas.

Vassilis Bourdakis, Dipl. Arch. NTUA (Athens), PhD on building evaluation and performance (University of Bath). Worked and taught at the Centre for Advanced Studies in Architecture (CASA) at the University of Bath until 1998. Since 1999, he is working at the University of Thessaly where he's currently an associate professor. He has researched and written extensively on the topics of 3d modelling and applications in large scale digital city models, use of computers in architectural design, virtual reality focusing in communication of ideas / designs, development of interactive virtual environments and their use as a visualisation tool for urban planning and policy making and lately intelligent environments their design implementation and implications. While in CASA, he was the prime investigator and developer of the first, virtual reality, architectural oriented model of the Georgian city of Bath as well as the London's Map of the Future.

Domenico Camarda is Assistant Professor in Spatial Planning at the Technical University of Bari, Italy. He has a Master degree in Civil Engineering, a Master of Science in Economic Policy and Planning, a PhD in Spatial Planning. His interests include Governance and environmental planning, Scenario building, Participation and multiagent systems in planning, Decision support systems, that are subjects of his recent research activities and published papers in international journals. He has coordinated Italian and international meetings and seminars, in many of which he participated as lecturer as well as speaker, presenting about 40 papers. He published about 30 scientific papers and a dozen scientific-informational articles, on those topics, in which an increasing importance was given to the linkages between planning and information and communication technology, today his main object of research interest.

Antonio Caperna is Senior Lecturer and member of MetaUniversitys' team at postgraduate Master Course in Interactive Sustainable Design and Multimedia (Università di Roma Tre). He has a PhD in Sustainable Urban Planning, MA in Urban Planning, BA in Architecture, PGCert in Inclusive Design and PGDip. in Interactive Sustainable Design and Multimedia. Current research and writing focuses on numerous aspects related to sustainability, the role of Information Communication Technology (ICT) and, in particular, the application of the most exciting scientific developments of the past decade, such as fractals, complexity theory, and artificial intelligence to understand the fundamental processes behind urban design and able to create a new human-oriented architecture. In 2008, He was invited at XXIII U.I.A. world congress of Architecture (Turin) contributing as guest lecturer and co-tutor during the international workshop "Transmitting the sustainable city".

Yun Chen was awarded PhD title in ePlanning Systems Development by the University of Salford (UK) in 2007. During her PhD period, she held the highly prized Overseas Research Scholarship (ORS) from UK government (i.e. Universities UK). Currently, Dr. Chen is a research assistant in the University of Salford, working on a UK government EPSRC funded project called Sustainable Urban Regeneration (SURegen) (£2.5 Million, 2008-2012) to design Regeneration Simulator Workbench (RSW). She was working on an EU-funded project called Virtual Environmental Planning systems (VEPs) (INTEREG 3B, 4.7 million Euros) from 2005 to March 2008, contributing to design and implement ePlanning systems. She also achieved the 'Certificate of Professional Development' in consultation services and techniques from the Consultation Institute in the UK in 2008.

Maria Manta Conroy is an Associate Professor of City & Regional Planning at The Ohio State University. She received a B.S. (Systems Engineering) from the University of Pennsylvania, Masters of Science (Systems Engineering) and Planning (Environmental Planning) from the University of Virginia, and a Ph.D. in City & Regional Planning from the University of North Carolina – Chapel Hill. She is a member of the Association of Collegiate Schools of Planning, and the American Planning Association. Dr. Conroy's primary research area is evaluating sustainable development as it relates to local comprehensive planning. Her research extends to technology enhanced public participation, especially focused on impacts on the planning process.

Alex M. Deffner, Dipl. Arch. NTUA (Athens), MSc Urban & Regional Planning (London School of Economics and Political Science: LSE), Ph.D. Planning Studies (LSE). Since 2008 he is Associate Professor of Urban and Leisure Planning, and since 2003 he is Director of the Laboratory of Tourism Planning, Research and Policy at the Department of Planning and Regional Development, University of Thessaly, Volos, Greece. His research experience includes: parks and culture - museum marketing - city networks - urban green - place marketing of cultural heritage - participation in planning - marine parks - sustainable tourism planning - strategic and sustainable planning and development - integration, planning and metropolitan development in Southeastern Europe - observatories of spatial planning - European Cities of Culture - Atlas of Greece - second homes. He has been scientific co-ordinator of the various programmes relating to cultural heritage. His publications in English focus on urban cultural and time planning, as well as city marketing.

Jürgen Döllner studied mathematics and computer science at the University of Siegen, Germany. He got his Ph.D. in computer science from the University of Münster, Germany, in 1996; he also received here his habilitation degree in 2001. In 2001 he became full professor for computer science at the Hasso-Plattner-Institute at the University of Potsdam, where he is leading the computer graphics and visualization research division. His research interests include 3D real-time rendering, 3D non-photorealistic rendering, geovisualization, virtual city and landscape models, and software visualization.

About the Contributors

Jennifer Evans-Cowley is an Associate Professor and Section Head of City & Regional Planning at The Ohio State University. She received a B.S. (Political Science) from Texas A&M University, Master of Urban Planning from Texas A&M University, Master of Public Administration from University of North Texas, and a Ph.D. in Urban and Regional Science from Texas A&M University. She is a member of the American Institute of Certified Planners. Dr. Evans-Cowley's research interests include information technology in planning, public participation, and zoning policy.

Mikael Granberg holds a PhD in political science. Granberg is a member of the Centre for Urban and Regional Studies (CUReS) and the School of Humanities, Education and Social Sciences (HumES), Örebro University, Sweden. He is also a member of the Network of Excellence DEMO-net funded under the European Commission's sixth framework program.

Andy Hamilton is the Director of the Virtual Planning Research Group in the Research Institute for the Built and Human Environment, (BuHu). BuHu is the top rated Research unit in its field in the UK. Andy is an EC appointed senior expert in Urban Information Systems and, in this capacity, has recently advised Yantai City, China on development of systems to manage sustainable urban change. He is currently technical director of the project Sustainable Urban Regeneration, SURegen, (UK Government EPSRC funded, £2.5 Million, 2008-2012) and has been the technical director of European projects such as Virtual Environmental Planning systems (VEPs - INTEREG 3b 2005 -2008) and INTELCITIES (FP6 IP 2004-5). He has published over 50 papers.

Liisa Horelli, PhD, is adjunct professor of environmental psychology at the Centre for Urban and Regional Studies of the Helsinki University of Technology. She has conducted action research on participatory planning and evaluation with children, young people and women for over two decades. She has also coordinated the EuroFEM - Gender and Human Settlements network and is an active member of the international Women in Defense of Place group. She is currently interested in community informatics-assisted time planning and development.

Angela Howell was raised in Maryland and obtained an Associate Degree in Computer Programming from Prince George's Community College and her Bachelor's Degree in Information Systems Management, Summa Cum Laude, from the University of Maryland Baltimore County. She is currently pursuing her Master's Degree in Information Technology with a concentration in Information Assurance from the University of Maryland University College. Following graduation, Ms. Howell was employed 23 years at Xerox Corporation in various technical, sales, marketing and managerial positions. Her main focus was as a technology engineer responsible for the analysis of customer requirements to design, develop and implement integrated and custom document management and workflow systems. She has worked with Fortune 500 companies throughout the world to improve business processes through technology and workflow advances. During her time at Xerox she received numerous awards for her outstanding achievements, dedication and customer satisfaction. In 2008 Ms. Howell was presented with a wonderful new career opportunity at the Department of Defense where she is currently employed. At the Department of Defense she is a Technical Project Manager in the field of Information Assurance.

Markus Jobst, born in 1972, graduated 2003 at the Vienna University of Technology with his thesis on "Multimedia Technologies in Map-related Depictions". From 2003 to 2007 he worked as research scientist at the Institute of Geoinformation and Cartography, where he received his PhD in 2008 with the topic "A semiotic model for the cartographic communication with 3D". He followed with a Post-Doc grant at the "Research School on Service-Oriented Architectures" at the Hasso Plattner Institute at University of Potsdam in the field of geovisualization. Besides several multimedia-cartographic businesses and the management of various scientific projects, Markus Jobst owns funded knowledge of digital photography, x-media cartography and Service-Oriented geoinformation. His main interest lies in the field of geocommunication, multimedia 3D cartography and geo-process management. At the moment Markus Jobst manages the implementation of the EU-directive INSPIRE at the Federal Office of Metrology and Surveying in Austria.

Jens Klessmann works, since January 2008, as a scientific officer in the Work Group High Performance Portals at the Fraunhofer Institute for Open Communication Systems in Berlin, Germany. During the years 2006/2007 he was a project manager at the Information Office d-NRW. There he was responsible for developing and operating the "Model Environment eGovernment". Beginning in 2005 Jens was a staff member at the office of the Task Force for Innovation and High Technology of the State of North Rhine-Westphalia at the State Ministry of Economics. He graduated as a Dipl.-Ing. for Spatial Planning (Urban and Regional Planning) at the Technical University of Dortmund. His thesis was on strategic economic development in the areas of cluster development and local economies. It was funded in part by a grant from the EU Cost Action 26.

Herbert Kubicek was born in Cologne, Germany, in 1946. He studied Business Administration at the University of Cologne (diploma 1971; Ph.D. (Dr. rer. pol) 1974). 1978-1988 he was professor of business administration, in particular organization theory, at the University of Trier. Since 1988, he is professor of applied computer science in the Department of Computer Science and Mathematics of the University of Bremen, Germany, and since 2003 director of the Institute for Information Management Bremen (ifib) as well as scientific director of the Digital Opportunities Foundation. He was consultant to the German Federal Parliament and to Federal Ministries on e-participation and digital divide. His research covers practical development of web-based systems as well as analytical and empirical studies in e-government and e-democracy with particular focus on usability and usefulness (more: http://www.ifib.de).

Olaf Lubanski, born in 1975, in Poland, studied landscape architecture at the University of Natural Resources and Applied Life Sciences, Vienna. His diploma thesis is about real time virtual scale models. Since 2003 he works at RaumUmwelt Gmbh as 3d-artist and CAD-designer.

Arun Mahizhnan is Deputy Director of the Institute of Policy Studies (IPS). Until June 2008, he was also an Adjunct Professor at the Wee Kim Wee School of Communication & Information at the Nanyang Technological University. Mr Mahizhnan's research interests at IPS include policy issues relating to information communication technology (ICT) and mass media; development of Singapore as an Information Society; and arts and cultural developments in Singapore. He has edited books and contributed articles on the above topics. He has also co-authored policy-oriented reports such as "Future of Broadcasting"; "Singapore as a Renaissance City"; "Developing Creative Industries in Singapore"; and "Partnership among Public, Private and People Sectors in Developing the Arts and Culture." His

About the Contributors

co-authored/co-edited books include "Impressions of Goh Chok Tong Years in Singapore," "Broadcast Media in ASEAN"; "Singapore: Re-Engineering Success," and "Selves: State of the Arts In Singapore."

Vanessa Morris is a strategic leader who has a wealth of global and corporate experience in the Human Resources, Health & Safety, Renewable Energy and Program Management arenas. She has a B.S. degree in Business Administration, a B.A. in Management and is currently pursuing a dual MBA & Master of Science in Information Technology degree. In 2005, Vanessa joined the Baltimore Gas and Electric Company (BGE), a Constellation Energy company where she has partnered with several Vision 2020 initiatives including Workforce Planning and Knowledge Management. These initiatives were designed to help define BGE's strategic planning to attract, develop and retain a future workforce. Most recently, she is working in Utility Training where she has developed a new program to support testing qualification for technical and constructions positions at BGE. In addition, Vanessa is an Adjunct Professor at the Community College of Baltimore County and serves on several Board of Directors for non-profit organizations. She currently resides in the Maryland area.

Pilvi Nummi is an architect (M.Sc. in architecture), born in 1976. She is specialized in urban planning and user-centered design. She was working as a researcher in OPUS project in Helsinki University of Technology years 2006-2008. Her main focus in the research project was the usability of ePlanning tools. She will continue her post-graduate studies in the fields of user experience design and ePlanning. From the year 2008, she has been working as a detail planner in the City of Järvenpää, which is a growing city near Helsinki in Southern Finland. In her present work she has an opportunity to apply and test interactive web based tools as a planner in real life situations in detail planning projects.

Les Pang is a Program Director at the University of Maryland University College and teaches courses on database technologies, information technology foundations and homeland security. He is a former professor at the Information Resources Management College, which is part of the National Defense University in Washington, D.C. He has taught military and civilian leaders in the areas of data management, enterprise applications, multimedia, simulation, the Internet, and software technologies. He received a PhD in engineering from the University of Utah and a Masters in Business Administration (concentrating on Management Information Systems) from the University of Maryland College Park. He is the 2004 recipient of the Stanley J. Drazek Teaching Excellence Award at the University of Maryland University College.

Heli Rantanen is an architect (M.Sc. in architecture), born in 1963. For over ten years, she was developing and studying Internet-based tools for participative land use planning as a researcher in the Helsinki University of Technology, Finland. Her main interests are the production and utilization of formal and informal knowledge in local land use planning using web applications and forums, and the web based interaction between the citizens and local governance in general. She is working on her dissertation which deals with the combination of the Internet, urban planning and local knowledge. From the year 2008, she has been developing web based feedback systems and interactive mapping applications in the City of Helsinki as a project manager in the IT Division of Economic and Planning Centre.

Darren Reed is a sociologist who has worked within Human Computer Interaction and Science and Technology Studies for nine years. Working in Computer Science at York and then the Science and

Technology Studies Unit (SATSU), he undertook analysis of a large-scale bus management and information system in York called BLISS. Given his breadth of experience, he undertook both a computer science research fellow role and complemented this with a sociology fellowship in the second half of the project. Since that time he has worked in the area of the inclusive design of technology to support older people in various guises. First as a Research Fellow on the iDesign project, and later as a leading member of the Inclusive Digital Economy Network

Thomas Reichhart (Dipl.-Ing., Dipl.-Ing.) holds a master degree in landscape planning and landscape architecture from the BOKU - University of Natural Resources and Applied Life Sciences, Vienna as well as a Master degree in architecture form the Technical University of Vienna. He is specialised in 3d-visualisation and digital architecture as well as tourism and recreation planning. In his academic work he combines the very technical aspects of modern computer technologies with environmental and social important issues. Beside this he works as project manager and architect for tourism projects and contributed to a row of important Austrian tourism hot spots. His professional and research interests include outdoor recreation, leisure behaviour, 3d-graphics, social and ecological carrying capacities, and the development of regional tourism concepts.

Alexandre Repetti got his Engineering degree and his Doctorate from Ecole Polytechnique Fédérale de Lausanne (EPFL). He was a postdoc researcher in GIS and Environmental Planning at University of California Berkeley. He is currently the head of an Environnmental Planning company, mainly active in Switzerland. He is also a scientific consultant at Cooperation@epfl, the EPFL group specialized in enabling technology for developing countries. Alexandre Repetti has a solid experience of environmental planning, in Switzerland and in Africa. He works on the potential of ICTs for improving the sustainability of urban planning and natural resources management. He has published about 10 journal papers and book chapters on the subject. His work won him the 2004 Lausanne Research and Innovation Award.

Aija Staffans is an architect and D.Sc. (tech.), born in 1957. She is an expert in collaborative land use planning and sustainable urban development. She has been the forerunner of developing and researching local, participative methods in urban planning in Finland. During the last ten years the main focus has been on the exploitation of Internet in the processes of land use planning. She is the manager of the Laboratory of Urban Planning and Design in the Helsinki University of Technology, Department of Architecture, and the initiator and leader of the research group working with the local internet forums. During the years of 2005-2008 she was the project manager of the OPUS project. Since 2004 she has also been the chair of Helka, which is a civic organization of 76 neighbourhood associations in Helsinki.

Koray Velibeyoglu is an Assistant Professor at the Department of Urban and Regional Planning, Izmir Institute of Technology, Izmir, Turkey. His teaching interests and methods cover project-based courses in urban planning and urban design as well as urban ICT policy for planning. The main foci of his research are urban design, urban ICT policy-making and knowledge-based development processes. He is an expert in understanding networked urbanism and the impacts at the metropolitan and local level and the role of ICTs in sustainable urban development.

Sirkku Wallin M.Sc. (planning geography) is a researcher at the Centre for Urban and Regional Studies at Helsinki University of Technology. She has been involved in research projects on European

About the Contributors

structural and regional policies, evaluation of regional development and planning, and development of participatory methods in planning. During last five years, she has worked with a local action research project that applies the coordination of time and space from the everyday life perspective and ubiquitous technology in practise.

Andrew Webster is Director of the Science and Technology Studies Unit (SATSU), and Head of Department of Sociology at the University of York. He was Director of the £5m ESRC/MRC Innovative Health Technologies Programme, is member of various national Boards and Committees relating to innovation and was Specialist Advisor to the House of Commons Health Select Committee. He was a member of the Royal Society's Expert Working Group on Health Informatics. His most recent book is *Health, Technology and Society: A Sociological Critique* (Palgrave Macmillan) 2007. He was elected a Fellow of the Academy of the Social Sciences in 2006.

Bridgette Wessels is Lecturer in Sociology at the University of Sheffield. She has worked on European projects that have addressed information and communication technologies in public services, everyday life, and in the public sphere. She was expert on EU Fifth Framework IST programme, OST/DTI and Royal Society cybertrust programmes. Her books include 'Inside the Digital Revolution: policing and changing communication with the public' (Ashgate, 2007); 'Information and joining up services: the case of an information guide for parents of disabled children' (Policy Press, 2002); 'Understanding the Internet: a socio-cultural perspective' (Palgrave, 2009). She publishes in journals including The Information Society, New Media and Society, Journal of Computer Mediated Communication and the European Journal of ePractice. She is book reviews editor for the International Journal of Media and Cultural Politics.

Tan Yigitcanlar is a senior lecturer in urban and regional planning at the Queensland University of Technology in Australia. He has a multi-disciplinary background and extensive work experience in private consulting, government and academia. The focus of his research is promoting knowledge-based and sustainable urban development. He has been responsible for a wide variety of teaching, training and capacity building programmes on urban planning, environmental science, policy analysis and information and communication technologies in Turkish, Japanese and Australian universities. He is the Editor-in-Chief of the International Journal of Knowledge Based Development.

Rustam Khairi Zahari was trained as an urban planner and did his undergraduate studies at the Arizona State University and the University of Louisiana. This was followed by graduate studies at the California State University at Fresno. He then joined the International Islamic University Malaysia as a project manager at the University's Development Division. He later did his doctorate studies at the University of Nottingham with a focus on urban hazards. Upon obtaining his PhD, he re-joined the University as a lecturer. He is now an Assistant Professor and also the Head of the Department of Urban and Regional Planning at the University.

Index

Symbols

3-D visualizations 106, 113
4D 40

A

Aarhus convention 171, 180, 181, 193, 255
accessibility 273, 275, 283
active tags 390, 393, 399, 403
adaptability 199, 211
Advanced TransEuropean Telematics Applications for Community Help (ATTACH) 292, 293
Agenda 21 340, 343, 352, 356
aggregation 121, 134
algorithms 121, 122, 128, 129, 132
ambiguity 376
analog process 260
analysis agents 207, 210
analytical approach 385
analytical models 406, 419
architecture 367, 370, 371, 373, 376, 377, 385
artificial agents 195, 196, 207, 209, 213
attitudinal survey 238
au pair 196
Auto-ID 389

B

basemap 144, 150, 156, 157
best GIS 150
binary logit regression 108
brainstorming 197, 198, 199, 201, 202, 203, 208, 215, 216
bureaucracy 325, 327
bus drivers 366, 368, 370, 373, 374, 376, 379
Bus Location and Information SubSystem (BLISS) 365, 368, 369, 370, 371, 372, 373, 374, 375, 376, 377, 378, 379, 380, 381, 382, 385, 386
bus operators 368, 374, 382

C

cartography 121, 122, 125, 127, 129, 130, 131, 134, 135, 138, 139
chronotope 62
chronotopic 61, 62, 68, 72, 73
citizen control 256, 263, 266
citizen participation 218, 219, 227, 232, 235
City Beautiful movement 2
city planners 218
civil engineering 137
civil society 325, 336, 338
classic urban planning 323
cognitive appropriation 379, 380
cognitive mapping 126
collaboration environment 259
collaborative planning 269
collaborative spatial decision support systems (CSDSS) 20
collective memory 91
combinatorial explosions 201
common core 268, 274, 281
common good 247
communication agents 207
communicative planning 1, 2, 3, 237, 242, 245, 246, 247, 367
community-based Internet GIS approach (CIGA) 23, 24, 25, 26, 27, 29

Index

community knowledge 84
community planning 271, 281, 285
compatibility 275
complete data sets 150
complex coevolving systems 64
comprehensive plan 236,
comprehensive planning 367
computer-aided 199, 203, 215
computer-aided design (CAD) 20, 273, 275
computer games 103, 104, 105, 106, 107, 108, 109, 113, 114
computer scientists 146
computer supported collaborative work (CSCW) 19, 20, 24
conceptual models 43
constructivists 202
consultation 256, 257, 260
contextual analysis 58, 59, 69, 78
contextual architecture 370, 371
contextualization 205
continuous feedback 195, 196
control stations 239
conurbation 313
convergent 289
coordination agents 206, 209
cost-efficiency 39

D

data accuracy 153, 156
database management 405, 419
database management system (DBMS) 405, 419
data collection 388, 389, 394
data-driven planning 150
data management 146, 406, 419
data mining routines 195, 196
data source level 96
dead values 121, 133, 134, 135
decision making 15, 16, 17, 18, 19, 20, 21, 22, 23, 24, 25, 27, 28, 29, 31, 32, 34, 35, 80, 82, 83, 86, 93, 94, 95, 98, 340, 351, 353, 404, 405, 406, 408, 409, 419
decision support system (DSS) 15, 17, 20, 22, 24, 31, 34, 443, 447, 449, 452, 453
delegated power 256

deliberative consultation 173, 175, 179, 180, 191, 193
Delphi survey 200
departmental GIS 148
Department of Town and Country Planning 437, 450, 453
design methodologies 286, 300
developing cities 306, 307, 309, 310, 311, 312, 313, 314, 315, 316, 318, 319, 320, 322, 323
development forum 86, 87, 88, 91, 92, 93, 95, 96, 97, 101
development plan 435, 437, 438, 439, 440, 441, 442, 443, 444, 445, 446, 447, 448, 449, 452, 453
digital citizens 80, 82, 93
digital citizenship 58, 59, 81
digital divide 41, 223, 224, 227, 228, 229, 236, 241, 288, 297, 305, 324, 325, 326, 331, 332, 333, 335, 338
digital elevation model (DEM) 154
digital models 279, 280
digital outreach teams (DOTs) 299, 300
digital services 255
digital terminology 59
direct democracy 93
disseminating 22, 28
distance learning 363
distributed interaction 195, 196

E

ecoware 434
e-democracy 60, 169, 181, 183, 192, 220, 221, 222
e-governance 59, 60, 71, 336, 338
e-government 60, 218, 219, 220, 221, 222, 223, 224, 225, 226, 227, 228, 229, 230, 231, 232, 233, 235, 236, 253, 262, 266, 286, 287, 296, 297, 301, 305
e-government action plan 1 (eGAP1) 326
e-inclusion 288, 292, 297, 298, 299, 301, 305
elite democracy 240
emoticon 236
empirical study 106, 114

513

engineering frame 365, 367, 382, 385
enterprise GIS 149
entry point 258, 266
environmental democracy 181, 193
environmental planning 104, 114, 195, 196, 197, 212, 213, 214
environmental sustainability 195
Environmental Systems Research Institute (ESRI) 21, 30, 32, 158, 159
e-participant 218, 219, 225, 226, 236
e-participation 168, 169, 170, 176, 177, 179, 180, 181, 182, 183, 184, 185, 186, 187, 188, 190, 191, 192, 193, 194, 237, 238, 239, 241, 242, 243, 244, 245, 246, 247, 252, 257, 258, 259, 261, 262, 263, 266
e-petitions 179
epidemiology 344
epistemological 2, 14
e-planning 365, 366, 378, 379, 382, 404, 405, 406, 407, 408, 409, 410, 411, 413, 419
e-planning model 143
e-Planning systems 36, 37, 38, 39, 40, 41, 42, 43, 44, 47, 48, 49, 52, 57
equal opportunities 269, 270, 282
e-services 286, 287, 288, 289, 290, 291, 292, 293, 295, 296, 297, 299, 300, 301, 302, 305
e-training 268, 273
evaluation gap 170
exacerbated 311, 314
executive information systems 453
executive planning 238
expert agents 196, 208, 212

F

feasibility 39, 42, 43, 44
ferroelectric random access memory (FRAM) 391, 401
field of view (FoV) 133
finware 434
flaming 222, 236
flexibility 199, 202, 271, 275
forecasting 62

framework 340, 341, 342, 343, 344, 347, 349, 351, 353, 356, 363
free resources 241
future-making assessment 58, 59, 61, 66, 71
future-making assessment approach (FMA) 58, 59, 60, 61, 65, 66, 70, 71, 72, 73, 74, 78
future workshop 197, 198, 199, 203
fuzziness 59

G

Garden City movement 2
geographical information system (GIS) 15, 16, 17, 18, 19, 20, 21, 22, 23, 24, 25, 27, 28, 29, 30, 31, 32, 33, 34, 35, 177, 178, 182, 193, 225, 226, 230, 236, 241, 273, 275, 313, 315, 317, 318, 320, 404, 405, 406, 407, 409, 410, 411, 412, 413, 414, 415, 417, 418, 419, 435, 436, 438, 439, 440, 441, 442, 443, 444, 445, 446, 447, 448, 449, 450, 451, 452, 453, 454
geographical mediation 257
geographic information systems 37, 53, 54
geometric distortions 133
geospatial 15, 28, 120, 121, 125, 126, 127, 130, 131, 134, 135, 136, 137, 139, 142, 143, 144, 146, 147, 148, 150, 153, 158, 159, 160, 161, 163, 164
geospatial analytic ability 39
geospatial database 143, 144, 147, 148, 150, 153, 158, 159, 160, 163, 164
geospatial technologies 15
geovisualization 130, 139
GIS-based systems 447, 448, 449, 453
global positioning system (GPS) 241, 394, 397
glocal 63, 72, 74, 76
glocal networks 63
Google Earth 28
government-to-business (G2B) 327
government-to-employees (G2E) 327
granularity 127
grassroots 324

greenwaving 377
groupware 17, 19, 32, 200, 209, 216

H

Hajj 143, 144, 145, 146, 147, 148, 149, 150, 151, 152, 153, 160, 163, 165, 166,
human agents 195, 196, 200, 206, 207, 210, 213
human computer interaction (HCI) 40
hybrid tree model 124

I

informality 311, 323
informational era 325
information media 39
information pools 252, 255, 258, 259, 262, 266
information poverty 227
information security management 405, 412, 415, 419
information society 27, 81, 367
information system development methodology (ISDM) 37, 42
information systems 313, 315
information transmission 126, 127, 130, 131, 139
infrastructure 238, 239, 242, 243, 244, 287, 290, 298, 303, 307, 308, 309, 310, 311, 312, 313, 314, 315, 316, 317, 318, 319, 322, 323, 345, 347, 348, 352, 354, 364
instrumental rationality 239, 240
integrated mobility services 365
intelligent transport system (ITS) 366, 379
intelligent vehicle-highway system (IVHS) 396
interactional theory 49
interactive tools 88, 101
interactive tracking 289
interactive whiteboards 63
interconnectedness 371
interconnections 104, 113
intermediate agents 196, 207
Internet mapping 21, 22, 29
interoperability 39, 42, 43, 44
interpretive turn 367

interpretivism 43
interrelationships 202
inventory forum 86, 87, 101

K

knowledge agents 204, 207
knowledge building 81, 83, 85, 87, 93, 98, 101
knowledge management 91, 95, 96, 97, 100
knowledge management level 95, 96
knowledge structuring 120, 125, 126

L

ladder of participation 255, 257
land use 343, 364
land use models 15
linear system 62
lobbying agents 207, 208
local agencies 286, 305
local authority 343, 364
local content 88, 89, 98
local e-government 422, 434
local government 347, 348, 352, 364
local knowledge 82, 83, 84, 85, 86, 89, 90, 98, 100, 101, 102, 172
local knowledge map 86, 89, 90, 101
local plan 364
local planning 313
local planning authority 437, 438, 453
local policy 349, 350, 354, 355, 356, 364
location based services 28
longitudinal study 218
low earth orbit (LEO) 152

M

managerial model 220, 221, 223, 226, 229, 231
maneuvering space 367
manipulation 256
map accuracy 154
MeetingWorks 200, 201, 204, 216
meta-design 290
mobility 365, 366, 376, 378, 379, 380, 382, 385

modelling & matching methodology (M&M) 36, 37, 38, 42, 43, 45, 46, 47, 48, 52, 56
monolithic 18
motility 379, 380, 382, 385
mullti-agent system (MAS) 195, 196, 197, 202, 210, 211, 212, 213
multi-agent interactions 195, 196, 197, 215
multichannel marketing 93
multidirectional interactivity 221
multi-level agents 208
multi-methodology 42
multitexturing 123

N

National Computer Board (NCB) 326
networking capability 391
networking tools 95
new media 298, 299, 302, 303, 304
nonlinear dynamics system (NDS) 342
nonlinearity 342
non-photorealistic rendering (NPR) 131, 132

O

objectivist 43
one-stop government 254, 255, 266, 267
online discussion platform 27
online GIS 147
online participation tools 218, 220, 225, 226, 227, 228, 229, 230
online planning 16, 17, 18, 22, 28, 33
operating frequency 391
OPUS 82, 83, 84, 85, 86, 88, 89, 90, 91, 92, 93, 94, 95, 98, 99, 101
organizational knowledge 163
orgware 431, 434
orthorectified 155

P

panchromatic 155
paradigmatic 197
participant design 290, 301, 305
participation portals 252, 253, 258, 261, 262
participatory democracy 237, 240, 242, 245, 246, 247
participatory design (PD) 290
participatory e-planning 59, 60, 61, 64
participatory planning 61, 64, 65, 72, 75, 195, 197, 212, 214
participatory turn 240
participatory web 176, 194
passenger information panels (PIP) 367, 368, 373, 376, 377, 380, 381, 386
passive tag 390, 403
Pentagon-Prism model 434
photogrammetric 275
photogrammetry 121, 122
photomontages 104, 106
photorealistic 126, 128, 131, 132, 133, 135, 137
pilot study 25, 27
placelessness 81
planning association 260, 261
planning decision support systems (PDSS) 153
planning department 220, 225, 236,
planning forum 86, 101
Planning Inclusion of Clients through E-Training (PICT) 268, 269, 271, 273, 274, 276, 281, 282, 284
planning knowledge 195, 196
planning methods 195, 196
planning support system (PSS) 17, 315, 405, 406, 407, 408, 409, 410, 412, 414, 419, 425, 426, 433, 434
planning theory 1, 2, 3, 6, 7, 8, 9, 10, 12, 13, 14, 239, 240
plurality 228
policy clients 82
policy making 237, 238, 240, 242
political equality 256
polygons 124, 129
portal 252, 254, 255, 258, 259, 260, 261, 262, 263, 265, 266, 267
pragmatism 240
prioritisation 370, 377, 378, 380, 381
process tree 89, 90, 101
professional communication 40
project perspective 66, 68, 70
public green spaces 103
public opinion 269

public participation 16, 17, 18, 19, 20, 21, 22, 23, 26, 28, 32, 34, 170, 171, 172, 173, 175, 180, 252, 253, 255, 257, 258, 259, 260, 261, 262, 263, 266, 268, 269, 270, 278, 281, 282, 283, 285, 324, 338
public participation ladder 173
public participatory geographical information system (PPGIS) 19, 22, 28, 31, 40, 54, 182
public sector 252, 253, 254, 255, 258, 261, 262, 263, 287, 288, 291, 298, 301, 305
punctuality 380

Q

qualitative 42, 49
quantitative 42, 49
quick and dirty GIS 149

R

radio frequency identification (RFID) 388, 389, 390, 391, 392, 393, 394, 395, 396, 397, 398, 399, 400, 401, 402, 403
radio frequency (RF) 390
rapid development methodology 43
rational planner 245
rational planning 313
rational theory 2, 3, 4, 6
realpolitik 313
real time information (RTI) 368
reassembling 366
recommendation machine 67
redevelopment 273, 280
regional planning 252, 260, 261
regression analysis 244
regulation zones 309, 323
representational validity 105
representation technologies 40
representative democracy 239, 242, 244
responsiveness 176, 178
rich language 209
road-mapping 62
running board 372

S

scenario analysis 198
scenario building 198, 199, 200, 201, 202, 203
schema 269
scratchpad 203
script 63, 64, 69, 70, 73, 79
security policy 406, 411, 419
self-organization 342
semiotic structure 130, 131, 132, 139
semi-passive tags 403
sensory networks 59, 63, 79
service center 255, 259, 260, 263, 266, 267
service cluster 255, 261, 263, 267
short message service (SMS) 329
slums 306, 307, 308, 311, 312, 313, 314, 318, 319, 323
smart approaches 121, 130
smart card 289
SMURF 309, 317, 322
social formation 291, 295, 296, 300, 301
social frame 365, 367, 382, 385
social inclusion 325, 326, 327, 329, 338
social media 59, 63
sociotechnical constituencies 291
sociotechnical perspective 66, 68, 70, 71
soft systems methodology 43, 45, 56
spatial accessibility 380
spatial analysis 146
spatial data 143, 421, 422, 429, 434
spatial decision support system (SDSS) 422, 434
spatial development 364
spatial division 262
spatial information 28, 35
spatial knowledge 120, 121, 125, 126
spatial planning 126, 195
spatial-temporal data 152
stakeholder 45, 47, 48, 51, 52, 364
stakeholder group 47
stakeholders 2, 3, 4, 5, 7, 8, 14, 144, 162, 199, 200, 201, 202, 203, 204, 208, 237, 239, 269, 306, 307, 308, 309, 313, 314, 315, 316, 317, 319, 320, 323

statutory development plan 435, 436, 437, 438, 440, 441, 443, 445, 446, 447, 448, 449, 454
strategic decisions 24
strategic planning 354, 364, 434
subdivision 258, 259
survey data 406, 419
sustainable development (SD) 340, 342, 343, 344, 347, 348, 349, 350, 351, 352, 353, 355, 357, 359, 361, 363, 364
sustainable tourism 341, 342
SWOT analysis 430, 434
synthetic roofs 122
system development 37, 41, 42, 43, 45, 46, 48, 52, 57
system failures 379
systemic interrelations 323
systemic perspective 66, 68
system theory 1, 2, 3, 4, 6

T

tacit knowledge 85, 93
taxonomy 197, 206, 214, 215
technical agents 206, 209
technologies interoperability 39
technology acceptance model (TAM) 220
temporal flexibility 380
terms of references (TORs) 160
texture baking 123
theory of planned behavior (TPB) 220
theory of reasoned action (TRA) 220
timetable 372, 380, 382
timing points 372, 381
top-down 60
topography 120, 122, 134
toxicology 344
tracking services 28
traffic congestion management system (TCMS) 373
trail scenarios 103, 104, 107, 108, 110, 113
transformation process 46
transmitting media 120, 121, 122, 125, 129, 131, 133, 134, 142
transparency 172, 175, 178, 179, 186, 188, 256, 259

transportation routes 152
triangulated irregular networks (TIN) 122

U

ubiquitous computing 59, 62, 63, 72, 366
UNIX 158, 159
unwrapping 123
urban context 106
urban governance 269, 281, 285, 306, 308, 313, 317, 320, 323
urban information system (UIS) 97, 422, 434
urban integration 308
urbanisation 41
urbanism 2, 3, 9, 11, 12, 13
urbanization 307, 310, 323
urban management 310, 313
urban model 275, 276, 278, 279, 281
urban planning 1, 2, 3, 4, 5, 6, 7, 8, 9, 10, 13, 14, 15, 16, 17, 18, 19, 21, 22, 23, 24, 25, 28, 29, 30, 32, 37, 38, 39, 41, 42, 43, 44, 48, 49, 50, 52, 57, 80, 81, 82, 83, 85, 88, 95, 97, 98, 100, 101, 103, 120, 128, 137, 168, 169, 170, 171, 172, 174, 176, 178, 179, 181, 182, 183, 185, 187, 190, 237, 238, 239, 252, 285, 388, 419
urban planning paradigm 1, 2
urban planning theory 367
urban policy 434
urban public space 103
urban regeneration 47, 57
UrbanSim 36, 50, 52, 53, 55, 56, 57
urban sprawl 307, 322
urban structure 365
user-centered design 290
user-friendliness 199
user-generated content (UGC) 176

V

value sensitive design (VSD) 36, 37, 38, 42, 48, 49, 50, 51, 52, 57
videoconferencing 289, 363
view angle 134
virtual cities 21

virtual environment
 104, 105, 106, 111, 113, 277
virtual reality modelling language (VRML)
 20, 21
virtual reality (VR) 16, 37, 131, 133, 135, 273, 275, 276, 277, 279, 280, 281, 282, 285,
volunteered geographical information (VGI) 85
vulnerability 411, 412, 413, 419

W

weak signals 62
Weltanschauung 46
wiki 236